# JUDGING
# EDWARD TELLER

Edward Teller in the early 1950s. Courtesy of Los Alamos National Laboratory.

Istvan
Hargittai

# JUDGING
# EDWARD TELLER

A CLOSER LOOK AT ONE OF THE MOST INFLUENTIAL
SCIENTISTS OF THE TWENTIETH CENTURY

**Prometheus Books**

59 John Glenn Drive
Amherst, New York 14228–2119

Published 2010 by Prometheus Books

Inquiries should be addressed to
Prometheus Books
59 John Glenn Drive
Amherst, New York 14228–2119
VOICE: 716–691–0133
FAX: 716–691–0137
WWW.PROMETHEUSBOOKS.COM

14  13  12  11  10     5  4  3  2  1

Library of Congress Cataloging-in-Publication Data

Hargittai, Istvan.
    Judging Edward Teller : a closer look at one of the most influential scientists of the twentieth century / by Istvan Hargittai.
        p.  cm.
    Includes bibliographical references and index.
    ISBN 978–1–61614–221–6 (cloth : alk. paper)
    1. Teller, Edward, 1908–2003. 2. Physicists—United States—Biography.
3. Scientists—United States—Biography. 4. Hungarian Americans—Biography.
5. Atomic bomb—United States—History. I. Title.

QC16.T37H37 2010
530.092B—dc22

2010022918

Printed in the United States of America on acid-free paper

For victims of totalitarian regimes.

The author acknowledges with gratitude a generous grant from the Alfred P. Sloan Foundation, which made his research in American archives and some of the writing possible in the preparation of this book.

# ALSO BY THE AUTHOR

With Magdolna Hargittai and Balazs Hargittai, *Candid Science I–VI: Conversations with Famous Scientists* (Imperial College Press, 2000–2006).

*The Road to Stockholm: Nobel Prizes, Science, and Scientists* (Oxford University Press, 2002; 2003).

*Our Lives: Encounters of a Scientist* (Akadémiai Kiadó, 2004).

*The Martians of Science: Five Physicists Who Changed the Twentieth Century* (Oxford University Press, 2006; 2008).

*The DNA Doctor: Candid Conversations with James D. Watson* (World Scientific, 2007).

With Magdolna Hargittai, *Symmetry through the Eyes of a Chemist*, 3rd ed. (Springer, 2009; 2010).

With Magdolna Hargittai, *Visual Symmetry* (World Scientific, 2009).

# CONTENTS

# FOREWORD

## *Peter Lax*

Sixty years ago, Edward Teller was the center of one of the biggest science public policy controversies—the building of the hydrogen bomb. In the next forty years he managed to involve himself in more science–public policy controversies. This full-length biography, a scrupulously fair-minded and evenhanded account, is a timely review of the life and influence of this fascinating scientist.

Teller was one of a remarkable group of Hungarian scientists, affectionately referred to as "the Martians," who made outstanding, world-class contributions to their fields: physics, mathematics, computer design and computational methods, chemistry, and aerodynamics. They made equally important contributions to the defense of the United States and of the free world.

That so many Hungarians, out of a nation of ten million, became leading scientists of the world calls for an explanation. One is the linguistic isolation of Hungarians in a sea of Slavs in Eastern Europe. There was a general feeling that to maintain their identity they had to achieve extraordinary accomplishments. Another factor was that Teller and the other "Martians" were Jewish, and to overcome anti-Semitic prejudices prevalent in Hungary they had to be special, even among Hungarians.

Hargittai describes Teller as the victim of three exiles. In 1926 at age eighteen, he left Hungary because as a Jew he could not be assured of a university career. He went to Germany, studied in Karlsruhe under the great polymer chemist Herman Mark, then under Werner Heisenberg in Leipzig. Next he held a postdoc position in Göttingen.

This idyllic arrangement came to an abrupt end in 1933 when the Nazi gang came to power; Teller went on his second exile, ending up in the United States at George Washington University.

Hargittai describes in detail Teller's involvement in nuclear energy, which led to his wartime service at Los Alamos, and his growing preoccupation with thermonuclear weapons—and his clash with J. Robert Oppenheimer. Teller correctly perceived that if the Soviet Union developed thermonuclear weapons and the United States failed, the Soviets would dominate the world, so he devoted all his energies to convincing the US government to embark on a program to build thermonuclear weapons. Then he devoted his energies to the design of these weapons.

Calculations indicated that Teller's original design would fizzle. I was present at a lecture on the subject by George Gamow at Los Alamos, in which he illustrated this conclusion by an experiment. He crumpled a piece of paper, put a piece of petrified wood on top of it, lit the paper and exclaimed, "Look, no ignition!"

Subsequently Teller devised a new design of an H-bomb that was successfully tested. Other people contributed ideas, but it is entirely appropriate to call Teller the father of the hydrogen bomb.

In 1954, in a sinister conspiracy, Oppenheimer was accused of disloyalty and branded a security risk. Teller gave damaging testimony, and Oppenheimer was deprived of his security clearance. Most (not all) of the physics community turned on Teller, refusing even to shake his hand. Hargittai correctly calls this ostracism Teller's third exile.

After his great success with the hydrogen bomb, Teller's judgment deserted him and he was on the wrong side of a series of important issues, some political and some technical. He vigorously opposed the prohibition of nuclear tests in the atmosphere. He minimized the dangers of radioactive fallout, whose hazards were emphasized by proponents of the test ban. Linus Pauling was their leader; one of the dangers he pointed to was genetic damage due to radiation. In one of the lighter moments of the debate, Teller pointed out that wearing pants increases the temperature of the testes, which can also lead to genetic damage. Therefore, to be consistent, proponents of the ban of atmospheric tests should also insist that men wear kilts.

Teller was wrong to insist on unlimited testing; it would have been inconsistent with asking other nations to refrain from developing nuclear weapons.

Teller was equally wrong in opposing the Strategic Arms Limitation Treaties. He could not imagine that after the death of Stalin the Soviet Union was changing.

During his presidency, Ronald Reagan announced a program, called Strategic Defense Initiative, to "make nuclear weapons obsolete" through countermeasures. Most defense scientists thought this aim could not be achieved, but Teller thought otherwise. One of his proposals was to use X-ray lasers, another was called "Brilliant Pebbles." Neither was realistic. Some claim that the mere idea of the Strategic Defense Initiative was such a challenge to the Soviet defense establishment that it contributed to the downfall of the Soviet Union. Maybe.

Still, Teller was far from being an archconservative. In 1946 he joined Leo Szilard and other physicists to help push through legislation that placed nuclear energy in civilian, rather than military, hands. He strenuously opposed a loyalty oath imposed by trustees of the University of California on the faculty. When his friend Stephen Brunauer was falsely accused by Joe McCarthy of being a security risk, Teller came vigorously to his defense. Moreover, he was against secrecy; he liked to point out that in nuclear weaponry, where secrecy is highest, the Soviets are on par with the United States. On the other hand, in the relatively open field of computers, the Soviets are nowhere.

No other scientist engendered so much emotion as Edward Teller. He had hosts of admirers and an equal number who regarded him as evil personified. Hargittai's aim, admirably achieved, is to set the facts of his life straight.

Teller was a very complicated person; Hargittai says that there were at least three Tellers. We shall not see the likes of them again.

# PREFACE

No part of a book is so intimate as the Preface.
—CHARLES W. ELIOT[1]

The crowd at the corner of Honvéd Street and Szalay Street in downtown Budapest was unexpectedly large mid-morning on January 15, 2008, the centennial anniversary of Edward Teller's birth. It was the unveiling ceremony of a memorial plaque on the wall of an imposing, large apartment building that had been Edward Teller's boyhood home. Physicists, teachers, and whole classes of high school students came, as well as many others who had happened to learn about the poorly advertised event. It was initiated by a clerk in a secondhand bookstore in the building who had accidentally learned about its distinction, and he made it his avocation to mark this event.[2] The plaque was impressive. It said in Hungarian and in English:

EDWARD TELLER (JANUARY 15, 1908, BUDAPEST—SEPTEMBER 9, 2003, STANFORD, CALIFORNIA) LIVED IN THIS HOUSE BETWEEN 1913 AND 1926. HE WAS A WORLD-FAMOUS SCIENTIST AND MEMBER OF THE HUNGARIAN ACADEMY OF SCIENCES, AND WAS RECOGNIZED, AMONG OTHERS, BY THE CORVIN CHAIN (HUNGARY 2001) AND THE PRESIDENTIAL MEDAL OF FREEDOM (USA 2003).

There were official events to commemorate the Teller centenary as well. A scientific session at the Hungarian Academy of Sciences was held, where scientists and historians alternated in lecturing about Teller's life and oeuvre as well as twentieth-century history. The Hungarian National Bank created a limited-edition silver coin with the denomination of HUF (Hungarian Forint) 5,000 (about twenty-five American dollars); and the Hungarian Post Office issued a memorial stamp of HUF250 denomination. Both the coin and the stamp depicted an aged Teller and an image of the nuclear reaction representing fusion, the basis of the hydrogen bomb.

I was involved in the commemoration of the Teller centenary due to my recent book about the five eminent Jewish–Hungarian–American physicists who are also known as the *Martians of Science*.[3] They included Theodore von Kármán, Leo Szilard, Eugene P. Wigner, John von Neumann, and Edward Teller. They each made important discoveries in their respective fields, in aerodynamics and quantum mechanics, in developing the modern computer and molecular biology, in nuclear chain reaction and game theory, and elsewhere. What made them a distinct group was their commitment to contributing to the defense of the Free World in the most profound ways, even if they had to risk their scientific careers.

The Martians designation traces back to the Manhattan Project, when Enrico Fermi wondered about the origin of the many talented Hungarians. According to one version, Fermi mused about the probability of other civilizations existing beside the one on Earth. When their original planets got overpopulated, they started colonizing other planets in other galaxies, eventually finding our Earth, as "these highly exceptional and talented people could hardly overlook such a beautiful place as our Earth." But then, Fermi posed the question, "Where are they?" In this version, Szilard's answer was "They are among us, but they call themselves Hungarians."[4] Teller dedicated his *Memoirs* to the Martians, and added, "No accolade gives me so great a pleasure as that I was counted among them."[5]

I did not plan to write much more about them than my *Martians* book, but during the preparation for the Teller centenary I realized that we should know more about Edward Teller. It was a long road for me before writing a book dedicated solely to him, although by the time I embarked on the project it seemed inevitable. I have been interested in the deeds and lives of outstanding scientists. When I was a teenager, I read voraciously, always preferring nonfiction to fiction. Later, my professor of organic chemistry in Budapest sprinkled his lectures with human stories, which drew me to his class despite the subject's hardly interesting me. Upon graduation from university, the editor of popular science at Radio Budapest encouraged me to do some interviews with visiting scientist luminaries. But I soon left this type of activity to focus on my experiments and I limited my reading to my research.

I had narrowly focused for about twenty-five years during which I created my niche in science—in structural chemistry, at the interface between chemistry and physics. But without actually noticing it, I started to broaden my interest once again. Among other things, I developed an interviewing project, which soon became a family undertaking. My wife, Magdi, and I recorded hundreds of interviews with prominent scientists.[6] Edward Teller was one of our interviewees.

I read everything about the Martians I could get hold of, and found the least satisfactory and the most controversial material about Edward Teller. I also found him in some ways the closest and most similar to me. He was born the same year as my mother. He came from a family that, not unlike my own, was Jewish, and we both had ascendants for whom German (not Yiddish) was as much their mother tongue as Hungarian. Teller's father was a well-established Budapest lawyer as was my father; another lawyer in his family, his brother-in-law, was killed as a forced laborer by the Nazis, as was my father.[7] For the members of his family who stayed in Hungary, the ordeals did not end when World War II was over, as they did not for us. I felt that with my background, I might have some advantage in understanding Teller's character and attitude and the conditions under which he grew up.

Teller's childhood was hardly influenced by the direct experience of communism. At the time of the short-lived communist dictatorship in Hungary in 1919 he was eleven years old and spent most of the four-month period far from Budapest. On the other hand, his adult life was greatly motivated by his response to it. Communism, however, is more complex than is often projected. For me, it meant liberation from a Nazi concentration camp, but following that, discrimination. Due to my ostensibly capitalist roots, I was meant to be excluded from not only higher education but even from secondary education in the 1950s.[8] When we returned from the concentration camp in 1945, my mother, whose exemplary altruistic behavior in the hellish camp became legendary,[9] joined the Communist Party. She felt obliged to fight Nazism and regretted that she did not know about the communists before the Holocaust (as we call it today). By the time the Communist Party in Hungary transformed the fledgling democracy into a one-party dictatorship in 1948, she had been summarily kicked out of the party as a former member of the capitalist class.

The vastly conflicting opinions about Teller in American society have intrigued me. I received a taste of this conflict in the 1980s, when I edited an international science journal. I observed that the publisher and my co-editor were upset when the guest editor for a special issue got Teller to write an introduction.

During the 1980s and 1990s, I spent years as a visiting professor at the universities of Connecticut, Texas, Hawaii, and North Carolina. My family and I developed an interest in all affairs American. I learned to think about many things in more complex ways than I had before; more in shades than in stark black or white. I was impressed by the greater tolerance of Americans compared with my general experience elsewhere. So I was puzzled when this tolerance appeared to dissipate as soon as any discussion about Teller came up, regardless of whether I talked with his admirers or his critics. Most of my American colleagues spoke about Teller in extremes. Interestingly, I spent four academic years at Moscow State University in the 1960s where I earned my master's degree, and Teller figured there as an obvious enemy. On my more recent encounters in Moscow, however, I observed that he has become a hero.

My deep interest in American affairs in general and in Teller in particular has made me wonder about the black-or-white approach to him even by some outstanding minds. My principal motivation for writing this book was to counter such a one-dimensional approach and create a portrait regardless of any preconceived image about Teller. Years of reading and writing about Edward Teller have brought me close to him, and I have learned to be appreciative of his virtues and to be conscious of his flaws. I have grown to feel that his life has become an open book for me, which is surprising, considering I met him in person only once, when he was eighty-eight years old, and the only meaningful correspondence we had was during the last months of his life.

Our meeting took place in the Teller home in Stanford on February 24, 1996. Magdi and I went to record a conversation with him. We spent the whole afternoon with Teller and his wife, Mici. Teller was convalescing after an illness, and as we started talking, he was obviously in a bad mood. This we managed to turn around, but the start was slow. We asked if he would tell us about his family background, and he responded with four sentences he sandwiched between "I am not terribly eager to do so" and "Otherwise I cannot tell you anything of great interest." It was not the lowest point though. Our next question was about the situation of intellectual Jewish families in Budapest at the time of his youth and about the conditions that forced him and countless others to flee. He said that Hungary was mildly anti-Semitic when he was young. We were visibly startled, so he quickly added a qualification. "Mildly," he said, "because I call Hitler not mildly." At one point our conversation seemed to deteriorate so much that Mici whispered to him in Hungarian (our conversation was in English), "Don't be so unfriendly!"

The first sign of thaw came when we asked him about the widespread opinion that his anticommunism originated from his having suffered under the communist regime in Hungary. We knew this could not have played an important role since he was a child at the time of the short-lived communist regime in 1919, and Teller and his family did not personally suffer from it in any extreme way. He told us that he was "not exposed to the problems of communism in a direct way," and added that he "was exposed in the school to anti-Semitism" that came afterward. He went into detail discussing his relationship to communism, and the stage was set for a more meaningful exchange. It was further facilitated when I volunteered that I had heard that Arthur Koestler's *Darkness at Noon* was an important ingredient in his formulating a view of the world.

His responses lengthened, and he started visibly enjoying the conversation. It lasted long and included extended treatises on scientific topics, such as Teller's contribution to interpreting Linus Pauling's theory of resonance; the story of the discovery of the Jahn–Teller effect; and the philosophical implications of Heisenberg's uncertainty principle. Dinnertime was approaching, and we wound up the recording. At this point Teller wanted to have a chat in Hungarian, and he made some complimentary remarks about Hungary's then socialist prime minister, a former communist.

Magdi told him about having found experimental evidence for the Jahn–Teller effect in the geometry of a gaseous molecule—a first in the literature. Teller gave some useful pointers that Magdi acknowledged in her paper the following year. The visit became a success. At times I felt almost mesmerized. He spoke with such an enthusiasm about some well-known scientific theories as if they were the latest discoveries. As we discussed symmetry, a topic that was obviously of great interest to him, we promised to give him the recent second edition of our book *Symmetry through the Eyes of a Chemist*,[10] with which we returned later in the evening.

My last correspondence with Edward Teller took place in the summer of 2003. It was via e-mail, and he wrote me in Hungarian, thanks to his assistant, Margit Grigory, who had impeccable Hungarian as well as English. I sent Teller the Hungarian version of my book *Our Lives* (the English version appeared the following year) and Margit read it for him; Teller was blind by then. The plan was that he would send me his comments after having listened to a chapter or two at a time. He lived long enough for only one such message dated August 13, 2003, and an addendum dated August 17.

In this letter, among others, he explained why he did not give a response to my question about success in science. It sounded almost like a summary of his views on several important issues. It is worth quoting (in my translation from the original Hungarian):

> In my life's work, I loved science a thousand times more than its applications. I agreed to do the latter because I took the dangers of war very much to my heart. I hope you know that I was always against our becoming the first to deploy the hydrogen bomb. I only wanted to have the possibility of the H-bombs as a deterrent for wars, and this has worked so far.
>
> I am convinced that the business of scientists is exclusively science itself. The application of science is the business of politicians and consequently that of the voters. I had problems with my fellow scientists, especially with those according to whom we shouldn't have worked on anything like the hydrogen bomb.
>
> Incidentally, in this question we bitterly differed with Oppenheimer. Similarly we had different positions with Enrico Fermi, but with him our friendship did not suffer from this difference. The same can be said about Leo Szilard, who was the most gifted in treading on other peoples' corns, but he never bored anybody.
>
> However, let's speak about science. Very few performed true science and I knew two such people, Einstein and Bohr. I would be curious to know what Einstein thought when he received the Nobel Prize definitely not for relativity. . . . Things like relativity and quantum mechanics far surpass all other intellectual activities.
>
> For me it is important that the same four letters describe the DNA of all living creatures. This may bring us closer to the understanding of what life is.

A few days later, I received a shorter note from Teller, a continued response to my question about success in science, which seemed to strengthen what he said in his previous letter. I quote it here (again, in translation from the original Hungarian):

> Your question is a difficult one, but my answer is easy. I was not an unsuccessful scientist but my scientific research suffered from my work on weapons. Also, in part this happened when I was at the peak of my energies.
>
> In addition, as witnessed by my productive work, I liked to cooperate with others, and in this our disagreement with Oppenheimer caused a lot of damage.

This was his last message. He died on September 9.

If we consider that Teller was in the ninety-sixth year of his life, his letters were remarkable. They also convey the impression that he was not quite at peace with himself and the world, and that he was readying himself for the judgment on his oeuvre by later generations. This book may be a step in that process.

The emphasis in this book is on extracting the essence and the main lessons of Edward Teller's life and oeuvre rather than following his life day by day. There has been a lot of misleading information about Teller in the literature, much of it based on superficial errors and some not so superficial factual ones. It is also remarkable how such erroneous information propagates from one piece to another. I seldom point out such erroneous statements with references; rather, I tried to do my best to communicate everything from reliable and independent sources and after proper checking.

There were some significant points in Teller's life, and I build my account around them. There need be no special justification that his childhood and youth affected his later life. He spent his early years in an entirely different world from the one he operated in during his later life. He acquired many of his traits in Hungarian society where he spent his first eighteen years. It was a turbulent time, and his environment was by and large hostile to him. The first chapter covers this period with historical background. The second chapter shows Teller's road to science in Germany and the impact of his mentor, Werner Heisenberg. The third chapter covers the broadest domain geographically as Teller moved from Germany to Copenhagen to London and to the United States. Here we also get a glimpse into his science, which involved chemistry and physics.

The fourth chapter reflects on Teller's life as it was blending with world events and in particular with the creation of the atomic bomb. The uncertainties of the immediate postwar era are presented in the fifth chapter, as is Teller's increasing involvement with the dilemma over the hydrogen bomb. The development of the hydrogen bomb and how Teller reflected on it subsequently is presented in the sixth chapter. The seventh chapter introduces the forming of the second weapons laboratory, which was the penultimate step in Teller's breaking with his peers. His testimony in the Oppenheimer hearing and how it sent him into his third exile is the topic of the eighth chapter.

The ninth chapter shows him as a formidable debater in his fierce opposition to test bans. The tenth chapter deals with a plethora of issues illustrating that he was a "monomaniac with many manias." Teller's role in the Strategic Defense Initiative is presented in the eleventh chapter. The chapter of final thoughts brings him back to the land of his birth and to Russia, and assesses his achievements and his legacy.

I have examined everything about Teller in context and looked at him and his deeds against the backdrop of the historical circumstances of his time, bearing in mind how his peers acted during the same events. The twentieth century was a tragic period in the history of humankind from the point of view of the terrible wars fought and of the long reigns of two totalitarian regimes. It was also the time when democracy and science gained tremendous momentum and achieved so much.

Fans and foes of Edward Teller probably agree that he was an extraordinary person in our immediate past. As time passes, we become increasingly able to view and judge him with a diminishing amount of bias. Aware of the strong emotions he still generates, I feel the following caveat is justified. To those who are very sensitive to any criticism concerning their hero, Teller's oeuvre was significant enough to afford the luxury of viewing everything about him with a critical eye. After all the criticism has been waged, there will still be plenty to appreciate and admire. To those for whom Edward Teller personified evil itself, we owe it to objectivity to examine his life and oeuvre, filtering out emotions, because no good comes from blind condemnation. During more than the past half century, the world has survived a long period that was pregnant with the possibility of a calamity of unprecedented proportions. Let us see what role Edward Teller played, if any, in this survival.

A special note is warranted on the usage and spelling of Hungarian names. I use the versions of the names as their bearers used them in the United States, but I will indicate their original versions where context requires. As an example, Edward Teller was born Ede; he changed his first name to the German Eduard for a short while, and then to Edward in the United States.[11] Although some would prescribe using "he" or "she" rather than "he" in general formulation, I settled on the latter for the sake of simplicity. Likewise, it is not contrary to political correctness that I refer to Teller as "Teller" and to his wife as "Mici." I have equal respect for them both.

# Chapter I

# TELLER IN HUNGARY

## *Origins and Background*[1]

The eyes of childhood are magnifying lenses.
—EDWARD TELLER[2]

*Teller spent his first eighteen years in Hungary during which he experienced: the "happy peace time" of the Austro–Hungarian Monarchy; a lost war; the dismemberment of Hungary; democratic revolution; communist dictatorship; counter-revolution and right-wing terror with vicious anti-Semitism; the first anti-Semitic legislation in post–World War I Europe; and a degree of stabilization. He was too young to truly appreciate the benefits of the "happy peace time," but it remained a point of reference for most people around him. His childhood and early teenage years were typical of the Jewish–Hungarian upper middle class, and his lifelong friendships originated from the encounters of his youth. Many of his values, but also some of his limitations, dated back to this period. Being a Jew in Hungary in the 1920s was an important factor in shaping his views and outlook over his lifetime.*

Teller was not born in the house where the memorial plaque was unveiled on January 15, 2008. The building of his birthplace used to stand a few blocks away; it was typical of the architecture of fin de siècle Budapest. A few years ago, the building was demolished to give way to a modern glass-and-steel office complex. Teller was five years old in 1913

Edward Teller's family lived in this house at the time of his birth. The house was located at 3 Gorove (today, Kozma) Street. The building has been demolished to give space for a modern office complex. Drawing by Gyula Széles, courtesy of László Kovács, Szombathely, Hungary.

Edward Teller and his family lived in this house from 1913. The house was on 3 Szalay Street. Photo graphics by and courtesy of Magdolna Hargittai, 2009.

when he and his family—his father, mother, and sister—moved into the house in Szalay Street.

He was often taken to play at the big square of the Parliament building a few hundred yards away. The neo-Gothic complex dated back to the prosperous last decades of the nineteenth century and was large even for the Kingdom of Hungary, which was not a fully independent country but part of the Habsburg Empire at the time. The building was hugely out of proportion with the much smaller Hungary at the time of Teller's youth. The two lions standing guard in front of its main entrance remained imprinted in the child's mind forever. When decades later Nelson Rockefeller asked Teller if he would like anything from Budapest, Teller mentioned the image of these lions.[3]

It was from his boyhood house that Teller traveled to Germany in 1926; there, his boyhood home, is where he returned to recuperate after his trolley accident in Munich, and where he and later he and his wife came for periodic visits—the last in 1936—first from Germany, then from Denmark and from England, and, finally, from the United States.

The main entrance to the Hungarian Parliament
with the two lions somewhat enlarged by the artist
Gyula Széles, based on the author's photograph.
Courtesy of László Kovács, Szombathely.

## BROADER BACKGROUND

Teller's nephew, Janos Kirz, remembers how his grandfather, Edward's father, even in the late 1940s, talked continually about the good old days of the period before World War I. In Janos's words, "The period prior to World War I was a tremendous formative time that bordered on illusion. Edward lived through a period of well-being and comfort and security—false as it turned out to be—and a sense of Hungarian identity that went with it, something that people who were born in that period kept yearning for."[4]

The Budapest of the "happy peace time," 1867 to 1914, was a uniquely fertile time for promoting talent. The first date, 1867, refers to the so-called Compromise between the Habsburgs and Hungary, and the second, 1914, to the outbreak of World War I. In the Compromise, the Habsburgs and the Hungarians came to an agreement following the crushed Hungarian Revolution and war for liberation of 1848–1849 against Austria. The Habsburgs alone could not defeat the Hungarians, so they had to call in the mighty Russian czar to their rescue. A ruthless period of terror followed, but in a few years' time, Austria, weakened by lost foreign wars, could no longer live with a rebellious nation under her rule.

The Compromise created a dual monarchy of Austria and Hungary and a personal union under Franz Joseph I. In 1867, he became the king of Hungary in addition to being the emperor of the rest of the Empire, which he had been since 1848. The Austrians and the Hungarians henceforth reigned over smaller nations. The Kingdom of Hungary, for example, included Croatia and Slovakia, which today exist as independent nations. The monarchy had Vienna and Budapest as its twin capitals, and the two cities often competed against each other, which was beneficial to both. Budapest was born from uniting Buda, Pest, and Óbuda (ancient Buda) in 1873, with Buda and Óbuda on the west bank of the Danube and Pest on the east. Buda is hilly, and Pest is considered to be a plain because its elevation is gradual; its outer districts reach the altitude of the conspicuous Gellért Hill of the Buda side.

In 1867, the door opened to unprecedented progress in Hungary, and Budapest became one of the fastest-growing cities in Europe. Immigration was encouraged. At the end of the nineteenth century, Budapest was one of the main destinations of Jewish immigration in the world; second only perhaps to New York City. According to some, the more well-to-do Jews congregated in Budapest and the poorer went to New York.[5] The large class of the Hungarian nobility monopolized the political bureaucracy, the judiciary, and the military. The chairman of the first congress of the Hungarian chemists in 1910 lamented: "We are first of all a nation of lawyers and all who aim at dealing with and improving the fate of our fatherland have the mistaken belief that only a law degree enables us to do so."[6] He called for turning the attention of aspiring young people toward studying science and technology.

The one-sided interest of the Hungarian nobility left wide open other intellectual trades, which became vastly popular among the recently emancipated Jews as well as among the Germans and other minorities in Hungary. Jews even graduated from law school in large numbers, but they remained lawyers and only seldom became members of the judiciary. Such division of labor between Hungarians and Jews did not slow the intensifying Jewish assimilation. The political elite welcomed this assimilation because the Hungarian population was in the minority in large areas of this multiethnic country, and the Jewish Hungarians could be counted on as loyal Hungarians.

There are opposing views as to the origin of the great possibilities for Jews in fin de siècle Hungary. According to one view, it was because the country was so liberal. The opposing view maintains that it was exactly Hungary's backward feudalistic regime that brought about those unique opportunities for Jews. In any case, their numbers were swelling because the absence of persecution attracted immigration. They burst onto this scene after centuries of having been excluded from higher professions, so they were looking for new opportunities.

Budapest of the early 1900s has been the subject of scrutiny and admiration because of its extraordinary production of gifted scientists, artists, composers, and playwrights.[7] One-fifth of its population at that time was Jewish, a group consisting of native city inhabitants, incomers from the provinces of Hungary proper, and immigrants from all directions. My own ancestors came from the northwest on the paternal side and from the southwest on the maternal side. They spoke German, and soon learned Hungarian. Many came from the East, from Galicia, and they spoke Yiddish. Interestingly, Galicia has produced an extraordinary number of first-rate scientists, often going through Hungary. This fact is little known because often these outstanding contributors to world culture lost their Eastern origins at some point and are now regarded as Austrians. An example is the great American physicist Isidor I. Rabi, a Nobel laureate, whose autobiography states that he was born in Raymanov, Austria.[8] His biographers write more precisely that he was born in Rymanow, Galicia, "then a province of the Austro–Hungarian Monarchy, now in Poland."[9]

The Hungarian–Jewish coexistence prospered in the "happy peace time," and the Jews in Hungary spoke Hungarian and German; relatively few spoke Yiddish. Conversion to Christianity helped Jewish careers, but in many professions nonconversion did not constitute a barrier. Even though some professions did not open up for Jews between 1867 and 1914, there were numerous new opportunities for the Jewish middle class, and even for recently arrived immigrants. This was not unlike the modern American scene where recent educated immigrants may find their fastest road to success by becoming scientists, doctors, and engineers.

The outcome was spectacular. Its impact was magnified in the decades to come because much of this talent would leave and find self-fulfillment abroad, initially in Germany, and ultimately in the United States. Teller and other scientists were not unique in this respect. In addition to scientists, Hungary was the launching pad for many significant playwrights and composers, artists and film directors, mathematicians and economists, and others who would make their mark on the international scene.[10]

For prominent Jews the ultimate sign of assimilation in Hungary was to become a member of the hereditary nobility. And what could have been better evidence of the liberal atmosphere of the Habsburg Empire than the fact that even nonconverted Jewish families could acquire such a distinction? In the early 1900s, even a few members of the Hungarian government came from Jewish families, though they had names that sounded genuinely Hungarian. But World War I, known also as the Great War, brought an end to this seemingly ideal situation. The most forward-looking Jewish families had sensed that the peaceful conditions could not last forever. So they made sure that their children received a good education that would help them survive in any part of the world. To further this end, they stressed the cultivation of modern languages and practical professions. The Teller family was typical in this respect.

In 1867, new legislation guaranteed civil and legal equality for Jews in Hungary. There was also budding anti-Semitism, to be sure, that would grow into a formidable force after World War I and the ensuing revolutions. The anti-Semitic outbursts during the period from 1867 to 1914, however, did not disturb the peaceful atmosphere, and, if anything, the response showed the resilience of society by resisting it. There was, for example, the case of a young Christian peasant girl allegedly killed by the Jews as part of their religious ritual in Tiszaeszlár (a village in eastern Hungary). A trial followed, with a divisive impact on Hungarian society at large, but the accused were acquitted. "Respectable opinion, including most of the aristocracy and the gentry, rejected anti-Semitism."[11] Thus Hungary fared better on this occasion than France did in the Dreyfus affair.

Fin de siècle Budapest was conspicuously attractive. "Foreign visitors arriving in that unknown portion of Europe, east of Vienna, were astounded to find a modern city with first-class hotels, plate-glass windows, electric tramcars, elegant men and women, the largest Parliament building in the world about to be completed."[12] This description is characteristic at more than one level and conveys the perception of ambition as well as pretension. Budapest was like a younger brother to Vienna. The period witnessed prosperity in many areas. And Budapest became a banking center for the broader region.

Considerable technological innovation and large-scale industrialization took place in Hungary, with Budapest leading the rest of the country. Modernization was more conspicuous in Budapest than in most other great cities in Europe because in many areas it was not just the new gradually replacing the old, but the new being created where nothing had existed. Having lacked industry in many areas before, it could start with the most modern concepts and constructions.

The Elizabeth Bridge was completed in 1902, and at that time it was the largest single-span bridge on Earth. It stands today but completely rebuilt because it was—as all the other bridges of Budapest were—blown up by the retreating Germans in 1945. The city became a world-class metropolis by around 1896—the Hungarian Millennium—and one hundred years later it was still the progress of that earlier period that made Budapest attractive. Typical of this progress is the Castle Hill, which is recognized by UNESCO as a World Heritage site. The first subway on the European Continent was built and running in Budapest. The Ganz factory developed and built the first electric railroad engine in the world. Soon after the telephone system was introduced in Budapest, a telephone service for news and entertainment was initiated—a forerunner of radio.

There was unprecedented growth in educational and cultural opportunities, never approximated since. New buildings went up for the University, the Technical University, the Music Academy, and many other institutions. During the period from 1867 to 1914, Budapest was the fastest-growing city among the major cities in Europe, and may have been the fastest growing in the world, save for Chicago. Budapest had close to one million inhabitants by the time World War I began. That meant that roughly one in every twenty people in Hungary lived in the capital city, which was considered to be a healthy proportion.[13] Part of the progress was the development of the school system; parents expected their children to surpass them in all facets of life. In addition to the old landed aristocrats, a new upper class emerged, the financial aristocracy, but there was a well-defined division of influence between them. This division was bridged on occasion, however, by intermarriage between aristocratic sons and the daughters of rich, often Jewish, members of the financial aristocracy.

By the end of the nineteenth century, the gentry class had lost their land and largely congregated in Budapest, gaining positions in the civil service regardless of their qualifications. They would have found it beneath their pride and dignity to deal with commerce, finance, industry, or business of any kind. There is some resemblance in this behavior to the English attitude reflected in Jane Austen's novels, but the Hungarian version was being played out a hundred years later. The Hungarian gentry formed the so-called Christian gentlemanly middle class, which remained characteristic through World War II. The label stressed Christians to signify the exclusion of Jews from their ranks, and this echelon would foment future anti-Semitism. The sons of this class went preferentially into the army and law and seldom studied to become medical doctors and engineers. The division between the two strata was the "division between the urban and the populist, between the commercial and the agrarian, between the cosmopolitan and the nationalist," that is, between the Jewish Hungarian and the non-Jewish Hungarian culture.[14] In the period of growth and prosperity, the animosities seldom came to the surface, but this would change later.

Although the great majority of the Jewish population was not politically active, there was a vocal and visible layer of it. After World War I, this group was conspicuously among the former prisoners of war returning from Soviet Russia who wanted to change the social system and took their example from Lenin's revolution. The bourgeois democratic government of Count Mihály Károlyi came about at the end of the war, in October 1918. However, it soon folded under the pressure of the Western Allies' demands for the dismemberment of Hungary. Fearing an attack from neighboring countries, Károlyi handed over the government to the communist Béla Kun and his comrades in March 1919. An ill-fated and ill-managed communist rule of 133 days followed. Kun was Jewish, as were the majority of his people's commissars.

By the time Admiral Horthy entered Budapest on a white horse, the relationship between Hungarians and Jews had changed drastically. The peace treaty of June 4, 1920, signed in Trianon (a palace in Versailles), dismembered historic Hungary. It gave independence to Croatia and Slovakia in the new Yugoslavia and Czechoslovakia, respectively. It also carved out large chunks of the country to add them to Romania and even some territory to Austria. There was no longer any need for assimilated Jews to enhance the Hungarian population. In Hungary proper, there were hardly any sizeable minorities, whereas Hungarians were trapped by the millions in the neighboring countries.

The period between 1920 and 1944 is usually called the Horthy era after the head of state Miklós (Nicholas) Horthy. There were significant differences between the Hungary where Teller spent his childhood and the one of his high school years. The education in Hungary at the time of the monarchy was outward-looking, introducing reforms to modernize instruction. In the Horthy era, education was inward-looking; it placed emphasis on nurturing extreme nationalistic feelings, keeping alive the idea of Greater Hungary, and vowing to replace the ostensible Jewish presence with true Hungarian aspirations.

The main ambitions of educational policy were pronounced in a 1921 publication of a leading official, which set the tone for the whole Horthy era, including the entire period of Teller's high school studies.[15] Its main goals were (1) "positive nurturing of the sense of a nation"; (2) "protecting the minds of young people against the spirit of internationalism"; and (3) "re-hungarianizing the nation's intellectuals, or what might term their hungarianization in place of their judaization."[16]

What followed World War I, the failed revolutions, and the humiliating Trianon Peace Treaty was the White Terror and vicious anti-Semitism of the early 1920s. Horthy became the regent of Hungary in the absence of a king in the Kingdom of Hungary, and so he remained until October 1944. Legislation was introduced in 1920, known commonly as *numerus clausus*. The Latin expression means "closed number," and it severely limited the number of Jewish students admitted to higher education. This law (XXV: 1920) had the dubious distinction of being the first anti-Jewish legislation after World War I in Europe.[17] What a contrast it was to Law XII: 1867, which made it possible for Jews to become high school teachers and even university professors.

Some ascribe the explosion of excellence in Hungarian mathematics to the beneficial impact of the 1867 law. Many of these mathematicians studied in Germany but returned to Hungary afterward and enriched the country's intellectual and university life. The weight of the 1920 legislation could be appreciated if we compare the estimates of the Jewish proportion of the population at about 5 percent and the Jewish proportion of the country's medical doctors and some other professionals at well over 50 percent. The consequence of *numerus clausus* was a severe depletion of qualified professionals. Toward the end of the Horthy era, *numerus clausus* practically became *numerus nullus*, parallel with Hungary's increasing alignment with Nazi Germany. However, during the second half of the 1920s, there was some relaxation in the application of this legislation.

Compared with the extensive industrialization and modernization in Hungary prior to World War I, stagnation characterized the period between the two world wars. The difficulties in restarting the country after Trianon cannot be overstated. In spite of restrictive measures in higher education mentioned above, there was also progress. There were attempts to bring back outstanding scientists who had left the country. The most conspicuous case was Albert Szent-Györgyi, who returned from Cambridge to the University of Szeged. He became the only scientist who traveled from Hungary to receive his Nobel Prize in Stockholm, in 1937. In the sciences, all other Nobel laureates of Hungarian origin would be based in foreign lands when they were awarded this most prestigious of all prizes. The worldwide economic crisis signaled by the crash at the New York Stock Exchange impacted Hungary as well. The ensuing years brought

Gyula Gömbös to the position of prime minister in 1932 (until his death in 1936, the year of Teller's last pre-war visit to Hungary). Gömbös's views were very close to fascist and Nazi ideologies.

How and to what extent the Hungarian experience affected Teller's life and the lives of others in similar predicaments is an interesting question. The physicist George Marx prepared a graph in which he plotted the school years of outstanding scientists and noticed that the peaks appeared during the years of the world wars. He suggested that hard times facilitated the occurrence of creativity, at the same time that many of the traditional values lost their meaning.[18] The anticipated nature of the next regime and its ideology directed people in deciding whether to try to adapt or escape. This applied to everybody, but given the circumstances, Jews felt it the most. Living through drastic changes provided a unique Jewish experience, according to Marx, which promoted their getting adjusted to different conditions whether at home or in exile.

Von Neumann ascribed the great production of talent in the region to "a coincidence of some cultural factors . . . an external pressure on the whole society of this part of Central Europe, a feeling of extreme insecurity in the individuals, and the necessity to produce the unusual or else face extinction."[19] This is an important circumstance that added much to the feeling that emigration was a way out. According to the Hungarian–Swedish tumor biologist George Klein: "You either became successful or you were going to end up in the gutter."[20] The Nobel laureate discoverer of holography, Dennis Gabor, simply stated, "Innovate or die!"[21]

Among his Martian friends, Teller stayed in Hungary for the longest period after World War I and the revolutions. His high school years coincided with the time of the most atrocious anti-Semitism in Europe of the 1920s. Alluding to the anti-Semitic discrimination during his youth in Hungary, he remarked that he would "have to be better, much better than anyone else" in order to survive and succeed.[22] This atmosphere had to have a profound effect on him, though clearly he persevered and succeeded.

Eugene Wigner, Nobel laureate, mused toward the end of his life about the question that he had been asked repeatedly, "Why was his generation of Jewish Hungarians so brilliant?" Wigner gave credit to "the superb high schools in Budapest" and ascribed an even greater share in this to forced emigration. "Emigration can certainly be painful, but a young man with talent finds it stimulating. Outside your own nation, you lack a ready place. You need great ingenuity and effort just to find a niche. Hard work and ingenuity become a habit. Often they are enough to earn you a place above natives of your adopted country quite as talented as you."[23]

Walter Laqueur wrote about the success of most of the young Jewish refugees from Nazi Germany in their new countries. He noted that for them it was "swimming or sinking." They "had to start from scratch, because there was no helping hand, no money, no connections, no safety net."[24]

Between the two world wars Hungarian public life was soaked with irredentism. Following the Trianon Treaty, a competition was announced for a national prayer and a national slogan expressing the feelings of Hungarians. Teller and his classmates had to recite the winning prayer every morning: "I believe in one God, I believe in one fatherland. / I believe in one divine eternal justice, / I believe in Hungary's resurrection! Amen." The winning slogan was displayed and pronounced ubiquitously: "Truncated Hungary is no country; complete Hungary is heaven."

In time, Hitler brilliantly played on Hungarian irredentism and the fact that Hungarian foreign policy focused on regaining the lost territories. It was a small price to pay to introduce German-modeled anti-Jewish legislation and to send hundreds of thousands of Jews to be murdered at Auschwitz. There was also homegrown anti-Semitism, to be sure, which fueled the compliance with German–Nazi efforts. The state machinery of Horthy's Hungary, and in particular the Gendarme, often outdid the Germans in their zealousness and eagerness to get rid of the Jews.[25] There was then an additional component—a material interest in expropriating Jewish property. The anti-Jewish laws of the late 1930s and early 1940s benefited many in direct material and financial sense in Hungary, where the Jewish population amounted to about 5 percent of the whole population. This is also why the few surviving Jews, as they were returning from the concentration camps after the war, were met with animosity or worse by the people who had benefited from the loot of the deported and murdered Jews.

## ORIGINS

It is now time to take a closer look at Teller's roots and his first years. We cannot go back many generations in the Teller genealogy, because most records have been lost due to the wanderings and persecution of Teller's forebears. What follows is a brief summary of what we know about his immediate ancestors.[26]

They lived in regions of the Kingdom of Hungary that are today parts of other countries. His paternal grandfather was a merchant in Érsekújvár in northwestern Hungary (today it is Nové Zámky in Slovakia). Though a

## THEODORE VON KÁRMÁN (1881–1963)

There were five Martians of science, von Kármán, Leo Szilard, Eugene P. Wigner, John von Neumann, and Edward Teller. The last four were closer to each other in age than von Kármán, but his path paralleled the rest to a remarkable degree. All of them were born into well-to-do upper-middle-class Jewish families in downtown Budapest during the "happy peace time" of the Austro–Hungarian Monarchy. Von Kármán went to one of the famous high schools, the one his father, Maurice Kármán, had founded and where Teller went two decades later. The father was elevated to hereditary nobility in 1907; hence the "von" in front of their surname.

Theodore studied engineering at the Budapest Technical University and spent a few years at the start of his career there after graduating and before he left Hungary for Germany. He did his doctoral work in Göttingen and became full professor at and director of the Aeronautical Institute at the Technical University in Aachen. Then he accepted an invitation from the California Institute of Technology (Caltech) and moved to the United States a few years before Hitler came to power in Germany.

He founded the US Institute of Aeronautical Sciences from which the Jet Propulsion Laboratory evolved in Pasadena. Von Kármán was as interested in defense matters as the others, but he differed from Teller and Szilard in that he worked behind the scenes and did not publicly lobby for any of the programs he found important. He was instrumental in modernizing the Air Corps of the US Army prior to and during World War II, and in establishing and developing the US Air Force after the war.

He received recognition in the United States and internationally, and was the first recipient of the National Medal of Science from President John F. Kennedy in 1962.

thriving regional center, Jews could not stay overnight in the town until the early 1800s; still they were active in trade on market days. When they finally gained the right to settle in the town, they brought prosperity to it. Érsekújvár developed early into a railway center.

Teller's father, Miksa (Max) Teller (1871–1950) went to elementary school in Érsekújvár and to high school in Pozsony (today, Bratislava, the capital of Slovakia). Miksa had a difficult youth; by the time he went to law school in Budapest, he had lost his parents and had to finance his own studies. In addition, he had to help his sisters who had also moved to Budapest. He was hardworking and successful and after graduating from

## LEO SZILARD (1898–1964)

Leo Szilard was born in Budapest into an upper-middle-class Jewish family. After high school, he started his university studies at the Budapest Technical University, and continued them at the Technical University of Berlin from 1920. He gradually shifted to physics at Berlin University and received his doctorate there under Max von Laue, in 1922. Decades later, his thesis work was recognized as the starting point of information theory. After the Nazis' accession to power in Germany he spent a few years in London and moved to the United States in 1938. During the period of 1933–1934, he originated the ideas of nuclear chain reaction and critical mass.

After the discovery of nuclear fission, he engineered Albert Einstein's letter to President Franklin D. Roosevelt, which called attention to the possibility of Germany's developing the atomic bomb and the need for America to develop the bomb first. Szilard thus initiated the American efforts for nuclear weapons. He participated in the Manhattan Project but then tried to prevent the deployment of atomic bombs without prior demonstration against Japan. During the cold war, he carried out extensive activities to promote negotiations and peaceful coexistence with the Soviet Union, disarmament, and banning nuclear explosions for testing.

He always conducted his efforts at the highest level, writing letters to Stalin and exchanging his views with Khrushchev. Some of his innovative schemes that initially sounded absurd were eventually adopted. He was a most colorful personality who left valuable marks in the form of appreciated or unsolicited advice in numerous scientific fields, including molecular biology.

law school, he opened his law practice in Budapest in 1895. He met Ilona Deutsch (1884–1977) in January 1904; he was thirty-three and she was twenty years old.

The Deutsch family lived in Lugos in southeast Hungary (today, it is Lugoj in Romania). The town had a substantial Jewish community originating from Germany and Austria starting from the end of the seventeenth century and from Galicia starting from the end of the eighteenth century. Ilona's father was a well-to-do banker and he became Teller's favorite grandparent. The language of the Deutsch family was German, but Ilona learned several other languages as well, including Hungarian. She enjoyed playing the piano, which she did very well.

Miksa and Ilona became engaged within a few weeks after they met and married within a few months. They settled in Budapest and had two children, Emma (born in 1905) and Edward, whose original Hungarian name was Ede (born in 1908). After a brief first marriage in the mid-1930s, Emma married a lawyer, András Kirz. He had been an office worker and completed his law studies in night school. Following graduation he stayed in the bank where he had worked as a clerk and became involved with the bank's legal matters. He was killed in a concentration camp in 1945. Miksa Teller continued a thriving practice in spite of various restrictions. He was a prominent member of the legal community in Budapest and at one time co-editor of the law review *Jogtudományi Közlöny*. Anti-Semitism prevented him from becoming a judge.

## EUGENE P. WIGNER (1902–1995)

Eugene Wigner was born in Budapest into an upper-middle-class Jewish family and went to a Lutheran high school famous for its outstanding teachers and pupils. He started his university studies in chemical engineering in Budapest, moved to Berlin in 1921, and continued at the Technical University there. He did his diploma work under Herman F. Mark and his doctoral work under Michael Polanyi. He attended the famous physics colloquia in Berlin and started his university career in Germany.

He moved to Princeton, first part time, then, from 1933, full time. Wigner had excellent graduate students at Princeton and through their contributions he did more than anybody to create modern condensed matter physics in the United States. Wigner spent a short period of time at the University of Wisconsin in the mid-1930s and during World War II worked for the Manhattan Project. He utilized his thorough knowledge of materials from his chemistry studies in his work for the atomic bomb. His contribution was especially important in developing the first nuclear reactors, and he was the world's first nuclear engineer. After the war, he served as director of research and development at the Clinton Laboratory in Oak Ridge, Tennessee, which is today the Oak Ridge National Laboratory. However, he preferred teaching and independent research. He moved back to Princeton and stayed there to the end of his life.

Wigner did pioneering work in applying symmetry considerations in atomic physics in the 1920s and 1930s, for which he received the Nobel Prize in Physics in 1963.

## CHILDHOOD

In spite of the warm and caring family environment, Teller was very lonely as a child. He did not start speaking until he was three or four years old (according to different sources), which caused anxiety in the family. But when he started speaking, he did so in full sentences immediately. He first started playing with numbers when he was four years old, and this became one of his earliest memories.[27] He built up his own inner world in which thinking about numbers had an important place. Afraid of the dark, he played with numbers in his head to help him bridge the period between lights out and his falling asleep. He liked to count the seconds in a minute, the minutes in an hour, the hours in a day, the minutes in a day, the seconds in a day, and so on. He developed the ability to carry out complex arithmetical operations in his head. Teller ascribed his interest in numbers to the confusion of speech he had heard at home due to his family being bilingual. He was taught German and Hungarian at the same time, so words became confusing, whereas numbers were unambiguous.

He was not unique in his talent with numbers among the Martians. Von Neumann was a child prodigy as a mathematical wizard, and so was von Kármán. Teller's parents did not mind their son's developing such ability, and neither did von Neumann's parents. In contrast, Maurice Kármán, Theodore's father, who was an educational expert, asked his son to forget about mathematics—not because he did not care for his son's mental development but because he was afraid that his son would turn into a freak and be used merely for entertaining audiences. In pre-calculator times there were entertainers who made their living by performing as mathematical wizards.

From the age of eight, Teller learned to play the piano, and he liked it a great deal. But when he was made to attend courses at the Music Academy, he failed there. He preferred to play for his mother's enjoyment and his own, a pleasure he enjoyed to the end of his life. Also, to broaden the children's outlook, the parents took him and Emma to sites in Hungary and in neighboring countries. They engaged a governess, Magda Hesz, from the time Teller was seven years old, and she stayed with the family for seven years. She had grown up in Chicago where her Hungarian parents had immigrated, but when she lost them, her relatives sent her back to Hungary. This is how the Teller children learned English from an early age and how they learned American usage, which was unusual in Hungary, where the British version was generally preferred. Magda later returned to Chicago and after World War II served as a link between the American and Budapest wings of the Teller family.

Teller's mother was a worrying type; she worried about everything in connection with her children. Teller was not sent to regular school right away when he reached school age; rather, he was partly educated at home by his mother and the governess, and partly at a nearby private school for a few hours daily. At home he was both serious and contemplative, yet also a lively child. Many decades later his father's law clerk remembered him as "a little boy with very flushed cheeks, playing Mowgli" bursting into the law office.[28]

He started attending regular school, *gimnázium*, when he was nine years old, a little earlier than most children. Teller liked to immerse himself in his own thoughts from early childhood. On such occasions when others tried to speak to him, he asked them to leave him alone, saying that he was busy thinking. He suffered much in school from the teasing by his classmates, but once he learned to ignore it, his situation eased. Then he earned some respect by helping others in their studies, and started making friends.

Teller liked his maternal grandfather the best of the whole family. He even resembled him in appearance. Even the limp Teller would later develop after his trolley accident in 1928 was similar to that of his grandfather, who developed one after he got caught under a horse. The grandfather was likable and even-tempered according to Teller's description. Everybody in the family liked him, except his wife.[29]

During his summer vacations and the longer forced absence in the fall of 1919 in Lugos, Teller had interesting exchanges with his revered grandfather that provided life lessons for him. Teller asked his grandfather about the validity of laws because he had some doubts. He asked him whether it was right to take "an eye for an eye, a tooth for a tooth." He never forgot what his grandfather told him:

> Laws must be obeyed without exception. The law cannot make everyone a saint. Only a very few people are saints, and obeying the law must be possible for all people. If someone knocks your tooth out, you have a strong urge to hit back. The meaning of the law is that you must never take more than one tooth for a tooth. To forgive is much better. But the law cannot forbid the desire for revenge. It can only limit it by justice.[30]

Teller never accepted that personal responsibility might override what the law says, even if it worked against one's conscience. Judith Shoolery, Teller's longtime editorial assistant and the co-author of his *Memoirs*, noted in this connection:

He held firmly that, in a democracy, the obligation to obey the law was absolute. To disregard democratically established law at one's own convenience would lead to anarchy. If a citizen of a democracy considered a law wrong or unjust, the citizen had an obligation to work to get the law changed. He never answered my question as to what should be done under a fascist system.[31]

## GIMNÁZIUM EXPERIENCE

At the beginning of the twentieth century the Hungarian high school, the *gimnázium*, encompassed eight years of the student's life, from the age of ten to eighteen, so it had a profound impact with long-lasting consequences on its pupils. The gimnázium as an institution had a prominent role in Hungary's cultural life, and on a relative scale, the fin de siècle Hungarian gimnázium was far more advanced than the Hungarian university of that era. It was the single noteworthy influence on Teller in Hungarian life, outside of his family.

The Minta Gimnázium, then and now. Drawing by László Pittman, courtesy of László Kovács, Szombathely, Hungary.

The Hungarian gimnázium was modeled after its German counterpart, the gymnasium, which had been thoroughly studied by Hungarian experts of education before it was adapted. The Hungarian state education system was reformed following the Compromise, and the architect of the reform was Maurice Kármán, who studied the German system in Leipzig. Upon his return to Hungary in 1872, he initiated teachers' training at the University of Budapest and founded the Minta Gimnázium—a state school—to provide the teachers-to-be an opportunity to practice teaching. The name *Minta* means "model," and it was to be a model high school.[32]

The high school was a great equalizer among its pupils, and Jewish boys could enter non-Jewish denominational schools.[33] Thus, for example, John von Neumann (or as he was known at that time, János Neumann), who had not yet converted, attended the Lutheran Gimnázium. The future Nobel laureate in chemistry George de Hevesy (or, Georg von Hevesy) attended the Catholic gimnázium of the Piarist Fathers as a Jewish pupil.[34]

The gimnázium was an elitist high school providing an all-around humanistic education that included Latin and Greek, but its students usually excelled in mathematics and the sciences as well. The high level of gimnázium education was ensured by excellent teachers, a few of whom were elected members of the Hungarian Academy of Sciences and many of whom did independent research. There was interaction between high schools and universities, and it was not unprecedented for a gimnázium teacher to be invited to give a course at a university in a field in which he excelled. There was an overproduction of high-level university graduates and many of the best minds landed in gimnázium positions.

The world-renowned physicist Loránd Eötvös was an important contributor to elevating the level of math and science education in Hungary. He was briefly minister of education in 1894 and president of the Hungarian Academy of Sciences for over fifteen years (1889–1905). The *Középiskolai Matematikai Lapok* (High School Mathematical Periodical) was initiated at that time. It has survived (augmented with physics) to this day, along with an elaborate system of high school competitions. Loránd Eötvös took the Minta Gimnázium under his patronage.

Theodore von Kármán called the Minta "the gem" of his father's educational theories and lamented that it was little known in the West.[35] A visiting journalist of the London *Observer* called it a "nursery for the elite" and compared it to Britain's Eton and France's Le Rosey. In addition to von Kármán and Teller, the school's famous pupils have included medical doctor turned physical chemist turned philosopher Michael Polanyi; two British economists, Thomas Balogh and Nicholas Kaldor, who eventually became

# JOHN VON NEUMANN (1903–1957)

John von Neumann was born in Budapest into an upper-middle-class Jewish family. His father was conferred nobility, hence the "von" in front of his surname. Von Neumann graduated from the same high school as Wigner. Then he studied mathematics at Budapest University and chemical engineering in Berlin and Zurich, and received his PhD in mathematics in Budapest in 1926. He had university positions in Germany, first in Berlin, then in Hamburg, and later moved to Princeton in 1930.

He became one of the first professors at the Institute for Advanced Study (IAS) in Princeton. He solved numerous problems in a great variety of scientific and defense areas, and during the war served as a consultant at the Los Alamos Laboratory and for other laboratories. In addition to his works in pure mathematics and quantum mechanics, he co-authored a book on game theory and economics. After the war, he was increasingly active in defense matters and was aggressively anti-Soviet, but he did not clash with his colleagues.

Von Neumann's visibility increased after 1948, when he acted as consultant in the US armed forces as well as for private companies. His principal occupation remained his developing the modern computer with stored programming capabilities. He supported Teller's fight for the development of the hydrogen bomb and helped him in the actual work. He was a member of the General Advisory Committee of the Atomic Energy Commission (AEC) in 1952–1954, and from 1955, he was a commissioner of the AEC. He received the first Fermi Prize in 1956 and the Presidential Medal of Freedom from President Eisenhower in 1957.

As chairman of the nuclear weapons panel of the Scientific Advisory Board of the US Air Force, he had a great impact on American defense policy. The panel determined that it would be possible to develop long-range missiles and equip them with nuclear warheads. The United States developed a strong missile force over the years primarily due to von Neumann's activities.

Von Neumann possessed an exceptionally quick mind. He was famous for his wit and humor, but he did not have a very warm personality and kept a distance even from his closest friends. According to Wigner, there was only one genius among the Martians, and that was von Neumann. Others noted that von Neumann was a very special species and when, sometime in the future, there is a higher version of human existence, it will be like him.

barons in Great Britain; Oxford physicist Nicholas Kurti; and the 2005 Abel Prize winner, mathematician Peter Lax of New York University.[36]

Maurice Kármán believed that the best way to teach was by showing connections with everyday life. Even in teaching Latin, he did not start with the rules of grammar; rather, he and his students walked around the city and collected Latin expressions wherever they found them, on buildings, shop signs, everywhere. They also looked for similar words, tried to figure out their meanings, and looked them up in the dictionary. This way they built up a vocabulary in a playful manner and learned a lot about Hungarian history. Latin was especially important because it had been the official language in Hungary until 1844. All the speeches in the Hungarian Parliament used to be given in Latin. The language was taught as if it was alive and it came to life in the classroom.

Mathematics was taught in the same vein. Students looked up various statistics about production in Hungarian agriculture, set up tables, constructed graphs and located minima and maxima on the graphs, and learned about the rates of change. This way they got close to calculus rather early on in their studies. Such studies proved useful not only for enhancing their mathematical skills but also for augmenting their knowledge in geography and social studies as well. The approach to education at the Minta did not favor memorization of rules; rather, the pupils were encouraged to develop the rules themselves. Learning inductive reasoning in practice, that is, deriving general rules from specific examples, proved to be especially useful in von Kármán's later life.

The Minta was innovative in other aspects as well. The encounters with practicing teachers who were university students and only a few years older than the high school pupils brought teaching closer to the students. They also learned to form judgments of good and bad teaching. The "grown-up" teachers of Minta were encouraged to mingle with their pupils during the breaks in the hallway and were allowed to shake their hands if they met outside class. They could talk not only about school matters but about other topics as well.

The Hungarian gimnázium had some negative elements, however. Lukacs noted the rigidity of its requirements. "In most classes the hour began with recitation, meaning that each student had to be ready for testing and questioning each day." As a consequence, "students were haunted by the fear of being suddenly called on, of being inadequately prepared, and of receiving a consequent poor or failing mark at the end of the semester." Lukacs allowed that this approach may have contributed to self-discipline, but he warned that the "almost impossible demands also brought forth . . . an early and youthful realization that cutting corners and disregarding rules, that clandestinity and prevarication were inevitable conditions of survival in a world with rigid, categorical, insensitive and often senseless rules."[37] Peter Lax went to the Minta just one generation after Teller, and he referred to his experience succinctly: "You recognized who your enemy was—the teachers—so maybe something like that could explain why it was so efficient. You had to fight for your life."[38] (On a personal note, our daughter went to the same school in the late 1980s, and found a much more pleasant atmosphere.)

Teller joined the Minta in 1917; he studied mathematics, history, Hungarian, German, and Latin, and participated in physical education. What he found to be an academically uninteresting program was not helped by his being a social outcast during his years at the Minta. Teller does not devote much space in his *Memoirs* to his high school experience; it was a painful topic for him. But it lasted eight long years, from September 1917 through June 1925, and he underwent some trials during those years. First, Teller had a very good math teacher (a communist) who taught the class things that Teller remembered even eight decades later, in a 1990 interview.[39] For instance, he learned the rule of nines from this teacher. There is a number, such as 243; if it is divisible by 9, then the sum of its figures, $2 + 4 + 3 = 9$, is also divisible by nine. The pupils liked this rule because it was simple and made sense, and they also liked that the teacher gave a proof to the rule, so it was not merely pronounced but proven. This was the first mathematical proof Teller ever encountered.

When the communist dictatorship was suppressed and the Tellers finally returned to Budapest, Edward had a new math teacher whom he labeled a fascist in the same 1990 interview mentioned above. This teacher cared only that the pupils could write the equations in a legible way. When Teller suggested an alternative solution to a math problem, the teacher wondered if he was taking the same class for the second time, implying that he must have failed the first time, which was an insult. When on another occasion Teller showed a simpler solution to the one the math teacher gave on the blackboard, the teacher called him a genius and added ominously that he did not like geniuses. This teacher's rigid disciplinarian approach to teaching Teller's favorite subject "set me back several years," he commented later. In 1984, he said, "If I ask myself who in my life is the person who did me most damage, it was this mathematics professor and it took years until I recovered."[40]

"Challenging students to explore ideas was not a common aim at the Minta." This was Teller's summary judgment, although he describes some positive experiences with other teachers.[41] Though he referred to his studies of physics there as "painful and slow," he made a point in his *Memoirs* that his studies at the Minta served him well later at the University of Leipzig.[42] So it seems that on balance the Minta provided a strong foundation of knowledge but did not tend to the individual needs of exceptional students. Elsewhere, Teller stated that his playing ping-pong at the Minta helped him more in his later life than anything else he learned there.[43]

According to the yearbooks of the Minta Gimnázium, Teller entered the school as a high school freshman, one of forty-five pupils, in the academic year 1917–18. The distribution of religions among the pupils showed that the majority of pupils were either Roman Catholic or Jewish, each numbering eighteen pupils, or 40 percent.[44] In these data, the number of Jewish students corresponded to students of the Jewish religion. In addition, a few of the pupils were converted Jews, including one of Teller's best friends and the brother of his future wife, Ede Schütz-Harkányi. Even if the Jewish pupils constituted a minority, it was not a small minority, especially at the beginning of Teller's school career. As for the social distribution of the Minta pupils at Teller's time, the bulk was upper middle class.

The yearbooks during Teller's time there record outstanding student activities but seldom mention Teller's name. His name appears, however, among a list of diligent pupils, and he was one of the two pupils who spoke at a cemetery commemoration of two former directors of the Minta. Teller was graded invariably "excellent," which was not very common; Schütz-Harkányi was graded "good." Of the thirty-one pupils who took the final exams at the end of the eight-year studies, five finished "excellent," including Teller and Schütz-Harkányi. Schütz-Harkányi's sister, Auguszta (nicknamed Mici), eventually became Teller's sweetheart and in 1934 his wife.

Teller laments in his *Memoirs* his poor fate during his first years at the Minta. He was so unhappy there that he wanted to transfer to another school. He mentions a friend, Pál Virágh, with whom he had interesting philosophical arguments. He was two years Teller's senior, and his family lived in the building next to the Tellers; Virágh's father was also a lawyer. Virágh was Catholic and attended the gimnázium of the Piarist Fathers.[45] His description of his school sounded like a dream to Teller as compared with the Minta. Virágh invited Teller to join him for excursions organized by the Boy Scouts at his school. According to the *Memoirs*,[46] Teller convinced his parents to apply for him to get accepted by the Piarists, but "unlike Paul, I was not Catholic. I was turned down." Virágh's records can be found in the archives of the Piarists.

The Piarists have been known for tolerance, so Teller's experience seemed puzzling. Combing through eight years of Teller's high school time in the Piarists' archives, where every piece of paper was logged meticulously, I found no trace of correspondence with the Tellers. It was also of interest to see the other side of the puzzle, that is, whether or not the Piarists accepted Jewish students; in other words, not just converted Jewish students but students of the Jewish religion. Here the findings were positive: the Piarists had pupils of the Jewish religion during all the years Teller was a high school student. It seems that the attempt to transfer to the Piarists was only a thought experiment on Teller's part rather than real action.

Incidentally, Teller was the only one among the five Martians who never converted, although conversion was another response, apart from emigration, to the heightened anti-Semitism in Hungary. There was a much stronger movement of conversion among Hungary's Jewish population in 1919–1920 than before World War I, and for good reason. The antagonism grew worse.

The high school years remained an unhappy time in Teller's memory, and the subsequent decades did not mellow his antipathy. When in 1949 he wanted to explain his fear of something, he reached back to the time when he was preparing for his final examinations, "the matura," at the end of his studies at the Minta. "I was convinced that all my sins and deficiencies [would] become evident on that day."[47]

## HIGHER LEARNING BEGINS

Teller's father taught his son that he, as a Jew, had to excel just to keep abreast. He further said that one day he would have to immigrate to a country where conditions were more favorable for minorities, and therefore he had to acquire a profession that would be valued internationally. Emigration was forced upon young people like Teller not only by anti-Semitism but also by the lack of job possibilities at home. These young people typically went to Germany, whose economy had also suffered from a lost war but was at the time a flourishing democracy, known as the Weimar Republic. Many followed this path, and their destinations included other West European nations as well. (My future stepfather was among this group.)[48]

While Miksa Teller wanted his son to study abroad, his wife had a different opinion. She was concerned with Teller's immediate physical well-being and could not imagine him safe anywhere except near her. A compromise was reached according to which Teller would stay with the family until he was eighteen. After that, he could go where his father would direct him. He graduated from high school in 1925, at the age of seventeen, rather than the customary eighteen. Some of his few positive experiences came from tutoring others at that time, and he determined that the best of all professions was being a professor.

Although Teller gave the impression in his *Memoirs* that his relationship with his father was not very warm, Miksa Teller emerges as a caring parent through his actions. When Teller was ten years old, his father, knowing Edward's affinity for numbers, took him to meet Leopold Klug, a retired professor of projective geometry in Budapest.[49] Professor Klug was the only grownup Teller had ever met in his childhood who obviously enjoyed what he was doing. However, aiming at a professor's career was not a practical goal for Teller in the Hungary of the 1920s. As a teenager, he knew that Budapest could not be a place for his career and for his future life. Over the years, "anti-Semitism had, if anything, grown worse."[50]

By all signs, Miksa was a caring father, but Teller's *Memoirs* give the impression that Miksa was not very close to his son. There was an incident that seems to have come between father and son. Teller describes their rivalry in chess, which he used to enjoy but which abruptly ended after he had beaten his father. Without trying to read too much into this incident, it might be taken as an example of Teller's developing doubts and distrust even toward those who were close to him.

Teller received the greatest pleasure in his high school years from meeting "three young men from the Jewish community in Budapest": Eugene Wigner, Johnny von Neumann, and Leo Szilard.[51] They were his seniors by five to ten years and all were doing physics in Germany. Teller met with them during summer vacations, and they represented role models for him. Teller had doubts in himself, which were alleviated at least partially when he shared the first prize with two others in mathematics and won the national physics competition alone in 1925.

The three winners in mathematics were Rudolf Fuchs, Ede Teller, and László Tisza. It was a relatively small group of twenty-eight students, because at that time only those who had completed high school studies were allowed to participate.[52] Tisza ended up in the United States, just like Teller, although he took a more adventurous route. Fuchs stayed in Hungary and was killed in a Nazi labor camp in 1944. Fifteen students competed in the physics competition; Fuchs and Teller solved all three problems, but the committee supervising the competition decided to give Teller the first prize and Fuchs the second.[53]

Because of his mother's worries, it was decided that Teller should start his studies in Budapest at the Technical University. The question of what to study had already been decided upon. Teller would have opted for mathematics, but it was as impractical at the time as physics was. A few years before, Szilard, who was interested in physics, came to the conclusion that "there was no career in physics in Hungary."[54] Physics was not a common profession in other countries either during the first decades of the twentieth century. The future Nobel laureate American physicist and co-inventor of the laser, Charles Townes, for example, did not know that physics could be a profession when he was choosing his direction of study in the 1930s.[55] So Teller settled for chemistry—a convenient compromise for him as it had been for others: Eugene Wigner and John von Neumann had also started in chemical engineering.

Teller did not need to be concerned about the restrictions of the *numerus clausus*. He was eligible to attend university due to his exceptional performance at the national high school competitions in mathematics and physics. However, the Jewish students were regularly beat up at the entrance to the Technical University. This practice began in 1919 and is what physically prevented Leo Szilard and his brother from entering the university when they wanted to resume their studies after the war and revolutions. Szilard thought that if he showed his documents of conversion to the nationalists they would leave them alone, but he was mistaken. We know about these beatings also from Teller's friend Laszlo Tisza. However, the temporary relaxation by the mid-1920s had its influence on the attitude of the nationalist students as well, and for a few years the beatings stopped. To Tisza's luck, and Teller's, this happened when they attended the Budapest Technical University.[56]

According to the archives of the Budapest University of Technology and Economics, Teller started his studies in the division of chemical engineering of what was then called the Hungarian Royal Palatine József (Joseph) Technical University in September 1925.[57] The same year Teller started, Laszlo Tisza was also a freshman. The two became friends when they met as winners of the national competition.

The documents show that Teller took courses on relativity and vector analysis in addition to the traditional ones prescribed for chemical engineers. Even as early as his years at the Minta, he had been interested in the theory of relativity and had acquired a book on it. Since he could not understand the theory, he turned to his physics teacher. The teacher was known to carry out independent research in addition to his teaching, but it was in experimental physics. He could not or did not want to explain

Edward Teller's transcripts at the Budapest Technical University with a listing
of his subjects and grades. Note the entries of vector analysis and theory
of relativity, rather unusual for a student of chemical engineering.
Courtesy of the Archives of Budapest University of Technology and Economics.

what Teller asked him about. Instead, he "borrowed" the book for the duration of Teller's high school studies.[58]

Teller left Hungary for Germany two weeks short of his eighteenth birthday. Officially, however, he did not leave the Budapest Technical University until November 2, 1928, according to the archival documents. From this point on, Teller's only connections with Hungary were his visits during his vacations. That he maintained some connection with the Budapest Technical University is indicated by the fact that when the future Nobel laureate Robert S. Mulliken and his wife visited Budapest in 1928, or a little later, Teller acted as their host along with Professor Béla Pogány of the Technical University. Mulliken and Teller would meet again as professors of the University of Chicago after the war. The Japanese physicist Yushio Fujioka was another visitor to Budapest for whom Teller played host. When, thirty years later, Teller's mother and sister were arriving in Vienna after they had finally received permission to leave Hungary, Fujioka—working at the International Atomic Energy Agency—met them and helped them with their transition.

There was nothing unusual in the fact that young Hungarians aiming to become members of the professional class went to complete their education in Germany. It was only that in the 1920s the circumstance forced them to do so, and it was mostly a one-way movement. In the "happy peace time," it was also Germany and German culture that attracted the young, aspiring Hungarians and particularly the Jewish youth to go there. Theodore von Kármán's life paralleled the lives of the other Martians, but he was there one generation before Teller. The pupils of the Minta got their share of German literature and were well versed in the German classics, which, for Teller, were judiciously augmented by his mother teaching him German literature.

The attraction of Germany and German culture, and more generally of the West and the culture of the West, was not unique for Hungarians. There was a general migration from East to West, which only intensified with the emergence of totalitarian and semi-fascist regimes. Teller's first emigration was not very painful because of his having already been immersed in German culture and language and because of the compatibility of the Hungarian and German educational systems (this kind of compatibility is nowadays being revived in the European Union). It could also be argued that by the time Teller embarked on his emigration from Hungary to Germany, he had already experienced what we might call internal emigration: the experience of Jewish families preparing their children for new environments, new conditions of life, in order to get accepted in their surroundings.

In spite of the hardship imposed on them by official Hungary and its active complicity in the genocide of 1944–1945, Jewish Hungarians felt a loyalty to their country. Teller's Hungarian patriotism, no matter what, provides an excellent example. He never equated Hungary with its fascist and communist regimes. It is also true that during his last years he received an almost unreserved admiration in Hungary, the likes of which was never afforded to him in the United States. After the political changes of 1989–1990, he basked in the exceptional treatment he received in Hungary from the authorities, regardless of their political affiliations. It is also true that by the time he returned, after more than half a century, none of his friends were around; most had perished in German concentration camps or were killed in Budapest by Hungarian Nazis.

## FAMILY FATE

Although Teller was absent from Hungary starting in 1936, his parents and his sister and her family lived through the horrors of the coming two decades. His sister Emma and András Kirz had one child, János. Teller's parents, Emma, and her son survived the period of anti-Jewish laws and the war in Budapest. Amid suffering and persecution, a small but significant incident for Miksa occurred when he—along with all other Jewish lawyers—was stripped of his membership in the Hungarian Chamber of Advocates in 1944.[59] His elation by the liberation in 1945 was tempered by fear of the future under Soviet rule. Miksa died in 1950, and the communist authorities deported the rest of the family from their Budapest home to the countryside in 1951, ostensibly for the Tellers' "capitalist" past. Despite Teller's absence from Hungary at that time, the treatment of his family in Hungary must have had an effect on his view of communists.

There is a mistaken notion in the literature that the authorities waited with the deportation until Miksa Teller died because he was well protected by his professional standing and contacts.[60] The communist authorities were not that subtle, not in this case, and the deportation of the Teller family was part of a large operation. Its excuse was to remove the belligerent capitalist elements of society from Budapest and the major cities. They also wanted to induce fear among the rest of the population. Moreover, homes emptied in this manner could be given to members of the new elite. This was simpler than carrying out the necessary construction of new homes.[61]

Miksa was no longer well connected during and after the war, and his family was not protected; everything that he had worked for his entire life had disappeared. He continued his practice as a lawyer into his seventies, working until the month he died. Under the communist regime, his profession lost its respectability, as well as its lucrative character. It was a painful period in his life, but he remained a caring and gentle man who took his grandson for long walks in the Buda hills.[62] He may have changed in his old age, because Teller's *Memoirs* give the impression that he was not close to his son. It may also be that Teller did not fully perceive that his father deeply cared for him. The father may have had a reserved way of conducting himself, but he clearly looked out for his son's well-being. Miksa gave ample testimony of this in his actions if not in their direct personal interactions.

The Teller family was deported to a farmer's house without running water in the small village of Tállya, in eastern Hungary, the most backward region of the country. There was no proper school in the village for Janos, but he succeeded with his studies because a friend in his class sent him weekly reports of their lessons.[63] Janos had been a good student and excelled especially in mathematics, according to the secret police documents. When he applied for permission to take his examinations or to study at least in a vocational school, he was turned down.[64]

The Teller family learned that the immediate justification for their deportation to Tállya was Miksa's alleged board membership in the "Hungaria" rubber factory; in fact, he had only been their solicitor. The family sent appeals against their deportation and applications for permission to allow them to return to Budapest or at least to move to a nearby city where Emma could find employment and Janos could continue his studies. At some point during January 1952, Emma and Janos submitted an application for permission to immigrate to Israel.

By this time, the case of the Teller family had gained added significance because in mid-1951 the Hungarian secret police, called "the authority of the protection of the state" (Államvédelmi Hatóság, ÁVH), opened a broad-based investigation of Edward Teller.[65] He was given the code names "Kárász" and "József Kerekes." He was a "target," and targets as well as secret agents were assigned code names. The long-range plan was to make contact with him and engage him one way or another. The secret police were at this point considering releasing the Teller family from deportation but did not know how to do it without attracting attention. They made plans for using Teller's sister, though they did not know yet in what way. Another consideration was the health of Teller's mother. The authorities were afraid that she would die during deportation, which would make their work in engaging Teller much harder.[66]

They understood that Teller was a highly respected and important sci-entist who had participated in the development of the atomic bomb. The purpose of their interest was taking shape to see whether "he could be used for scientific intelligence," obviously for the Soviet bloc.[67] First, they col-lected all the information about Teller and his family and friends.[68] There was an ever-broadening net around people whom the secret police consid-ered worthy of their investigation. Their interest over the years included close and distant relatives, Teller's classmates in school, his fellow students at Leipzig, the people who attended various courses of foreign languages that Emma Kirz had given at a variety of foreign trade companies, and the like. They sent agents to Leipzig in East Germany to investigate Teller's past during his doctoral studies. Their meticulous efforts extended into the mid-1950s. Eventually, they managed to put together a detailed and in many respects realistic portfolio about Teller's background.

Initially, however, there was scarcely any information about him and his views, and he was described repeatedly by adjectives such as "leftist," "progressive," and "democratically thinking," all indicating that he was expected to cooperate with the Hungarian secret police. Furthermore, it was learned that Teller had sent word to his father that "answering to his conscience, he [was] no longer working on the development of atomic weapons; rather, he [was] working on defense against atomic weapons and [had] already achieved results in his new line of work."[69] This only whetted the appetite of the Hungarian security organs to gain the trust of the family and to learn as much as possible about Teller.

Finally, the termination of the deportation of the Teller family was decided on November 9, 1952. The date is significant because the bulk of the deportees were released only after Stalin's death in March 1953. Once the release of the Tellers was approved, the authorities wanted to act with utmost speed. The family's ordeal at Tállya ended in December 1952. However, the authorities wanted to recruit Teller's sister, Emma, as an agent to help net Teller. Engaging Emma as an agent nonetheless seemed difficult since they recognized her lack of trust in them. She did not reveal her previous employment at the Budapest US Embassy in the immediate post-war years nor the fact that her brother was living abroad. In the meantime, the collection of information about the family continued, and it was discovered that the link between Teller and his Budapest family was their former governess who was living in Illinois as Mrs. Jacob Schutz. Uncovering this was all the more challenging because Mrs. Schutz did not send Teller's letters directly to the Tellers but to someone else's address. Agents were dispatched to Chicago to investigate her.

Upon the termination of their deportation, the Tellers returned to Budapest. As they had lost their home, they huddled in a room of a shared apartment. Further hardship followed. Emma became very much a target for the secret police: one of the most notorious personalities of the secret police, Vladimir Farkas, was helping to plan the strategy on how to deal with her and through her, to get to "Kárász," that is, Edward Teller.

The secret police finally communicated their success in engaging Emma. It happened on May 13, 1953, but the report did not sound promising for the authorities.[70] Apparently, they gained Emma's consent "by pressure." They held her for days with repeated interrogations, and in spite of the intention of the officers who wanted to entice her to become a secret agent, the guards had roughed her up. She was given the code name "Mária Zsoldos." In reports about her engagement, complaints were made that she did not produce usable information.[71] When reading these reports, there is a question of who was using whom, because Emma repeatedly managed to get hard-to-come-by medication for her mother from her holders.

At this time the reports about Teller became thorough and informative, although it is questionable whether they were of any use for the Hungarian secret police. There was ample information about the Oppenheimer hearing with a complete translation of Teller's testimony. According to the commentary added, Teller's "shrewd" testimony was a typical capitalistic approach in which someone wanted to push away a competitor and place himself in his stead in the coveted leading position.[72] In this report the secret police formed the opinion that Teller was a dedicated fighter on behalf of the United States who would not let the Soviet Union overtake it in the arms race.

Still they continued to try to utilize Emma and devised an alternative action plan. This included "marrying her" and then letting her and her would-be husband—obviously a secret police agent—to leave the country. The plan was to use Emma to make her brother return to Hungary and continue his activities in the interests of peace—an obvious euphemism. At one point they set a barely two-month deadline for finding a husband for her and it seemed they might have had a candidate for this purpose.[73] But the supervising higher officer made a note on this report that he did not find the deadline realistic.

We know that the secret police did not succeed in "marrying her out." They also became increasingly disappointed in her persistent lack of cooperation in providing any useful information to the secret police. Alas, there are no more documents in the secret police archives about their "handling" of the Tellers between the spring of 1956 and their final departure from Hungary. There must have been further documents, but they disappeared or have been destroyed. What we do know is that Janos Kirz left Hungary in the wake of the suppressed revolution of 1956, reached his uncle's family in Berkeley, California, studied physics at the University of California at Berkeley, and eventually became a professor at the State University of New York at Stony Brook. As we will see, Teller's mother and sister succeeded in leaving Hungary legally in early 1959.

Here it is of interest to note that the secret investigation of Teller's affairs extended beyond Hungary. Romania was involved in examining Teller's past in Lugos and East Germany at least to the extent of the years Teller spent in Leipzig. It might be considered strange that there seems to be no trace of direct Soviet involvement in the archives, considering the weight of the nuclear program in which Teller was a key player. On some of the top-secret documents though, there is a cryptic sentence: "Give it to Comrade for translation."[74] Two important pieces of possible information are missing in this sentence: the fact that "Comrade" is capitalized but not accompanied by a surname probably referring to someone important and highly secret, and the missing designation of the language into which the document must be translated. It is a guess, but highly probable, that the translation was to be made into Russian. Hungary was not involved directly with nuclear weapons, and the Teller operation by the Hungarian secret police could not have been carried out without Soviet participation; it must have been guided by Soviet interests.

Teller made various attempts to get his family out of Hungary. He sent an affidavit for his sister and nephew for this purpose.[75] Apparently Teller's father and mother preferred to stay in Budapest. After the death of Teller's father in 1950, the situation changed in that from this time on, his mother also wanted to leave. Lewis Strauss, then chairman of the Atomic Energy Commission, informed Teller in the spring of 1956 about a conversation with CIA director Allen Dulles and deputy under-secretary of state Robert D. Murphy on his behalf. These officials considered an open request to the Soviet authorities in Budapest to gain permission for their departure.[76] Nothing came of it. In the fall of the same year Teller wrote to the Hungarian–Swedish Nobel laureate George [von] Hevesy in Stockholm. Teller was considering immigration of his family to Sweden and asked Hevesy for support. Teller's mother had by then received permission to immigrate to Israel, but she did not want to leave without her daughter and grandson.[77]

Szilard had long wanted Teller to visit the Soviet Union, but when he raised the issue in a conversation with him in Washington, in the spring of 1958, Teller declined. He explained to Szilard that with his family still living in Hungary he might be subject to blackmail. He would not even let Mici visit Vienna out of fear that she might take the wrong train and end up in Hungary, where she would be detained and never return.[78] Szilard told Teller that he would try to do something about his mother and sister, but Teller was skeptical because of his prior experiences.

During the next Pugwash meeting, in September 1958 in Kitzbühel, Austria, Szilard raised the issue with a member of the Soviet delegation. Szilard had played a pivotal role in initiating the Pugwash movement, which was a forum for intellectuals from both the Western and the Eastern blocs to meet and informally discuss questions of peaceful coexistence. It was no secret that whereas independent people came from the West, those from the East were under strict government control. For Szilard trying to help Teller, this was just as well, since he could then expect the Eastern delegates to be close to their governments. The Soviet delegate talked with Lajos Jánossy, a well-known physicist member of the Hungarian delegation. At the time, Jánossy was one of the top administrators of the Hungarian Academy of Sciences, and in 1958 he was elected to the central committee of the Communist Party (under the name of Hungarian Socialist Workers' Party in Hungary's one-party system). Jánossy sought out Szilard during the same meeting, and Szilard called Teller right away for information about his mother and sister. The next time Szilard heard from Jánossy, it was in a letter on Christmas Eve with the good news that Teller's mother and sister had received permission to leave Hungary.[79] They arrived in California early 1959. Teller was relieved.

# Chapter 2
# GERMANY
## *Road to Science*

The researcher's art is first of all to find himself a good boss.
—ANDRÉ LWOFF[1]

*The significance of Teller's German period is twofold. First, he received state-of-the-art training in chemistry, mathematics, and physics. He spent these years in the center of the scientific universe and became a member of the international community of scientists. Second, his first experience as an immigrant in a foreign land provided excellent training for his second immigration, to the United States.*

## WHY GERMANY?

Germany was the primary destination for the Hungarian intellectual exodus. Many Jewish–Hungarian scientists and engineers went there after World War I and the revolutions, during the White Terror and the first decade of the Horthy era. Germany might have seemed an odd destination: the country suffered defeat in World War I and the Allies imposed heavy reparations on it. The Kaiser abdicated. The country shrunk. And there was staggering inflation. However, there was a new democratic constitution, enacted in the small literary city of Weimar in 1919 (hence the name the Weimar Republic). An important difference between Germany and Hungary was the liberal atmosphere and the lack of state-sanctioned anti-Semitism in Germany. It also boasted a tremendous concentration of intellectual power.

The constitution of the Weimar Republic was not only very democratic, it was also very complicated. This is why Leo Szilard referred to it as the "entropy of governance."[2] It was fragile, but as long as it lasted it was progressive. At this time, Germany was the scientific capital of the world and the German language was the unofficial language of science. One of the attractions of Germany was its institutionalized effort at the national level to stay at the apex of science with the establishment of the Kaiser Wilhelm Institutes. (After World War II, there would be a similar attempt by the United States in its creation of the network of national laboratories.)

German science was an international magnet, and the best minds from all over the world congregated there, especially those from German-speaking territories in Central Europe, where German was the intermediary of world culture and progress. In other countries, like Austria, the lack of scientific depth drove aspiring young people to Germany. One of Teller's future colleagues and co-authors, Victor Weisskopf, had studied physics at the University of Vienna. After his first two years, his mentor, Hans Thirring, told him that there was no reason for Weisskopf to continue in Vienna and he should go to Göttingen or Munich—Vienna was nowhere near as advanced in modern quantum mechanics as they were.[3]

Teller had his best time in Germany, rivaled only by the time he would spend at George Washington University in Washington when he first went to the United States. However, that is not to say that the period he spent in Germany, 1926–1933, was ideal. It is a tribute to his resilience that he hardly ever complained about the negative conditions of his life in Germany, because that was the period that led to Hitler's rise to power and Teller's forced emigration from Germany. It was, of course, not only resilience; it was also a will to adjust. Teller was so much taken by his opportunities to learn and become part of real science that he tended to overlook what he did not like. All his Martian friends behaved the same way.

It was especially lucky for Teller and the other Hungarian émigrés that Germany was a superpower in science and that German was the language of science. Through the 1920s, the Americans had to learn German if they wanted to be informed firsthand about the latest progress in science. This would all change after Hitler's accession to power, when the center of gravity of science would move to America, and English would become the language of science. The United States and Great Britain were the two major, almost exclusive recipient countries of the refugee scientists, and this worked to their mutual benefits. However, these countries could hardly have been destinations for those leaving Hungary in the early 1920s. The English language was not at all widespread, and immigration to the United States was hindered by the American legislation of 1921 that established quotas on the basis of "national origin"; the Reed–Johnson Act of 1924 made immigration even more restrictive.[4]

In the late nineteenth century and during the first third of the twentieth century, Germany was the home of great science. Wilhelm Conrad Röntgen and his X-rays; Max Planck and his epoch-making quantum theory; Fritz Haber and nitrogen fixation; Werner Heisenberg and Max Born and their quantum mechanics; and above all, Albert Einstein and his general theory of relativity were all discovered in Germany, and they all lived and worked in Germany. Almost 60 percent of all Nobel Prizes in physics and chemistry were awarded to German scientists through 1933, which is very impressive even if there was considerable bias toward German science by the Swedish awarders at the time.[5] Of course, the bias had developed initially due to the excellence of science there. Then politicking got mixed in with the Nobel Prize institution. After the Norwegians had awarded the Nobel Peace Prize to the incarcerated peace activist journalist Carl von Ossietzky in 1936, Hitler forbade German scientists to accept Nobel Prizes.

Hitler's action was misdirected because the Swedes awarded the science prizes and not the Norwegians, and the Swedes showed no sign of anti-German sentiments. On the contrary, it was controversial when the Nobel Prize in Chemistry was awarded to Fritz Haber just as World War I ended. Haber's name was sadly linked to the first poison-gas attacks on the battlefront during World War I. A quarter of a century and another war later it was similarly tactless to award Otto Hahn a Nobel Prize for the discovery of nuclear fission. The award was bestowed right after the dropping of the atomic bombs over Japan, the scientific roots of which went back to Hahn's discovery. Then there was a question of the omission from the award of Lise Meitner for her contribution to the discovery of nuclear fission, which thereby enhanced the injustice she had suffered when she was forced to flee Germany in 1938.[6] It took an effort for the Nobel Prize institution to regroup and move away from their German orientation following World War II, but it succeeded soon enough.

The tremendous success of science in Germany in the first three decades of the twentieth century had built on tradition. Chemistry was the most distinguished area of science, and its applications flourished in Germany. This included the discovery of the therapeutic effects of aspirin, the discoveries of dyestuff, the invention of Novocain as an anesthetic, and

Salvarsan for the treatment of syphilis. By the eve of World War I, "Germany had become the international Mecca of science."[7]

The German educational system encouraged the study of science and also the applications of the achievements of science. There was unprecedented symbiosis between science and industry in Germany, which in modern terms we call "technological transfer." At that time, it was being carried out without special offices fostering it, as is more customary nowadays. This interaction was characteristic of all the technical universities (or institutes of technology, as they are called in America). Teller started his university studies at the Budapest Technical University, which was created and organized after the model of the German Technische Hochschule. When he continued at the Karlsruhe Technische Hochschule, only the language of instruction was different, which caused no difficulty for him.

In spite of the international character of science in Germany, when World War I broke out, fierce nationalism took hold of German scientists, as it did their counterparts in the allied countries. After the German defeat in World War I, there was international boycott of Germany by scientists of the former allied countries, but it did not last very long. German science prevailed and kept thriving. Thus international students and scholars congregated in Germany, again, up to Hitler's accession to power in 1933. Both Imperial Germany before World War I and the Weimar Republic in the 1920s supported science. They organized the Notgemeinschaft der deutschen Wissenschaft, the Emergency Society for German Science, to provide funding for basic research.

The informal way that they dispersed their support for research would be the envy of young researchers today, but, of course, science was operating on scales many orders of magnitude smaller then. Germany would fund what appeared to be innovative ideas. According to the late biochemist Erwin Chargaff, who had experience with both the German funding system of the 1920s and later the American peer-review system, "... most important is to get the general behavior, the general way of thinking of the person rather than to decide that this is a marvelous problem. [Ideas] are not marvelous except in lucky hands, in very gifted hands. The hands you can't look at in a proposal."[8]

Notwithstanding Germany's prominence in science, it, too, had an undercurrent of anti-Semitism during Teller's time there. This was perhaps not conspicuous then because anti-Semitism existed elsewhere as well, notably, in American academia. However, it manifested itself more as discrimination rather than persecution in both Germany and the United States. This was not something that Teller and the others would have found too unusual at the time. Besides, there was a difference between Hungary and Germany in that Jewish students were not limited in getting a university education in Germany. Yet another difference was that there were protests against discrimination however ineffectual they might have been.

Richard Willstätter's was the most notable among such rare protests. He was a Nobel laureate organic chemist, a professor at the University of Munich. In 1924 he announced his resignation at the age of fifty-two as a gesture against increasingly anti-Semitic hiring practices at his university. He received attractive offers after his resignation, but when the University of Heidelberg considered him for a chair, Nazi students voiced their strong opposition. Willstätter never came out of his retirement; he lived in Munich until 1938 when in order to escape from getting arrested by the Gestapo, he fled to Switzerland.[9]

The dismissal of Jewish scientists had begun before Hitler came to power in Germany in 1933, as part of the anti-Semitic atmosphere that was increasing in scope. A case in point was what happened to Herman F. Mark, who had been Teller's favorite professor in Karlsruhe. Mark conducted pioneering research of polymers at I. G. Farbenindustrie in Ludwigshafen, the huge chemical concern, I. G. Farben, in short. His activities exemplified the cooperation of academia and industry. He made discoveries in the most fundamental fields of structural chemistry in addition to his applied work.

Mark did not think much about his partly Jewish roots: his surgeon father, born in Budapest, was Jewish. He describes in his autobiography how on a summer day in 1932, he was asked to the office of his director and was told that in view of the anticipated Nazi takeover of Germany they decided to let him go.[10] Mark returned to his native Vienna and assumed a university professorship there. He held it until the *Anschluss*, the German annexation of Austria, when he became a German citizen automatically, and had to move again. At that point he and his family immigrated to the United States, and to the end of his distinguished career he taught at Brooklyn Polytechnic. Mark's case is of interest also because many could not have believed even at the moment of Hitler's accession to power that the Jewish scientists would lose their jobs. Apparently, at I. G. Farben, they had better foresight. The company soon after became infamous for its exploitation of slave labor in Auschwitz and elsewhere during the war. It has been called "the industrial demon of two world wars . . . the jackal to Hitler's lion," and its twenty-four top executives were tried at Nuremberg for war crimes.[11]

## KARLSRUHE AND MUNICH

Teller moved from Budapest to the Technical University of Karlsruhe in January 1926. He was majoring in chemical engineering, but he took courses in mathematics in addition to chemistry. Karlsruhe was one of the favorite destinations for Hungarians interested in chemical engineering. Michael Polanyi went there first as a chaperone of another student in 1912. This was Polanyi's first opportunity to learn from leading authorities in the field of his interest, which was the adsorption of gases on solid surfaces.[12] In about a quarter of a century, Teller would achieve one of his most important results in the same area of research. In Karlsruhe, there was also the Hungarian chemistry student Ferenc Kőrösy, two years Teller's senior, who later became a leading professor in Israel. Teller referred to Kőrösy as his "first effective teacher."[13]

Teller signed up for Herman Mark's course on wave mechanics, which was almost as new for the young professor as it was for his students. Mark wrote that on occasion he was not able to clarify it for his audience, and in such situations Teller would say, "'I think, Professor Mark, it must be quite clear to all of us what you have just now explained, namely . . .' and then . . . he gave a completely correct and lucid version of [Mark's] somewhat obscure presentation . . ."[14] This is quite remarkable, for the course was Teller's introduction to quantum mechanics. Their meeting at Karlsruhe was the start of a friendship that would continue for Teller with Mark's son, Hans Mark.

Teller's world was greatly interconnected. Michael Polanyi had graduated from the Minta and went to Karlsruhe; he was interested in and became very successful in adsorption studies. At one period of his life he was in charge of the section at the Kaiser Wilhelm Institute in Berlin where Herman Mark worked. Polanyi supervised Wigner's doctoral work in Berlin. Wigner was Teller's friend, and Mark was Wigner's thesis supervisor in Berlin. Herman Mark and Wigner were closer to each other in age than Mark and Teller. Yet for Wigner in Berlin, Mark seemed to be rather aloof, so Teller may have contributed to opening up Mark to their interactions.

The American chemist Linus Pauling visited Mark at I. G. Farben, and he was so taken by Mark's and Raimond Wierl's invention of the gas-phase electron diffraction—a new technique for molecular structure determination—that he decided to start a group dedicated to it in Pasadena. Mark welcomed the idea because this research was too academic for an industrial laboratory, so he gave Pauling the blueprints for the experimental apparatus. Teller and Pauling did not meet at that time, but Pauling would figure in Teller's later life, primarily as his main adversary in the test ban debates. Still, their interests overlapped in science.

Initially Teller was not too enthusiastic about chemistry, but in time, his knowledge of the subject would serve him well. In his subsequent career, he would make important contributions to chemistry. He discovered effects and set up equations—always in cooperation with others—and his achievements in that field would have sufficed for a whole career of a most distinguished scientist.

Teller's interest in physics proved to be useful in his chemistry studies. Once he had to analyze an unknown substance in his chemistry course in Karlsruhe, and he supposed it was a salt. He knew what chemical approach he should be taking by preparing the necessary solutions to carry out the analysis. However, instead of going through this technique of "wet" chemistry, he decided to make use of a spectroscope that he noticed in the corner of the laboratory. A spectroscope would display a characteristic colored flame, depending on the nature of the metal in a salt. Table salt would have colored the flame yellow, and Teller's unknown sample colored the flame red, the characteristic color of lithium. Thus he solved the problem with the simplest possible test, circumventing a more elaborate analysis.[15]

In addition to Mark, Teller encountered another remarkable visiting lecturer at Karlsruhe, Paul P. Ewald from Stuttgart. Ewald's principal contribution to science led to the invention of X-ray crystallography, which became one of the dominant tools in twentieth-century science. Since Ewald had some Jewish ancestry and his wife was Jewish, he left Germany in 1937 and the rest of his family left in 1938; first they lived in Great Britain and then immigrated to the United States in 1949. One of Ewald's daughters, Rose, married Teller's good friend, Hans Bethe.

Other experiences also came Teller's way in Karlsruhe. He noticed the rude behavior by a laboratory assistant toward him. The man had lost a leg in World War I, and Teller's fellow students explained to Teller that because of this misfortune, he was an anti-Semite. By then Teller knew that he could encounter anti-Semitism for no rational reason, but this was the first time that such belligerence caught up with him in a personal way.[16] Otherwise, it is remarkable how oblivious he was to anti-Semitism amid German academia up to Hitler's accession to power. It may also be that he had already been sufficiently hardened not to let it get to him.

Teller was not entirely happy in Karlsruhe with his chemistry and mathematics, whereas physics increasingly occupied his imagination. However, he could not make a move independent of his parents, not only because they supported his studies in Germany, but also because parental consent was a more important consideration at that time than it would be today. Fortunately, Teller's father was understanding and had in mind not only his son's job prospects but his happiness as well. He consulted a distant relative, a physics professor in Vienna, Felix Ehrenhaft, who encouraged Teller's changing to physics. Teller's father also consulted some of Teller's professors in Karlsruhe; as a result, Teller finally had his father's consent to move away from chemical engineering to physics. Teller's mother did not play a role in these considerations because for her it was trial enough not having her son under her wings in Budapest; the question of the subject of his studies was irrelevant to her.

Once Teller had the opportunity to study physics, he decided to leave Karlsruhe. There were several famous places for physicists in Germany, but as far as teaching was concerned, Arnold Sommerfeld at the University of Munich was the most sought-after professor. He was a legend, distinguished not so much for his discoveries but for his pedagogy; the brightest youths from all over the world flocked to his laboratory. Teller moved to Munich in the late spring of 1928, and started attending Sommerfeld's lectures. Alas, he did not feel comfortable with the formal and pompous professor; he missed the personal approach he had experienced with Mark. The only good thing Teller could say about Sommerfeld was that "he was not as bad as my mathematics teacher [in Budapest]."[17]

Teller did not spend much time at Munich University. The semester at Karlsruhe ended in April 1928, and he suffered a terrible accident in Munich on July 14. Between his arrival in Munich and the accident, he went away to visit Mici. She was spending some time as an instructor at Odenwald School, not far north of Heidelberg, a considerable distance from Munich.[18] So Teller attended classes at Munich for at most two months, probably less. The accident happened as he was preparing to meet his friends for a hike—a favorite form of recreation for him. He was hurrying to the meeting place and missed his stop while traveling on a trolley car. When he noticed his mistake, he jumped off the platform, lost his balance (probably because of his backpack), and fell beneath the trolley car, whose iron wheels severed his right foot.

Teller writes about the accident in a detached way in his *Memoirs* and hardly ever complained about it during his long life.[19] He had a limp, and in his later years he carried a stick, which became his trademark. As an older man, he would often use this walking stick to rap the floor for emphasis as he spoke. Otherwise, he seldom gave any sign of being bothered by wearing a prosthesis. Donald Glaser told me that when he and his wife had Teller visit their home, which was on the hillside, Teller walked the eighty stairs without complaint. Glaser learned only later that he walked with a prosthesis.[20] Teller skied in Copenhagen in the winter of 1934, but in Los Alamos, he wore snowshoes when he joined the others who went skiing.[21]

During his convalescence Teller despised his mother for showing off her grief. When he noticed that the painkillers he took interfered with his thinking, he stopped using them. From then on, when he felt pain he swallowed plain water and made himself think that he took a painkiller and forced himself not to think about the pain. It worked. This was not the only example of how serious trials made him stronger and more determined in achieving his goals.

His surgeon in Munich, Dr. von Lossow, discussed with Teller the available procedures for his injury. The one they chose consisted of fusing the remaining part of his heel with the two bones of his lower leg. This procedure meant less radical surgery, but it necessitated a more sophisticated prosthesis. By the time the first version was constructed, Teller was already recuperating in Budapest after yet another operation there. The first prosthesis did not work to his satisfaction, and so another one was fitted, but by then he was back in Germany. This second prosthesis proved to be perfect and worked well for the rest of his life. In his old age Teller joked that it performed better than the other parts of his body. Dr. von Lossow made a good impression on Teller, and he was very grateful to the surgeon.[22]

The recuperation period was slow, giving Teller much time to contemplate about his future. He did not see much point in returning to Munich. Sommerfeld was no longer a big attraction for him, and the professor was going away for a whole year anyway. Teller chose Leipzig, where a young professor of international fame, Werner Heisenberg, seemed to be the most promising mentor for him. Teller's move to Leipzig proved to be a decisive step in his scientific career.

## LEIPZIG AND GÖTTINGEN

As Teller was preparing for Leipzig, he started thinking about a dissertation. The German system of education in physics and mathematics at the time was different from the American system. It was also different from the German system in other fields, such as chemistry. In physics, the students took courses, but there was no official end to their studies unless they

## WERNER K. HEISENBERG (1901–1976)

Werner Heisenberg was born in Würzburg, Germany, into a German intellectual family. He studied physics under Arnold Sommerfeld at the University of Munich and completed his doctoral studies in 1923 in theoretical physics, in which he showed exceptional talent and inventiveness. He had already started working at Max Born's institute at the University of Göttingen in 1922, and spent some time doing research at Niels Bohr's institute at the University of Copenhagen. He was appointed professor of physics and put in charge of theoretical physics at the University of Leipzig in 1927.

By then, he had already published seminal papers on the foundations of quantum mechanics and on the discovery of the uncertainty principle, which was to bear his name. In 1933, he was awarded the Nobel Prize in Physics for the creation of quantum mechanics and for the discovery of different modifications of hydrogen. There has never been any doubt that Heisenberg deserved the Nobel Prize, but the omission of Max Born from the award for the creation of quantum mechanics was noteworthy. Born was later awarded the prize in 1954.

Upon the Nazis' accession to power in Germany, Heisenberg hoped to avoid politics and restrict his activities to physics, but such separation proved impossible. He was considered an opportunist by Nazi scientists because he had refused to condemn physical theories. The Nazis condemned physical theories that had been created by Jewish scientists, e.g., Einstein's theory of relativity. Heisenberg was even labeled a "white Jew" and was fiercely attacked in the Nazi press. However, he had supporters, notably, the infamous SS-Führer Heinrich Himmler, whose interference resulted in Heisenberg's being left in peace and his career being unhindered.

In 1941, he was appointed professor of physics at the University of Berlin and director of the Kaiser Wilhelm Institute for Physics there. He was in charge of the German project of developing an atomic bomb, which, as we know now, failed. Whether and to what extent there was a German project may have been the subject of the Bohr–Heisenberg conversation on the occasion of Heisenberg's visit to Copenhagen in September 1941. This was depicted in Michael Frayn's acclaimed play *Copenhagen*. Heisenberg's and Bohr's views on the subject of their conversation differed in retrospect. Heisenberg subscribed to the myth after the war that the German scientists could have developed an atomic bomb but did not due to moral considerations.

At the end of the war, Heisenberg was detained by the Western Allies along with nine other leading scientists who might have participated in the German nuclear program, although not all did. After months of isolation, they returned to Germany. In 1949, Heisenberg was appointed president of the newly established German Research Council in West Germany. He continued his research in physics, expanding his interests to include philosophy. After some reservations, he was again afforded international recognition, awards, and lectureships; however, without the authority that he had wielded in the early 1930s.

В. ГЕЙЗЕНБЕРГ

Werner Heisenberg, drawing by physicist Dmitrii D. Ivanenko in the Russian edition of W. Heisenberg, E. Schrödinger, P. A. M. Dirac, *Modern Quantum Mechanics: Three Nobel Reports* (Leningrad and Moscow: Technico-Theoretical State Press, 1934), p. 13.

decided to pursue a doctorate. In contrast, the order in professional qualifications for chemistry was introduced earlier, including the Diplom-Chemiker degree, which corresponds to the American master's degree. Being a chemist or a chemical engineer was a popular profession, whereas being a physicist was not, and there was no lower degree in physics than the doctorate.[23]

Teller could go directly for his doctorate, since it was at the discretion of his supervisor when his work would be considered worthy of satisfying the requirements, which were rather vague. Many of the great physicists of the twentieth century earned their doctorates the same way Teller did. There is no indication that the vague requirements lowered the quality of their dissertations in comparison with today's doctoral degree, which usually takes many years longer to earn.

Until Teller formulated his plan of getting his doctorate, he did not need to make his studies official. Now, with an eye on earning a doctorate, his affiliation with the University of Leipzig had to be formalized, and so he asked for his transcripts from the Budapest Technical University in November 1928. He spent less than a year and a half in Leipzig before he became Dr. Teller. It should be noted here that the official papers in the Archives of the University of Leipzig list him as "Ede Teller" even after he got his doctorate, but his dissertation carries the name "Eduard Teller." His research papers from England would carry this name. There is no indication that he had at this time contemplated changing his citizenship, and when he was applying for his American visa in London, he did so as a Hungarian citizen. Some among his closest friends changed their citizenship after they had left Hungary. Polanyi took Austrian citizenship in 1920[24] and Leo Szilard took German citizenship in 1930.[25]

Teller became a member of the physics community in Leipzig. His revered mentor was not only a great scientist but he was also close to his young disciples in both age and interests. Heisenberg participated in the revolution of modern physics by being one of the principal creators of quantum mechanics, and he was the consummate young star. He also had a competitive spirit toward whatever endeavor in which he was engaged. Though Teller, despite his infirmity, could initially beat Heisenberg in table tennis, it was not for long. When Heisenberg went for a trip to the Orient, he returned with much more skillful abilities in this sport, and from then on he invariably beat Teller.

Teller made important friends in Leipzig. The group around Heisenberg was stellar and included two future Nobel laureates, the American Robert Mulliken and the Swiss Felix Bloch. There was also the Japanese Yushio Fujioka, who interacted with Teller. Another Japanese visitor in Leipzig, Yoshio Nishina, may not have met with Teller; he would later be in charge of the Japanese attempts to build an atomic bomb. Those gathered at Heisenberg's seminars discussed physics in broad terms. The discussions often extended to questions bordering on philosophy, such as, what did understanding something in theoretical physics mean?[26]

Two other young physicists in Heisenberg's group played important roles in Teller's life. They were the Russian Lev D. Landau and the German Carl Friedrich von Weizsäcker. Of the two, Landau was the greater physicist, and his creative years were compressed to short periods of time during his turbulent life. Teller knew him as a dedicated communist. In time,

Landau became utterly disillusioned with the Stalinist regime. But Teller and Landau never met again. After World War II, they worked on opposite sides of the world's political divide: Teller as a fierce advocate of nuclear armament in the United States, and Landau—by his own characterization—as a "learned slave" in the Soviet nuclear program. Teller counted the brutal imprisonment and treatment of Landau in 1938 as one of the reasons he became anticommunist.

Von Weizsäcker's interest, in time, shifted toward philosophy. He came from a distinguished aristocratic family; his father, Ernst, would become a high official in the Nazi administration, and his younger brother, Richard, would become president of the Federal Republic of Germany. Whereas after World War II his father was sentenced to years of incarceration for war crimes, his brother, while he was occupying the highest office of the land, apologized for the Nazi war crimes. In the early 1930s, Teller noticed that his friend Carl Friedrich was attuned to the political developments in Germany; he also noticed his sympathy toward the Nazis.

Teller arrived in Leipzig in the late fall of 1928 and had to overcome considerable red tape to be accepted as a bona fide student of Leipzig University.[27] There were considerations of his being a foreigner, and he had to produce a list of documents, which included a photograph; his high school graduation certificate; his foreign passport; a police certificate testifying his good conduct; and the departure certificates from his previous universities. Finally, he was allowed to sign up as a regular student. The formal admission of students was an elaborate ceremony, and Teller was unreservedly happy because he recognized that he had finally arrived at the place of his dreams. Hans Bethe called Leipzig Teller's "intellectual home."[28]

For his dissertational research Teller had to do a lot of numerical calculations on a noisy mechanical calculator. His office was just beneath Heisenberg's bedroom, where the single professor lived at the institute. To make matters worse, Teller was a night owl; however, Heisenberg did not once complain about the noise Teller generated.[29] Instead, one night, sometime in the late fall of 1929, Heisenberg came downstairs to see Teller and suggested that he write up his results for his thesis. Such an arrangement was not unusual, even though Teller was relatively new at the school. John von Neumann already had a draft of his doctoral dissertation on the axiomatization of set theory in 1922, even though he had just completed his high school studies at age eighteen in 1921. However, he did not defend his doctoral dissertation until 1926 in Budapest. But in the meantime he had also completed his university training in chemical engineering in Berlin and Zurich. There were other similar examples.

Teller's principal professors were Heisenberg, Peter Debye—a future Nobel laureate—and Friedrich Hund. Teller attended several courses in physics by Debye, courses in thermodynamics, electrodynamics, applied quantum mechanics, and optics by Heisenberg, and another course in the theory of molecular structure by Hund. There were seminars, colloquia, and laboratory exercises for advanced students in physics, materials structure, thermodynamics, electrodynamics, optics, and molecular structure in which the same three professors were heavily involved. Debye was soon joined by Erich Hückel, who had previously co-authored with Debye important theories, and whose simple yet efficient approaches to quantum chemistry have been much used ever since. The atmosphere around Teller in Leipzig was eminently conducive to broadening his knowledge in physics and his outlook on the world of science.

It may have been a short period during which Teller prepared his doctoral work, but it was also a very efficient period. First, Heisenberg gave him a paper by Eugen Wigner (as he was known then) to study and to report on. The paper was about the application of group theory in quantum mechanics. This was to become Wigner's leitmotif for his theoretical physics, and it was the topic of his new monograph in 1931.[30] Although Teller was versed in group theory—not a small feat for a recent student in chemical engineering—he did not understand Wigner's paper. Here came Heisenberg, the pedagogue, who showed Teller how to decipher such an article, how to find its essence, and how to master the rest. In the paper, Wigner demonstrated that symmetry considerations—this is what the group-theoretical approach represented—could simplify the handling of even very complex systems.

Heisenberg was satisfied with Teller's performance, so he assigned to him a research project, again, with a great pedagogical sense. Teller was to decide between two conflicting solutions for the same problem. An American and a Danish researcher each published his results for the most stable state of the simplest possible molecule, consisting of two protons and one electron. This is a molecular ion: a molecule, because there are two hydrogen atoms linked in it, and an ion, because one of its two electrons is missing. The formula of the molecule would be $H_2$ and the formula of the molecular ion would be $H_2^+$. Teller possessed the necessary mathematical skills to solve the problem and decided who was right and who was not and why.

His next assignment was to determine the excited states of the hydrogen molecular ion, that is, when its only electron moves to orbits of higher energy. Such states may exist in infinite number—in theory, that is.

# Über das
# Wasserstoffmolekülion

## Dissertation

zur

Erlangung der Doktorwürde einer
Hohen Philosophischen Fakultät
der Universität Leipzig

eingereicht von

**Eduard Teller**

Leipzig, den 6. Februar 1930

Verlagsbuchhandlung Julius Springer in Berlin
1930

Title page of Edward Teller's doctoral dissertation.

There was no practical importance to the question, only the academic interest, but it was a good project for a doctoral dissertation. It also involved a lot of calculations. It then depended on the supervisor to determine how far the doctoral candidate should be required to go before considering the work sufficient for the degree. Heisenberg was not too demanding, and rightly so, because after the first few excited states the work becomes a routine procedure.

As of January 1930, Teller stopped doing course work and concentrated on his writing. On February 6, he presented his research before the Faculty of Philosophy of which physics was part. Hund and Heisenberg acted as referees of his dissertation. According to Hund, Teller showed independence in carrying out his research and great talent in finding the most suitable mathematical approach for solving his task. Heisenberg stressed that Teller's work demonstrated that he was in complete control of the physical problem he had to solve and of the mathematical techniques he chose to apply. Both Hund and Heisenberg gave Teller's dissertation the grade "very good" (*sehr gut*)—a high grade, but not the highest.[31] The oral examination took place on February 28. In addition to physics, his major, for which Hund was the examiner, Teller had to be examined in his two minors, mathematics and chemistry. All three examiners gave the highest grade, "excellent" (*ausgezeichnet*), for his performance. The certificate of his doctorate was issued on May 19, 1930; four months and four days after his twenty-second birthday.

Even before Teller completed his dissertation, he was offered an assistantship at the University of Leipzig in the fall of 1929. He was Hund's assistant until the spring of 1931.[32] The assistantship did not involve research with Hund, only correcting students' papers. Teller carried out independent research and published papers, but neither with Hund nor with Heisenberg. The Leipzig years and his interactions with Heisenberg were nevertheless of paramount importance for Teller, and their impact on him lasted a lifetime. Heisenberg took him along for trips to Berlin and elsewhere to attend meetings, among them the famous Berlin colloquium at the University of Berlin and a lecture at the Kaiser Wilhelm Institute of Physical Chemistry and Electrochemistry.

At the colloquia Teller could listen to such luminaries as Albert Einstein. Actually, he became rather depressed when he did not understand Einstein's presentation, but he was fortunate to have Wigner at hand, and they mutually consoled each other for their shortcomings. When Teller moved to Göttingen, he did some work with Heisenberg's former mentor, Max Born, although they were not linked in any formal way. Yet Born referred to Teller as one of his "closest collaborators" in his memoirs.[33] Both Heisenberg and Teller participated in the celebration of Born's fiftieth birthday.

Teller was formulating his own direction for research when he returned to chemistry for some suitable projects. Most of his papers from the work during this period appeared after he had left Germany. The application of modern physics to problems of chemistry was a fruitful choice of topics, similar to what Linus Pauling was doing. Pauling devoted his entire creative career to this approach, whereas Teller took only some initial steps in this direction. But even these steps were to prove of lasting significance. There would later be some overlap between the works of Teller and Pauling in which Teller supported some of Pauling's disputed claims. This was more remarkable when considering how fiercely they would be opposing each other in politics in the late 1950s and after.

Obviously, Teller was the sole author of his dissertation. It was unusual, however, that Heisenberg did not figure in as co-author of the publications that reported the main results of the dissertation. Thus, Teller's first two publications, one each in 1930 and 1931, carry his name alone. Teller, nonetheless, started involving others in joint work during his brief Leipzig period. It should be stressed that Teller invariably proved to be generous in assigning co-authorship of published papers and giving weight to the contributions of his colleagues.

His very first co-author was Laszlo Tisza, whom he involved in his molecular physics and urged him to come to Leipzig and join Heisenberg's group even if only for a few weeks. This brief sojourn in Leipzig proved very fruitful for Tisza. They were engaged in describing the vibrations of molecules. Molecules are not rigid bodies, and there are displacements of the positions of the atoms relative to each other all the time. This motion is very rapid and goes back and forth in the order of a trillion times in the duration of every second. The motion is an integral part of the structure of molecules; it can be calculated and it can be measured. The first task was to describe this motion in the clearest and simplest way, but when Teller and Tisza succeeded in doing that, they learned to their dismay that others had already solved the task. One can take such a disappointment either of two ways. The optimist is happy that he came to the answer that had been proven correct. The pessimist considers the whole exercise superfluous. Teller took the optimist's approach.

He also knew that their work prepared them for solving further problems that had not yet been solved. In this case they knew this because they—however belatedly—checked them in the literature. The work involved the internal rotation of molecules. For this, they not only had to be versed in the literature but also to be aware of what other laboratories engaged in similar research were doing. One can carry out research that is so hot that no other researchers are yet involved in similar work. On the other hand, one can carry out research that is so hot that there are others— competitors, in a good sense of the word—who are working on the same or related problems. The internal rotation in molecules was such a problem in which others were involved as well, and it was to everybody's advantage to be informed and to inform the others. So that is what they did.

Teller felt secure under Heisenberg's wings in Leipzig and comfortable at Heisenberg's institute. But the time came, and it came fast, when an opportunity opened up to gain a little more independence, and Teller, wisely, took it. Arnold Eucken, professor of physical chemistry at the famous University of Göttingen, invited Teller in the spring of 1931 to become his assistant. No documents could be found about Teller's stay in Göttingen in the University Archives, but the records at the City Archives show that he had an assistantship at the Institute for Physical Chemistry and that he was a resident in Göttingen from May 5, 1931, through September 4, 1933.[34]

In terms of present-day American usage, Teller was a postdoctoral fellow in Leipzig, and the position Eucken invited him for was similar to an assistant professorship, though less independent. Eucken was an ambitious physical chemist who wanted to keep pace with the developments in atomic physics. He was interested in interpreting his results employing the new quantum mechanics. Teller's main task was to watch out for possible blunders and to do so with tact, without hurting his professor's feelings. Teller had heard from his fellow Minta pupil Nicholas Kurti that less careful assistants did not last long with Eucken; Teller did.

The role of selecting the correct ideas from among someone else's research suited Teller at this stage of his career. Apparently, he had the ability to show when an idea was of no use in a manner that did not damage anyone's ego. He performed impeccably to Eucken's satisfaction, so word got around in Göttingen about his abilities in both science and attitude. Before long, there was another call.

The physics Nobel laureate James Franck was eager to understand molecular chemistry by applying the latest achievements of physics to it, so he also turned to Teller for assistance. This was another fortunate encounter for Teller, because Franck was not only an outstanding physicist, wielding enormous authority among his colleagues, but he was respected for his personal qualities as well. In 1965, soon after Franck had died, Teller remembered their interactions in the early 1930s: "Professor Franck was one of the two or three people who had the deepest influence on my own scientific development. I learned from him not only that most of the important things in physics can be described in a nonmathematical language but also that mathematics is being used all too often to obscure the essentially simple character of underlying ideas."[35]

Upon Hitler's accession to power, Franck would also leave Göttingen and Germany and would become professor of physics at the University of Chicago. He and Teller would see each other there during subsequent years. With Eucken, Teller had only one more meeting after 1933; it was when Teller visited Europe for the first time after the war, in 1948. He visited Göttingen briefly and found Eucken the most changed among his old acquaintances; the professor was in a depressed state and committed suicide not long afterward.

Göttingen provided another exceptional environment in science for Teller in the years from 1931 to 1933. In addition to its stellar professors, it boasted several future luminaries. There was Gerhard Herzberg, whose lifelong interest was spectroscopy, a field in which Teller had already demonstrated his skills. Teller worked jointly with him and also with the Czechoslovakian George Placzek. The latter had a complicated attitude toward Teller; he cared for him but could also be nasty to him. Teller had not experienced such a belligerent attitude since his early days at the Minta. Again, it was a combination of Placzek's teasing him and Teller's taking it badly that further incited Placzek to keep teasing him. It was not terribly serious; for instance, Placzek would address him as *Herr Molekular-Inspektor*, referring to Teller's interest in the structure of molecules. It could have been considered a compliment, and even if it was not, Teller should have let it go—as is easy to say in hindsight—but apparently he could not. This is interesting in light of how much antipathy he had ignored around him growing up.

That Placzek was not completely malevolent toward Teller was shown by his suggesting new books to read and by his advising him to stay at the Hungarian Academy, the *Accademia d'Ungheria*, when Teller was planning to visit Enrico Fermi in Rome. Fermi wrote an appealing recommendation on Teller's behalf to the Hungarian authorities to let Teller get room and board at the Hungarian Academy, which was granted.[36] Fermi was a few years older than Teller and had arrived on the scene of physics in time to solve some of the fundamental problems of twentieth-century science. But it was not primarily the timing that made the difference. In Rome, Fermi built up his own very strong group, composed of excellent physicists and chemists. They worked in a concerted effort with Fermi as their undisputed leader, who already had made important discoveries.

It was a couple of years after Teller's visit that Fermi could have made another pivotal discovery, but he failed on this occasion. The consequences impacted Teller's life, and not only his. Fermi and his group bombarded various elements with neutrons in a systematic way. In the course of these experiments Fermi discovered that slow neutrons were much more efficient to cause nuclear reactions than the fast neutrons. In 1934, it was thought that the experiment in which uranium was bombarded produced two new transuranium elements. Although Fermi was always very cautious about communicating new findings, his mentor, the politically savvy Orso Mario Corbino, announced the discovery of the two new elements. The event was greeted in the media as another great success of fascist Italy.

Fermi had been warned by a German chemist, Ida Noddack, that they may have split the uranium nucleus in their experiment, but he dismissed her suggestion. Had he not, he might have discovered nuclear fission four years earlier than Hahn and his colleagues, and many things might have developed differently in physics and politics. To put it simply, the quest for the atomic bomb might have started much sooner and Germany might have been a major player in this quest. Though Szilard had initiated the idea of nuclear chain reactions, he had failed to search systematically for the right element to accomplish it.

Teller interacted in scientific projects not only with Eucken and Franck but others as well. Though he was only in his early twenties, both his scientific acumen and his personality manifested themselves strongly. Teller's help was very useful to Max Born in writing one of the chapters in his optics book. The chapter was about a recent achievement, the Raman Effect, and Teller was already versed in it. Born noted in his memoirs written at the end of his long life that "the collaboration was not easy because his way of thinking was rather different from mine, and he insisted on his formulations with the same stubbornness as he now insists on his political concepts."[37]

In 1933, it became obvious that Teller had to leave Germany. He noted that "the unique and wonderful community that was German physics in its golden years was [to be] destroyed."[38] His mother urged him to return to Hungary. However, Teller did not see safety there either, in the long run, even if the situation was milder than in Germany for the time being. It was good foresight. His additional argument was that in Germany, at least, he was in touch with the world of science, whereas going back to Hungary would mean turning back from what he was aspiring toward. Once he tasted being in the center of the scientific universe, he could not give it up.

Teller realized that he was forced to leave Germany without a clear understanding of where to go. This was a more dire situation than the one in 1926 when he was leaving Hungary without particular haste; at that time he knew his destination and he was master of the language of his target country. He could have returned home to continue his studies. Now he was no longer a student but an established scientist who needed a good job that would pay enough for him to start a family. He was in a much different stage of his life at the age of twenty-five from the high school graduate of eighteen.

He was now Dr. Teller, with an internationally known name and a cosmopolitan outlook, much of which was due to Heisenberg, who helped Teller to become a member of the international physics community. This world was now expanding, but it was also closing in on Teller. Teller gave numerous evidence of his devotion to his mentor over the years, whereas on Heisenberg's side there were no reciprocal expressions. The essay he wrote for a volume honoring Teller's sixtieth birthday sounds staunchly impersonal.[39]

## TELLER AND GERMANY

Teller left Germany before the atrocities began. Whatever suffering his and Mici's family members had gone through during the war years was inflicted by Hungarian Nazis rather than Germans. Teller's closest family member, who had fallen victim to Nazi persecution, was his friend and brother-in-law, Mici's brother. The pain caused by this loss was so acute, especially for Mici, that they hardly mentioned his name. Wendy Teller does not remember a single time that persecution in Hungary or the Hungarian Holocaust ever came up in their conversations, not once![40] Wendy explains this by her father's tendency to look ahead and never behind.

On the background of Teller's fierce anticommunism, his opposition to Nazism was rendered secondary. Teller felt that Hitler was a madman and that however terrible he may have been it was easier to get rid of a crazy person than to overcome the cold and calculating leaders of the Kremlin. This attitude came up in family conversations and in Teller's writings. In 1983, he expressed his opinion in the following way: "The events of the past thirty-five years have demonstrated that while the danger from a ruthless adventurer named Hitler was more immediate, the danger from the patient, unrelenting leaders in the Kremlin is in reality greater."[41]

Werner Heisenberg was one of the greatest physicists of the twentieth century; he also had weaknesses as a human being that may have appeared exaggerated because of his eminence as a scientist. Here our aim is not a universal evaluation, and our interest in Heisenberg is limited to two aspects: one is his impact on Teller's studies and his position as a top-notch physicist in Germany. The other is how Heisenberg's image formed after the war, and how Teller talked and wrote about him then and in later years.

Teller idolized his former mentor to the degree that he did not question Heisenberg's actions during Germany's Nazi period. Even in the quiet of his home, he insisted that Heisenberg sabotaged the efforts of building an atomic bomb for Hitler. To bolster this argument, Teller also posed the question, why did Heisenberg never state that he committed sabotage during the war? He found the answer in Heisenberg's character: "Admitting [the sabotage] before the end of the War would have been equal to suicide. Talking about this after the War would have seemed self-praise."[42] This does not mean that Teller never thought of his teacher in less than the most complimentary ways. In 1929, Heisenberg complained to Teller about the stagnation of physics due to all the interesting problems having been solved. Teller's audible response was polite, but his unspoken thought was, "You were there when the really great events occurred. You turned physics upside down. What more do you want?"[43] But, apparently, this was the most rebellious thought that Teller ever had in connection with his revered teacher, and he mentioned this in his *Memoirs* seven decades after the event.

Hitler did not realize what science had meant for Germany or what it could have meant. It is well known that when Max Planck pleaded with him against expelling the Jewish scientists, his response was, "Our national policies will not be revoked or modified, even for scientists. If the dismissal of Jewish scientists means the annihilation of contemporary German science, then we shall do without science for a few years."[44] What Hitler would have done with an atomic bomb had he possessed one is a worthwhile question. London comes to mind before anything else. The Nazi minister of armaments, Albert Speer, noted Hitler's unbounded enthusiasm when he watched planes diving toward Britain, saying, "What I want is annihilation."[45]

The expulsion of Jewish physicists from Germany weakened science in Germany. Whether this made it impossible to carry out a successful program for the atomic bomb in Germany is difficult to ascertain. The atomic bomb program in Germany may never have succeeded, but the remaining contingent of physicists in Germany was not a negligible group. They had a strong scientific tradition, and nuclear fission had been discovered in Germany. Moreover, Heisenberg became the head of the German atomic bomb project during the war. Although he was a theoretical physicist, his appointment to this position is not surprising, because he was the undisputed leader of physics in Germany. In hindsight, this is even less strange because of the excellent scientific directorship of the American theoretical physicist Robert Oppenheimer at Los Alamos, where the final stage of the American atomic bomb project took place.

Heisenberg could have been under terrible pressure had the Nazi leadership really expected a bomb from him. However, all indications point to their lack of understanding of the project. They seem to have assessed that whatever it was, it was not for them during the current war. Thus it may be said that Heisenberg and his associates toiled on the atomic bomb project without the pressure of the Nazi leaders expecting them to perform. Heisenberg and his colleagues worked on the bomb project *in spite* of the disinterest of the Nazi leadership. This renders the often emphasized claim meaningless that Heisenberg deliberately and bravely applied a brake to the program. Yet Teller writes in his *Memoirs*, "I believe the idea of putting the power of an atomic bomb into Hitler's hands was consciously or unconsciously repellant to many of the scientists involved, but most especially to Heisenberg."[46] From Heisenberg's errors in estimations connected with the atomic bomb, Teller concludes that Heisenberg constructed "mental barriers to making progress on the atomic bomb."[47] Then comes his ace, provided by no other than von Weizsäcker: "Carl Friedrich, who was very close to Heisenberg, once told me that 'Heisenberg died without regrets.' To me, that statement makes it clear that Heisenberg never worked for the Nazis in a real sense."[48] To say the least, Teller was most magnanimous toward Heisenberg.

The first contact between Teller and Heisenberg after the war took place when Teller received a short note from his former mentor some time early in the summer of 1946. Heisenberg wrote about *where* he had been during the previous years, but Teller noticed that he did not write anything about *how* he had been. Heisenberg asked Teller to send him the journal *Physical Review* in exchange for German publications.[49]

Teller and Heisenberg never had a conversation about Nazism, the war, or anything else concerning politics when they met in 1948, after fifteen years of having not seen each other or on later occasions. About their meeting in 1948, Teller noted only Heisenberg's "weary reserve." Teller's visit to Germany was part of a trip to Europe. On the same occasion he also saw von Weizsäcker, who was more talkative than Heisenberg. He told Teller about the "injustice" being committed by the Allies against his father, who were trying him for alleged war crimes.

Teller wrote with great sympathy about the elder von Weizsäcker's fate. He was sentenced to seven years of prison, of which he served only one and a half due to poor health. Von Weizsäcker told Teller that he would have found it justified "if the Americans had come in and shot every tenth German," but he did not think their persistent hunt after the war criminals had anything to do with justice.[50] Incidentally, the Nazis frequently carried out indiscriminate executions of the innocent for acts committed by others who had escaped capture. One of their favorite methods was decimation: killing every tenth person.

In his account to Maria Goeppert Mayer, Teller condemned the occupational policies of the Allies. Further, he wrote that he was convinced that "everything we heard about Weizsäcker being a Nazi is complete nonsense." Teller though notes that: "It is true that Weizsäcker did make compromises. But does anyone of the people who accuse him know why he made them?" Teller ends his comments, "To judge him on the evidence is a scandal."[51] As the topic of Weizsäcker's visit to the United States was coming up, Teller found it necessary to issue a warning: "It is not necessary to avoid the subject of Nazism. One can even disagree with them [German scientists] quite violently." This is an interesting point; why should it be necessary to disagree with Weizsäcker, and quite violently at that, if he was no Nazi and did not represent Nazi ideals?

In February 1949 Teller learned that he was appointed to a committee on von Weizsäcker.[52] Then, on November 22, 1949, Teller and von Weizsäcker met with Samuel Goudsmit, the physicist who led the Alsos Mission, which was charged with finding out about the German atomic project.[53] Teller defended his German friend against accusations concerning his actions. Teller's main argument was that von Weizsäcker's behavior was forced upon him by the fact that his father was a high-level official in the Nazi administration, an undersecretary in the Foreign Ministry. Among other deeds, the father signed the deportation order of the French Jews.[54] After their joint meeting, Goudsmit told Teller "that the general theme you followed, namely that the position of his father forced him to act as he did, is utterly ridiculous. . . . I was glad that you were present at our interview. . . . You can't be completely blind."[55]

Goudsmit was very critical of Carl Friedrich von Weizsäcker, whom the Nazis had named as a professor at Strasbourg University (made part of Germany by their war conquest). Goudsmit had investigated the documents in von Weizsäcker's office in the framework of the Alsos Mission. Among the documents he found were von Weizsäcker's reports on the pro-Nazi and anti-Nazi sympathies of his foreign colleagues and on the work of his German colleagues, on behalf of the infamous SS. This seems to have amounted to spying on his colleagues to find out about their loyalty to

Nazism. When he met with Goudsmit, von Weizsäcker tried to convince him of the similarities between the idealism of the German knighthood, Christian ethics, and the Nazi ideology. In a most despicable way, von Weizsäcker tried to discredit Goudsmit's criticism by ascribing his belligerence to the fact that Goudsmit's parents were murdered by the Nazis.[56]

In 1945, the Allies collected and detained ten leading German scientists in a manor house in Cambridge, England, called Farm Hall. The German detainees were a mixed lot; they included active supporters of the Nazi regime and known anti-Nazis. The most dramatic exchanges there occurred after they had heard the announcement of the Hiroshima bombing on the radio. The first reactions were disbelief turned into anger and accusations, and finally a "version" of history was concocted. According to this "version," while there were various reasons for the Germans' failure to produce an atomic bomb, such as the Allied bombings, the principal reason was that they found it morally wrong to provide Hitler with such a terrible weapon. Von Laue called it *Lesart*, meaning a certain way of reading something.

Apparently von Weizsäcker initiated the "version," but Heisenberg most convincingly adopted it for the rest of his life. Unbeknownst to the German detainees at the Farm Hall, their conversations were taped, and their conversations were most revealing. Teller wanted to protect Heisenberg's and von Weizsäcker's reputations and attempted to prevent the Farm Hall recordings from becoming public. He wrote about this in no uncertain terms later to von Weizsäcker. In the same letter he disparaged Goudsmit and criticized him for "insensitivity."[57] Teller did not succeed, and the conversations were eventually published. Teller's whitewashing Heisenberg of any wrongdoing and accepting his "version" of what really happened appears to be a sad example of Teller's opportunism.

Teller never questioned Heisenberg's behavior during the Nazi era, but he noted that Heisenberg bitterly complained to him about the American bombings.[58] On another occasion, Teller complained that Heisenberg's "*apparent* cooperation with the Nazis" was misunderstood by many (emphasis by Teller).[59] Teller's leniency toward Heisenberg may have been out of deference for his former teacher, although it is difficult to fathom how he explained and even praised Heisenberg's behavior, ascribing to it the great value of knowingly sabotaging the German atomic project. But just as Teller's allegiance to Hungary was unshaken, so, too, was Teller's loyalty to those he admired or had a friendship with. The way he functioned throughout his school years—ignoring the bad—was how he functioned in some of his lasting friendships—through denial. It enabled him to survive and move ahead.

To appreciate how unrealistic Teller's attitude was, let us examine Heisenberg's behavior. Before doing so, however, we need to make two points. One is that no shred of evidence has ever been found to show that Heisenberg was a Nazi; he was not. The other is that Heisenberg never made any unprincipled false statement concerning physics in order to belittle the importance of any of its teachings that would diminish the significance of the contributions of Albert Einstein or other Jewish physicists. If this sounds trivial, one would see it otherwise if one were to examine the records of many German scientists during the Nazi era. It is true, though, that while defending theoretical physics, Heisenberg was willing to let its creators disappear from public consciousness.

Heisenberg could have left Nazi Germany, but he did not. Teller could have stayed in Nazi Germany, at least for a while, under the protection of his Hungarian passport, but he did not even consider staying; none of his Jewish–Hungarian friends—the Martians—did. George von Hevesy, the future Nobel laureate chemist, is an interesting case. He worked in Freiburg im Breisgau in southwest Germany as a revered and popular professor, where it was not known that he was Jewish. Nonetheless, he did not stay but left in 1934, moving to Copenhagen soon after the proclamation of the racial laws. When it became dangerous to stay in Denmark, he moved to Sweden.

A conspicuous case of a Jewish scientist staying in Nazi Germany under the protection of a foreign, in this case, Austrian, passport was Lise Meitner. She was an associate of the Kaiser Wilhelm Institute and left only when she found herself a German citizen upon the *Anschluss* in the spring of 1938. At that point she fled to Sweden via Holland; it was a dangerous escape. After the war, Meitner regretted that she had stayed in Nazi Germany for those additional five years. She wrote to Otto Hahn in 1945, "I know today that it was not only stupid, but a great injustice that I didn't leave immediately."[60] A later letter in 1948, again to Hahn, is yet harsher toward herself: "[T]oday it is very clear to me that it was a grave moral fault not to leave Germany in '33, since in effect by staying there I supported Hitlerism."[61]

Meitner's opinion hardened as she observed the ease with which her German colleagues tended to forget the horrors of Nazism, to ascribe the crimes against humanity to a select group of top Nazis, and to absolve the rest of Germany from them. The very German scientists who never became Nazis themselves, such as Max von Laue and Otto Hahn, and who had the moral right to speak out, caved in, ostensibly for the good of the nation. Hahn became the head of the Kaiser Wilhelm Society, soon to be renamed the Max Planck Society. It then took half a century before the society decided to investigate its past during the Nazi regime.

Teller was not alone in excusing the negative deeds of his former German colleagues. Here is a minor example. When Lise Meitner was asked by a former Nazi, G. von Droste, to testify that he had never been a real Nazi, Meitner's immediate reaction was that she called von Droste's claims "lies from beginning to end." Yet she complied with von Droste's request and wrote a letter, which might have helped exonerate him.[62] To be sure, there were exceptions, such as when Pascual Jordan, a "rabid Nazi" physicist, asked Max Born for a testimonial that he, Jordan, protected Jewish contributions to theoretical physics from the radical Nazi side. Born, in his declining response, expressed his distaste for Jordan's request and former behavior by listing the names of friends and relatives killed by the Nazis.[63]

From a humanistic point of view, one of the lowest points in Heisenberg's behavior came about during a pre-war encounter with his former mentor, Max Born, at the beginning of Born's forced immigration to England. Heisenberg came to Cambridge to give a lecture, after which they talked. Heisenberg mentioned to Born that he was authorized to ask Born to return to Germany and take up a research position—teaching was excluded. Born was sufficiently disoriented to show interest and inquired about bringing his family. No, Heisenberg informed him, Born's family was not included in the offer.[64] Considering Born's contributions to German physics, this was especially egregious. In the period of 1921 to 1933, he attracted eight future Nobel laureates and at least as many other luminaries in physics to Göttingen, and he counted Teller among them.[65] Born was astonished by Heisenberg's attitude; he noted in his memoirs, "We could not understand how Heisenberg, whom we knew as a decent, humane fellow, could have agreed to convey such a message to me."[66] And this was long before the war and before any considerations for using nuclear physics for weapons. How could Teller or anybody judge Heisenberg's behavior during the Nazi times on the basis of having known him only before 1933?

Heisenberg's behavior has been characterized as "redolent of shameful compromise, that he acted weakly and confused moral purpose with his own private comfort, that he was a profoundly opportunistic character, that he failed his duty as a human being—all these severe judgments seem to him not only offensive, but utterly incomprehensible."[67] But his behavior under the Nazis was topped by his behavior following the end of the Nazi era: "The situation is made all the worse by Heisenberg's glib attempts to rationalize the indefensible in an often absurd manner, and the all too ready resort to half-truth, evasion, and self-delusion."[68] After the war Heisenberg remarked to Francis Simon, a German refugee scientist who lost family members in Nazi extermination camps, "The Nazis should have been left in power longer, then they would have become quite decent."[69]

Apart from his trip to Copenhagen in 1941 and the famous puzzle of his conversation with Bohr, Heisenberg made numerous other trips to occupied countries during the war. Heisenberg let himself be used as the flag carrier of the "Culture" of Nazi Germany. It has been alleged that his trips were at least in part a repayment for Heinrich Himmler's many-sided support of Heisenberg, which included protection from the attacks of Nazi physicists and recognition by the Nazi regime in high positions and awards.[70]

The most distasteful trip was perhaps Heisenberg's visit to Poland, where his school acquaintance Hans Frank, called the butcher of Poland, reigned and had invited him for the visit. This was one of the examples that demonstrated how Heisenberg "turned a blind eye to what was happening."[71] Frank was later tried in Nuremberg and hanged in 1946 for his crimes against humanity.

Heisenberg made a questionable visit to Holland in October 1943. According to G. P. Kuiper, a Dutch–American member of the Alsos Mission, the Dutch physicist Hendrik B. G. Casimir had a private talk with Heisenberg in Leiden during this visit. From their conversation Casimir learned that Heisenberg knew about the concentration camps and about the German atrocities in occupied territories, yet "he wanted Germany to rule."[72]

According to another source, "There is no evidence that Heisenberg ever regretted his wartime role, and every evidence that he falsified his memory of it."[73] Even the mild-mannered James Franck said of Heisenberg, "I have neither forgiven him nor forgotten what he did."[74] Goudsmit criticized Heisenberg because Heisenberg "doesn't seem to be willing even now to condemn the Nazis openly."[75]

In contrast, we look to what the head of the Japanese atomic bomb project, Yoshio Nishina, said. We have come across his name before as one of the visitors to Heisenberg's laboratory. The Japanese project was a small one that suffered from the rivalry between Japan's army and navy. They had excellent physicists there, though, if only considering the later Nobel laureates, Hideki Yukawa, Sinitiro Tomonaga, and the then student Yoichiro Nambu, in addition to the project leader Nishina.

The Japanese physicists never did so much as even hint—hiding behind their failed effort—that they might have been morally superior to the American scientists who made up the Manhattan Project.[76] Nishina exclaimed upon having heard of the success of the Americans in having created the atomic bombs, "U.S. scientists won a great victory over [us] Japanese Riken scientists. In the end, they have had higher morale than us."[77] Riken was the Institute of Physical and Chemical Research in Rikagaku-kenkyujo. The attitude of the Japanese physicists is even more noteworthy if one considers that Nishina made this statement despite Japan's reluctance to face the true history of its role and deeds in World War II.

In Germany very few could survive the Nazi regime and not become tainted to a small or great extent—and the scientists were no exception. But the differences of involvement were enormous. It was not the same to accept an empty chair after a Jewish professor had been expelled from it as to denounce a Jewish professor in order to take over his position. Of the scientists staying in Nazi Germany, there have been doubts about the behavior of most. Those with known Nazi activities tried to whitewash themselves, and they often succeeded. Many factors in both the emerging West Germany and East Germany enabled former Nazis to avoid accounting for past behavior. Different people met different fates, and their treatment by no means was always proportional with their deeds.

One of Teller's professors in Leipzig, the Dutch citizen Peter Debye, left Germany in 1939 and spent the rest of his life as a great, revered scientist at Cornell University. It was already under the Nazis that Debye had sought and gained administrative prominence in Germany, becoming director of the Kaiser Wilhelm Institute of Physics in Berlin and president of the German Physical Society. He carried out the anti-Jewish measures that were expected of him in his positions. Having moved to America, Debye never had to answer for his deeds, whereas his former colleague, the less important Hückel, had to go through the so-called denazification process—proving that he had not been a Nazi—after the war. Debye should have at least explained his controversial behavior and actions in the period of 1933 to 1939 after the war, but this never happened. Instead he was honored with the chairmanship of his department; lectures and scholarships were named after him, and his bust stands at Cornell University, where he spent the last twenty-six years of his life. Recently, in Holland, books were published about Debye's anti-Semitic and Nazi past.[78] Vigorous discussions followed in Holland and at Cornell University. After some soul-searching, the Department of Chemistry at Cornell decided not to change anything in the way Debye's name is commemorated. The expedient decision seems to be an exercise in self-justification for earlier carelessness and also, likely, inertia. There was only one dissenting voice, by the Holocaust survivor and Nobel laureate Roald Hoffmann.[79]

How far-fetched is it to bring up cases like Debye's in connection with Teller's relationship to Heisenberg? The anti-Nazi alliance was fast giving way to the cold war after 1945. And the United States found it expedient to rely on various components from Nazi Germany in its struggle against the Soviet Union. Consider just the example of Wernher von Braun, whose rockets pounded London during World War II. He became a respected member of the American defense establishment. Given this backdrop, it is less surprising that Heisenberg could not have appeared to Teller or Debye could not have appeared to the Cornell faculty as someone whose past deeds and attitudes should be scrutinized and should have been made to account for.

As for Teller, he remained devoted to Heisenberg. In his article he wrote honoring Heisenberg's seventieth birthday, he said, "For me Heisenberg lives as two persons. One is a demigod whom I revered in my youth. The other is a Man who understood his call and saved a lot of values from a terrible shipwreck of world history."[80] Teller maintained to the end that "[t]he most important part of my education was my study with a great man whom I admired and whom I consider a wonderful person, and not only a wonderful scientist."[81] Noting Heisenberg's efforts after 1945 "to re-create" German physics, Teller expressed regret in 1980 "that all of us did not lend more of our thoughts and efforts to the purpose of re-establishing what used to be a great intellectual community in Germany."[82]

# Chapter 3

# TRANSITIONS

Emigration is in many ways very stimulating.
—EUGENE P. WIGNER[1]

*However painful it was for Teller to leave the golden age of physics in Germany, he showed remarkable adaptability and resilience in finding himself at home first in Copenhagen, then in London, and finally in Washington, DC. His scientific output did not diminish, which is the more remarkable because he was making a transition not only between countries, languages, and continents, but also from chemical physics to nuclear physics. When his transition from a European university assistant to an American full professor was complete, he had also transformed himself from a somewhat unsure young man into a popular and respected member of the community of physicists. His Washington period turned out to be the last stint of a worry-free life in which he had built up considerable capital in both scientific achievement and human relations.*

Wigner's remark about the benefits of emigration was applicable to Teller, but there were differences between Teller's departures from Hungary and from Germany. In 1926, he left Hungary with the promise of a better future, or, even more accurate, to have a future, yet it was not a forced departure. After the Nazi takeover in 1933, he had to leave Germany. This departure also meant leaving behind both languages in which he was fluent and both cultures in which he grew up.

## COPENHAGEN AND LONDON

Teller spoke English, but nowhere near as well as he spoke German, and he was hardly versed in British or American literature and culture. On his first visit to London, his host, George F. Donnan of University College, suggested that he read Lewis Carroll's two classics *Alice in Wonderland* and *Through the Looking Glass*[2] to make up for some of his deficiency. Teller became very fond of Carroll's writings and started to read Carroll's stories and poems to his two-year-old son, according to Laura Fermi, who met the Tellers in Los Alamos in 1944.[3]

The British organized a rescue mission for Jewish scientists who were being kicked out of Germany and also recognized the great opportunity for British science to strengthen itself. It was a beautiful example of sensible and humane behavior with mutual benefits. Teller met Donnan when the London professor came on a rescue-recruiting mission to Göttingen, as did Frederick A. Lindemann of Oxford University. Teller's friend Szilard was active in the rescue operations, and his schemes initiated the Academic Assistance Council (AAC). It was founded by Lord Beveridge and inaugurated with a meeting at the Royal Albert Hall in London on October 23, 1933, where Albert Einstein was among the speakers. The AAC helped thousands of refugee academics, and it has survived to the present day.[4]

In 1933, and throughout subsequent years, finding jobs for refugee scientists in the thick of an economic depression was not a trivial endeavor; jobs and opportunities in academia were scarce. The oldest fellow of the Royal Society, who was a notorious protester, criticized Michael Polanyi's appointment to the Chemistry Department of Manchester University, lamenting, "How are we to make progress if we do not employ those we train?"[5] In a previous letter to the media, he protested "against the appointment . . . of a gentleman who is not an Englishman nor in any way connected with us." He added that "[t]he introduction of a foreign outlook into Manchester is most undesirable."[6]

Expressions of reservation cropped up from the United States as well. One of the leading mathematicians voiced his fear that refugee scientists would take precious academic jobs away from the Americans, who, in turn, "may become hewers of wood and carriers of water."[7] He suggested that refugee scientists should be content with lesser positions than their qualifications called for, which was the case in most instances anyway. However, the voices of dissonance were rare, and the assistance of the British and the Americans can only be even more appreciated when looking at it in this context. Teller and his Martians friends, von Neumann and Wigner, were privileged exceptions because they gained better positions upon coming to the United States than they had in Germany.

Teller's stay in London was helped by the AAC and funded by Imperial Chemical Industries.[8] He had friends among the many refugees, including Hans Bethe. During his German period Teller was hardly interested in politics beyond discussions with colleagues. Curiously, Michael Polanyi, who had a much deeper interest in politics at the time, was slow in recognizing the importance and consequences of the Nazis coming to power in Germany. This is the more surprising because he had accurately assessed the Hungarian situation some years before. It was Szilard who had the foresight to appreciate that Hitler and the Nazis were there to stay, and he recognized the utilitarian approach of the German population to politics that made that possible. Szilard ascribed the developments in Germany not so much to the strength of the Nazi movement and Hitler, but to the weakness of the resistance to them.[9]

Despite the difficulties of the emigration from Germany as compared with the departure from Hungary, there were advantages. Teller came to Germany as a student and left as a scientist with a doctorate. Moreover, he was not just a scientist but one with excellent training; with some important papers under his name and with unique connections in the world of science—yet he was barely twenty-six. This was a time for decisions at more than one level. As soon as he heard that Mici was back in Budapest from the United States—after two years of graduate studies in sociology and psychology—he rushed to Budapest to renew their friendship. It had been a painful two years during which they didn't even exchange letters, but the two years strengthened their resolve to spend the rest of their lives together. When Teller proposed to her, she did not need much deliberation before she said yes. They planned the wedding for Christmas of 1933.

Both Donnan and Lindemann interviewed Teller in Göttingen. Both meetings were a success and both issued him an invitation. Teller felt closer to Donnan, who urged him to come to London for a quick exploratory visit. Donnan was a physical chemist who had worked with the pioneers of the field, including the Latvian–German Wilhelm Ostwald and the Dutch Jacobus van 't Hoff. When Teller went to London, he liked what he saw, and to sweeten the deal, Donnan offered him great flexibility with his appointment. This was meaningful to Teller, who had applied for a Rockefeller Foundation fellowship because he wanted to spend some time at Bohr's institute in Copenhagen.

It would be interesting to speculate what might have happened had Teller preferred Lindemann's offer over Donnan's. Lindemann—Lord

Cherwell—was Winston Churchill's friend and scientific adviser, and he played a significant role during the war in matters involving science. After the war, in 1948, Teller would dine with Lord Cherwell in London on the occasion of a trip to England.[10] Regardless, Teller chose Donnan, who became a caring and understanding mentor and friend. His offer included a salary that exceeded what Teller earned in Germany.

Teller had applied for a fellowship at the Rockefeller Foundation, and he had a good chance of getting it. However, the conditions stipulated that the applicant have employment to which he could return upon the completion of his fellowship. His Göttingen professors, Eucken, Born, and Franck, promised him a joint appointment, but this would not satisfy the Rockefeller Foundation. Showing good foresight, the Rockefeller people were afraid that the situation in Germany would deteriorate and the foreigners there would be dismissed. A representative of the Rockefeller Foundation, a Mr. Miller, visited Göttingen at the beginning of December 1932 and advised Teller to get assurance from Budapest that he could continue his career in science there. Teller asked his father for help with the Hungarian authorities.[11]

Miksa Teller, ever the dutiful father, took immediate action. He visited the chairman of the Department of Theoretical Physics of Budapest University, Rudolf Ortvay, and asked for his intervention. Ortvay knew Teller and his work because Teller had met with him and given talks at the department during his visits to Budapest. That same day, Ortvay wrote a letter to Kálmán Szily, the state secretary of the Ministry of Religion and Public Education, asking for support in the form of a statement that Teller would have a position in Hungary. Ortvay attached Teller's handwritten curriculum vitae and a few reprints of his papers to his letter.[12]

Miksa Teller's next visit was to his former fellow student Elek Petrovich, who at the time was the director of the Museum of Modern Art. On December 13, Petrovich sent a letter, presumably also to Szily, on Teller's behalf. He testified that Miksa Teller was an honorable man and that his colleagues held him in high esteem. Further he stated that the young Teller aimed at carrying out *scientific* work (the emphasis was in the original).[13]

The question arises whether Teller might have contemplated at this point to return to Hungary after a stint at Bohr's laboratory, which was presumably the goal of his application. Teller made it clear in his letter to his father that he would rather stay in Germany than go to Budapest. But he understood that had he not made a commitment to return to Budapest, he could not have expected a promise from the Hungarian authorities. It is doubtful whether the Hungarian authorities would have welcomed his return, and Ortvay made it clear to the state secretary in his letter of December 19 that there was no hope in the near future for employing Teller. At the same time, he warmly recommended sending a letter of support to the Rockefeller Foundation.

The Rockefeller Foundation finally granted Teller the fellowship. The story is of interest because it shows yet another example of Miksa Teller supporting his son and doing so very efficiently. It also shows that Teller reached back to Budapest for support in this critical point of his life. The exchanges cited above took place less than two months before Hitler came to power in Germany.

This was a significant time for Teller for yet another reason; he and Mici had decided to get married on Christmas. However, the Rockefeller Foundation frowned upon their marriage plans, citing another Hungarian who had recently used a Rockefeller fellowship for an extended honeymoon. Teller talked it over with Mici, and she agreed to postpone the marriage. Teller immersed himself in his work and in interactions with colleagues in Copenhagen. When he wrote about his life to his old friend James Franck—who had already left for the United States—Franck protested the stipulation concerning marriage, not only with Teller, but also with the Rockefeller Foundation. The foundation headquarters suggested a scheme that would allow Teller to marry while still ensuring that their European office, which had objected to his marriage, would save face.

Things would have resolved themselves, but Teller wanted to include Bohr in his decision. Bohr agreed that there should be no problem with Teller's getting married, and this set things into motion.[14] The way Teller handled the whole affair, and especially his involving Bohr, says something about how much importance he placed on getting the consent of his superiors for his actions. We again see how strongly Teller felt about following the wishes of authority.

Teller and Mici were married in a civil ceremony in Budapest on February 24, 1934; thus the delay with respect to their original plans was not considerable. Only a few close friends and family members attended the event, with Teller's sister, Emma, and Mici's brother, Ede, acting as witnesses (according to the Registry of District V in Budapest). Teller writes in his *Memoirs* that the reason they opted for a civil ceremony was that he was Jewish and Mici was Calvinist. It also might be that Teller did not care for having a religious ceremony and that in Hungary in 1934 it was more expedient to skip the issue of religion.

Mici came from a Jewish family of long tradition; both her parents' ancestors had settled in the southwestern Hungarian town of

Nagykanizsa.[15] The imposing building where her father's family lived at the end of the nineteenth century still stands today largely unchanged. The family had been made up of well-to-do builders and iron factory owners. Mici's father, Ede Harkányi (a Hungarianized surname), had a law degree, but he became known as a writer interested in sociology. This interest may have brought him and his wife, Ella (according to other sources, Gabriella) Weiser together, because by then she had made her name in social science. The house of the Weiser family in Nagykanizsa was designed by a Viennese architect. It has gone through many changes, but its internal framework has remained more or less intact.

Mici's father, Ede Harkányi, died at a young age, leaving behind his wife with two small children. Her father was a convert to Catholicism, and her mother was a convert to Calvinism, which is why their son Ede became Catholic and Mici became Calvinist. Ella Weiser's (Mici's mother's) second husband, Aladár Schütz, a pediatrician in Budapest, adopted both children, hence their double surname, Schütz-Harkányi. Mici's stepfather was Jewish and her two half-brothers, István and Gábor, born during her mother's second marriage, grew up in the Jewish religion.

István Schütz immigrated to Australia, but Gábor stayed in Hungary. There he was incarcerated as a forced laborer during the war, escaped and fought against the Nazis as a partisan, and after the war became a prominent communist. He was soon disillusioned and joined the reformers under Imre Nagy's leadership. After the suppression of the 1956 revolution, he left Hungary and lived in Switzerland.[16] Gábor Hungarianized his surname from Schütz to Magos. The Archives of the State Security Services holds a detailed interview with him about Teller and Mici, conducted on March 24, 1955.[17] What he said about Mici is especially interesting, as he must have known her quite well and there are few sources about Mici's youth. Also, what he said appears consistent with information from other sources.

According to Magos, Mici was a determined, even defiant girl with a lot of contrariness; some in the family considered her impulsive. She often had disagreements with her parents, especially with her stepfather. A typical episode that illustrates her defiant behavior occurred when one of Mici's girlfriends, Ágnes Birki, was arrested for communist activities, and Mici's parents wanted to ban Mici from Birki's circle of friends. Mici responded that she could not care less for such bans. Magos considered Mici oppositional, but not leftist; rather, progressive with no connections to the Left. She was against the Horthy regime and was an excellent student.

Even though the Tellers did not have a large or extravagant wedding, their union was for life. However, Teller was anxious to show the Rockefeller Foundation the opposite behavior to their previous experience with a Hungarian grantee. So, rather than going on a honeymoon, Teller and Mici headed back to Copenhagen with only a brief stopover in Leipzig. Teller's friend Lev Landau was taken by surprise observing how serious Teller was about his marriage. Although Landau liked Mici, he asked Teller how long they intended to stay married.[18] When it came to marital issues, Landau's views were shaped by his upbringing in revolutionary Soviet society and so he argued that "only a capitalist society could induce its members to spoil a basically good thing by exaggerating it to that extent."[19]

Copenhagen had become a great meeting place for physicists, and in 1934 there was the additional urgency of a gathering place for departing physicists from Germany. They felt uncertain about the future but wanted to continue their work without slowing down, even temporarily. Hans Bethe, who had become Teller's friend during his Munich days, was in Copenhagen, as was George Placzek. For Lev Landau, Copenhagen was his last station in Western Europe while he was preparing to return to Soviet Russia. There were others from Germany, not victims of Nazi persecution but just spending time in Bohr's institute, including Carl Friedrich von Weizsäcker, who was the closest among them to Teller. Von Weizsäcker had an affinity for philosophy and so did Bohr, and this environment influenced Teller's thinking.

Teller was very much taken by questions of determinism, stemming from quantum mechanics. He tried to formulate what he thought about it: "Although we cannot change the past, we can know it. But according to the new theory of atoms, the future is undetermined and therefore truly unknowable. *We can change it* [the future], *but we cannot know it*, at least not completely"[20] (italics added). In a later formulation, Teller would state that we cannot lie about the future because it is impossible to know it; in other words, there is no way to disprove in advance what one might promise about it. This later became a trademark of his approach in his bidding wars in weapons programs, to the consternation of both his competitors and his colleagues. The latter were supposed to fulfill what he had promised in his negotiations.

Considerations about the future were not the only thing that influenced Teller's thinking during his Copenhagen period. He traced his love for exaggeration to what he had ostensibly learned from an episode with Bohr. The story went like this: Bohr was discussing a problem in molecular physics, and Teller had to correct something Bohr had stated. Bohr accepted his correction and added that, of course, Teller knew a hundred times more about molecules than he, Bohr, did. This was a polite way of

Bohr acknowledging Teller's contribution to the discussion, but Teller took it literally and called Bohr's polite statement an exaggeration, which it was—so obviously so that it did not need to be pointed out.

What may have been the reason for Teller's commenting on Bohr's exaggeration is difficult to imagine; it may have been to extend the pleasure of his little triumph of having corrected Bohr. But Bohr said in response to Teller's mock modesty that Teller was right because Teller knew only ninety-nine times more about molecules than Bohr did. Bohr then claimed his right to exaggerate because he would be unable to talk if he could not use hyperbole. Teller considered this story a justification for his future exaggerations and referred to it frequently, although it was a dubious excuse for the kinds of exaggeration that Teller used in the future.

His first meeting with Bohr took place in Copenhagen back in 1929, when he was seated next to the great man at an event. Teller did not feel shy, in fact, he initiated a conversation about the future—the future of matrix mechanics, that is. He probably did not earn much appreciation from Bohr, because Teller expressed his hope that "contradictions will disappear from the orderly discussions of science," whereas Bohr thrived on paradoxes. Teller himself quoted Bohr, saying on another occasion, "We will never understand anything until we have found some contradictions."[21]

Teller was much taken by the complexity of Bohr's thinking and by the way Bohr expressed himself, but he never felt he fully understood Bohr. This was very different from his relationship with Heisenberg, with whom he felt on the same wavelength. Teller compared the complexity of Bohr to riding a bicycle: "Its relatively slow progress, based on a balance of dynamic variables that one cannot explain adequately in a few words, is, I think, a good introduction to the character of Niels Bohr."[22] If one finds this comparison too simplistic, Michael Polanyi's discussion of the complexity of riding a bicycle is a convincing counterargument.[23] Another expression of Teller's covert criticism of Bohr's expressing himself in complicated ways is his statement "I must apologize for at least one deficiency. Some of my sentences are short, and this conflicts with Bohr's way of using any language."[24] Teller says in his *Memoirs*, "In one important way, Copenhagen was a disappointment. Niels Bohr had turned out to be an embodiment of the paradoxes of science."[25]

Teller and Bohr did not interact directly in science, and Teller never felt fully comfortable during his stay in Copenhagen. For him the most valuable part was his interactions with other scientists. He co-authored a paper with Landau, but, according to Teller, his contribution to Landau's work was minimal. Landau published their joint report in a German-language periodical in the Soviet Union.[26] There was a more significant exchange between Teller and Landau, though it did not result in a joint publication; it had some connection to the Jahn–Teller effect.

A much-admired paper that came out of Teller's Copenhagen period was one he co-authored with a Danish physicist, Fritz Kalckar.[27] Kalckar was two years younger than Teller, a close co-worker of Bohr's, and his untimely death in 1938 was a heavy blow for Bohr because he liked working with him.[28] The subject of the Kalckar–Teller paper was catalysis. Catalysts are substances that make it easier for chemical reactions to happen while the catalysts themselves are preserved during the process. Teller called them "chemical matchmakers"—an ingenious choice of expression and much to the point. He considered his work with Kalckar to be part of his transition from molecular physics to nuclear physics, though he had not taken a special interest in nuclear physics as yet during his stay in Bohr's institute.[29]

It was in Copenhagen that the Tellers became known for their legendary hospitality, and their first dinner guest was Werner Heisenberg, who was on a visit from Germany. In Copenhagen, Teller also renewed and strengthened his friendship with George Gamow. They had first met in Copenhagen in 1930 when Teller was visiting there and Gamow was spending some time with Bohr. The next time they met was in 1931, again, in Copenhagen, during another of Teller's visits. On this occasion they spent more time together as Gamow took Teller along for a grand tour of the Danish countryside on his motorcycle. They had a good time and discussed physics a great deal. Their next meeting, again in Copenhagen, took place in 1934, when they were both married. Their wives also took to each other. This was the last encounter between the Gamows and the Tellers in Europe, but their close interactions would soon continue in the United States.

The Tellers moved from Copenhagen to London in September 1934. One of the benefits of the London environment was that Teller was surrounded mostly by chemists and physical chemists at University College. He became involved in research problems in which he could utilize his knowledge of and interest in both chemistry and physics. Another benefit of the London stay was a closer interaction with Szilard, who appreciated Teller and used him as a sounding board for his often unorthodox ideas. Teller excelled in such exchanges and often found himself on both sides. He could be the sounding board for "crazy" ideas, as he was for Szilard and would be later for Gamow in Washington. He could also be the one firing away one idea after the other, letting his colleagues decide which idea was crazy and which was the one that was worthy of further development.

Teller shared Szilard's excitement about the possibility of nuclear chain

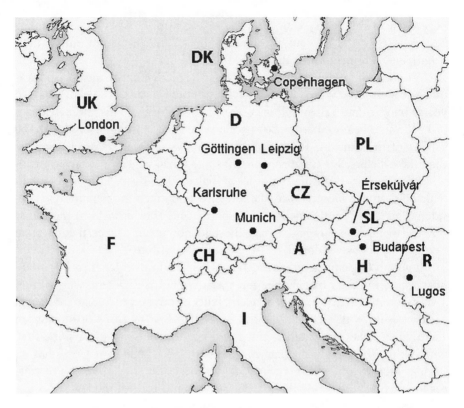

Teller in Europe (with current political borders): Budapest (Hungary); Karlsruhe, Munich, Leipzig, and Göttingen (Germany); Copenhagen (Denmark); and London (UK). Also indicated are his parents' birthplaces, Érsekújvár (Slovakia) and Lugos (Romania).

reactions and their possible culmination in an explosion. Szilard's concept was so revolutionary that, upon hearing it, Ernest Rutherford had become terribly upset. This reaction was surprising, considering the fact that Rutherford should have been the most appropriate person at the time for assessing Szilard's new ideas. After all, it was Rutherford who first carried out experiments in which atomic nuclei were transmuted—the old dream of alchemists. Opinions are divided as to whether Rutherford truly could not have imagined what Szilard proposed or whether he rendered it impossible because he sensed the danger in the realization of Szilard's suggestion.

Teller gave a course for students in London; this was his second attempt at lecturing in English, and it went very well. The first test of his English had been in Copenhagen, but it was a single presentation, and it was not in an English-speaking country; besides, it was for an audience of physicists. Although Teller's job in London was temporary, he and Mici

## GEORGE GAMOW (1904–1968)

George Gamow was born Georgii A. Gamow in Odessa, Russia (now the Ukraine) to Russian parents. For higher education, he attended Petrograd State University, while at the same time working as an instructor at an artillery school of the Red Army.

Gamow was introduced to the theory of relativity by the Russian mathematician Alexander A. Friedmann. Although the professor died in 1925, he had kindled Gamow's interest in theoretical physics for a lifetime, an interest that was strengthened by his friends Lev D. Landau and Dmitrii D. Ivanenko at the University of Leningrad (as it was called at that time). Gamow started publishing valuable papers in the mid-1920s. He was sent for a summer to Göttingen in 1928, followed by periods of time in Copenhagen and Cambridge, UK. His scientific acumen was recognized by such greats of the era as Max Born, Niels Bohr, and Ernest Rutherford. Early on, Gamow's interest turned to the application of the new quantum theory in nuclear physics, and his work augmented Rutherford's recent experiments.

His results were praised in the Soviet media as the achievements of the new order, but Gamow found a distasteful atmosphere on his return home in 1931, in which science was being increasingly looked upon as a weapon in the fight against capitalism. When Gamow and his wife had the opportunity to attend the forthcoming Solvay Congress on nuclear physics in Brussels in 1933, they simply did not return to the Soviet Union. Gamow was soon appointed to the Physics Department of George Washington University. There, he recommended Edward Teller for a second professorship. Their collegial interactions during the Washington years proved exceptionally fruitful for their research.

During the next decades Gamow made major contributions to both astrophysics and genetics. His most remarkable theory concerned the origin of the universe, which his critics ridiculed as the Big Bang—by which name it has become generally accepted. Gamow also became involved in the quest for cracking the genetic code by calling attention to the problem of information transfer from the nucleic acids to the proteins. During the last decade or so of his life, Gamow was a professor at the University of Colorado at Boulder.

Gamow was a visionary of science: a flamboyant and colorful person with a great sense of humor. In addition to his pivotal research contributions, he became a highly sought-after author of popular science books. He received the Kalinga Prize from UNESCO, his only recognition—despite his scientific achievements—and it was for his books. There is a charming autobiography by Gamow, *My World Line*, published posthumously, describing mostly his life prior to living in the United States.[30]

were so confident of staying in London that they committed themselves to a long-term lease of an apartment. But uncertainties persisted because there was no guarantee for long-term employment. Donnan was supportive and enthusiastic about having Teller around, but he could not have given him a firm promise for a permanent position. At this point, Teller received two prestigious invitations from the United States.

## NEW WORLD

Both offers that came Teller's way from America were attractive. One was from Princeton, initiated by Wigner, but it was only for another temporary position. The other, from George Washington University (GWU), was engineered by Gamow, who had moved there and needed a partner in building up his research environment. The ambitious president of the university, Cloyd H. Marvin, had tremendous foresight in how to expand and improve the university itself and lift its academic level. He had established a fund for creating a strong physics department. GWU had an outstanding law school and a similarly prestigious medical school, but the school did not excel in the area of science. Marvin learned from his adviser, Merle Tuve, that even the sum of $100,000 (an exorbitant amount at the time) he had amassed would not suffice to fund a department strong in experimental physics. Tuve suggested building up a powerful group in theoretical physics for which only scientists, travel money, paper, and pencil would be needed. Upon Tuve's recommendation, Marvin invited Gamow in the fall of 1934.

In 1935, George Washington University offered the twenty-six-year-old Teller a full professorship in physics. But getting American visas for the Tellers was more complicated than expected. There were quota visas assigned to different nationalities and non-quota visas to professionals, including teachers. Ignoring the simpler route of applying for quota visas, Teller went for a non-quota visa. But for that visa, the applicant had to demonstrate previous practice of sufficient time in the given profession. Teller did not have the prescribed time in teaching, so his and Mici's visa applications were denied. However, the more straightforward way of applying within the allotted quota for Hungarians was available, and they acquired the visas that way. In their dealings with the visa problem the Tellers were assisted by former Minta pupil Thomas Balogh, who had become a prominent economist in Britain.

In the Tellers' consideration of where to continue their lives, the political situation did not play much of a role. They might have thought they'd be more secure in the United States than in Europe, where the Nazi danger was looming and an increasing number of signs were pointing toward a war. Yet these considerations did not enter their equation in weighing their options, and so it was to some extent accidental that they ended up in the United States. To the extent it was not accidental was probably Mici's love for America, where she had spent two wonderful years in Pittsburgh. As for Teller, he was surrounded by friends, mostly from his German period both in Copenhagen and in London. Their sailing to the United States brought him and Mici to a new world, but many of their friends were also headed to the United States.

It happened that they were crossing the Atlantic Ocean together with Hans Bethe. George Gamow met them soon after their arrival. In spite of the political upheavals that forced Teller to cross borders in Europe and transverse the Atlantic Ocean, his life showed continuity. If there was any change, it was for the better; his status among his peers was growing and his positions kept improving. His full professorship at George Washington University signaled that he had reached the pinnacle of academia. There could hardly be a better job that provided job security and financial security, and also complete independence, than a university professorship. Teller's disposition reflected the happiness of his position. They lived in a small house, not far from the university—and not far from the White House, for that matter. They enjoyed entertaining both friends and visitors; the visitors often became friends as well.

Gamow and Teller developed a fruitful and pleasant collaboration in which the flamboyant Gamow regaled Teller with new ideas every morning; that is, as soon as Teller got up, which was not very early. Teller methodically weeded out the least worthy ideas, kept the few good ones, and together they developed them into joint projects and subsequently into joint papers. Decades later, when Paul Dirac wanted to refer to Gamow's and Teller's work, he asked Teller about their respective contributions, which he saw as inseparable.[31]

The two friends participated actively in the organization of the annual conference on theoretical physics. This was not a big meeting, just a few dozen scientists. The meetings became very popular, because until then, there were no such gatherings of theoretical physicists in the country. Teller's interactions with Gamow began to color his scientific outlook, and he gradually developed from a molecular physicist into an atomic physicist, becoming increasingly interested in nuclear physics. However, for the first years in his American life, he continued his research at the interface between chemistry and physics.

## "MOLECULE INSPECTOR"

Teller was not one of the great creators of the new physics; for such a contribution he was late by a few years and perhaps was not suited for it by temperament. He characterized his kind of physics as "low-brow."[32] When George Placzek called Teller "Molecule Inspector" in Göttingen, Teller resented the label. Eventually he warmed to it; indeed, it aptly describes the main thrust of his scientific research in the 1930s. It is curious that Teller, who had started out as a chemistry student in Budapest and Karlsruhe, had to become a theoretical physicist before returning to chemistry in his research. Part of this paradox is in how chemistry versus physics is viewed.

Today theoretical and computational studies have equal footing in chemistry with the so-called wet chemistry, which is done in the laboratory. But in Teller's time chemistry was mostly done in the laboratory, which was not quite his kind of chemistry. Legend has it that his "effort created a fine filigree of scar tissue all over the digits of his hands, a living memorial which lingers through the present day to the huge amount of chemical glassware which died in his service."[33] Teller's help in transferring C. N. Yang, a quarter century later, "from experimental to theoretical work of an analogously afflicted graduate student" at the University of Chicago has been ascribed to his own laboratory experience.[34]

Although Teller would never attack any scientific problem with the fervor that would eventually characterize his dedication to the hydrogen bomb, he was fully devoted to a research project while he was working on it. It was also characteristic of his scientific oeuvre that he moved from one problem to another in quick succession. There are two main types of researchers: drillers and diggers. A driller spends long years, often decades, his entire life dedicated to one particular problem. The digger solves many problems and does not stick to any for too long. We find drillers and diggers among the greatest scientists. It is not a question of scientific strategy promising success; rather, it is a question of temperament. Teller certainly belonged to the latter group. Even among the diggers Teller distinguished himself with an unusually large variety of scientific projects. He could best create in collaboration with others, typically with one associate, and he changed his associates frequently.

Among the many co-authors of Teller's scientific papers there are world-renowned names. Gerhard Herzberg is one of them. He was a German spectroscopist who emigrated from Germany when the racial laws prevented him from teaching because his wife was Jewish. He became a leading scientist in Canada and was awarded the Nobel Prize in Chemistry in 1971. While he was at the University of Darmstadt, Germany, he attended a meeting on molecular structure, where he became acquainted with Teller. More than fifty years later, Herzberg still remembered vividly his interactions with Teller, and his words shed light on how Teller approached problems and interacted at that time.

> My function was that of a midwife: Teller had the ideas, which I tried to get out of him by describing the experimental results to him and by drafting a tentative form of the paper, which he then corrected. Teller had an extraordinary reservoir of ideas in this field (as well as in other fields) and was always ready to share his knowledge. Working with him was an experience that I shall never forget. Although the ideas came from him, he insisted that on the title page we follow the alphabetical order of the authors.[35]

Almost all of Teller's scientific achievements are somewhat complex to present to a general reader because they deal with specific aspects of molecules and atoms rather than their most fundamental features. Of the rich collection of Teller's scientific contributions in the 1930s, I've selected four very different ones in character.

The first example concerns the internal rotation of ethane, when one part of the molecule rotates relative to the other parts. The ethane molecule can be imagined as two methyl groups connected to each other, $H_3C–CH_3$. There is thus a central bond between the two carbon atoms. The $CH_3$ groups may be turning around this central C–C bond. The question then arises as to whether these $CH_3$ groups can rotate freely, that is, without any restriction or whether they need to overcome some energy barrier in order to be able to turn around. When chemists started thinking about such rotations, they first imagined them as not being restricted at all. It took not only some imagination to consider a barrier to such rotation, but also the observation of some experimental facts that pointed to the notion that such rotation may not be completely free.

At the Chemistry Department of University College London, Teller joined forces with a tenured colleague with a strong research record, Bryan Topley. Together they studied a simple chemical reaction in which ethane figured.[36] They were interested in particular in the changes of the heat content during the reaction. There had been conflicting reports about this heat content from different laboratories. We might recall that investigating controversies was not new for Teller; the first research project he worked on at Heisenberg's request was a case of conflicting results from two different groups.

The gist of the problem in the present project was that the different results implied very different barriers to internal rotation in the ethane molecule. Looking at their work, it appears amazingly daring how Teller and Topley gradually considered higher and higher barriers to approximate the experimental results for the internal rotation of ethane. When they supposed 1 kilocalorie, which was much higher than previous researchers had suggested, they almost coyly noted that such a barrier would be "highly improbable." Yet they further increased the supposed barrier up to 3 kilocalories, because only at such a high value would they find consistency with the experimental measurements. Today, the accepted value for the barrier to restricted rotation in ethane is close to 3 kilocalories and it has been confirmed many times over with the most diverse techniques and in numerous laboratories.

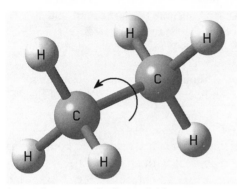

Internal rotation of ethane, $C_2H_6$.

Teller and Topley's result was a supposition, but it must have taken some courage to come up with and communicate such a value, when even one-third of it would have been considered improbable at the time. They directed attention not only to the fact of restricted rotation but to the need of accurate knowledge of the barrier to rotation for a comprehensive understanding of how chemical reactions happen. Today we know that hindered internal rotation is often crucial in biological activities. Further discoveries of some aspects of internal rotation were recognized by a Nobel Prize in Chemistry.[37] Teller did not pursue this research further. Today only very few chemists are aware of Teller's contribution to this area of research, because eventually many scientists got involved in it and the initial results have been superseded by the new findings. This is an example of a pioneering contribution that was important when it was made but that has disappeared into oblivion.

A second kind of research contribution produced a result that has proved very useful and that has remained in the same form as had been originally communicated. This is the famous BET equation describing the multilayer adsorption of gases on solid surfaces. "BET" refers to Stephen Brunauer, Paul H. Emmett, and Edward Teller. This piece of research was done entirely in the United States with all three co-authors working at

American laboratories, and its report appeared in 1938 in the flagship journal of the American Chemical Society.[38]

Emmett was from Oregon, where he attended college together with Linus Pauling at Oregon Agriculture College, today known as Oregon State University. They both then went to the California Institute of Technology, where they did joint research and published their work without a supervisor. A few years after he received his doctorate, Emmett went to work at the Fixed Nitrogen Laboratory in Washington and eventually hired an ambitious young chemist by the name of Stephen Brunauer. Brunauer was interested in broadening his knowledge and signed up for Teller's course on quantum mechanics at George Washington University. Teller gave this course late afternoons to accommodate his students, most of whom had jobs and were "older" than the average student and had already earned their degrees a few years before. Quantum mechanics, however, had not been in their curriculum, and they wanted to catch up with this new development in physics. The schedule fit Teller's work habits perfectly; he seldom went to bed before midnight or got up before ten in the morning.

Teller became friendly with Brunauer and began discussing research projects with him. Brunauer brought up the question of multilayer adsorption. (Adsorption is the process wherein molecules from a gas stick to a solid surface.) In a typical "Tellerism," Teller remembered later that "[a]t the time, my ignorance of adsorption was superlative."[39] In order to understand multilayer adsorption, they had to understand what was going on layer by layer as the molecules were forming subsequent layers in the process. First, a layer of molecules covers the bare surface. The adsorption of this first layer of molecules had already been described by the famous scientist Irving Langmuir. He was awarded the Nobel Prize in Chemistry in 1932 "for his discoveries and investigations in surface chemistry."

The nature of adsorption of the second layer of molecules was different from the first because the molecules of the second layer came into contact with the molecules of the first layer rather than with the original surface. All subsequent layers then behaved like the second layer. Brunauer had recognized that starting from the second layer, long-range interactions played a decisive role, whereas in the case of the first layer, it was short-range interactions. The phenomena of adsorption were directed by the gas pressure surrounding the surface rather than by the properties of the gas; hence, it was possible to establish a description of the multilayer adsorption by a general formula. When Brunauer and Teller came up with their result, they turned to Emmett, whose knowledge of surface chemistry was

## STEPHEN BRUNAUER (1903–1986)

Stephen Brunauer was born into a Jewish family as István Brunauer in Budapest. He completed his high school studies in Hungary before immigrating to the United States. In New York City he attended City College and Columbia University. He had a double major, chemistry and English, and continued with graduate studies in chemistry and engineering, earning a master's degree at George Washington University.

While he was a student, he joined the communist front organization, the Young Workers' League in New York City (such front organizations were run by the Communist Party, but they were made to appear independent of it). Brunauer soon left this organization, and his later views have been characterized as liberal anticommunist. He became an associate of the Fixed Nitrogen Laboratory in Washington, DC, in 1929. He married the similarly liberal-leaning Esther Caukin, and from 1931 they lived in Baltimore. Brunauer completed his PhD studies at Johns Hopkins University in 1933, and his wife earned a PhD in history from Stanford University.

Brunauer sought application of his knowledge and experience for defense-related work and joined the US Navy. He was soon appointed to be in charge of a research group dealing with high explosives at the Bureau of Ordnance. He developed an important research project, which initially employed a small group that eventually grew into more than a hundred people. He involved well-known scientists in his projects, most notably Albert Einstein, George Gamow, and John von Neumann.

By the end of the war he was a lieutenant commander of the US Navy, but he soon transferred to civilian employment. In the meantime Esther Brunauer joined the State Department, rose to the rank of minister, and represented the United States in founding UNESCO. She was prominent among those attacked by Senator McCarthy in 1950 in his famous accusations against State Department employees. Although she was eventually exonerated, she was still considered a potential security risk because of her husband's prior involvement in communist front organizations, and her job was terminated. She died in 1959.

Brunauer resigned from the navy, found a job with the Portland Cement Association in Chicago, and became an international expert in his new field. Later, he joined Clarkson University and served as chairman of its Chemistry Department until his retirement. He married Dr. Dalma Hunyadi in 1961. His widow now lives in Hungary.

legendary, and the three jointly produced an expression, which became known as the BET equation.

The amount of adsorption depends on the surface area, which increases as the material is being subdivided. The wide applicability of the BET equation came from its utility in determining surface areas. Just to give an example, the total surface area in one handful of soil can be compared to the area of a football field.[40] To illustrate the importance of the problem, note that there is a lot of methane gas trapped in coal deposits that may be liberated and exposed as the fresh surface area increases enormously in the process of mining. The process often leads to an explosion and the death of miners. Many previous attempts had failed to determine the effective surface areas of porous materials and other finely divided powders.

The need for the technique could not be demonstrated better than by the fact that it was one of the top-cited papers during a fifty-year period, and it is still being widely used in its original formulation. For many physical chemists, Teller's fame is related to the BET equation more than anything else. There have been attempts to improve on the BET equation; some were advanced by Teller himself, and another set of papers in this direction was published by the future Nobel laureate Gerhard Ertl (Chemistry, 2007). But "for most situations, . . . the BET equation provides as good a model, both from the practical and the philosophical viewpoints, as any of the models advanced to date."[41] The paths of the three co-authors of the BET equation diverged after their joint work. The three scientists were awarded honorary doctorates in 1969 from Clarkson University, Brunauer's workplace at the time, and they met again for a quiet reunion in San Francisco in 1979.

An entirely different kind of discovery, considering its consequences, was the Jahn–Teller effect.[42] Not only has it been frequently cited, it has taken off and developed in new directions without losing the significance of its origins. This kind of discovery lives its own life. The original discoverers might not even have fully appreciated the impact of their discovery, but their fame is increasingly enriched by the new developments. The discovery of the effect has an interesting history that involves two additional persons. One was Rudolf Renner, Max Born's student in Göttingen, for whom Teller served as consultant. The other was Lev Landau. The phenomenon is rather involved, and it is not our purpose to go into its intricacies here. Suffice it to note that it was a difficult problem to handle and it was something over which Landau and Teller had had a friendly argument. Teller won, to his great satisfaction, because, as he observed, it was unusual for Landau to lose a scientific argument.

In studying the interaction of electronic structures and vibrations of molecules, the possibility arose that certain very symmetrical molecules with fewer than the possible number of electrons might not be stable. In such molecules there is a mismatch between the symmetry of the ensemble of the atomic nuclei and the symmetry of the electron cloud surrounding it due to the smaller number of electrons than what there would be room for in the molecule. This causes some of the nuclei to move from their original positions into ones that match the symmetry of the electron cloud, and thus the symmetry of the molecule decreases. This symmetry-lowering happens in many molecules, except some very simple ones, like carbon dioxide, $CO_2$, which Landau actually used as his example. The symmetry of the carbon dioxide molecule can be depicted as O=C=O: triatomic, linear, and symmetric. The structure of such molecules was the subject of Renner's thesis and subsequent article.[43] The symmetry-lowering in the carbon dioxide molecule would mean that the three atoms would no longer be aligned along a straight line. Renner showed that there are certain situations when even such a structure may undergo distortion. This was originally referred to as the Renner effect.

At some point, Gerhard Herzberg started using the term Renner–Teller effect instead of the Renner effect, perhaps because he was familiar with Teller's role in discovering it. Rudolf Renner was born in 1909, in Schweidnitz in Silezia, Germany (now Poland). He studied first in Breslau, then, from 1929, in Göttingen, where he was awarded his doctorate in 1934.[44] His former supervisor, Max Born, and his consultant, Teller, by then had left Germany. This might have been the reason Renner published his paper under his name alone, but it might also have been the case that Born and Teller would have urged him to publish alone.

Teller did not know anything about Renner for a long time, until 1980, when he received a letter from his former pupil. As the physics program in Göttingen disintegrated, Renner obtained a teaching certificate but could not find employment. He entered the weather service and worked as a meteorologist until the end of the war. His wife's family owned a pharmacy, and the need arose to run it, so Renner studied pharmacy, took over the business in 1950, and continued as a pharmacist for the next thirty years until his retirement. He wrote that he followed Teller's career with great interest, and Teller responded with a kind letter. The correspondence was in German.[45] Renner died in 1991, and he never spoke to his family about his "first career."[46]

Landau's choice of the linear symmetric molecule was not fortunate for discussing what would later become the Jahn–Teller effect, because it

## HERMANN ARTHUR JAHN (1907–1979)

In contrast to Edward Teller's fame, the other eponym of the Jahn–Teller effect is hardly known. Teller referred to him only as a refugee from Germany.

Hermann Jahn was of German descent. His father left Germany and settled in England in 1890. He was born in Colchester and earned a BSc degree in chemistry at University College London in 1928. He was drawn to quantum mechanics and went to Leipzig to continue his studies, which he completed with his doctorate in 1935 under the supervision of Werner Heisenberg and Bartel L. van der Waerden. Upon his return to London, he worked at the Royal Institution, then, during the war, at the Royal Aircraft Establishment, followed by a stint at Rudolf Peierls's department in Birmingham. From 1949, he spent the rest of his career at the Department of Mathematics of the University of Southampton, from which he retired in 1972.

His interaction with Teller occurred during the short period Teller spent in London at University College when Jahn was at the Royal Institution. Their paper about what soon became known as the Jahn–Teller effect appeared as the first of a two-paper miniseries; the second of which was authored by Jahn alone. Jahn's other publications carry the marks of a well-established, independent researcher who made serious contributions to the structure of molecules, especially regarding their vibrations. During the war, Jahn moved from molecular vibrations to studying the vibrations of larger structures. As a result, he probably learned more about the vibrations of airplanes than he had wanted, and "he vowed never fly in an aeroplane—and he never did," according to his obituary.[47]

In an ironic quirk of history, in 1947, Jahn was interviewed for a job by Klaus Fuchs, the German refugee turned Soviet spy, who was then the head of theoretical physics at the British nuclear project in Harwell. Jahn expressed his doubt as to whether he, having German ancestry, might be employed by an establishment involved in classified research. Fuchs reassured him, saying that he himself had German ancestry and had full access to classified materials.

In Jahn's obituary, his successor in the chair of their department made this observation: "It is surely true that personal modesty and aversion to self-advertisement can also lead to a certain lack of contemporary appreciation by the scientific community."

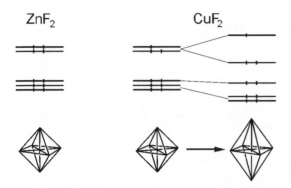

Illustration of the Jahn–Teller effect: The short vertical bars represent electrons populating energy levels represented by bold horizontal lines. For zinc difluoride, $ZnF_2$, there is no effect, due to the completely filled energy levels by electrons. The energy levels of copper difluoride, $CuF_2$, are incompletely filled, and the regular octahedral shape of the molecule elongates into a less symmetrical one—a manifestation of the Jahn–Teller effect.

did not apply to such systems. Nonetheless, Teller never missed an opportunity to give credit to Landau for having initiated their discussions. He even suggested calling the effect the Landau effect or the Landau–Jahn–Teller effect.[48]

The Jahn–Teller effect has become famous, whereas Jahn seems to have disappeared into oblivion. The paper was published at a time when Teller had produced numerous papers with a host of co-authors. One might assume that during the ensuing years Teller might have become curious as to what had happened with his co-author; apparently, he never did. On one occasion, he referred to Jahn as a fellow refugee,[49] which Jahn was not. Teller liked to talk about the Jahn–Teller effect, and on the occasion of one of his visits to Israel he chose the Jahn–Teller effect as the subject of his lecture. The choice surprised his hosts, but it might have been a preference for a noncontroversial topic, for a change.

The afterlife of the Jahn–Teller effect includes the emergence of some derivatives of the originally postulated effect, such as the pseudo Jahn–Teller effect. Monographs are published about the Jahn–Teller effect and international conferences are held regularly. A conspicuous application of the Jahn–Teller effect was its use as a starting point in the quest for high-temperature superconductors by J. Georg Bednorz and K. Alex Müller at an IBM research laboratory in Switzerland. They made their discovery in 1986 and were awarded the Nobel Prize in Physics in 1987; this

## LEV D. LANDAU (1908–1968)

Lev Landau was born into a Jewish–Russian family in Baku, where his oil-engineer father was sent for his work. His mother was a gynecologist. Landau went to Petrograd State University at the age of fourteen and published his first paper when he was eighteen. He started his research career in Abram Ioffe's Roentgen Institute in Leningrad.

Landau spent some time in Western Europe in the late 1920s—in Leipzig, Berlin, Cambridge, Zurich, and especially in Copenhagen—and worked with Werner Heisenberg, Edward Teller, George Gamow, Wolfgang Pauli, Ernest Rutherford, and especially with Niels Bohr, whom he considered his teacher.

In 1932, Landau moved to Kharkov, where he worked at the Ukrainian Physical Technical Institute (UFTI), which was fast becoming an internationally recognized center of experimental and theoretical physics. Sadly, by about 1938, UFTI was destroyed by the arrest of several of its leading scientists during the madness of Stalin's purges for alleged crimes of anti-state or anti-party activities. This was the fate of Landau as well, who by then had transferred to Peter Kapitza's Institute for Physical Problems of the Academy of Sciences of the USSR in Moscow.

Some of the Kharkov physicists—most notably the exceptionally gifted experimentalist Lev V. Shubnikov—were executed. Landau was incarcerated for one year and was freed because Kapitsa had pledged personal responsibility for Landau's behavior. It is noteworthy that Landau's accusations of anti-Soviet activities were not withdrawn upon his liberation from prison; they were finally dismissed long after Landau's and Kapitza's deaths and years after the dissolution of the Soviet Union.

Landau became indispensable to the Soviet nuclear program, but he always considered himself an educated slave who participated in the nuclear program under duress. After Stalin's death, he resigned from the nuclear program. One of his most significant contributions to world science was the textbook series he authored jointly with Evgenii M. Lifshits, titled *Theoretical Physics*.

Landau was awarded the Nobel Prize in Physics in 1962 "for his pioneering theories for condensed matter, especially liquid helium." Sadly he could not truly enjoy and appreciate this distinction. On January 7, 1962, nine months before the Nobel announcement, he was the victim of an automobile accident in which his skull and brain was damaged. After the accident he lived for six more years without fully recovering the legendary capacity of his mind. Boris Gorobets of Moscow recently published a trilogy about Landau's life and science.[50]

was one of the shortest time periods between a bestowal of an award and a discovery in the history of the Nobel Prizes. In their joint Nobel lecture the laureates paid ample tribute to the Jahn–Teller effect and illustrated it in two attractive figures.[51]

Teller ascribed great significance to the discovery of the first high-temperature superconductor (see above), and he could have not missed its relationship to the Jahn–Teller effect. He also stressed the importance of other pieces of his research for high-temperature superconductivity.[52] He had modestly noted before that: "[w]hat we failed to do turned out to be of much greater importance than what we actually did accomplish."[53]

Teller made a pointed reference to the strategic noteworthiness of the discovery of high-temperature superconductors in an article titled "Deterrence? Defense? Disarmament? The Many Roads toward Stability" in 1988.[54] In the section on the technological advantage of the United States vis-à-vis the Soviet Union in the 1990s, he singled out three areas, namely, electronics, space, and lasers. At the end of the section he stated that the technological picture may greatly change by the year 2000, and he ascribed the change to the potentials of high-temperature superconductors. In 1991, when mentioning an example to illustrate his statement that "[m]olecules are able to produce miracles, or at least unpredicted phenomena," he again referred to the new high-temperature superconductivity.[55] All these references could be justly considered as highlighting the importance of the Jahn–Teller effect.

Two Soviet scholars wrote a monograph about the Jahn–Teller effect in the late 1980s. However, the Soviet authorities forbade them to use the term "Jahn–Teller effect" in the title of the book. They did not want to see Teller's name on a book cover; he was considered by then to be strongly anti-Soviet. The authors had to resort to the generic expression.[56] That politics could never be too far away in Soviet science will be seen again in the concluding example below.

The covalent bond between two atoms is at the core of the foundation of chemistry. (A covalent bond is one or more pairs of electrons to which the two participating atoms each contributes equal numbers of electrons.) It was first suggested by Gilbert N. Lewis in 1916 and its quantum mechanical bases were worked out subsequently by two physicists, Walter Heitler and Fritz London. These covalent bonds may consist of two electrons being shared, which is called a single bond, designated by a line, and four electrons being shared, which is called a double bond, designated by a double line, for example.

In the 1930s, Linus Pauling wanted to explain why the structure of the

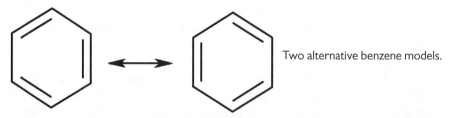

Two alternative benzene models.

benzene molecule could be equally well described by two equivalent structures in which single bonds and double bonds alternated. This pair of structures corresponded to the hypothesis of the German organic chemist Friedrich August Kekulé, who pioneered it for benzene back in the nineteenth century, well before chemists understood the nature of chemical bonding.

Pauling suggested that the two separate structures in the description of the benzene molecule might better be imagined as being in a continuous resonance of going back and forth between the two. Hence the name of the theory of resonance, for which benzene is only one of many examples. The critics of this theory argued that two different structures could not have coexisted and that the benzene molecule should be described by a single structure. Pauling took great pains to stress that his description was merely a model, a mathematical construction, but that did not pacify the opponents of the resonance theory.

Teller was intrigued by the problems of the description of the benzene structure and used his expertise to clear up the situation by providing spectroscopic evidence for Pauling's theory. Again, the details would stray far from the origins of the issue, but Teller and two co-workers published a paper in 1940 in which they showed that it was meaningful to speak about the benzene molecule in the way Pauling described it.[57] Incidentally, sixty-eight years later the Nobel laureate and physicist Philip Anderson also found modern means to give an interpretation to the resonance structures.[58]

According to Soviet ideologues, Pauling's description of the benzene structure was not only meaningless but it was contrary to dialectical materialism—the philosophical foundation of communism. A meeting was organized in Moscow in 1951, where a group of vocal though scientifically weak chemists rejected the ostensibly harmful activities of the proponents of the theory of resonance. Some of those condemned were internationally renowned scientists who were forced to engage in humiliating self-criticism during the discussion. A thick volume of the minutes of the meeting was soon published.[59] Some outstanding Soviet chemists lost their jobs as a consequence of this witch hunt. The chemists fared better than some of the biologists who lost their lives as a result of similar yet more fierce ideological battles. At one point, ideological attacks against the Soviet physicist community had also been considered but were decided against when in the postwar years the Soviet leaders understood that such actions might cost them nuclear weapons.

An ironic quirk in the Soviet controversy about the theory of resonance was that Pauling had leftist leanings and friendly sentiments toward the Soviet Union, and this was broadly known in the United States.[60] However, it took some time for the Soviet apparatchik to learn about it and to dampen at least the personal attacks against Pauling. There is parallelism between the witch hunting in the ideological struggles in the Soviet Union and McCarthyism in the United States. It is true though that the consequences of the former were much heavier than those of the latter, but the former occurred in a totalitarian state and the latter in a democracy.


## ENTER NUCLEAR PHYSICS

The Tellers enjoyed their life in Washington in the latter half of the 1930s, a period to which only his life in Germany in the late 1920s could be compared. Yet they were not without worries about the events in Europe. When they visited Hungary in 1936, they traveled through Germany under the protection of their Hungarian passports. When they considered a meeting with von Weizsäcker, he preferred to see them in Switzerland rather than in Germany. Von Weizsäcker reassured Teller that there would be no war.[61] All signs pointed to the reverse of such a prediction; one of them being the *Anschluss* in March 1938, which brought the Nazi empire to the borders of Hungary.

Teller and Mici had planned another visit to Budapest for the summer of 1938, but now they postponed it. They would not return to Hungary for more than half a century. Nonetheless, the European events seemed far away, and Teller kept both himself and Mici busy. He was much involved in organizing the annual Washington conferences on theoretical physics and gradually he became its chief organizer, since Gamow had lost interest in the day-to-day running of the meetings. Teller did not participate in the first conference in the spring of 1935, but the second, in 1936, carried his unmistakable mark. He selected the topics of chemical bonding, reaction kinetics, magnetism, van der Waals forces, molecular vibrations, and isotopes. About sixty people attended the lectures, among them nine Nobel laureates or future Nobel laureates. Most were Teller's friends, present or future co-authors, future friends, and future friends-turned-future-adversaries.

When there was no conference, the Tellers still continued their active social life, holding informal gatherings at their home for visiting scientists. Many of them came from faraway places, others from the vicinity of Washington. Hans Bethe was a most welcome visitor. Teller never forgot that Bethe, who was two years his senior (a fact that counted when they were students), showed up at his hospital bedside when he was recuperating from his injury in Munich. Teller and Mici happily played the role of matchmaker for Bethe and Rose Ewald in Washington; Bethe and Rose had known each other from Germany. The Tellers took them on a trip to the West Coast in 1937, and the two couples became the best of friends. Bethe and Rose married soon afterward.

The trip to the West Coast was memorable for meeting other physicists, too. The Tellers visited Berkeley, where Edward and J. Robert Oppenheimer met for the first time. Oppenheimer was an impeccable host, inviting the Tellers for dinner at a Mexican restaurant and also inviting Teller to address his group at the university. For Teller, though, their first meeting was not without its problems.[62] He found Oppenheimer's personality overpowering and the Mexican dishes so spicy that he lost his voice. His loss of voice might also have been attributed to nerves, as he had not yet become the experienced speaker he would be in time.

Teller's approach to research projects was often to use dialogues with other scientists to find out about unsolved problems. He thrived on these challenges. Either his partner in the dialogue or another scientist would then become his co-author of the emerging report. This approach explains why he worked with such a diversity of topics and co-authors, though usually only one or two for most papers. His projects were often improvements or perfections of previous discoveries by others. "Improvement" is something to be appreciated; even Alfred Nobel's one-page Testament, serving as the foundation of the institution of the Nobel Prize, contains the expression of improvement in addition to discovery and invention.

One of Teller's joint works with Gamow aimed at such improvement. The work was related to Enrico Fermi's discovery of beta decay and it was eventually shown to have more general applicability than the original discovery.[63] Fermi proposed his theory of beta decay in 1934. During beta decay the atomic nucleus emits an electron and a neutral particle while its positive charge increases by one. Without going into the details of the description of the process, Fermi established some selection rules, supposing no change in nuclear spin for beta decay. However, Gamow and Teller determined that there may be spin flips accompanying the phenomenon, which complemented Fermi's theory.[64]

Their findings pointed the way for progress in nuclear physics such as the discovery of parity violation, the theory of fundamental weak interactions, and the standard model of the origin of the universe. No one knows whether Teller would have arrived at any of these discoveries had he continued the work, but the fact is that he did not, and this was typical of him. One may also argue that his versatility would have suffered had he anchored his efforts to one or a few particular research projects. Half a century after the publication of the Gamow–Teller paper, the Livermore physicist Stewart D. Bloom wrote, "The fertility of the Gamow–Teller concept of spin flip in reaction processes was and still is a profoundly impressive example of the power of 'simple' ideas in our understanding of an enormous range of physical phenomena."[65]

Teller collaborated with Julian Schwinger (another future Nobel laureate), Eugene P. Wigner, John A. Wheeler, and others in describing various nuclear processes. When in 1937 Teller visited Columbia University in New York, he became interested in Schwinger's work related to the scattering of neutrons and suggested that he develop it further and write up the results. Two joint papers in *Physical Review* grew from their interactions.

Teller's nuclear physics soon brought him to selected problems of astrophysics, since the two disciplines share a natural interface. Hans Bethe developed his theories of how nuclei were synthesized. Together with one of Gamow's and Teller's students, Charles Critchfield, Bethe realized the basic importance of the proton–proton reaction in the sun, and this reaction corresponded to the Gamow–Teller description of nuclear reactions. Teller and Gamow determined the temperature-dependence of thermonuclear reactions. This was again a revision of an earlier theory about thermonuclear reactions, which are reactions involving nuclei at very high temperatures.[66] This was Teller's first involvement with the topic that would eventually play a central role in his projects for developing the most powerful weaponry.

Teller and Gamow continued their synergistic interactions and remained on the best of terms. Even the discovery that Gamow was strongly anti-Semitic did not disturb their relationship. Gamow's anti-Semitism came to the surface during a trip the two couples took to Florida, where Gamow was upset by the presence of so many Jews in Miami. The Tellers and the Gamows "solved" the problem by leaving Miami. Again, Teller's penchant for ignoring disturbing details about those he admired enabled him to move forward and meet new challenges. Clearly, he had hardened himself against displays of anti-Semitism.

Under Gamow's influence, Teller became increasingly interested in nuclear physics, the science about the tiny center of the atom where much of its mass is concentrated. This was the hot area in the 1930s in which new discoveries were emerging. He became involved in the field before the epoch-making discovery of nuclear fission at the end of 1938 and the beginning of 1939. If we look back, it seems as if he was getting ready for the developments to come in a carefully choreographed way. The topics of the Washington conferences included nuclear physics; molecular physics; elementary particles (twice); stellar energy; stellar evolution and cosmology; low-temperature physics; and the science of the interior of the earth.

Gamow's and Teller's interest in the nuclear processes in the stars gave them the idea of organizing the next conference in 1938, which would focus on the energy production in the stars. Bethe had attended their previous meetings, and three decades later declared that these meetings were the "most stimulating that I have ever attended."[67] He attended the 1938 meeting at Teller's explicit urging, and the topic then became one of his major interests and research areas. In 1967, Bethe was awarded the Nobel Prize in Physics primarily for his discoveries concerning the energy production in the stars. Gamow had done pioneering work in this area, too, and the Swedish presenter mentioned Gamow's contribution in his introduction to Bethe's work. It is not often that the Swedish presenters mention contributions by living non-Nobel laureates in their presentation speeches. In fact, the contribution was a joint paper by Gamow and Teller.

Gamow then co-authored a paper with Bethe in 1948 about the origin of the chemical elements in the universe. The paper has a bizarre history. Gamow and his doctoral student, Alpher, wrote a paper on the basis of Alpher's dissertation to which Gamow—ever the great joker—added Bethe's name for euphony. He claimed that it was "unfair to the Greek alphabet to have the article signed by Alpher and Gamow only."[68] Thus, the paper was published by Alpher, Bethe, and Gamow.[69] It did not help the paper to be taken seriously, especially since it was published on April 1, but that was not by design. The paper was, nevertheless, a step toward a model of the universe, and many people referred to it as the alpha–beta–gamma model.[70] Curiously, Bethe gave his consent to have his name appear on the paper without having contributed to it. When Gamow first suggested adding Bethe's name to their paper, Alpher was rather apprehensive about it, lest his name not be recognized in the shadow of the two well-known names.[71] Incidentally, in 1937, when Alpher had begun his studies at George Washington University, Teller was his freshman physics professor.[72]

The overriding principles of the micro-world of nuclear physics and the macro-world of cosmology are similar, and Gamow would become a major player in the latter as well, after the war. But already one of his joint papers with Teller, which they worked out in 1938, was concerned with a cosmological phenomenon, the origin of great nebulae.[73] They specified the conditions under which galaxies could be formed by condensation. Although some of their conclusions had to be modified in time, their main ideas withstood the test of rapid progress in cosmology.[74] Wigner noted that Teller's collaboration with Gamow working on the behavior of matter in the stars was an important ingredient in Teller's developing interest in the properties of substances under unusual conditions.[75]

When Teller left Germany in 1933, he left the German language behind just as he had left Hungarian behind when he moved from Budapest to Karlsruhe in 1926. His last two papers in Germany appeared in 1934 in the form of considerable reviews, one on molecular spectroscopy and the other on crystal spectroscopy. From 1935 on he published in English with the curious exception of a paper with Landau, mentioned above. A compilation of Teller's publications lists fifty articles by Teller and his associates in the years from 1935 to 1943.[76] The end of this period is already the completion of yet another transition toward defense-related and mostly classified work.

Writing fifty papers in this time period means that Teller published between five and six papers annually on average, but the distribution is uneven. First, only two or three papers were published; from 1937 to 1939, up to nine papers appeared, then, slowly, the production decreased. Only on two of the fifty is he listed as sole author. He published five papers jointly with Gamow. Among his co-authors are five Nobel laureates or future Nobel laureates, and his other co-authors are also well-respected scientists. If looking at the otherwise often fuzzy division between what might be labeled as molecular physics and nuclear physics, the number of papers of the latter grows steadily during this period at the expense of the former. However, Teller never entirely abandoned his interest in the science of molecules.

If we look at Teller's life in its totality, the first half decade of his post-German life is hardly distinguishable from the preceding years in terms of his demeanor and his scientific production. This could have been a very difficult transition period for him, but it was not. This is remarkable and tells a lot about his resilience and dedication but also reveals a great deal about the countries, institutions, and people with whom he and his wife came into contact during these years. At the time, Teller was characterized in the most positive terms, which must have bolstered his spirits. His *Memoirs* quotes from a letter in response to a query from the University of

Chicago, which was looking for a physics professor to hire. Unfortunately, the letter could never be found, although Teller himself would have liked to find it; apparently, he only saw the letter but never had possession of it. In an interview in 1984, he quoted from it:

> What he [Tuve Merle] wrote about me was, "Now, if you want to get a genius, don't get Teller, get Gamow. But geniuses are a dime a dozen. Teller is much better than a genius. He is a man who gets along with everybody, who helps everybody. He has talked with every physicist in this area, and helped with their work, and never got into disagreement with a single person." I was recommended as the paragon of the uncontroversial figure.[77]

Even though the letter has never been found, and Teller quoted from it differently at different times, its gist sounds realistic. The comment in the last sentence is an example of his self-deprecating humor. This was in 1984 and he knew how far his image had departed from the one described by Tuve.

The question arises, when did this period end? There were disturbing signs from Europe that the world was nearing a tremendous calamity. This concerned the Tellers most directly because their families lived there. Still, they canceled their planned visit in 1938 because of the frightening conditions there. At that time, America tried to follow a path of isolationism, as if it could have protected itself from any potential war on distant continents. The immigrant scientists had the double burden of foresight of a bleak future. Not only did they understand the nature of rogue ideologies and forces that were taking over Europe, they were also keenly sensitive to the scientific discoveries that might be harnessed for destruction. The emergence of nuclear science with the enormous energies it might unleash carried the potential for unimaginable danger.

# Chapter 4

# ATOMIC BOMB QUEST

The first atomic bomb destroyed more than the city of Hiroshima. It also exploded our inherited, outdated political ideas.
—Albert Einstein[1]

There is no evil in the atom; only in men's souls.
—Adlai E. Stevenson[2]

*Teller became entangled in the story of the atomic bomb right from the beginning, partly by accidental circumstances—being one of Leo Szilard's close friends—and partly due to his increasing involvement in nuclear physics. It was no accident that Teller participated in the Manhattan Project, but it was in Los Alamos that, for the first time, Teller's negative traits dampened his friendly relationships with some of his peers. His stance against communism was taking shape in this period. He did not oppose the deployment of the atomic bombs against Japan, even though he gave a different interpretation of his stand later in his life.*

World War II broke out on September 1, 1939, when Germany attacked Poland, and soon Great Britain and France declared war on Germany. For quite a while nothing happened on the western front, hence the label of the "phony war." By the time the war started, the scientists involved with nuclear physics had been in turmoil for the better part of a year. In December 1938 in an experiment by Otto Hahn and Fritz Strassmann in Berlin, uranium was seen to produce two known elements

123

of about half its weight. This was interpreted by two refugee scientists, Lise Meitner and Otto Frisch, to be nuclear fission.[3]

When Szilard learned the news, he realized that uranium was the element he had been looking for in his thought experiments on a nuclear chain reaction. The discovery opened the possibility for liberating enormous amounts of energy from the atomic nucleus, something that the great Ernest Rutherford declared impossible only a few years before. Given the combination of the possibility of an explosion of heretofore unprecedented force and the Nazi menace, the danger for democracies appeared huge and imminent. The Roosevelt administration was very cautious in war preparations lest it provoke the isolationists who did not wish to get involved in yet another "foreign conflict." Though initially the events followed at a snail's pace, Teller was transformed from a ringside viewer into an active participant.

## IDYLL ENDING

At the beginning of 1939, Teller was busy organizing for January 19–20, the next Washington conference whose topic was low-temperature physics. Two especially prestigious attendees were expected, Niels Bohr from Denmark and Enrico Fermi from Italy. Teller had had friendly encounters with each and had enjoyed their hospitality. Fermi arrived in the United States directly from Stockholm, where, on December 10, 1938, the king of Sweden had handed him the Nobel Prize in Physics. He took up the job of professor of physics at Columbia University, which had been arranged for him. Bohr, who came to Princeton for lecturing, was accompanied by one of his sons and by his co-worker, Leon Rosenfeld. Bohr always needed someone around him to play the role of sounding board.

Before his departure for America, Bohr learned from Otto Frisch, who was a visiting member of his Copenhagen institute, about the nuclear fission discovery. This news preoccupied Bohr's mind during the trip, but he said nothing about it to Fermi, who met him at the pier in Manhattan. Bohr respected Meitner's and Frisch's priorities and knew that they were still in the process of working out the theory of nuclear fission. Bohr, however, failed to warn Rosenfeld, and while Bohr stayed in New York for a brief visit, Rosenfeld told John A. Wheeler about nuclear fission and gave a seminar on this topic at Princeton. Gamow learned about it from Bohr when the Danish scientist arrived in Washington on the eve of the opening day of the conference. By then Bohr could speak freely, as the Princeton seminar had broken the silence. Gamow called Teller, who saw at once the far-reaching implications of the discovery. It did not escape his attention either that the discovery was made in Nazi Germany, which threatened the world with war.

Although the news did not spread as quickly as it would today, word caught on fast. Wigner was not present at the Princeton seminar; he was hospitalized with jaundice, but as soon as he learned of the news, he informed Szilard. Gamow and Teller had rearranged the program of the first day of their conference to give Bohr the opportunity to report on the new development. Fermi had a meeting with navy representatives in Washington, trying—unsuccessfully, as it happened—to spark their interest in nuclear research. Szilard also came to Washington, and Teller, Fermi, and Szilard got together to discuss secrecy. Ironically, it was Szilard, the most free-thinking member of the group, who insisted on the need for secrecy, because he had the most acute foresight about the possible consequences of the fission discovery. Teller joined Szilard in urging for secrecy, but Fermi did not yet see its necessity.[4]

Szilard had contacted Lewis Strauss, who would become a crucial friend and ally to both Szilard and Teller. He was a financier, a wealthy self-made man, who was interested in science and defense matters. Szilard had informed him about the discovery of nuclear fission and that it "might lead to a large-scale production of energy and radioactive elements, unfortunately also perhaps to atomic bombs."[5] Szilard then founded the Association for Scientific Collaboration (ASC), which—so far—consisted of Szilard, Wigner, and Teller. Their intention was to provide a framework for collecting funds and carrying out research in nuclear physics to develop atomic weaponry.[6] However, funding was not forthcoming, which tormented the Hungarian trio, since they were looking for the best way to counter any actions Germany might be taking. Their naïveté and dedication was staggering. Teller's role in the association would also be to keep in contact with the Roosevelt administration. This was a rather interesting role, and one to which some of Teller's later actions might be traced.

At this time it was already known that of the two uranium isotopes, U-238 and U-235, in naturally occurring uranium, the rare lighter one—U-235—could participate in neutron production and thus maintain a nuclear chain reaction. The notation of U-235 (or, uranium-235) refers to the uranium isotope with mass number 235. In a more complete way, it is $^{92}$U-235, where 92 is the atomic number of uranium, meaning 92 protons in its nucleus. The atomic number 92 belongs to uranium and vice versa; any uranium isotope has the same atomic number, 92. On the other hand,

## ENRICO FERMI (1901–1954)

Enrico Fermi was born in Rome into the family of a clerk of the Italian State Railway system. He went to school in Rome, attended college at the famous Scuola Normale in Pisa, and received his doctorate at the age of twenty. Then he spent some postdoctoral stints in Germany and Holland as well as in Italy, and was appointed full professor of physics at the University of Rome at the age of twenty-five. Fermi was one of the last of the great physicists who made pivotal discoveries in both experimental and theoretical physics. In Rome, he built up a strong research group of young talented physicists and chemists, and his gift for creating schools manifested itself during his American career as well.

The discovery of slow neutrons had multiple consequences in science and beyond. Fermi and his associates investigated the phenomenon of artificial radioactivity: they bombarded various targets systematically with fast neutrons. When Fermi placed a piece of paraffin as a filter in between the neutron source and the target uranium, a much larger amount of radioactivity was detected than without the filter. Fermi understood that the hydrocarbon paraffin slowed down the fast neutrons, thereby making them spend more time in the target and enhance the efficiency of their interactions.

Fermi's wife was Jewish, and the Fermis felt increasingly threatened as Italy was rapidly adopting anti-Jewish legislation at the end of the 1930s. In 1938, Fermi was awarded the Nobel Prize in Physics, and the Fermi family did not return to Italy from the Nobel Prize ceremonies in Stockholm but sailed to New York, where Fermi had accepted a professorship at Columbia University. When the Metallurgical Laboratory was established at the University of Chicago with the purpose of developing the atomic bomb, Fermi moved there and played a pivotal role in creating the world's first nuclear reactor. He was instrumental in the assembly stage of the atomic bombs at the Los Alamos Laboratory in the framework of the Manhattan Project.

After the war, he returned to Chicago to what later became the Enrico Fermi Institute. Fermi abhorred the prospect of developing the hydrogen bomb, yet when President Truman decided on the program, he returned for extended periods to Los Alamos to assist in its successful completion. Fermi was one of the few of Teller's many friends who remained his friend to the end.

the mass number is not a unique characterization of an element. It is the sum of protons and neutrons in the atomic nucleus. Both U-235 and U-238 have 92 protons in their nuclei, but there are 235 − 92 = 143 neutrons in the nuclei of U-235 and 238 − 92 = 146 neutrons in the nuclei of U-238. The separation of the two uranium isotopes is a formidable task because there is only one U-235 atom for every one hundred and forty U-238 atoms; besides, their masses are very close to each other, and this makes their separation even more difficult.

Niels Bohr said that the whole country would have to be turned into a laboratory to produce a sufficient amount of U-235 for a bomb.[7] Of course, there was also a question of scale. By the time World War II was ending, a workforce—comparable to that of a small country—was working on projects related to the atomic bombs in the United States, with greater expenditures than the budget of a small country. Bohr and Wheeler worked out the details of nuclear fission; they understood why U-235 was easily fissionable and could predict other nuclei to be similarly fissionable. Plutonium, Pu-239, would be a candidate, but it did not yet exist.

Merle Tuve at the Carnegie Institution had also taken interest in the fission experiment. During the Washington meeting, he rushed back to his laboratory as soon as he heard Bohr's announcement. He quickly set up a demonstration of the fission experiment and invited the conference attendees to view it. On February 17, Teller informed Szilard in a letter about neutron-producing fission experiments at Carnegie. He mentioned the "chain-reaction mood in Washington. [He] only had to say 'Uranium,' and then could listen for two hours to their thoughts."[8]

Szilard, Wigner, and Teller kept each other informed about every step of progress, or lack thereof, in nuclear matters. Szilard liked to delegate manual work, but he made exceptions when he found it absolutely necessary. He was allowed to operate at the Physics Department of Columbia University and set up an experiment to confirm recent French findings about the extra neutrons being produced during the fission of uranium. When he completed his measurements, he told Teller on the phone, "I have found the neutrons."[9]

Secrecy continued very much to be on Szilard's and Teller's minds. At one point even Bohr agreed to secrecy if Fermi would too. And when Fermi let himself be persuaded, it was yet another small victory for them. In the meantime, Fermi and Szilard, each with a small group of associates, confirmed experimentally the possible reality of an atomic bomb. In this period, Szilard and Fermi were the major players; their names will be linked forever in nuclear physics. They co-designed the world's first nuclear reactor (called "pile" at the time) and had a joint US patent issued publicly in 1955, by which time Fermi had died.

Fermi's and Szilard's personalities contrasted sharply: For Fermi, science was everything; for Szilard, it was only a means to save humankind.[10] If we consider their differences, it is remarkable that when it was critically important, they could rise above them and join forces to accomplish historic deeds. Due to their very different personalities, their working relationship at Columbia University was an uneasy one, and they both agreed that they needed help. This is why Teller had been invited to Columbia to play the role of a friendly catalyst. Teller and Mici counted on staying in Washington full time and to set up only a temporary home in Manhattan. They rented an apartment within walking distance of the Physics Department of Columbia University. Teller's acceptance as mediator by Fermi and Szilard was another example of how friendly and trusted he was during this period.

Szilard had the broadest vision of the happenings in the world and the ability to analyze events from the point of view of how they might affect the approaching war. The Czech uranium ores came under the rule of Nazi Germany in March 1939, and the Germans immediately banned the export of uranium. Szilard anticipated that the Germans would be attacking Belgium at any time, a prediction that exhibited greater foresight than the Belgian government showed at the time.[11] The Hungarian physicists were afraid that with the fall of Belgium, the Belgian Congo with its vast uranium ore resources would also fall into German hands. This only added to their concerns that Germany might be headed toward possessing atomic weapons. In a parallel development, it was also realized in the Soviet Union that uranium ore supplies might be the bottleneck in their quest for atomic weapons. Hence, expeditions were organized by their Uranium Committee at the initiative of the famous geologist Vladimir I. Vernadskii.

Szilard and his colleagues felt that they should warn the Belgians of the strategic importance of uranium. The discussions of these plans revolved about whom to address, how to transfer the information, what it should be, whether they should contact first the American authorities, and so on. Gradually, the direction of their concern changed, and they realized that it would be even more crucial to alert the American administration to the possibility of the atomic bomb and the danger represented by Germany possessing it. Albert Einstein's involvement had already come as they were scheming about informing the Belgians because of his connection with the Belgian royal family. There were, however, reservations about contacting the Belgians because foreign relations are a prerogative of the State Department (the Logan Act of 1799 prohibited private diplomacy), and Szilard, Wigner, and Teller did not want to break any laws.

In the descriptions of how Einstein was contacted there remains some benevolent discrepancy between Teller's *Memoirs* and elsewhere in the literature. The documents reproduced and commented upon in the volume *Leo Szilard: His Version of the Facts*[12] are a good guide through this maze. Here is what happened. On July 12, 1939, Szilard and Wigner drove to Peconic, Long Island, where Einstein was vacationing at the time. To be exact, Wigner drove his Dodge; Szilard never learned to drive. They knew only the general direction and headed for Patchogue rather than Cutchogue.[13] At some point, they changed direction, and arrived at Peconic, but they did not have Einstein's exact address. Finally, a child gave them proper directions because he recognized the man they were looking for based on their description of his peculiar appearance.[14] The story had become a legend, and Teller described the same adventure as if it had happened to him.[15] When Szilard had to return to Einstein on August 2, Wigner was not available, as he had left for California. Szilard asked Teller to drive him to Peconic and on this second trip Szilard knew where they were going.

On the occasion of the July 12 visit, Szilard and Wigner first had to explain to Einstein how nuclear fission could lead to energy production and to an atomic bomb. Einstein had known of Szilard's acumen for practical things, although many others considered him more of a dreamer than an engineer. However, Szilard had had some technical training at the Budapest Technical University. He and Einstein had filed a series of patents for refrigerators of a new kind in Berlin in the 1920s. As for the fundamental physics of making an atomic bomb, it was all based on the famous relationship between energy ($E$) and mass ($m$), which Einstein had discovered: $E = mc^2$, where c is a constant, the speed of light. Yet until Szilard told him about it, Einstein had not believed that nuclear energy would be utilized in his lifetime. Now, he merely said, "I haven't thought of that at all."[16] Although Einstein understood the concept quickly, he predicted that "[i]t will be a hard thing to put this across to the military mind."[17]

There are variations in the descriptions about the extent to which Einstein personally participated in writing the letter that would go with his signature to President Roosevelt. In Szilard's narrative, on the first visit there was no immediate plan at the start of their discussion with Einstein; rather, it was being formulated as they spoke. When they left, the Belgium version was still on the table. Einstein had dictated a draft in German, addressed to the ambassador of Belgium; Wigner took it down; and when they returned to Princeton, Szilard translated it from German into English. They thought of a second letter as well, which would be sent to the US secretary of state, in which Einstein would be asking for his thoughts about a letter to be sent to the Belgians. Eventually, their plans changed and these letters never materialized. Soon after the first visit to Einstein, Szilard mailed a letter to him, on July 19, 1939, in which he unfolded a different plan of action, according to which they should inform the American president rather than the Belgians.

August 2nd, 1939

F. D. Roosevelt
President of the United States
White House
Washington, DC

Sir:

Some recent work by E. Fermi and L. Szilard, which has been communicated to me in manuscript, leads me to expect that the element uranium may be turned into a new and important source of energy in the immediate future. Certain aspects of the situation which has arisen seem to call for watchfulness and, if necessary, quick action on the part of the Administration. I believe therefore that it is my duty to bring to your attention the following facts and recommendations:

In the course of the last four months it has been made probable—through the work of Joliot in France as well as Fermi and Szilard in America—that it may become possible to set up a nuclear chain reaction in a large mass of uranium by which vast amounts of power and large quantities of new radium-like elements would be generated. Now it appears almost certain that this could be achieved in the immediate future.

This new phenomenon would also lead to the construction of bombs, and it is conceivable—though much less certain—that extremely powerful bombs of a new type may thus be constructed. A single bomb of this type, carried by boat and exploded in a port, might very well destroy the whole port together with some of the surrounding territory. However, such bombs might very well prove to be too heavy for transportation by air.

Szilard reported to Einstein that he had discussed the new plan with Teller, who "has been in the conspiracy with us from the beginning." Teller agreed with the new plan of action. Then Szilard added that he would like to take Teller with him for the next visit "not only because I believe his advice is valuable but also because I think you might enjoy getting to know him. He is particularly nice."[18] He did not mention that he also needed Teller because of his 1935 Plymouth and because Teller could drive. The letter, addressed to President Roosevelt, was finalized on their visit and carried the date of the visit, August 2, 1939. Teller liked to joke about his

The United States has only very poor ores of uranium in moderate quantities. There is some good ore in Canada and the former Czechoslovakia, while the most important source of uranium is the Belgian Congo.

In view of this situation you may think it desirable to have some permanent contact maintained between the Administration and the group of physicists working on chain reactions in America. One possible way of achieving this might be for you to entrust with this task a person who has your confidence and who could perhaps serve in an inofficial [sic] capacity. His task might comprise the following:

a) to approach Government Departments, keep them informed of the further development, and put forward recommendations for Government action, giving particular attention to the problem of securing a supply of uranium ore for the United States,

b) to speed up the experimental work, which is at present being carried on within the limits of the budgets of University laboratories, by providing funds, if such funds be required, through his contacts with private persons who are willing to make contributions for this cause, and perhaps also by obtaining the co-operation of industrial laboratories which have the necessary equipment.

I understand that Germany has actually stopped the sale of uranium from the Czechoslovakian mines which she has taken over. That she should have taken such early action might perhaps be understood on the ground that the son of the German Under-Secretary of State, von Weizsäcker, is attached to the Kaiser-Wilhelm-Institut in Berlin where some of the American work on uranium is now being repeated.

Yours very truly,
A. Einstein

role, saying that he was merely the chauffeur and that Einstein was cordial because he invited the chauffeur into the house. Einstein's letter has become a famous document of history.[19]

The letter explicitly mentions the possibility of a bomb. Another interesting point in the letter is the suggestion to entrust a person to serve as a link between the scientists and the administration in an unofficial capacity. This shows that Einstein and the instigators of his letter did not fully appreciate the enormity of the task they were pursuing. In Szilard's words, they "were all green."[20] They did not know their way around in America; how to do business; or how to embark on a project they had in mind. They were conditioned by their middle-European upbringing and had too much faith in government bureaucracy, which still intimidated them. In the United States, the less-hampered private foundations might have provided a better starting point for their project. At the same time, they correctly sensed that a proper channel for interaction between the government and the scientific community was lacking. Future developments in presidential advising would show that the suggestions by Szilard and his friends were most forward-looking in this respect.

A technical memorandum by Szilard was added to Einstein's letter about the peaceful and military applications of nuclear power. On August 15, the two documents were handed over to Alexander Sachs, a Russian-born economist and a friend of the president, who had an interest in scientific matters. A frustrating period followed because they heard nothing from Sachs for quite some time. When the war broke out in Europe on September 1, the United States remained neutral and continued to be so for two years. However, the US Congress repealed the heretofore existing arms embargo and made it possible for belligerent countries to purchase arms from America.

Sachs managed to deliver the Einstein documents personally to Roosevelt only on October 11. It was not an easy task to capture the president's attention, not only because he had no knowledge of nuclear physics but also because he was preoccupied with matters of national and international importance. Besides, the proposal in the letter might have seemed too esoteric to be taken seriously. On the other hand, Einstein's name carried weight, and the issue sounded too grave to be ignored. It was clear that the letter raised the possibility that the Germans might blow up the democracies, and that the president's action was required.

An immediate result of the letter was the establishment of an Advisory Committee on Uranium, called also the Uranium Committee, or the Briggs Committee because it was headed by Lyman J. Briggs, the director of the

National Bureau of Standards (NBS). He was a soil physicist and had worked in the Department of Agriculture prior to his appointment to the NBS. Commander Gilbert C. Hoover of the Navy Bureau of Ordnance and Colonel Keith F. Adamson of Army Ordnance were made members. The first meeting took place on October 21, 1939; Sachs, Szilard, Wigner, and Teller were present, in addition to others. Szilard's account[21] and the official history of the Atomic Energy Commission[22] are fully consistent in describing the meeting, which had two highlights.

One highlight was the discussion of funds, although the three Hungarian physicists had decided not to ask for money. They only wanted to use the meeting to get the "blessing of the government" and then go to private foundations for support. The meeting, however, took a different direction. It fell to Teller to report what Merle Tuve, who was invited but could not come, had told him to convey. According to Tuve, who had more experience than Teller and the other scientists in such matters, it would be impossible to spend more money than a total of $15,000 on all the research involved in the project. After some discussion, the military representatives promised $6,000, which took months to arrive. Teller also spoke for Fermi, who could not come either and had asked Teller to represent him. Teller liked to say that by this time, he had been promoted from chauffeur to messenger boy.

The other highlight of the meeting was a most remarkable exchange between Colonel Adamson and Eugene Wigner. The colonel was very skeptical about the feasibility of the project. Even if it succeeded, he said, "It usually took two wars to develop a new weapon, and it was morale, not new arms, that brought victory."[23] At this point Wigner suggested cutting the army's budget by 30 percent if indeed weapons were secondary in winning wars. Teller cites further colorful details of the meeting. Colonel Adamson told them about a goat tethered to a post at the Aberdeen Weapons Proving Grounds, and how the army offered a $10,000 award to anyone who could produce a lethal ray to kill the goat. The animal was still there, unharmed.[24] This story was the colonel's way of telling them that he did not take them seriously, and he wrote himself into the history of the atomic bomb with this performance. The Briggs Committee initiated various studies; one of them was aimed at investigating the fission by fast neutrons. When Teller saw the report, he estimated that a uranium sphere of thirty tons would be necessary for an explosion, which would make that goal unrealistic.[25]

No work was going on in connection with the uranium project in the United States between June 1939 and the spring of 1940. Szilard again mobilized Einstein, and again, through Sachs, they contacted the president of the United States. Roosevelt then suggested another meeting, which Briggs called, but he was reluctant to include "foreign" scientists. Fermi, Szilard, and Teller were not yet US citizens (Wigner was at that time). Briggs said, "Well, you know, these matters are secret and we did not think that they should be included."[26] There are, of course, multiple ironies in this. First, the secret was what Fermi and Szilard had invented. Second, the "foreign" scientists had pushed for secrecy to prevent the Germans from learning about their results. Third, they, and Szilard in particular, had pushed the American authorities to do something about their inventions. Eventually though, the atomic bomb project became operational, and it maintained the necessary secrecy without the exclusion of Szilard, Fermi, Teller, and the other "foreigners."

In the spring of 1940 it also happened that Louis A. Turner suggested the use of neutrons to convert uranium-238 into the highly fissionable plutonium-239.[27] This element was discovered in the work of two California scientists, Glenn T. Seaborg and Edwin M. McMillan, and their associates. What they accomplished was what Fermi had thought happened in the 1934 experiments at the University of Rome, when the production of two trans-uranium elements was announced. First, McMillan and Philip Abelson made uranium capture a neutron and, following beta-emission (the ejection of an electron from the nucleus), the element of atomic number 93 was formed. They called it neptunium, Np, after the planet Neptune orbiting next, outward, after Uranus. Following McMillan's departure for other defense-related research, Seaborg and his colleagues took over the project. They detected the next trans-uranium element, formed by another beta-emission; it had atomic number 94. It was given the name plutonium, Pu, after Pluto, orbiting next outside Neptune. The nuclear reactions are depicted here in shorthand notation. The n represents a neutron and the $\beta$ represents an electron emitted by the atomic nucleus.

$$^{92}U\text{-}238 + n \rightarrow \,^{92}U\text{-}239$$
$$^{92}U\text{-}239 \rightarrow \,^{93}Np\text{-}239 + \beta$$
$$^{93}Np\text{-}239 \rightarrow \,^{94}Pu\text{-}239 + \beta$$

Szilard's next new idea built onto Turner's suggestion. It was for a new type of reactor, eventually called a breeder, in which fast neutrons from a radioactive core would bombard a surrounding uranium blanket, turning uranium-238 into plutonium-239.[28] Plutonium was the second fissionable material used for atomic bombs. Turner submitted his manuscript to *Physical Review* and asked if he should have it withheld for secrecy. His finding had tremendous military value, but there was still no mechanism in place to decide about withholding manuscripts from publication except by voluntary action. Turner agreed to withhold publication but was unhappy to see that important papers might be handled on a "catch-as-catch-can" basis.[29] Briggs set up a new committee, the Advisory Committee on Nuclear Physics, with Szilard, Wigner, Teller, and Gregory Breit, the editor of *Physical Review*, among its members. Its task was to decide questions of withholding papers from publication.

About one year after the Hungarian physicists had first worried about the fate of Belgium, the Germans finally attacked and quickly conquered Belgium and the other Low Countries; the inaction on the western front came to an end. Teller felt increasingly that he was coming to a crossroads and would have to choose between academia and preparing for war. He writes of this dilemma: "I was uncertain whether I wanted to remain a bystander or become a participant."[30] One may tend to doubt Teller's words knowing the history of the rest of his life, yet Laura Fermi corroborates Teller's words: "Despite his discussions with Fermi, Edward Teller had not joined the uranium project by the spring of 1940. He could not make up his mind. Was it right or wrong for science to serve war?"[31]

It was at that point that Teller, along with many other scientists, received an invitation to a meeting at which President Roosevelt was scheduled to address them. It was the Eighth Pan-American Scientific Conference, and the president gave a speech for the hundreds of scientists from the Americas. He spoke about human rights and democracy, and addressed more immediate issues of the war. The United States was still neutral in the conflict, but barring armed participation, it was helping those resisting the Nazi conquerors. The president mentioned that in the modern world, distances had shrunk and the United States could no longer consider itself to be out of reach for any foreign power just due to its location. He also mentioned the responsibility of scientists. He called upon them "to use every knowledge, every science we possess, to apply common sense and above all to act with unanimity and singleness of purpose." Roosevelt, once again, declared his pacifism, but it was obviously not his take-home message. The crescendo of the speech was when he said, "I believe that . . . you and I, if in the long run it be necessary, will act together to protect and defend by every means at our command, our science, our culture, our American freedom and our civilization."[32]

Teller supposed that he may have been the only scientist present who was aware of the atomic bomb project and the unprecedented power of nuclear weapons. He felt as if the president was speaking directly to him and personally charted his future activities.

## WAR-FOOTING

The Roosevelt administration was slow in organizing the heretofore unknown nuclear science to provide the necessary support for defense. However, research and development for radar had already been going on with considerable force. As noted, the president's speech had a profound impact on Teller, and, moreover, the events in the European war strengthened his resolve. After the Nazi invasion of France had ended with the defeat of the French in June 1940, Teller and Bethe got together and decided that they should do something for the war effort. They turned to Theodore von Kármán for advice, and the famous aerodynamicist suggested that they investigate the properties of shock waves. Both Teller and Bethe found use of their prior experience and expertise, and their joint efforts resulted in a report in 1941. The study never reached the usual scientific literature, but it was widely circulated until finally it was printed in a compilation of Bethe's selected works in 1997.[33]

The atomic bomb project was a latecomer among defense-related efforts, but it did not lack scientists: it benefited from the wave of European refugee scientists. The initially slow beginning of the nuclear program was followed by rapid actions of increasing efficiency from the summer of 1940. The National Defense Research Committee (NDRC) was set up in June 1940 with the purpose of linking government and science. This was essentially what the Einstein letter of August 2, 1939, had suggested doing, except Einstein—that is, Szilard—had thought that this could be done by a single individual in an unofficial capacity. The NDRC was headed by Vannevar Bush. He did research for the navy during World War I; taught at MIT; was president of the Carnegie Institution in Washington, DC, before World War II; and acted in various capacities as scientific adviser to the American government. The chemistry professor and president of Harvard University, James B. Conant, was the vice head of NDRC. When Bush created and became head of the Office of Scientific Research and Development (OSRD) to oversee the development of all new weapons, Conant was named head of NDRC. As head of the OSRD, Bush reported to the president and Conant reported to Bush.

The Uranium Committee was reorganized and became part of NDRC. The "foreign" scientists became consultants to the committee. The NDRC held a meeting on December 6, 1941, on the eve of the Japanese attack on Pearl Harbor (December 7), and decided to accelerate work on the nuclear chain reaction. On the day after Pearl Harbor, the United States officially became an active force in World War II. Arthur H. Compton, the Nobel

# HANS A. BETHE (1906–2005)

Hans Bethe was born in Strasbourg, Germany (now France). His father was Protestant and his mother Jewish, and both were from professorial families. Bethe had a sheltered childhood; he studied physics, chemistry, and mathematics at the University of Frankfurt and did his doctoral work under Arnold Sommerfeld in Munich in from 1926 to 1928. His project was the translation of Paul Ewald's theory of X-ray scattering into the theory of electron scattering. He served as assistant to Ewald in Stuttgart for a short while, then to Sommerfeld, again, in Munich. Then he spent a year each in Cambridge and in Rome.

He was already at Tübingen University when the racial laws of Nazi Germany forced his university to dismiss him. He held a temporary position in Manchester, England, and then moved to Cornell University, where he started as assistant professor in 1934, and where he remained for the rest of his life. He left for varying periods of defense-related work during and after World War II. First he participated in developing microwave radar, and then he was in charge of the theory group at Los Alamos in the Manhattan Project. In 1947, Bethe was suggested for the post of his former professor, Sommerfeld, at Munich. It was a rare occasion when a former refugee from Germany was asked to return to a position that was worthy of the scientist. In an eloquent letter to his former mentor, Bethe explained why he found it impossible to accept the offer.

Bethe worked on the hydrogen bomb at Los Alamos in the 1950s, although he had fiercely opposed its development. Once the solution for the thermonuclear weapon became available, he helped to make it happen, despite his earlier stand on the issue.

He wrote comprehensive reviews of several theoretical fields in the 1930s that served the physics community for a long time, and his fundamental overviews were referred to as "Bethe's Bible." His main strength was problem solving. One of his former students at Cornell University, Freeman Dyson, wrote in his obituary: "He was not a deep thinker like Heisenberg and Dirac, who laid the foundation of modern physics in the 1920s. But he took their theories and made them into practical tools for understanding the behavior of atoms, stars and everything in between."[34] Bethe was awarded the Nobel Prize in Physics for 1967 for the discoveries related to the energy production in stars.

laureate physics professor, was named head of the atomic bomb project, and the Metallurgical Laboratory was established in Chicago.[35] One of Compton's first measures was the appointment of J. Robert Oppenheimer of the University of California at Berkeley to head a portion of the physics work.

As a final touch of the initial phase of the project, General Leslie R. Groves was put in charge of the whole atomic bomb program in September 1942. It became known as the Manhattan Project; its full code name was Manhattan Engineer District. Groves headed the project almost from its inception through 1946, when the new Atomic Energy Commission took over from January 1, 1947, as legislated by the US Congress. Initially Groves was made responsible for the construction works of the Manhattan Project—he had had vast experience in such work, which included the construction of the Pentagon building complex—but eventually he was charged with the whole project. Years after the war, the leader of the British team of physicists in the Manhattan Project, Nobel laureate James Chadwick, told Teller, "Without Groves, the scientists could have never built the bomb."[36]

One of the factors that may have stimulated the work on the bomb was the obvious difference in opinion and attitude between Groves and the scientists. As Groves expressed it, the scientists "don't know how to take orders and give orders."[37] Both sides brought their own experiences to the project and both believed that the other hindered efficiency of performance. Groves had industrial experience where people concentrated on their immediate tasks. The scientists had learned to benefit from the broadest possible interactions and from viewing the picture as a whole rather than in its details only.[38] According to Teller, "between 1943 and 1945 General Groves could have won almost any unpopularity contest in which the scientific community at Los Alamos voted."[39]

The Manhattan Project was under the jurisdiction of NDRC and it eventually had four principal sites for its subprograms:

1. The earliest was the Metallurgical Laboratory (Met Lab) at the University of Chicago, whose task was to build the first uranium-graphite pile and to show that plutonium would work. Compton was in charge (initially he was the head of the whole atomic bomb project). Two other university laboratories were also involved. One was at Columbia University under Harold C. Urey, working on a gaseous diffusion method of separating the uranium isotopes. The other was at the University of California at Berkeley under Ernest

O. Lawrence, working on an electromagnetic method of separating the uranium isotopes.
2. Clinton Engineer Works in Oak Ridge, Tennessee; its purpose was to separate U-235 from U-238.
3. Hanford Engineer Works in Hanford (close to Pasco), Washington, where scaled-up uranium-graphite piles were built in which U-238 was bombarded by neutrons to produce plutonium-239.
4. Los Alamos Laboratory in New Mexico, which was the secret site for assembling the first atomic bombs. J. Robert Oppenheimer was its scientific head; it collected brainpower from the Met Lab and elsewhere, uranium-235 from Oak Ridge, and plutonium from Hanford. Sometimes Oppenheimer is assumed to have been the scientific head of the whole Manhattan Project. However, he was the scientific director of Los Alamos, which was part of the Manhattan Project. Of course, when it became operational, it was the overwhelming sector of the project.[40]

The United States was gradually but surely getting onto war-time footing. It was as if President Roosevelt were waiting for a signal to join the antifascist coalition. That signal was Pearl Harbor. One important catalyst for getting the atomic bomb project under way was that the American officials learned that the British were making advances under much more severe conditions. These advances had been initiated by two refugee scientists, Otto Frisch in Liverpool and Rudolf Peierls in Birmingham.[41] They had shown that the atomic bomb was a possibility. The developing British experience greatly enhanced the American resolve to assign manpower and resources to the development of nuclear weaponry. While in October 1941, President Roosevelt gave no explicit instructions about nuclear research and ordered Bush not to proceed; the president in January 1942 formally approved the development of the atomic bomb.[42]

Various estimates have appeared that up to a whole year could have been gained had the American government and military recognized the possibility and the importance of the atomic bomb when it was first brought to their attention. Some even dismiss the significance of Einstein's letter in the initiation of the American nuclear efforts.[43] Such speculations, however, have no credible bases. A British contingent of scientists soon joined the Manhattan Project because a similar plan could not have been accomplished in war-time Britain. The contingent was led by Chadwick and Mark Oliphant, both former disciples of Rutherford. Frisch and Peierls were among the members, but so was Klaus Fuchs, who would be unmasked as a Soviet atomic spy after the war. He was a non-Jewish German communist who had left Germany for Britain.

Ever since President Roosevelt's speech, Teller was fired up to take part in the war efforts. However, he was disappointed to see that the Americans did not have the same anxiety about Hitler and Mussolini that he and the other immigrants did. A case in point was Robert Serber, Oppenheimer's former student, an outstanding physicist and later important participant in the Manhattan Project. But back in May 1940, Serber was not interested in the war because he thought it was "a clash between capitalist interests."[44] Serber told this to Teller when Teller dropped by for a visit at the University of Illinois at Urbana during a lecture tour organized by the Sigma Pi Sigma Society. It was at the time when the Soviet–German Non-aggression Pact was in effect, and many left-leaning Americans found themselves in the same camp with isolationists and worse from the other end of the political spectrum.[45]

Teller and Mici passed the first Citizenship Examination on November 11, 1940, and became American citizens on March 6, 1941, upon having fulfilled the five-year residency requirement.[46] It was lucky because after Pearl Harbor they would have become enemy aliens, as Enrico and Laura Fermi did. This did not prevent Fermi from working on classified government research, but it did make his travels in the country more difficult.[47] In the spring of 1941, the Tellers still considered Washington, DC, their home in spite of Teller's visiting professorship at Columbia University in New York City. Once again they made long-term plans and purchased a house in the Maryland suburbs. They never lived in this house though, because it was then that Teller was invited to be a full-time visiting professor at Columbia University.

Teller had already spent the summer session of 1940 at Columbia University and proved himself "a physicist of high standing and an exceptionally good teacher," according to Dean George B. Pegram, himself a professor of physics.[48] Columbia University needed Teller because several of its physics faculty members, including Isidor Rabi, had moved away to do war-related research. Teller's appointment as visiting professor of physics was from July 1, 1941, to June 30, 1942. The dean's letter about Teller's appointment was dated August 6, which indicated that Teller had taken up his duties at Columbia before his formal appointment.

In May 1942, Pegram sent another letter of recommendation to Columbia's President Butler for Teller's reappointment from July 1, 1942, to June 30, 1943.[49] In a subsequent letter, Pegram mentioned that Teller "had given a good deal of time to his subject," meaning "research in connection with one of the most important problems under the Office of Scientific Research and Development."[50] It turned out, however, that Teller's "good deal of time" did not satisfy the OSRD, which was concentrating the atomic project in Chicago, and it requested that Teller be released so he could fully devote himself to his war-related work. Columbia complied.[51]

The first two years of the war brought many victories for Nazi Germany just as it did for Japan, which frightened the European refugees in America. There is no information about whether the Tellers ever contemplated a plan of action in case of a German victory, but such considerations were not out of the question. Not only the possibility of a German invasion came to mind, but a Nazi takeover within the borders of the United States was also a possibility.

The Tellers joined the Metallurgical Laboratory in Chicago in the fall of 1942 and then soon moved from Chicago to Los Alamos. Teller never rejoined George Washington University. Clearly, the worry-free period of his life in the United States was over. His position at the Metallurgical Laboratory was his first assignment in a purely defense job. It was a testament that he, in Wigner's words, "felt a deep obligation to the country which provided a new home for him and wanted to serve it to the best of his ability."[52]

As we have seen, the United States becoming a belligerent in World War II changed the character of the uranium project. When it was decided to transfer the project to Chicago, Teller first stayed in New York. In the words of a member of the Metallurgical Laboratory, Leona Woods, Teller "was more or less disinvited to follow along to Chicago because he had relatives in Hungary and was considered, therefore, unclearable."[53] Oppenheimer found Teller's participation significant and managed to get clearance for him, which was not a trivial matter. Another respected refugee physicist, Maurice Goldhaber, authored important papers during the war, and the Manhattan Project used his results, but he could publish his papers only after the war. He was never invited to join the atomic bomb project; he did not have clearance, because his in-laws lived in Germany. By the time the Manhattan Project was under way, his in-laws had been killed in Germany, but Goldhaber and his wife learned about this only after the war.[54]

During this time, Teller developed a hernia. It was diagnosed by Hymer Friedell, a young doctor who joined the Metallurgical Laboratory from San Francisco and who was to be the first of the medical personnel assigned to the Manhattan Project. Teller was one of his first patients and in Friedell's words, had "a hernia the size of my fist."[55] During their house hunting in Chicago, the Tellers were staying with the Fermi family, then they rented an apartment near the campus of the University of Chicago. They bought their furniture at an auction at the Congress Hotel in Chicago. They also bought a Steinway concert piano, which traveled with them to all their subsequent homes. When we visited the Tellers in Stanford in 1996, they still had the hotel furniture and the Steinway.

The first nuclear reactor started up in Chicago in December 1942. Then in March 1943, the first settlers arrived in Los Alamos, including Teller. This was after Oppenheimer had become head of the weapons design group and he had suggested building a separate laboratory at Los Alamos. When the Tellers moved to Los Alamos, they did not have much choice in housing, since they arrived childless. Had Teller been among the top brass, then they would have qualified for a house on the so-called bathtub row, where the bathrooms were furnished with bathtubs. The Tellers' only luxury was their big piano. Teller, who was a night owl, would play at the most unexpected times, usually at night. The neighbors had a hard time deciding whether to enjoy his playing or get upset for the lack of rest.[56]

Oppenheimer assigned Teller the job of briefing new arrivals. Earlier Teller helped him recruit scientists for Los Alamos. Oppenheimer also charged Teller with organizing weekly seminars. From the very beginning, Teller was more interested in the possibility of a thermonuclear explosion than a fission bomb because the latter had already been solved from a theoretical point of view. There were stories about his wandering "about the Mesa projecting a confident, uncannily hypnotic assumption that only the thermonuclear reaction was worth working on."[57]

Teller's attitude at Los Alamos may be considered from two different angles. One is that the immediate task of Los Alamos was the creation of fission bombs and Teller did not sufficiently contribute to its development. The other is that his abilities could not have been best utilized for such work because at this stage the necessary theoretical work was mainly reduced to calculations, which was not his strong suit. It seems that Teller did not devote himself with the expected enthusiasm to whatever tasks arose to further the production of the fission bomb. On the other hand, one might argue that he was preparing the ground for later projects. Teller may also have been led by Oppenheimer to believe that the hydrogen bomb was to be an important part of their current work.

Hans Bethe was head of the theoretical group in which Teller worked. They approached physics differently. Bethe liked to work on what Fermi called "little bricks," and Teller characterized Bethe's work style as "methodical, meticulous, thorough, and detailed." How did Teller characterize his own style in physics? "Although I have made a few tiny little bricks, I much prefer (and am much better at) exploring the various structures that can be made from brick, and seeing how the bricks stack up." Here, Teller also characterizes Oppenheimer: "Oppenheimer also approaches physics in a manner more like a bricklayer than a brick maker."[58] It hurt Teller when Oppenheimer named Bethe to head the theoretical division. When Bethe then asked Victor Weisskopf to be his deputy, this led to a personal conflict. According to Weisskopf, "Edward maintained that he was the better physicist and should have been given the job," but Weisskopf thought that he was considered better at dealing with people.[59]

A collision came when Bethe wanted Teller to work on the detailed calculations for the implosion scheme. Teller did not want to be bogged down with detailed calculations, and he declined the task. Instead, Rudolf Peierls did the work, and his recollection of the episode, over four decades later, agreed with Teller's.[60] This conflict between Bethe and Teller was the beginning of the end of their friendship, according to Teller.[61] Oppenheimer perceived Teller's sensitivity; since he did not want to lose him, from then on he paid personal attention to him and assigned precious time for one-on-one consultations with Teller. In addition to the variety of programs Teller was involved with, a new one was added, travel. He was sent to discuss progress of the work with the people at Columbia University. Others were dispatched to Hanford, Oak Ridge, and Chicago. When Fermi moved from Chicago to Los Alamos, Oppenheimer created a separate unit for him called the F Division (F after Fermi). He lifted Teller and his small group out of Bethe's division and placed him in the F Division, thereby much improving Teller's situation. Both Oppenheimer and Fermi knew the right way to treat Teller.

For the first atomic bombs, two approaches were chosen at Los Alamos. The foundation of both was that from a so-called critical mass, the neutron multiplication would liberate such a large amount of energy that the process would lead to an explosion. In the so-called gun method, the critical mass would be reached by bringing together two halves—each less than the critical mass. The chain reaction would start when a detonation shot one part of fissionable material (uranium-235, in this case) into the other.

The other approach was for the utilization of plutonium. Since it was prone to spontaneous fission, plutonium had to be stored in small portions to avoid pre-detonation. The shock waves caused by high explosives would implode the fissionable material into the critical mass, which was a smaller mass at high pressure than at ordinary pressure. This is why the more sophisticated implosion design was applied to the plutonium bomb, whereas the straightforward gun method was satisfactory for the uranium bomb.

The implosion method was first suggested by Seth Neddermeyer, but his suggestion received little attention. Everything changed when John von Neumann showed up in Los Alamos at the end of the summer of 1943 on one of his visits as consultant. He made calculations and proved that the implosion method would be faster than the gun method and that it would produce higher pressures that would compress the fissionable material; thus less material would suffice. In the implosion bomb a hollow sphere of plutonium is surrounded by a layer of high explosives. When the explosives are detonated, they compress the fissionable material into a ball and thus produce the needed critical mass. Many credit von Neumann alone with recognizing that the compression of the fissionable material by implosion would have additional advantages.

Many years later, however, Teller remembered it differently. He dictated a twenty-page memorandum on September 20, 1979, after he had suffered a heart attack that he ascribed to the strained political atmosphere in the wake of the Three Mile Island nuclear plant accident.[62] The memorandum took the form of a taped conversation between Teller and George A. (Jay) Keyworth in Keyworth's office at Los Alamos. It discussed two events: one was the inception of the implosion method and the other was the sequence in which the idea of the workable hydrogen bomb came to life. (The second will be discussed in chapter 6.) Here we focus on how Teller viewed the story of the implosion approach by discussing, "How the idea of the implosion emerged."

According to Teller's story, when Neddermeyer proposed the approach of implosion, he had in mind only a faster assembly of the fissionable material and did not consider the advantages of compression. It was then that von Neumann came for his visit, which included a talk, after which Teller and von Neumann went to the Tellers' home. Von Neumann became interested in the implosion approach, and the two began calculations about the compressibility of solid materials. In school, we learn that solids are incompressible, but Teller remembered that geophysicists talk about compressed iron in the center of the earth under enormous pressures. Teller knew that the amount of the critical mass of the fissionable material would be considerably lowered if it could be compressed. The implosion could create enormous pressures, hence this approach would not only avoid pre-detonation but would also necessitate less fissionable material.

Teller and von Neumann realized that the geometry of the implosion should be carefully designed to ensure symmetry for maximum efficiency.

Already Neddermeyer's proposal contained the idea of surrounding the fissionable material with a shell of high explosives whose explosion would bring about the implosion. The necessary calculations were too involved even for von Neumann, but his qualitative estimates were very promising. This was one of the occasions when von Neumann strongly advocated the use of computers, although they had to wait for them for some time. Three years after Teller's memorandum, he wrote once again about the implosion story. The occasion was his nomination of Neddermeyer for the Fermi Award. Naturally, Neddermeyer's contribution received greater emphasis in the nomination, but Teller did not ignore von Neumann's and his own contributions in his letter.[63]

What is important here is that thirty-five years after the event Teller in his memorandum felt compelled to narrate a version of the history of the implosion technique that was successfully utilized in the Alamogordo test on July 16, 1945, in the New Mexico desert, and then in the Nagasaki bomb on August 9. Even in 1979, it would have been impossible to research whether or not Teller's memory was impeccable; today there is even less hope to do so. However, there is consistency in his description of how the idea of the compressibility of the fissionable material came about in what he wrote prior to 1979 as well as later. In Blumberg's and Owens's book, *Energy and Conflict*, there is an appendix submitted by Teller, which includes a brief review of his activities in Los Alamos during the war.[64] The review must have been prepared at the authors' request, and since the book was published in 1976, it could not have been written later than 1976; that is, at least three years before Teller's memorandum in 1979.

There are six issues narrated in the appendix. One is the indoctrination of the new arrivals at Los Alamos. The next is the specific question of compressibility at implosion, and this is given the most exposure. Here, carefully delineated, are von Neumann's and Teller's contributions. In essence, von Neumann estimated the pressures expected by using the high explosives planned for the implosion, and Teller remembered the compressibility of metals at enormous pressures. The implosion considerations demonstrated Teller at his best in bringing out new ideas and as one of the most imaginative scientists. When the practical calculations followed, and Bethe, nominally his boss, requested that he perform an unimaginative and repetitive work, Teller declined. Bethe, a person of a different mindset, took this very badly, but Oppenheimer, who by his nature was much closer to Teller than Bethe was, solved the situation as described above.

We mention the other aspects of Teller's research activities at Los Alamos only in passing. He dealt with the question of opacity, which refers to the resistance of materials to the effect of nuclear radiation. He succeeded in involving Maria Goeppert Mayer and her students in this work at Columbia University in New York. This was possible because Goeppert Mayer and her students did not need to know anything about the rest of the project. Sometime in 1945, though, Teller managed to actually bring Goeppert Mayer to Los Alamos.

Another issue was the examination of the possibility of a nuclear explosion in the industrial process of separating the isotope of uranium-235 by the gas diffusion technique at the Oak Ridge plants. It was feared that an explosion might occur in the process due to the accumulation of the fissionable uranium isotope in the apparatus, called diffusion cascades. This project contributed to developing Teller's keen interest in safety in nuclear industry during the subsequent decades.

Mention is also made of the "Super," the hydrogen bomb, in the appendix, but very little is said about the actual work. The final topic, which suited Teller eminently, was to consider whether there might be new laws of nature to be found from experience with the fission bomb. There were no such new laws, but it could have been imagined that there might be.

We now return to the question of why Teller discussed the emergence of the implosion concept in his 1979 memorandum when it had been covered in the appendix of *Energy and Conflict* in 1976. He may have forgotten what he had written about it or he may have wanted to be extra sure that his contribution was on record. Under scrutiny, we may notice two minor errors in this segment of the 1979 document. One occurs when he speaks about introducing newcomers to Los Alamos: he mentions that he especially enjoyed giving the introduction to Fermi, who arrived in Los Alamos a *couple of years* later. As people started arriving in Los Alamos in 1943, this would have placed Fermi's arrival no earlier than 1945, but Fermi was already there in 1944. The other is a Freudian slip—as Teller calls it—when he mentions the introduction of computers in Livermore and quickly corrects this to Los Alamos. This slip of the tongue may suggest that the memorandum following the disasters of the Three Mile Island accident and Teller's heart attack could not really add much in accuracy let alone scope about his activities at Los Alamos.

## ROOTS OF ANTICOMMUNISM

The Los Alamos period was important for Teller's political development because it was there, in 1943, that he read Arthur Koestler's *Darkness at*

*Noon.*[65] It was a "major milestone" in his thinking.[66] Teller says that as late as 1937, he was still open-minded about communism, at least open-minded to what might be the best course for Russia to take. The Hitler–Stalin pact in August 1939, which was to last until Hitler's aggression against the Soviet Union in June 1941, was a big blow not only to Teller but especially to communists worldwide. Teller wrote in 1970 that when they went to Los Alamos, both he and Mici "were inclined to be on the 'left.'"[67]

For Teller, the primary concern was defeating Hitler, but he also started viewing communism increasingly unsympathetically. As for Koestler's book, he declared, "I don't believe I have ever been more fascinated with a book than I was with *Darkness at Noon*." Teller stressed that his becoming an anticommunist was not an overnight event; rather, it was the result of a long and painstaking process. He considered himself an anticommunist "of the school of Koestler."[68]

Teller felt a deep kinship with Koestler. This was not just because of the similarity in their family backgrounds. There was a more profound reason: when Koestler changed from being a communist to becoming an anticommunist, he lost many of his dear friends. "Eleven years later— Teller writes about the period of 1943 to 1954—I began to understand that loss better, because, in a minor way, I went through a similar transformation."[69] His "in a minor way" is an understatement. Moreover, the parallel is flawed. Koestler changed his philosophy and lost many of his former friends. Teller testified against a fellow physicist—which is a less noble cause—and lost his former friends. But Teller ascribes his separation from many of his friends to his gradual but unambiguous transformation into an anticommunist at Los Alamos.

Although the location, characters, and everything else in Koestler's novel were fictitious, there could be no doubt that the events took place in the Soviet Union under Stalin's ruthless dictatorship. The main character was the Old Bolshevik Nicolas Salmanovitch Rubashov, who was a synthesis of several of the old-guard leaders persecuted by Stalin. He could be identified most conspicuously with Nikolai Bukharin, who was considered second to Stalin in the Soviet leadership at some point. Rubashov was a highly decorated former party leader and partisan commander. Now his confession of nonexistent crimes was demanded of him as his last service to the Communist Party. His confession, trial, and execution would ostensibly strengthen the party and the supreme leader. It was further promised that some time in the future, following the final victory of communism, his sacrifice would be made known.

## ARTHUR KOESTLER (1905–1983)

Arthur Koestler was born in downtown Budapest, within a one-mile distance from where Teller was born, and died in London in a double suicide with his wife.

He was a journalist and writer; a popularizer of science and a devotee of para-science. He was a dedicated communist and equally dedicated anticommunist; a Zionist Jew and a nationalistic Hungarian. Known to be a loyal friend, he could also be an abuser of friendship. In short, he was a most gifted and complex product of the twentieth century.

His family belonged to the upper middle class of Hungarian Jews, whose primary language at home was German. His mother was from Prague and grew up in Vienna; his father was from Miskolc, an industrial city in northeastern Hungary. Koestler was an only child and a lonely child, and in many respects Koestler's and Teller's youths were similar.

Koestler's later life was especially turbulent and extreme. He attended the Vienna Technical University, where journalism became his major avocation. He not only reported world events, he also lived them and participated in them actively. When he was a Zionist, he spent some time in Palestine. Then he became a communist, visited Moscow, and became disillusioned but remained a party member for years. He sided with the loyalists in the Spanish Civil War, was captured, and was to be executed by the fascists, but was freed by worldwide protest. His reportage and books helped greatly to keep the world informed.

The combination of his prison experience, his travel to the Soviet Union, and the information gleaned from his friends returning from the Soviet Union helped him write *Darkness at Noon* during 1938–1940. The first edition of the book appeared in England in 1940 and in the United States in 1941, and became an instant success. It was published at the time when the Nazi–Soviet Pact was in effect, but the book remained a focus of interest all through the cold war era. Although it was transformed in the public consciousness from a literary creation into a political device, it was considered to be one of the best novels of the twentieth century.

Koestler started his writer's and journalist's career as an editor of a German and Arabian weekly in Cairo. He achieved literary fame for his writings in German. Later, he switched to English and wrote many of his major works in English. He took great pride in the fact that his books were burned and banned both in Hitler's Germany and Stalin's Soviet Union.

Teller and Koestler never met apart from a possible introduction by John von Neumann, but Teller considered Koestler's *Darkness at Noon* a defining contribution to his becoming an anticommunist.

Confession was crucial for the communist show trials; this was the general pattern later as well, including similar trials in the satellite countries after the war. Rubashov understood the situation; he said, "History knows no scruples and no hesitation."[70] He should have known, because he had committed crimes at the instructions of the party. He and his similarly Old Bolshevik interrogator, Ivanov, thought alike; as Ivanov said, "History is *a priori* amoral; it has no conscience."[71] Thus Ivanov could not be surprised when he also fell victim to the regime, to which Rubashov's new, younger interrogator, Gletkin, added, "[M]ankind could never do without scapegoats."[72]

In spite of the powerful story, it is difficult to imagine that Teller's attitude was impacted by it to the extent he ascribed to it. His attitude in the immediate years after the war, for example, did not yet show signs of such a strong sense of anticommunism as later developed. It is also questionable whether Teller's anticommunism was more ideological or political. If his abhorrence toward communism was primarily ideological, the next question would have to be, which version of communism did he abhor? Did he dislike the utopianism of the early communists, and did he despise the slogan of all work according to their abilities and consume according to their needs? In the case of Nazism, he must have refused its initial teachings as much as its embodiment in their actual realization after the Nazis had come to power. There was consistency in what Hitler had promised in his book *Mein Kampf* and what he did between 1933 and 1945. With communism, things were more complicated.

This complexity is ably described in a conversation between a journalist and Italian Holocaust survivor chemist turned author Primo Levi. The journalist quoted Solzhenitsyn about the voice of protest against the authorities in a Soviet concentration camp, which ran like this: "You're not Soviets! You're not Communists!" In Levi's Nazi concentration camp, everything was consistent with the Nazi ideology, and the prisoners could only say, "You're perfect Nazis, you're the embodiment of your idea."[73]

Soviet communism may have been more an expression of Russian ambitions than of the original communist ideology, even if it appeared under its disguise. The former West German chancellor Helmut Schmidt characterized Soviet communism as 25 percent communist and 75 percent Russian. Teller must have had something similar on his mind when he wrote to Golda Meir, "Anti-Semitism is an old Russian tradition. It is not an old communist tradition. In fact, the advent of communism gave Jews a respite for a generation."[74]

It is not easy to second-guess Teller's anticommunism, but the impression is that his opposition was directed primarily against the Soviet Union and its military might and was less concerned with communist ideology. *Darkness at Noon* is one of the defining pieces of twentieth-century political literature, but it is more about how Soviet communism treated Soviet communists and the Soviet people as well as foreign communists— Rubashov was a great instrument in their executions—rather than about how it would threaten the United States. When Teller was subsequently pushing for the American hydrogen bomb, he was likely driven by the determination to have the United States prepared against a foreign power rather than to resist an alien ideology.

There have been some misconceptions about the roots of Teller's anticommunism. It has often been ascribed to his childhood experience during the communist dictatorship in Hungary in 1919. However, he was only eleven years old at that time and was not exposed to communism in a direct way. At the time and during the following years, he was more impacted by the anti-Semitism he experienced. Further, it has been said that Teller "fled the communists in Hungary,"[75] but Teller left an anti-Semitic and semi-fascist Hungary for Germany in 1926 (see chapter 2). During his life in Germany, he "had an open mind about communism." In 1996, he told us, referring to the time of the worldwide depression, when the impression was that capitalism had come to end: "The only people who had new ideas, so it was said, were the communists. Now I was not a communist, not at all, but I was genuinely open-minded."[76]

Teller mentioned his "devoted communist" friend, the Russian Lev Landau and another friend, the Hungarian Laszlo Tisza. According to Tisza, with whom I talked sixty years later, Teller was indeed "very open-minded and he had other leftist friends, too, but he didn't sympathize with the communist ideas even then."[77] Tisza did not elaborate as to whether he meant the philosophical foundations of communism or what at the time could be observed under its disguise in the Soviet Union. Tisza was one of the few people from whom Teller gained firsthand information about the Soviet Union. The fate of Teller's family in Hungary in the early 1950s might have added to his anticommunist sentiments. However, he did not have much information about it then or later, and by the time he could have been much concerned about the fate of his family in communist Hungary in the 1950s, he had already become strongly anticommunist. Generally speaking, he was not to be much impacted by personal experience, whether that of his own or of his relatives.

Nonetheless, it seems doubtful that Teller's political development reached completion after having just read *Darkness at Noon*. However shocking that book is—and it describes Soviet reality so astutely—Teller's future motivations cannot be interpreted exclusively by what he read. Besides, the Soviet Union had undergone considerable changes after 1953, and Rubashov's fate no longer characterized the treatment of its internal "enemies." It is also worth mentioning that the old guard, whom Khrushchev removed from power in the mid-1950s, did not mount the scaffold but continued in insignificant positions. Khrushchev himself became a pensioner when he was removed from power and lived out his life in peace. His fate symbolized the changes he helped introduce.

Teller's anticommunism could not have been determined primarily from his learning about how the Soviet regime treated its own people. Rather, his opposition toward the Soviet Union—the superpower he saw as threatening the free world—must have fueled his continuous crusade for ever-more sophisticated weaponry. These were the principal roots of his anticommunism. Nikolas Rubashov's fate may have been an important component in Teller's political development, but it would be an oversimplification to ascribe his political attitude in its entirety during the ensuing decades to Koestler's book.

Curiously, *Darkness at Noon* had a more sophisticated message than just a condemnation of Soviet-type communism. Its more general lesson can be illustrated by the following episode. During Rubashov's imprisonment, his friend and devotee, Michael Bogrov, a highly decorated commander of the Soviet Eastern Fleet, was being executed.[78] His "crime" was his *opinion* in the so-called submarine question, which referred to a disagreement about the desired tonnage of submarines that were to be built for the Soviet armed forces. Bogrov wanted large tonnage and long-range submarines to facilitate the continuation of world revolution. Other leaders favored the less-expensive small submarines of short-range action of which many more could be built and which would suffice for defense purposes. Many in the admiralty supported Bogrov, but the supreme leader and party line prevailed, and so Bogrov was liquidated.[79]

Because passages such as this are so realistic, the reader has to remind himself that this is in fact fiction. Ideologically, Bogrov's stand is consistent with Leon Trotsky's advocacy of permanent world revolution versus Stalin's teaching that socialism could be built in a single country. Trotsky was exiled and then murdered by Stalin's agents. For the current reader, Bogrov's story induces a feeling of déjà vu. Koestler could not have had in mind the hydrogen bomb controversy when he wrote his book in 1940. Yet the eventual disagreement between the US Air Force and the US Army concerning the desired magnitude of nuclear weapons comes inevitably to mind when reading about Bogrov's predicament. We will see later that Freeman Dyson ascribed the Teller/Oppenheimer controversy to interservice fights in the American armed forces. It might be argued that it would be an exaggeration to compare the two stories. However, in considering the Oppenheimer case (in chapter 8), at least some elements of similarity will be seen between the two stories.

Later, in 1998, Teller wrote an article for *Science* in which he returned to the origin of his anticommunism.[80] He wrote, "By age 11 I had had a none-too-sweet taste of communism in Hungary." This does not mean, however, that his experience with communism lasted for a long time; it was only 133 days, and the Tellers had left town well before it ended, so his experience lasted an even shorter time. Teller repeatedly expressed that he was more unfavorably impressed by the anti-Semitism in Hungary during the rest of his life there; that is, during six long years, than by what he experienced under the Soviet rule in Hungary.

Then he described his encounter with his friend Laszlo Tisza, who was arrested as a communist and after he left prison became unemployable. Teller recommended him to Landau in Kharkov, where Tisza had gone and earned a second doctorate under Landau. Tisza returned in a few years' time, not only disillusioned by Soviet communism but also with the news about Landau's arrest as a capitalist spy. Teller wrote, "The implication of this event was for me even more defining than the Hitler–Stalin Pact. By 1940, I had every reason to dislike and distrust the Soviets." Conspicuously, there was not a word about Koestler's *Darkness at Noon* in Teller's 1998 *Science* article, whereas he described the decisive influence of the book only two years before, and in great detail.

What he wrote in his *Memoirs* concerning the impact of Koestler's book on him was fully consistent with what he said in 1996. Koestler's book remained one of Teller's favorite books and it was second only to Goethe's *Faust* in his list of favorites, which he prepared at the request of a journalist in 2002.[81] The impression is that Teller's ascribing the factors in his becoming an anticommunist showed variations over time, and it would be a mistake to rely upon a single utterance of his when trying to understand his motivations.


## BOMB DILEMMAS

In April 1945, after reigning for close to identical lengths of time, President Roosevelt and Adolf Hitler died. Germany was defeated in early May, and the war ended in the European theater. For many of the scientists in the Manhattan Project the original purpose of the quest for the atomic bomb had ended, but only one scientist left the project; the rest continued.[82] In June 1945, preparations for the atomic bomb were at their fullest. Uranium-235 was scarce, so it was decided to use the uranium bomb without testing. For the implosion bomb of plutonium, one was to be tested and a second was to be used. It was at that point, at the end of June, that Teller received a letter from Szilard. Enclosed was a petition to the new president, Harry Truman, and Szilard asked for Teller's help to make it possible for the Los Alamos scientists to sign it if they agreed with its contents.

Szilard explained the motivations of the petition in his letter accompanying it.[83] He stressed that the petition was based on moral considerations only, and he did not exclude that the decision about using the bombs would have to be made on the basis of expediency. However, Szilard felt it important for the scientists to go on record with their opinion, especially in view of the American people being unaware of the existence of the atomic bombs. Teller discussed the circumstances of the petition and those of his response in detail in his *Memoirs*, but the text of the petition is not there. (See pp. 154–55.)

The petition is generally identified with James Franck's name, but Szilard wrote its final version, and sixty-eight scientists signed it. They warned of the inevitability of a nuclear arms race after the war and of the vulnerability of the United States due to its high concentrations of population and industry. They appealed to the president not to use the bomb against Japan without warning and not to ignore the moral aspects of the deployment of nuclear weapons. The style of the petition unmistakably bears Szilard's mark, as it instructs more than implores.

When Teller read Szilard's letter and the petition, they "made good sense" to him and he "could think of no reason that those of us at Los Alamos who agreed shouldn't sign it." This is how he remembered it in his *Memoirs*.[84] Back then, he decided to consult Oppenheimer, not only because he was the director but because he respected Oppenheimer's opinion. This was yet another example of Teller not wanting to act on his own but seeking the approval or guidance of his superiors. Oppenheimer brushed the petition aside. He told Teller that its authors did not see the complete picture and that the decisions should be left to the leaders in Washington, because they had all the information, and they should be trusted.

Again, in his *Memoirs*, Teller remembered that he "readily accepted his [Oppenheimer's] decision and felt relief at not having to participate in

A Petition to the President of the United States
July 17, 1945

Discoveries of which the people of the United States are not aware may affect the welfare of this nation in the near future. The liberation of atomic power which has been achieved places atomic bombs in the hands of the Army. It places in your hands, as Commander-in-Chief, the fateful decision whether or not to sanction the use of such bombs in the present phase of the war against Japan.

We, the undersigned scientists, have been working in the field of atomic power. Until recently we have had to fear that the United States might be attacked by atomic bombs during this war and that her only defense might lie in a counterattack by the same means. Today, with the defeat of Germany, this danger is averted and we feel impelled to say what follows:

The war has to be brought speedily to a successful conclusion and attacks by atomic bombs may very well be an effective method of warfare. We feel, however, that such attacks on Japan could not be justified, at least not unless the terms which will be imposed after the war on Japan were made public in detail and Japan were given an opportunity to surrender.

If such public announcement gave assurance to the Japanese that they could look forward to a life devoted to peaceful pursuits in their homeland and if Japan still refused to surrender our nation might then, in certain circumstances, find itself forced to resort to the use of atomic bombs. Such a step, however, ought not to be made at any time without seriously considering the moral responsibilities which are involved.

The development of atomic power will provide the nations with new means of destruction. The atomic bombs at our disposal represent only the first step in this direction, and there is almost no limit to the destructive power which will become available in the course of their future development. Thus a nation which sets the precedent of using these newly liberated forces of nature for purposes of destruction may have to bear the responsibility of opening the door to an era of devastation on an unimaginable scale.

If after this war a situation is allowed to develop in the world which permits rival powers to be in uncontrolled possession of these new means of destruction, the cities of the United States as well as the cities of other nations will be in continuous danger of sudden annihilation. All the resources of the United States, moral and material, may have to be mobilized to prevent the advent of such a world situation. Its prevention is at present the solemn responsibility of the United States—singled out by virtue of her lead in the field of atomic power.

The added material strength which this lead gives to the United States brings with it the obligation of restraint and if we were to violate this obligation our moral position would be weakened in the eyes of the world and in our own eyes. It would then be more difficult for us to live up to our responsibility of bringing the unloosened forces of destruction under control.

In view of the foregoing, we, the undersigned, respectfully petition: first, that you exercise your power as Commander-in-Chief to rule that the United States shall not resort to the use of atomic bombs in this war unless the terms which will be imposed upon Japan have been made public in detail and Japan knowing these terms has refused to surrender; second, that in such an event the question whether or not to use atomic bombs be decided by you in the light of the considerations presented in this petition as well as all the other moral responsibilities which are involved.[85]

the difficult judgments to be made."[86] He added three comments: He found that Szilard was right in assigning special responsibility to the scientists involved in creating the bombs; that Oppenheimer was right in that the scientists did not have sufficient information about the political situation; and that the Los Alamos scientists should have worked out the technical changes to make an atomic bomb suitable for a demonstration over Tokyo Bay. This last comment needs to be looked at closely, because the scientists could not have just decided to modify a bomb, and if they could have, it would have meant further delay, while American troops were in danger. Besides, after the test in Alamogordo, only two bombs remained; another plutonium bomb and one uranium bomb for which a test could not have been conducted.

Teller did not sign the petition, nor did he even inform his Los Alamos colleagues about it. He wrote a letter to Szilard, the essence of which could be summarized in three points. One was that if Szilard could convince Teller that Szilard's moral objections were valid, Teller would be ready to quit working (although Szilard did not expect him to quit, only to sign the petition). If that was not the case, Teller added, "I can hardly think that I should start protesting." The second was that "actual combat-use might even be the best thing" to make people understand how terrible a next war could be. Teller's third point was that just because they happened to work on the bomb, they did not have the responsibility of participating in the decision about how to use it.[87] Upon sending his letter to Szilard, Teller sent a copy to Oppenheimer.

Looking back, Teller disagreed with his own decision and kept implying that Oppenheimer misled him, because—as Teller found out eventually—Oppenheimer did participate in the decision about how to use the atomic bombs. In any case, he never stopped blaming Oppenheimer for his own decision. Incidentally, Hans Bethe also thought that Szilard's protest against the use of the atomic bombs was based on incomplete understanding of the situation. Bethe thought that even Szilard's foresight did not suffice to understand the consequences of taking or not taking certain actions in connection with the atomic bomb.[88]

In 1984, Teller was interviewed by Clarence and Jane Larson for their large-scale interviewing program with famous scientists and technologists.[89] Clarence was a former commissioner of the AEC. In this interview, Teller said that Oppenheimer "effectively and cleverly stopped Szilard's proposal for the demonstration before it [the bomb] was to be dropped." Teller thought that if "Oppenheimer and Szilard had gotten together on that point . . . they might have influenced Truman and the American policy in a direction which I believe would have been right." This sounds improbable, especially in hindsight. Teller then diminished his role by reducing it and further threw some indirect blame on Oppenheimer: "I was a go-between the two [Oppenheimer and Szilard], and was thrown out of Oppenheimer's office, incidentally with very kind, convincing, and tactful words."

Unbeknownst to Teller at the time, Oppenheimer was a member beside Arthur H. Compton, Ernest O. Lawrence, and Enrico Fermi of the Science Panel of the Interim Committee. The committee was a creation of Secretary of War Henry L. Stimson in May 1945 with the purpose of giving advice about nuclear issues that might come up following the war. Concerning the use of the first atomic bombs, the Science Panel considered two principal aspects of the issue. One was the possible consequences on subsequent international relations of dropping the atomic bombs, and the other was saving American lives by a rapid conclusion of the war against Japan.

They had three points in their recommendation. One was to consult about the new weapons with all allies of the United States, not only with Great Britain, before possible use. Another was that a technical demonstration would not suffice to end the war; they saw "no acceptable alternative to direct military use." In their third point, they stressed that they did not have proprietary rights in connection with the atomic bombs, but they thought about the problems of their utilization as only few could have thought about them. Nonetheless, they had "no claim to special competence in solving the political, social, and military problems which [were] presented by the advent of atomic power."[90]

The difficulty of making a recommendation of this nature is seen in this report. While the scientists did not consider themselves competent in solving political, social, and military problems, they understood that their unambiguous recommendation would have heavy political, social, and military consequences. The panel of the four scientists considered the two main alternatives: technical demonstration of the bomb and direct bombing.[91] Finally, they came out with recommending direct military use. This is what happened, and, following Hiroshima and Nagasaki, Japan announced surrender on August 15, 1945. Teller continued to regret that the bomb had not been demonstrated, but he did not regret his contribution to the bomb. He was asked many times whether he ever regretted that he had worked on the bombs, and his answer was always an emphatic no, as well as his own question: What if he hadn't worked on them? They did not know that the Germans were not seriously working on the bomb.

By the time he was writing his *Memoirs*, however, the weight of his argument had shifted—in a curious switch—from the Germans to the Russians: "Those who question the morality of working in wartime Los Alamos seem to have forgotten that the Soviet Union of the 1940s was controlled by Stalin."[92] In reality, the Stalin argument played well for the hydrogen bomb but not for the atomic bomb. Teller's reference to Stalin in this context was an example of projecting later history onto earlier events.

There could have been, however, a different question: What would have happened had the bomb been ready one year earlier? Teller contemplated such a question in his *Memoirs*, and blamed Groves for not letting Urey go ahead with the production of uranium-235 by the technique of centrifugation.[93] Teller said in his *Memoirs* that a bomb available by the summer of 1944, and deployed, would have meant, among other things, saving hundreds of thousands of Jewish Hungarians in Auschwitz. In addition, tens of millions of East–Central Europeans would have escaped Soviet domination for decades. Nevertheless, there would have been damages due to the atomic bomb in Central Europe.

The question about the scientist's responsibility with respect to the deployment of his own creation is an intriguing one. This was an especially difficult question when only the scientists were aware of the existence of the bombs, in addition to a small group of political and military leaders.

It is quite natural that the scientists, as learned members of society, would feel responsibility and would want to voice their opinions. In fact, the creation of the advisory panel of the four prestigious scientists was recognition by the US administration of the importance of their viewpoint.

In totalitarian regimes it is characteristic that the politicians and military leaders avoid letting the scientists meddle in what they consider to be their affairs. In August 1961, Soviet leader Nikita S. Khrushchev withdrew from the interim test ban, and the Soviet Union started testing huge nuclear weapons on a large scale. When Andrei Sakharov protested the decision, Khrushchev publicly humiliated the father of the Soviet hydrogen bomb. He explained to everybody that the scientists should confine their activities to the laboratory. It was the beginning of Sakharov's dissent, and within three years he was removed from the Soviet weapons program.[94]

In another historical example, Hermann Staudinger, polymer chemist and future Nobel laureate, questioned the propriety of conducting Fritz Haber's poison gas warfare. He discussed the matter at the time of the Versailles Treaty following Germany's defeat in World War I. He was afraid that the advancement of science might make war more terrible. Haber responded by warning that scientists should not engage in ethical considerations that concern the applications of scientific achievements in warfare; such considerations should be left to the political and military authorities.[95]

When one of the other scientist participants, Otto Hahn, asked Haber about the legality of gas warfare, Haber explained "that gas might promptly end the war, saving uncounted lives."[96] Incidentally, half a dozen future Nobel laureates participated in working out the "science" of gas warfare in Germany. Contrary to Haber's expectations, this did not end the war and it certainly did not end it in Germany's favor. By the end of the hostilities, Britain, France, and the United States had also been waging their own gas attacks. Recent studies suggest that "poison gas in the Great War [World War I] . . . was far less lethal than explosives, . . . it was better to be gassed than shot," though it left its fair share of fatalities.[97]

After the mid-July 1945 test, Oppenheimer reorganized Los Alamos and gave greater emphasis to the fusion bomb, putting Fermi and Bethe in charge. Teller welcomed this, as he saw in this a greater effort to develop the fusion bomb. Reading his *Memoirs* though, this does not sound convincing.[98] Teller had previously been disappointed when Bethe had been made the head of the theoretical division—essentially Teller's superior— and this time Bethe was put in charge of Teller's pet project. Still, Teller must have been satisfied that the thermonuclear project was finally gaining importance. His disappointment was all the greater then, when, after the

Japanese surrender, Oppenheimer declared that work on the hydrogen bomb should be discontinued.

Teller was often asked whether it was the right thing to do or whether it was a mistake—let alone a crime—to drop the atomic bombs over Japan. His succinct opinion expressed repeatedly was that: "[i]t was unnecessary and wrong to bomb Hiroshima without specific warning."[99] It is then instructive to consider how Teller's colleagues saw the same question. Some of their comments were made at the time of the war's end, others, later. Of course, as years and even decades have passed, the changing circumstances may have altered opinions as well.

In early 1944 Leo Szilard was a proponent for using the atomic bomb in World War II. He argued that peace could not be had unless the public fully understood the potentialities of the bombs. One of the actions to be taken in order to restrain the powers from developing and possessing nuclear weapons would be to control all mineral deposits that might be needed for such weapons. But, warned Szilard, "it will hardly be possible to get political action along that line unless high efficiency atomic bombs have actually been used in this war and the fact of their destructive power has deeply penetrated the mind of the public."[100] Was not this the kind of expediency so eloquently advocated against by the same Szilard in the petition quoted above?

It appears that a little more than a year later, Szilard changed his mind, hence the petition. In 1945, one of his concerns was that by dropping the atomic bombs, the United States would call attention to this new kind of arms. He had no idea about the Soviet espionage that was constantly tapping into the American nuclear project for crucial information. Subsequently Szilard realized that not deploying the bombs in 1945 would not have meant gaining any time for the United States in the forthcoming arms race due to Soviet espionage.

When in 1987 Luis Alvarez, former participant in the Manhattan Project, looked back to the ending of World War II, he noted:

> What would Harry Truman have told the nation in 1946 if we had invaded the Japanese home islands and defeated their tenacious, dedicated people and sustained most probably some hundreds of thousands of casualties and if *The New York Times* had broken the story of a stockpile of powerful secret weapons that cost two billion dollars to build but was not used, for whatever reasons of strategy or morality?[101]

The atomic bombings, given their horrific force, had a special place in the history of World War II and in public awareness. But there were other

American war actions, like the fire bombing over Tokyo, which had had at least as severe consequences as an atomic bomb, but which people tend to forget. As a consequence of a single bombing raid over Tokyo on March 9, 1945, eighty thousand people died, sixteen square miles of the city were devastated, and a million and a half people lost their homes.

When Eugene Wigner looked back to his contribution to the atomic bombs and considered the horrors they caused, he wished he could have said that he regretted working on the atomic bomb, just to please his questioners. In reality, however, he did not regret it, neither intellectually or emotionally.[102] He felt strongly that the bomb should have been created sooner, while the war was still going on with Germany, or at least in time to influence the Yalta meeting of the Big Three (Roosevelt, Churchill, and Stalin) to lessen the hardship of Eastern Europe and to ease Soviet domination. But Wigner never wanted to have the bombs dropped on Japan and he did not expect it to happen. However, he understood that the army developed the atomic bomb and the army was eager to use it to demonstrate its power. He thought that it probably saved a lot of lives and without it, many more American and Japanese lives would have been lost because America would have had to invade the Japanese islands. As Wigner remembered in 1984, there was a point during the Manhattan Project, when Germany had been defeated, and he thought

> that it was not necessary to continue the work on the bomb, but the government was not of that opinion. General Groves also wanted to continue and he said that we could use it against the Japanese and it would shorten the war.
>
> We proposed then to demonstrate the bomb in the presence of some Japanese scientists and military leaders. Groves once again disagreed and said that we should demonstrate it on a city. And that is what happened, but we were against it and were quite unhappy. We thought that many Japanese lives could have been saved if the bomb had been demonstrated on an uninhabited territory. But, apparently, I must admit, and I will admit, we were probably mistaken. Much later, I read in a book that the demonstrations in Hiroshima and Nagasaki may have saved many, many Japanese lives. Since I thought that a demonstration over an uninhabited territory in the presence of Japanese scientists and politicians could have sufficed, I went around and asked my Japanese friends about it. And with one exception, they said, "No, such a demonstration would have had no effect on the Emperor." According to all my Japanese friends, with one exception, "It would not have had the same effect; it was very good that you demonstrated it this way." Maybe that was the way to do it, but I did not think so at that time. Of course, they knew the Japanese politicians,

the Japanese Emperor, and the Japanese military leaders much better than we did. But I was very surprised. They thought that many Japanese lives were saved this way even though it led to the extinction of many Japanese lives. Apparently General Groves was right and the bomb had to be demonstrated the way it was.[103]

Wigner and Teller were among a few key figures in creating the atomic bombs who were asked in 1965 why they helped to make the bomb, and whether they would do the same again, knowing what they knew twenty years later.[104] Both said they would work on it again, and they justified their stand by what had been the looming danger of Germany getting the bomb first. However, they differed on whether dropping the bombs over Japan was the right thing to do.

In 1945, Wigner signed the petition opposing the use of the bomb, whereas, in 1965, he was "inclined to believe that the use of the nuclear bomb to terminate the war was a more humane way and led to less suffering and loss of life than any other way that was contemplated." Teller, on the other hand, declined to sign the petition in 1945, but in 1965, he stated that "I was positive then, and I am positive now, that we made a mistake dropping the bomb without a previous bloodless demonstration." This does not square not only with his having declined signing the petition, but with his saying in his letter to Szilard that "actual combat use might even be the best thing" (see above).

Over the years, Teller succeeded in creating the impression that, even as early as 1945, he was against dropping the bombs. Despite the absence of his signature from the petition, he was generally counted among those who had signed it. On the fortieth anniversary of the bomb, a *Washington Post* staff writer, who was preparing a CBS documentary about the topic, stated in an important article, "Some, including Dr. Edward Teller, wanted to stage a demonstration explosion over Tokyo Bay."[105] It was not a journalist's misunderstanding of the situation; rather, it was what Teller stated on various occasions; that is, his definite opposition to "dropping the bombs on Hiroshima and Nagasaki in 1945." This is what he told, for example, the President's Commission on Campus Unrest in Washington on July 24, 1970.[106]

For a reflection on the mood at the time of the bombing, we turn to John A. Wheeler, who was called the most versatile physicist of the twentieth century. He was a participant in the Manhattan Project, and he did not hesitate to name his contribution to developing the atomic bomb as his most important achievement. He recalled the many American troops who told him that those two bombs saved their lives. "All those American troops on the island of Okinawa in the fall of 1945 were ready to invade Japan—they knew that the Japanese were ready to die rather than to give in."

Wheeler often wondered "how many more lives could have been saved, had we done the bomb a year or so sooner." The question had a painful, personal relevance for him because his younger brother had been killed in action in Italy in October 1944. Shortly before his brother's death, Wheeler received a letter from him in which he wrote: "Hurry up!" Apparently, the brother guessed that Wheeler was involved in war work. Wheeler estimated that "had the war ended a year earlier; in mid-1944 instead of mid-1945; possibly 15 million lives would have been saved. A heavy thought . . ."[107]

In this connection, it is of interest to quote yet another Nobel laureate physicist, Philip Anderson, from a conversation in 1999. He had been vocal against the hydrogen bomb as well as against the Strategic Defense Initiative, but he did not take an apologetic attitude with respect to the atomic bombs of 1945:

> I'm not one of those who feel guilt about dropping them on Japan. The one thing that emotionally influences me is that I knew about something which most Americans don't, because, having been there, I knew about the fire bombing from my Japanese friends. The fire bombing of Tokyo was so close to genocide, killed so many people, that it seemed to me much more of a horror than the atom bombs. Another thing I was conscious of, and I don't know why so few Americans are conscious of it, is Nanking. Nanking and the Japanese behavior in China and Korea was a horrible thing, unbelievably savage. I don't think I have any complaint whatsoever about the atom bombs. And I'm not sympathetic to the Germans about Dresden. The old saying is absolutely right, "He that soweth the wind shall reap the whirlwind." That's what both the Germans and the Japanese did. The bombs left them with no illusions about being defeated.[108]

The Japanese emperor gave a radio speech a few days after the second atomic bomb had been dropped, announcing the Japanese surrender. Incidentally, there was not much news coverage in Japan about the first bomb; apparently, it took a second bomb to make the leaders of Japan realize that the war was over.

It was also after the second bomb, on August 10, that Japan officially entered a protest against the American use of the atomic bomb, citing a violation of the Hague Convention Respecting the Laws and Customs of War on Land. The protest was made not only in the name of the Japanese

Imperial Government but also "in the name of humanity and civilization." This protest was made despite the Japanese war crimes and indescribable atrocities during the previous decade. Apart from this letter, though, there was never any other protest by the Japanese government against the bombings of Hiroshima and Nagasaki.[109]

Not long ago, Teller figured in a "What might have happened if . . . ?" type of question in the August 14, 2007, issue of *Japan Times*.[110] The article concerns the Japanese atomic bomb project during World War II and was prompted by the recent publication of letters and documents of Yoshio Nishina, the project director. The article mentions a letter of April 21, 1933, by a German physicist to Nishina saying that "Edward Teller was hoping to stay in Japan after fleeing Nazi Germany." There is no mention of the name of the German physicist in the Japanese article and there is no hint of such considerations in Teller's *Memoirs*. The latter is no evidence by itself, because some other interesting aspects of Teller's life have been found missing from the *Memoirs*. It would be too easy to dismiss the suggestion of the letter, but it might have been a less impossible scenario than it had seemed at first glance.

The letter is dated years before Japan joined the Axis alliance (of Nazi Germany and fascist Italy), and refugee scientists had considered eastward immigration, including Turkey, the Soviet Union, India, and China. In the late 1920s, when the California Institute of Technology was persuading Theodore von Kármán to move there, he also received invitations from Japan. He was offered generous terms, and when he declined the invitation by requesting yet better remuneration, the Japanese accepted his terms. He went and later claimed (with false modesty) that "I do not wish to take too much credit—or perhaps in this case, blame—but I believe I was also the man who introduced Japan to metal airplane propellers."[111]

The journalist of *Japan Times* posed the question "If Teller . . . had joined Nishina's group, would Japan have been the first to produce the bomb?" The question is intriguing not so much for its verisimilitude, but for the fact that it was posed at all, and it might be worthwhile to attempt to answer it.[112] It is safe to say that Japan still would not have been the first. Even if Teller and others had joined the Japanese scientists around 1933, the bomb program could not have started until after 1938, that is, until after the discovery of nuclear fission. By then Japan had joined the alliance of Nazi Germany and fascist Italy, and, likely, no refugee scientist would have helped Japan build an atomic bomb. Besides, from 1939 to 1945, Japan's resources would not have sufficed to enrich uranium and create the bomb. Japan could have not competed successfully with the

United States, even if a few physicists of Teller's caliber had joined their program.

There is, however, an alternate scenario, supposing that nuclear fission might have been discovered years earlier than it was. In fact, Fermi and his group at Rome University with all probability performed the fission of uranium but misinterpreted the results of their experiments in 1934. Also, Szilard dreamed of the nuclear chain reaction, and had he made a systematic search for the fissionable element—something he was contemplating—he could have hit upon uranium as early as 1934.

If nuclear fission had been discovered in 1934, it would have given the Japanese more time to enter a race for the atomic bomb. It is unlikely that the Western democracies would have made as much of an effort or would have devoted as many resources to a similar program during the second half of the 1930s—that is prior to becoming participants in the war—as they did during the war. However, had either Fermi or Szilard discovered nuclear fission in the mid-1930s, it is probable that Nazi Germany would have seized the opportunity to build an atomic bomb, and, again, it would not have been Japan to succeed in it first. But would it have mattered whether Germany or Japan built the bomb first? Either outcome would have led to unimaginable calamity. This is why Szilard suggested many years later that Fermi and he (Szilard) deserved the Nobel Peace Prize, not for something they had discovered, but for something they had missed.

# Chapter 5

# NO CALM BEFORE THE STORM

Had we not pursued the hydrogen bomb, there is a very real threat that we would now all be speaking Russian.

—EDWARD TELLER

[We] were absolutely persuaded that our work was vitally necessary. . . . [T]o maintain peace it is inevitable to make such terrible things.

—ANDREI D. SAKHAROV[1]

*From the time Teller joined the Manhattan Project to the time he finally succeeded in establishing a second weapons laboratory, his activities were dominated by the fight for the development of the hydrogen bomb and by finding the solution to create it. This was his most all-consuming struggle—although it would not be the last—and he accomplished his goals at no negligible expense to human relations. Sandwiched between the projects of the atomic bomb and the hydrogen bomb, he spent a few years at the University of Chicago, where he produced yet another considerable scientific contribution. His dedication to nuclear matters manifested itself in his groundbreaking activities for reactor safety.*

The *New York Times* reported on October 4, 1988, that in a speech on the previous day, the eighty-year-old Teller vented his mocking "anger" at the hydrogen bomb. He addressed the bomb as if speaking to a child: "All these years I never received a Father's Day card."[2] Teller's joke was startling because he used to be upset when people called him "the father of the hydrogen bomb." He was not eager to claim the dubious title; however, he was proud of having earned it.

The story of the (American) hydrogen bomb began with a conversation between Enrico Fermi and Edward Teller in 1941 at Columbia University.[3] The idea of thermonuclear reaction came to Fermi because he worried about unwanted consequences of an atomic bomb explosion. Such an explosion would provide extraordinarily high temperatures of many million degrees, the likes of which had never been reached under terrestrial conditions. He was concerned that at such high temperatures a fusion reaction might occur; that is, the fusion of two light atoms into a heavier one.[4] For example, two deuterium (D) atoms might combine into one helium atom:

$$D + D \rightarrow He + energy$$

The term "thermonuclear" refers to very hot conditions. The energy—in large amounts—would derive from the mass difference occurring in such a reaction. (Thermonuclear reactions produce the energy in the stars. This had been discussed at the Washington conference of theoretical physics in the spring of 1938.)

The idea of setting up a thermonuclear reaction under terrestrial conditions first came up when Nikolai Bukharin, the Russian communist leader, offered support for the necessary experiments to George Gamow. The politician attended a lecture on energy production in the stars, which was given by the physicist in Leningrad in 1932. Bukharin was so impressed by what he heard that he offered to put the entire electric power of the Moscow industrial district at Gamow's disposal to carry out such experiments, but Gamow declined the offer.[5]

When Fermi first mentioned the idea about the thermonuclear reaction that might be initiated by the fission explosion, Teller deliberated for a while and then came to the conclusion that the explosion would not ignite a thermonuclear reaction. When Teller moved to Chicago in the early summer of 1942, and had some spare time, he and Emil Konopinski decided to put the negative possibility of the thermonuclear reaction to paper. However, their careful considerations led to a different result; they found that it might be possible to start a thermonuclear reaction with a fission explosion. The question came up again during the long gathering of theoretical physicists at Berkeley in the summer of 1942. The meeting was convened by Oppenheimer in preparation for intensified efforts for the production of the first atomic bombs.

Eminent physicists such as Fermi, Teller, Oppenheimer, Hans Bethe, Felix Bloch, Robert Serber, and John Van Vleck were present at the Berkeley meeting. They believed that the theoretical problems of the fission bomb had been solved. The question again arose as to whether the fission explosion might initiate fusion and ignite the atmosphere and the oceans, thereby causing a calamity of global proportions. Oppenheimer found the possibility so dramatic that he wanted to discuss it immediately with Arthur Compton, who at the time was in charge of the entire project. Compton was away at a resort in Michigan, so Oppenheimer visited him there. For the first time the scientists faced the dilemma of producing a weapon that might blow up the planet. Compton contemplated that it might be "better to accept the slavery of the Nazis than to run a chance of drawing the final curtain on mankind!"[6]

Subsequent calculations proved that the thermonuclear reaction was not a real danger as a consequence of the atomic bombs. On the other hand, the possibility of creating a thermonuclear weapon—called a hydrogen bomb, or super bomb, or simply, the Super—loomed over the nuclear weapons project from 1942 onward. However, the initial optimism about the problems of the atomic bombs having been solved gave way soon to the realization that much more effort was needed to accomplish these challenges and that they needed to be done in a timely fashion. Therefore, only a handful of theoreticians were assigned to work on the Super at Los Alamos, and Teller was responsible for keeping the project alive. He could not do it full time, however, as he was involved in other aspects of the work.

When the Alamogordo test showed that the atomic bomb project was a success, Oppenheimer renewed his interest in the Super. After the Japanese surrender, however, he had no desire to create a yet more terrible weapon. When Oppenheimer resigned the directorship in October of 1945, at his recommendation, the former Stanford physics professor and reserve naval officer Norris Bradbury was named the new director. He was a Berkeley-trained physicist, a quiet and cautious man, who had been in charge of the assembly team for the Alamogordo test of the device for the explosion in July 1945. Bradbury's immediate concern was to preserve the Los Alamos Laboratory and find the most appropriate projects for it. He did not consider the hydrogen bomb to be one of their short-term goals. He appreciated Teller's abilities and asked him to take over the leadership of the depleted theoretical group. Teller set some conditions for his staying at Los Alamos and stipulated that he would stay if either of two conditions would be fulfilled. One was a dozen tests of improved atomic bombs annually and the alternative was vigorous work on the hydrogen bomb. Bradbury could guarantee neither, and Oppenheimer refused to help lobby for the necessary support for Los Alamos. Thus Teller decided to leave, but he promised to return for visits to Los Alamos to help with their work.

## CHICAGO

The choice between Los Alamos and academia was not easy for Teller. For the first time in his life he had some real dilemmas. Up until this point, external events charted his path, and wherever he and Mici went, they had only the illusion of creating a permanent life. Teller had the best of memories of their time in Washington. It was academia; he was in a creative period; and he was in the midst of friends and colleagues with whom he enjoyed interacting and doing science. And now he had an even better opportunity: he was invited to join the University of Chicago where Fermi was organizing a strong institution called the Institute of Nuclear Studies. At Chicago, Teller could be part of a world-class physics community, which would be comparable to the pre-Nazi German centers of physics.

At Fermi's institute, "there was this unbelievable congregation of people—probably almost as great an assembly of people as at Los Alamos itself."[7] The latter was an obvious exaggeration, but, according to the outstanding physicist Valentine Telegdi, "[i]t was a place where you could be proud to be the dumbest one."[8] Among the faculty, graduate students, and frequent visitors at the University of Chicago, such names could be found as Fermi, James Franck, Leo Szilard, Harold Urey, Cyril Smith, Maria Goeppert and her husband, Joseph Mayer, Herbert and Jean Anderson, John and Leona Marshall, Owen Chamberlain, Emilio Segrè, Harold Agnew, Richard Garwin, Valentine Telegdi, Chen Ning Yang, Tsung Dao Lee, Willard Libby, Nicolas Metropolis, Jack Steinberger, and Jerome I. Friedman. There were eleven Nobel laureates or future Nobel laureates among them. Steinberger called the faculty "absolutely extraordinary,"[9] and for Friedman, getting his graduate education at the Institute of Nuclear Studies was "an exhilarating experience."[10]

The young British physicist Freeman Dyson was very impressed by Teller when they first met. He noted, though, with some ambiguity, that during the golden years of physics at Chicago, "Fermi was king and Teller was his court jester."[11] He elaborated on this: "They sat side by side at

Edward Teller's parents. Left: Ilona Deutsch. Right: Max Teller. Courtesy of Wendy and Paul Teller.

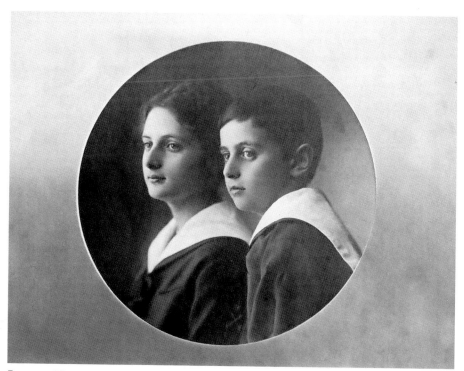

Emma and Edward Teller. Courtesy of Wendy and Paul Teller.

Left: Teller's high school graduation picture. Right: Teller's passport issued by the Budapest Police. Both courtesy of the late George Marx.

Edward Teller on the Pest bank of the Danube with a partial view of the Franz Joseph Bridge (today, Liberty Bridge). Courtesy of the late George Marx.

Left: Lev D. Landau in Moscow, 1937. Photo by and courtesy of the late David Shoenberg. Right: George Gamow with a cat. The handwritten note says, "Spinor or Scalar?" Courtesy of Igor Gamow, Boulder, CO.

Left: Hermann A. Jahn in the early 1930s. Courtesy of Michael Jahn and Margaret May, London. Right: Stephen Brunauer in navy uniform. Courtesy of Burtron Davis, Lexington, KY, and Dalma Hunyadi Brunauer.

Left: Enrico Fermi. Photo by Ed Westcott, courtesy of Oak Ridge National Laboratory. Right: Hans A. Bethe. Courtesy of Los Alamos National Laboratory.

Left: Maria Goeppert Mayer. Photo by Town Country Photographers, courtesy of the Special Collections Research Center, University of Chicago Library. Right: Arthur Koestler. Photo by Károly Forgács, courtesy of Petőfi Museum of Literature, Budapest.

Left: John von Neumann. Courtesy of Marina von Neumann Whitman, Ann Arbor, MI. Middle: Stanislaw M. Ulam. On the back of the photo is written "with love to my enemy, Yours, Stan." Courtesy of the late George Marx. Right: Leo Szilard in Cold Spring Harbor, 1953. Photo by and courtesy of Karl Maramorosch, Scarsdale, NY.

Left: Yulii B. Khariton. Courtesy of Alexey Semenov, Moscow. Right: Andrei D. Sakharov, 1955. Courtesy of Lyubov' Sakharova, Moscow.

Vitaly L. Ginzburg, 1940. Photo by Ivanov, courtesy of the Russian Archives of Documentary Films and Photographs, Krasnogorsk, Russia.

Lev Landau, George Gamow, and Edward Teller (on skis), and two of the Bohrs' children in Copenhagen. Courtesy of Igor Gamow, Boulder, CO.

Edward and Mici Teller relaxing. Courtesy of the late George Marx.

Some of the participants of the Manhattan Project at Los Alamos. Courtesy of Los Alamos National Laboratory.

American billboard toward the end of World War II. Photo by Ed Westcott, courtesy of Oak Ridge National Laboratory.

Left: Robert Oppenheimer receiving the US Medal of Merit from Secretary of War Robert P. Patterson in 1946. Right: The bomb over Nagasaki. Both photos courtesy of Los Alamos National Laboratory.

with the feeling of having done little or nothing."[21] Following his movements one can sense certain restlessness. There was a meeting at Los Alamos in the spring of 1946, a summer visiting professorship at Stanford University for the first half of the summer of 1946, and another stint at Los Alamos in the second half of the year. His and Mici's first child, Paul, was born in Los Alamos in 1943, and their second child, Susan Wendy (later known just as Wendy), arrived there in 1946. Mici stayed with the children in Los Alamos until Teller found a suitable house for them in Chicago. Their first home was uncomfortable, but very large, and it was nicknamed the Teller Hotel for the couple's hospitality and many visitors. Just as the Tellers stayed with the Fermi family when they first moved to Chicago in the fall of 1942, the Fermis stayed with them while Laura was house hunting in 1946.

There were conflicting opinions about Teller's research and teaching. Geoffrey Chew and Marvin L. Goldberger started out as Teller's doctoral students, but at one point Teller advised them to move to Fermi.[22] This could be interpreted as Teller's unwillingness to supervise them. However, Teller and Fermi had had a conversation in which Fermi expressed his intention to get a couple of students to direct in theoretical problems. Fermi had been involved in experimental work, but a new project caused a long break in his experiments—the building of a new accelerator—so he needed help. As it happened, Teller tried to expedite Fermi's request by steering two excellent students his way. He may have sensed, too, that Chew and Goldberger might have not been too happy working for him.

Another distinguished physicist, C. N. Yang, did complete his doctoral work with Teller at about the same time, but he required minimal supervision. Some time earlier, Teller and Konopinski communicated an unproven statement about the relationship between the angular momentum changes and the angular distribution of the products in nuclear reactions.[23] Yang produced a proof for this statement. He had arrived in the United States not long before, and one day he just walked into Teller's office and showed Teller his result. Teller suggested that he write it up immediately as a doctoral dissertation in theoretical physics. Yang left and returned soon with a three-page write-up. This was a little too short for a PhD dissertation, so Teller made some suggestions, and the paper grew to eleven pages. This seemed sufficient. Yang received his PhD degree and went on to a brilliant career, which included a Nobel Prize in Physics in 1957 shared with T. D. Lee, a fellow Chinese émigré and former University of Chicago doctoral student (under Fermi). Curiously, Teller referred to Lee as "my almost-student." He had helped Lee and Yang get

seminars; Fermi was the presiding figure and Teller was telling a
jokes. . . . If somebody asked a question, and Fermi answered it, tha
it; whereas Teller would give just some clever remarks; if you wan
straight answer, you'd go to Fermi."[12]

Telegdi provided another comparison: Fermi "was a very clear th
but not an exceptionally quick one—compared, say, to Teller o
Landau."[13] But in most comparisons between Teller and Fermi—wh
it was their performance at Los Alamos, their lecturing, or even
advising junior scientists—Fermi came out on top. Chen Ning (F
Yang mentioned in his Nobel biography that he wrote his doctoral
"under the guidance of Professor E. Teller," but for the same time p
"he came under the strong influence of Professor E. Fermi."[14] N
Goeppert Mayer, a favorite of Teller's, stated in her Nobel biograph
she "[owed] a great deal to very many discussions with Edward Telle
*in particular* with Enrico Fermi, who was always patient and hel
(italics added).[15]

Emilio Segrè remembered how actively Enrico Fermi and Teller
acted during their Chicago period after World War II. They had frequer
cussions, and "Fermi enjoyed Teller's unusual abundance of original
Fermi often developed these rapidly and far beyond the point reach
Teller—who sometimes missed the pleasure of nurturing his own
tures."[16] Another of Fermi's former doctoral students, Harold A
wrote, "Of all his colleagues of his vintage, Fermi's favorite for his in
tual ability was Edward Teller. He told me this and years later Laura
and his daughter confirmed this when I raised the question."[17]

Fermi was a great attraction and reason for Teller to be in Chi
But his feelings might have been ambiguous. He admired Fermi, b
was envious of him at the same time. In 1947, he wrote to Maria (
pert Mayer, referring to some trouble of his, "I do envy Enrico part
his brains but much more for his energy."[18] In another letter Teller
tioned some work on cosmic rays and the assistance he received from
itor by the name of Alfren. Fermi was also looking into the same n
and Teller noted: "You see how terrible this is: as soon as Enrico do
feel it is competition, when Alfren does it I feel it is help."[19] Later
lamented that he "did not start things rightly with Enrico."[20]

Chicago proved to be a short-lived interlude in Teller's life, one I
not quite enjoy for a variety of reasons. He complained about it to (
pert Mayer. He did not look back to Los Alamos with great satisf
either, saying, "I used to be so unhappy here," and when he contem
his return there, he expressed his apprehension: "I shall again come

accepted by the University of Chicago in spite of their lacking some of the prerequisites for admission.

John Firor, former director of the National Center for Atmospheric Research (NCAR) in Boulder, Colorado, was yet another graduate student at Chicago. He was an early voice in directing attention to the importance of human interference with the environment. Looking back to his student days, he remembered Teller as being very approachable to his students but also a very busy man. For example, Teller might be following a student's work on the blackboard while talking simultaneously with a uniformed officer. Firor found assistance from Teller especially helpful in an indirect way: Goeppert Mayer used to advise her students when they worked on a problem to ask Teller to guess the answer they should be expecting. Goeppert Mayer told them that they could save time because Teller had "such a great physical instinct" that his guess would be within a few percent of the solution. The students thus would have a good reference point in working out the problem. Teller welcomed the students posing a great variety of questions to him.[24]

As for Teller's research during his Chicago period, again, according to Goldberger, he did not carry out research at all; all he did was "[keep up] with physics and [interrelate] strongly with people and [discuss] problems with them."[25] This was quite typical of Teller. Still, the records show that his Chicago period was very fertile, and he thrived on actively interacting with others. He also enjoyed his interactions with scientists at Los Alamos, where he returned from time to time. During the five years between 1947 and 1951 he published fifteen research papers of which he was sole author of only two. His co-authors included Fermi (twice), Weisskopf, Maurice Goldhaber, W. G. McMillan (thrice), Konopinski, Feynman, Metropolis, Richtmyer, Maria Goeppert Mayer (twice), and Frederick de Hoffmann. Most of these names are well known even today.

Teller's participation in research projects often started with someone making an interesting observation or reading about some new finding in the literature and soon bumping into difficulties while looking for an explanation. Then the person, a graduate student or colleague, would turn to Teller, and a joint work and paper would follow. A similar process occurred on one occasion when Goldhaber read an intriguing paper on nuclear reactions. He thought of an analogy and then sought out Teller. Together they published a paper about nuclear giant resonance, which spurred research in the ensuing decades.[26] This phenomenon concerns the behavior of the principal constituents of the atomic nucleus: the protons and the neutrons. The collection of the protons has a tendency to move in one direction, and the collection of neutrons in the opposite direction, but the strong attractive forces between them generate a motion in the reverse direction. The result is a rapid resonance-like motion back and forth. Goldhaber and Teller proposed a model for the interpretation of the phenomenon, and forty years later, a historical review summarized the developments that their pioneering work had generated.[27]

It was in keeping with his emphasis on the importance of educating a broad audience that Teller offered a general physics course at Chicago and continued this course for the next forty years. Harold Agnew found Teller "a wonderful lecturer for two types of individuals: those who knew absolutely nothing—in basic freshman courses he was a superb lecturer; or for those who knew everything. But for the in-between individual, it was very rough."[28] The Nobel laureate physicist Val L. Fitch attended Teller's lectures at Los Alamos in 1945. When the war was over, and the group had not yet dispersed from Los Alamos, the academics started a small university program. Fitch signed up for Teller's course but "almost immediately took a dislike to his lecturing style; which was always directed toward a particular person in the audience, the son of a well-known physicist. He ignored all the others."[29] At one point Teller stopped giving lectures and had somebody substitute for him. Geoffrey Chew attended both Fermi's and Teller's lectures at Los Alamos as well as at Chicago, and found Fermi "a far more effective teacher than was Teller."[30]

Teller's nephew Janos Kirz attended several of his courses a decade or so later at Berkeley.[31] Though not necessarily an objective source, he was not alone in noting Teller's charisma and his ability to grab the attention of his students and always find the right level for his presentations. Janos attended his introductory physics course for nonscience majors and his graduate course for nuclear engineers; and according to him, the students found the lectures absorbing and fascinating. (These were the two levels to which Agnew referred in his evaluation.) Janos also attended Teller's lectures that were recorded for public television. His uncle asked him for feedback after each presentation; obviously, those lectures were carefully prepared.

By every indication, Teller cared a great deal about education. However, he was unable to fulfill his desire to teach in exactly the way he wanted because of his celebrity status. Ralph Moir, one of his future colleagues at Livermore, was an undergraduate student at Berkeley in the early 1960s. He signed up for Teller's graduate course Engineering with Nuclear Explosives in 1961. Only six graduate students and Moir had enrolled for the class, so he was astonished to find that the lecture was standing-room-only as he entered the auditorium. It turned out that the crowd was all media people with still and movie cameras. From the second lecture on, the seven students in the class were no longer disturbed.[32]

## LASTING FRIENDSHIP

Teller's friendship with Maria Goeppert Mayer had a long-lasting impact on him, and their most intensive exchanges happened during his Chicago period. Teller respected her for her achievements in physics, and they interacted in a most fruitful way. They wrote a seminal paper together on the origin of the elements in which they came to the conclusion that the light and heavy elements were produced through different processes. They discussed the origin of the heavy elements in detail and assumed that they were formed by a fission process from a nuclear fluid, which was rich in neutrons.[33] Eugene Wigner praised this work and noted that "the insights gained in the course of this work contributed significantly" to Goeppert Mayer's shell model of the atomic nucleus.[34] Goeppert Mayer's pioneering works earned her a share of the 1963 Nobel Prize in Physics. In 1983, William A. Fowler was awarded the Nobel Prize in Physics for his related work on the formation of lighter elements. Understanding the formation of elements was another area of research in which Teller could have continued but did not.

Teller's interactions with Goeppert Mayer were especially important to him on a human level, as he found in her a soul mate. She became his unique confidante to whom he could pour out his innermost thoughts, deliberations, hesitations, and doubts in himself. He almost seemed to be addicted to these interactions. He pleaded with her, "Please write—even if there is nothing to write about. I need it."[35] In February 1949 he wrote, "I should like to see you badly and talk with you a lot."[36] Later in the summer of 1949, again, "I am confused and want to see you badly."[37] By every indication, their interactions remained at the platonic level, though Teller to some degree seemed infatuated with her. Teller, who could not help but play to the galleries when people were around, found solace in showing himself barefaced to her with self-tormenting honesty. She was the one person before whom he felt he did not have to wear a mask or even code his words. On occasions when he was communicating something that was highly embarrassing, he asked her to destroy his letter.[38] She never did, and the letters, preserved by her, helped him write his *Memoirs*.

The letters had their own evolution. They started as letters about physics and about their joint research then developed into personal testimonies

## MARIA GOEPPERT MAYER (1906–1972)

Maria Goeppert Mayer was born Marie Göppert in Kattowitz, then Germany, now Poland (Katowice), and moved with her family in 1910 to Göttingen, where her father was appointed professor of pediatrics. She studied at Göttingen University from 1924, first majoring in mathematics, then in physics, and earned her doctorate under the supervision of Max Born. She had an engaging personality that invited trust, and Born found her lovely and industrious. Even as a young graduate student she won Born's confidence and provided spiritual support to him when he needed it.

After she married the distinguished American physical chemist Joseph Mayer, they moved to the United States, where his professional career took her to various universities. It was only after World War II at the University of Chicago that she began to gain recognition for her own contributions. She changed her specialty to nuclear physics during her stay in Chicago, primarily under the influence of Enrico Fermi and Edward Teller. Her intimate correspondence with Teller helped ease his isolation during a critical period of his life.

Her career trajectory is a telling example of the difficulties a woman scientist had to suffer in the world of science not such a long time ago. On top of everything, her theories were unusual and innovative, making it especially difficult to get them recognized. Her shell model of the atomic nucleus went against the accepted mainstream idea about the structure of the nucleus that was based on Niels Bohr's liquid drop model. She came up with so-called magic numbers for the amounts of protons and neutrons that she found characteristic of especially abundant isotopes.

Goeppert Mayer was awarded the Nobel Prize in Physics in 1963. Half of that year's prize went to Eugene P. Wigner and the other half was shared by Goeppert Mayer and J. Hans D. Jensen "for their discoveries concerning nuclear shell structure." Teller lamented that even after her Nobel Prize she did not receive much further recognition. The sad truth is that her deteriorating health prevented her from working much during the last years of her life. She died in San Diego, California.

without ever completely abandoning the topic of physics. During the war they were infrequent; there was an especially intensive exchange during the immediate post-war years; and in the 1950s they dwindled, but Teller's interest in her never completely vanished. Teller showed her his humorous side as much as his tendency for melancholy. Often his humor manifested itself in his applying terms of physics to everyday situations and human affairs. This kind of humor stems from what Koestler called bisociation. Whereas the connection in simple thought association is between thoughts on the same plane, bisociation refers to the connection of thoughts from different planes. According to Koestler, this is an important ingredient of both laughter and scientific discovery.[39] Their correspondence was in English with one exception, when Teller wrote her a letter in German in response to her having written to him in German—for no discernable reason.[40] Hardly any of Goeppert Mayer's letters to Teller remain.

Although they had met previously, their interactions really only began during the Manhattan Project. At some point, Teller suggested that Goeppert Mayer and her students at Columbia University (where she was teaching at the time) could do some calculations for him, as long as he didn't tell her their purpose (she and her students did not have security clearance). Teller went to New York to explain the job to her, and she did not ask any questions beyond what Teller volunteered. She and her students made an excellent contribution to the war efforts.

Teller's letters became increasingly personal, and in 1946 he gave Maria a detailed account of his and Mici's life in Los Alamos, including the great event of the birth of their first child, Paul. He mentioned a trip to New York that he would find painful, since she would not be there: Goeppert Mayer and her husband had moved to Chicago in the meantime. Teller mentioned to her that he would like to discuss his thoughts on the topics of the origin of the elements and of superconductivity when they met in Chicago. This sounds like an incentive to make sure that she would be willing to meet with him. An often-recurring topic in his letters was whether Teller would ever be able to return to physics. Teller concluded with a complaint: when he called her the day before, the conversation "sounded like a court-martial."[41]

Teller expressed his apprehension in another letter from the same time that Maria might be angry at him on account of what he had written to her. This was another recurring topic: his fear of her getting mad at him for some trivial reason. These high-school-like utterances were then mixed with serious scientific considerations about ongoing research and political topics such as the situation in postwar Germany. Apparently by this time she had told Teller about the magic numbers she had discovered in her studies of the shell structure of the atomic nucleus. His comments on and genuine interest in the topic must have greatly encouraged her at a time when many of her peers did not take them seriously. Yet another recurring theme was Teller's dilemma as to whether to return to Chicago from Los

Alamos. Teller almost begged her to write him: "[I]t would be good of you to write to me. The quantity does not matter so much at present but the quality might be different. Even very little would help."[42]

Teller wrote a lot about his son and about playing with him and teaching him. He wrote with self-deprecating humor that, as a physicist, he would be remembered "as . . . moderately lousy, but as a father he was hot stuff."[43] In his next letter Teller discussed their joint paper in-the-making about the origin of the elements; he mentioned Gamow's alpha–beta–gamma paper (see chapter 3), for which he did not care too much, and referred to the success in the production of liquid helium, which opened up the possibility of many further studies. He ended the letter pleading for Maria's letters.[44]

Goeppert Mayer must not have obliged his wish because Teller complained about her not writing in his next letter. He discussed their joint work; mentioned their work on a "new machine," a very probable and rare reference to the super bomb; and further discussed his dilemma about returning to Chicago or staying at Los Alamos. He felt that he was needed more in Los Alamos than in Chicago. He wrote, "When I am asked in Chicago, I am asked by reason of politeness. If I am asked here [Los Alamos], I am asked by reason of necessity." He found it frustrating being away from Maria and did not want to spend the coming year as he spent the previous one.[45] Teller implied on more than one occasion that Maria might be the main, if not the only reason why he would return to Chicago. He wrote to Maria in late January 1950, "There is really precisely one reason why I should want to come back. That reason you know. In the end you probably would not approve of it, if I decided on that reason."[46] Gradually, however, Teller's infatuation seemed to have diminished over the years, and his letters to her from the early 1950s were more about facts of life and science than about personal feelings.[47] This was also the time of Teller's intense involvement with the new laboratory at Livermore.

Teller clearly felt an emotional closeness to Goeppert Mayer, but there is no indication that he acted on these feelings. We cannot know whether their interactions meant the same for Goeppert Mayer as they did for him. What little can be learned about her letters from Teller's responses was that her interest was mostly scientific. She would also tease Teller, but whether there was any personal need on her part for their interaction remains unknown. Her close friendship, years before, with the much older Max Born back in Göttingen in the late 1920s, was imbalanced; Born seemed to need it a great deal and confided in her about his marital problems, and she lent a sympathetic ear.[48]

Teller retained his interest in Goeppert Mayer to the end of her life. He was genuinely delighted when he heard about her Nobel Prize in the fall of 1963. Earlier that year he nominated her for the much lesser, but still important Einstein Award.[49] In his nomination he mentioned that she was being proposed by others for the Nobel Prize. Teller described to the committee Goeppert Mayer's prize-worthy achievements that would then be the basis for her Nobel Prize later in the same year.[50] Teller made a statement about the scarce recognition she had received "because she [was] not one of the closely knit group of theoretical physicists whose opinions dominate the style of modern physics, appointments to the universities, and the political opinions expressed by the majority of the scientists."[51] This statement undoubtedly reflected Teller's personal experience with many of his peers in theoretical physics.

## MICI

Teller and his wife, Mici, seemed to form a perfect pair. By the time my wife and I met them, Mici had been in poor health for decades. Still, she was part of the group even if mostly silent. But this might have been anybody's role in Teller's presence. When she found him unfriendly, she had no trouble telling him so. My impression from various conversations with a number of people has been that she had full control of the Teller household from which he was often absent. She was a mathematician by training but may have not been very strong in it; nonetheless, she did not shy away from holding a job.

At the time of Teller's work at the Metallurgical Laboratory during their pre-Los Alamos period, Mici was already preparing for her position in the computer division at Los Alamos. Teller asked physicist Leona Woods to teach Mici "how to round off numbers, how to compute the right number of significant figures, how to estimate statistical accuracy."[52] She worked at Los Alamos and was given a certificate for it, signed by Secretary of War Henry Stimson after the war. She valued this document and kept it with their marriage certificate and naturalization papers.[53]

Mici excelled in dealing with family finances. Woods noted that she "[played] the stock market, making money where the pros barely [scraped] by."[54] Not even Mici's later illness would prevent her continued interest in the stock market or from augmenting their income with her investments.[55] She was knowledgeable about money, as opposed to Teller, who showed no interest in family finances. Their daughter learned from Mici everything about investing, buying a house, saving money, and filing income tax papers.[56]

Mici underwent surgery for a brain tumor in 1983. After the operation she could not drive. Through the local Hungarian community the Tellers found a helper whom she could call when she had to go somewhere. This helper, Judit Stur, was an exceptional woman, a recent arrival from Hungary, around fifty-five years of age, with a doctorate in chemistry and, in addition, an MD, though she never practiced medicine. Stur had worked at Szeged University instructing medical students in chemistry and she found it a challenge to obtain a medical degree. She did not work for the Tellers long; she attended a university course in San Jose, got a BSR degree, and became a nurse. After a few years she returned to Hungary. Stur speaks about the Tellers with devotion even after a quarter of a century. She was unhappy with how Edward Teller was treated in the hospital after a surgery. When he got home, she helped him with what she thought should have been the hospital's duty; the otherwise thrifty Mici surprised her with a big bonus.

Teller did not do much around the house, often neglecting even his own clothing. Mici tried to remedy the situation and on the occasion of her birthday in 1984, she asked him to go with her to an elegant shop of menswear, and in lieu of a present for her, he should buy three suits for *himself*. He complied.[57]

Teller did not write about the spirituality of his partnership with Mici. It may have existed, but he may have felt it too private to bring out into the open. And there are no sources available to attest to their relationship by any objective measure. By the time their children grew to an age at which they could appreciate their parents' interactions, Mici was already engulfed in an illness that would be difficult to describe even if it were proper to do so. According to every indication Edward was a protective and caring husband during these long, difficult years, just as Mici was a devoted and supportive wife, who kept a loving home and took care of all the essentials. In addition, she was neither ignorant of, nor uninterested in, his political and scientific activities.

## ENTERING POLITICS

Teller and Mici, both Democrats, became active in politics after World War II, during their Chicago period. The last months at Los Alamos changed Teller's attitude toward participation in public affairs. Mici had already gotten involved in politics, supporting the challenger in a 1948 congressional election. The incumbent was a member of the House Committee on Un-American Activities (HUAC). The committee existed from 1938 to 1975, and from the start it aimed at uncovering communist conspiracies against American democracy. Senator Joseph McCarthy would use many of the committee's tactics during his witch hunts of the early 1950s. While Mici was out canvassing votes, Teller was at home, babysitting. Her efforts were worthwhile: the challenger won the election.

Teller found it more effective to be part of political struggles at a different level. World War II in general and the atomic bombs in particular changed the status of scientists, including that of Teller's, as voices to be heard on political questions. He gradually got used to being asked about his opinion concerning various issues. He was invited to testify before a congressional committee at the suggestion of Stephen Brunauer, who was then a naval official. The testimony took place on February 1, 1946, before the US Senate's Special Committee on Atomic Energy, chaired by Brien McMahon, a Democrat from Connecticut.

It was impressive how deftly Teller on this first occasion expounded on an array of science-related politics—all in areas in which he would be waging his crusades in the years to come.[58] His dedication to these issues justified Fermi's labeling him "a monomaniac who has several manias."[59] The principal items were defense (with an emphasis on testing), nuclear energy and safety, the applications of isotopes, education, and the elimination of secrecy.

Teller stressed that a perfect defense did not exist, but it was better to have an unsatisfactory defense than no defense. He advocated the importance of defense against nuclear weapons at a time when the United States still enjoyed full monopoly of the atomic bombs. The general opinion was that the Soviet Union would not possess atomic bombs for many years to come.

Teller made a point that was to remain one of his permanent warnings: that the development of offensive weapons always preceded that of their defensive counterparts. He anticipated that more powerful and less expensive nuclear weapons "will dwarf the Hiroshima bomb in the same way that the [atomic] bomb has dwarfed high explosives." Further, he warned that the American military advantage might change quickly, and "[u]nless the possibility of a future war can be eliminated, we are going to live in a world in which safety no longer exists."[60] He saw a common interest in weapons tests between the military and the scientists; for the military, in improving weapons, and for the scientists, in gathering new knowledge about nature.

In the area of peaceful use of nuclear power, Teller warned that profound development should not be expected too soon. He hoped that energy production would become possible, but he excluded the utilization of small atomic machines, like the ones used to power cars and airplanes because of the necessity of heavy shielding. He found it more practicable to power large-scale vessels through nuclear technology. His insistence on efficient shielding already anticipated his growing concern about safety in the use of nuclear power.

He emphasized making use of isotopes in science and in particular in medicine at a time when such utilization had made hardly any progress. Especially noteworthy is that he mentioned the distribution of the isotopes not only within the United States but also internationally.[61] In the years to come this consideration would become a subject of controversy between his future ally Lewis Strauss and their common adversary-to-be Robert Oppenheimer. Oppenheimer did not see the harm in providing isotopes internationally, whereas Lewis was set on preventing their dissemination because he feared that the enemies of the United States might benefit from their possession.

Another prescient suggestion by Teller was his emphasis on education of the public to raise their science literacy. He expected not only a better informed constituency from such education but an increased interest on the part of scientists in participating in weapons development. He pointed out that money alone would not lure scientists back to weapons work. He called unnecessary secrecy a barrier to getting scientists to work on nuclear arms, as such secrecy would further isolate them from the rest of the scientific community. At the same time, he fully supported keeping secret various technical details and industrial know-how.

The members of the congressional committee must have been impressed by Teller's testimony on that February day. He had a persuasive style and his presentation was factual. He spoke with his characteristic accent, but by this time, a foreign accent was not a disturbing factor; on the contrary, it enhanced a scientist's persona. McMahon and his colleagues valued the expertise of their witness. They were involved in an important debate concerning the future of political and military jurisdiction over all matters nuclear.

In order to appreciate the significance of Teller's congressional testimony, we have to go back in time. Drafting of a bill on nuclear energy had commenced before the end of the war, and the bill was introduced late September 1945. It was to be known as the May–Johnson bill after its sponsors, Congressman Andrew May and Senator Edwin Johnson, and it had the full backing of the military and, specifically, General Groves.[62] This bill would have maintained full military control of atomic energy and atomic weapons development. It enjoyed the support of some of the scientists, most notably that of Enrico Fermi and Robert Oppenheimer; whereas others, especially the new Federation of Atomic Scientists, opposed it. The federation soon launched its periodical the *Bulletin of the Atomic Scientists of Chicago* ("of Chicago" was eventually dropped from its title). Teller published an article in one of the first issues welcoming the attempts at international control of atomic energy.[63] However, before any practical step could have been taken toward international sharing of secrets, the Americans had to sort out how they would be handling atomic matters.

First it seemed that the May–Johnson bill would have a smooth journey through the two houses of Congress, but most scientists resented that it would preserve wartime conditions for their future work. Matters were not helped by the military's attempt to hasten the bill through Congress. A conspicuous omission was the failure to invite relevant scientists to testify. The whole process was slowed down when a special Senate Committee on Atomic Energy was created, of which McMahon eventually became its chairman. The senator initiated hearings, which is when Teller, among others, got his invitation. In time, McMahon drew up his own bill aimed at transferring jurisdiction to civilian hands.

President Truman first hesitated to take a stand on either of the two bills, but eventually he came out in favor of the McMahon bill. It was during all the controversy that Teller gave his testimony. His words strengthened McMahon's position, which was, to use a more contemporary term, less "hawkish" than the May–Johnson bill. Both the Senate and the House passed the McMahon bill, and President Truman signed the Atomic Energy Act into law in the summer of 1946.

From the time of Teller's testimony, a good working relationship was forged between Teller and McMahon. It was Teller's entry into the corridors of power, and it was now possible for him to maintain his influence over the coming decades, even with no powerful political or administrative position. McMahon was an idealist who had initiated a movement for young Americans to disseminate the ideals of democracy internationally; it was a forerunner of the Peace Corps. His contribution to the legislation on atomic energy was probably his most remarkable achievement. His service in the US Senate was cut short by his early death at the age of forty-nine, in 1952.

On January 1, 1947, the US Atomic Energy Commission (AEC) came into existence.[64] Its five commissioners held equal power; the chairman's

Senator Brien McMahon on US stamp.

only distinction was being the spokesman of the commission. David E. Lilienthal, a Harvard-educated lawyer, became the first chairman. His parents were Jewish–Hungarian immigrants, and he had risen to prominence as head of the Tennessee Valley Authority under the presidency of Franklin D. Roosevelt. This assignment gave him experience in management as well as in energy affairs.

The birthplace of Lilienthal's parents became an issue at the congressional hearing of confirmation in January 1947. A senator tried to link the birthplace—at that time under Russian domination—with Lilienthal's ostensibly leftist political views.[65] This might seem an insignificant detail, but Lilienthal's response elevated it into a significant episode. It is worth quoting what he said, because it will help us understand the change in the American political scene from 1947, when Lilienthal's stand won enthusiastic support, to the McCarthy era in the early 1950s, when it would not have worked.

Lilienthal pointed out the importance of integrity of the individual as the fundamental principle of democracy. He counted among the tenets of democracy the protection of civil liberties "and a repugnance to anyone who would steal from a human being that which is most precious to him, his good name, by imputing things to him, by innuendo, or by insinuation."[66]

The attack on Lilienthal, even though unsuccessful, pointed to the precariousness of Teller's situation. It was not only that his parents had also been born in territories now under Russian control, but that they continued living there along with two other members of his immediate family—his sister and his nephew. Teller was not the subject of any appointment that would have necessitated congressional confirmation, but he must have felt as if he were being forewarned by Lilienthal's experience.

The commissioners of the AEC were usually prominent representatives of business—often from the energy sector—and politics rather than of science. Two other bodies augmented the AEC: the General Advisory Committee (GAC) and the Military Liaison Committee. The military committee ensured that the armed forces exercised proper oversight and control of any activities that concerned national defense. Under the cold war conditions, the bulk of the commission's activities comprised weapons development and production. The first chairman of the GAC was Robert Oppenheimer, who stayed in this influential position from its beginning until 1952.

Although the new director of Los Alamos made it clear to Teller that there were no plans to undertake vigorous efforts to develop the Super, some work continued in this direction. A meeting was convened in 1946, between April 17 and 23, to review the situation and chart further work.[67] Only five scientists among those who stayed at Los Alamos worked, at least in part, on the Super, but others, who had been involved though had since left, were also invited.[68] Their combined efforts had been going in two directions. One was research aimed at using a powerful fission explosion to ignite a small amount of fusion fuel to induce a thermonuclear reaction. This was called the booster project, referring to the fission explosion boosting the thermonuclear event. The other direction was to use a small fission explosion to ignite a large amount of fusion fuel and cause a powerful thermonuclear reaction. This was meant to become the super bomb.

No straightforward solution for the hydrogen bomb was forthcoming at this meeting, which was fortunate in retrospect, because one of its attendees, Klaus Fuchs, was later unmasked as a Soviet spy. He spied during the Manhattan Project and transmitted crucial information about the atomic bomb to the Soviets, who later followed up on it. He could not have transmitted much information about American progress in producing the hydrogen bomb, because not that much had been achieved. Of course, he must have transmitted information regarding whatever work, however scant, had been going on for developing the hydrogen bomb. There was one piece of information, however, that the American scientists did not seem to pay much attention to, but in retrospect it gained importance—the concept of radiation implosion.[69] As it was probably discussed at the 1946 meeting, it is probable that Fuchs conveyed this piece of information along with everything else to the Soviet Union. Radiation implosion was to become the pivotal idea for the workable hydrogen bomb in the spring of 1951.

The Soviet scientists, who had been compelled by their superiors to follow rigorously the American solutions for the atomic bombs, could now rely more upon their own initiatives for the Soviet thermonuclear weapons. The czar of Soviet nuclear weaponry, Igor V. Kurchatov, had charged Yulii B. Khariton with investigating the possibility of a thermonuclear bomb. A small collective prepared a report by December 1945, and in June 1946, a decision was made to initiate practical work on thermonuclear weapons in the Soviet Union.[70] Later the Soviets and then the Russians insisted on their independent invention of the Soviet hydrogen bomb. However, if Fuchs transmitted the information about radiation implosion, the Soviets may have paid more attention to it earlier than their American counterparts and benefited from it.

## YULII B. KHARITON (1904–1996)

Yulii Khariton was born into a Jewish family in St. Petersburg; his parents divorced when he was a child. His father graduated from law school, had a career as a journalist, and was exiled from the Soviet Union in 1922. His mother was an actress and left Russia in 1910. First she lived in Germany, then in 1933 she and her second husband moved to Tel Aviv. When she left the family, Yulii was six years old, and his father employed an Estonian woman as governess and housekeeper. She served as surrogate mother for Yulii and his two half-sisters and taught him to speak German fluently. By the time his father was exiled, Khariton was living an independent life. It was a miracle, and a testimony to Khariton's indispensable service, that, given this background, he was employed in a job that required the highest security clearance in the Soviet Union.

He graduated from high school in 1919 and started the Petrograd Polytechnic in 1920. He chose physics for his major under Abram Ioffe's influence, completing his studies in 1925. He had started working in 1917, began his scientific career at the Physical-Technical Institute in 1921, and earned his doctorate there. In 1931, he moved to the Institute of Chemical Physics, where he worked as head of the division of explosives. He investigated the oxidation of phosphorus at Nikolai Semenov's suggestion. Khariton's experiments eventually led to the discovery of branched chain reactions for whose interpretation Semenov was awarded the Nobel Prize in Chemistry in 1956.

The young Khariton was sent to study with Ernest Rutherford in Cambridge, where he earned a second doctorate. After two years in Cambridge, between 1926 and 1928, Khariton returned to Leningrad. Semenov in the meantime enhanced his research activities, and Khariton became head of a growing laboratory engaged in studying explosives in Semenov's institute.

When Khariton and another leading physicist, Yakov B. Zel'dovich, learned about the fission of uranium with slow neutrons, they realized that this was another occurrence of branched chain reactions. They recognized the possibility of creating powerful explosives on the basis of such reactions. Khariton and Zel'dovich did important work in the field of nuclear chain reactions. Eventually the Soviet Union wanted to enhance such activities within the framework of the Soviet Uranium Commission. The initial German military successes upon invading the Soviet Union in 1941, however, forced the Soviet leadership to close down uranium work for a while, and Khariton was delegated to work on traditional explosives.

Well before the end of World War II the work on nuclear weapons resumed. A secret installation was established, Arzamas-16—the analog of the Los Alamos Laboratory—in the town of Sarov. Khariton moved to the secret installation in 1946 and was made its scientific director. He was elected to the Soviet Academy of Sciences as corresponding member in 1946 and as full member in 1953. Khariton's wife, Maria, was a home-maker; their only child, a daughter, married Nikolai Semenov's son.

As a result of the work of Khariton and his associates, the world learned of the successful creation of the Soviet atomic and then hydrogen bombs. Khariton served for an unprecedented long period of time in his position, but for decades his name was hardly known outside a small circle of specialists.

After the war, conflicting views developed in the United States as to whether and to what extent to involve the international community in handling the issue of nuclear arms of heretofore unprecedented power. Teller was among those who advocated the necessity of establishing a world government to avoid future wars. Oppenheimer contributed to the proposal known as the Acheson–Lilienthal plan, which envisioned inter-national cooperation in nuclear matters. However, when the United States presented the plan to the United Nations, the Soviet Union rejected it. It has been argued that the American representative, Bernard Baruch, made the plan less attractive to the Soviets than Dean Acheson and David Lilien-thal had intended. However, it is doubtful whether Stalin would have agreed to any international cooperation. He was determined to create his own arsenal of nuclear weapons.

Teller's views on world government seemed rather vague even though he stressed the necessity of clarity in formulating its aims. He wrote, "We need a world government which makes war between the participating nations *technically* impossible" (italics added).[71] We can discern that as early as 1947, Teller already expected the solution of political problems by technical means. Another point in his thesis that he made numerous times over the years was that even if the world government would be formed initially without Russia, eventually it might be possible to induce her to join. His main message was that scientists in the United States should be relentless in continuing their efforts to perfect and enhance their nuclear weapons. He did not exclude the possibility that if the world government could not be established by agreement, it might be "after further bloodshed."[72]

In 1947, however, the United States was not yet prepared to adopt quite the kind of policy Teller would have preferred. The 1946 Los Alamos meeting called for only moderate efforts in producing the hydrogen bomb, lest the fission weapons suffer from disproportionate allocation of the limited resources. It was stated that an intensified effort could "be taken only as part of the *highest national policy*" (italics added).[73] So the Super did not receive the backing through 1949 that Teller had hoped for. In a characteristic "Tellerism," he found a way to express some optimism. When asked about recent progress, he said, "We still don't know if the Super can be built, but now we don't know it on much better grounds."[74]

By the time Teller said this in the spring of 1950, the situation had drastically changed. On January 31, 1950, President Truman made his decision and publicly instructed the Atomic Energy Commission to continue research on all atomic weapons, including the hydrogen bomb.


## REACTOR SAFETY

It was the task of the Atomic Energy Commission to address all issues related to the safety of nuclear reactors. Immediately after the war, there began a frenetic activity of building new reactors, which brought up the important question of safety. Soon after the General Advisory Committee had been formed, a subcommittee was set up to deal with the safety of nuclear reactors. This aspect of nuclear matters became one of Teller's long-term concerns. He, Richard Feynman, John A. Wheeler, and the chemist Joseph Kennedy became charter members of the subcommittee. Feynman did not stay long, but others were invited to represent the areas of chemical engineering, geophysics, meteorology, and public health and sanitary engineering. At their first session, the charter members elected Teller to be chairman of the Reactor Safeguard Committee, or RSC, as it became known.

He conducted his duties with the same zeal as he did everything else. He was dedicated to nuclear energy and he understood that it could gain widespread acceptance only if it could be regarded as safe and secure. He elevated the safety considerations to an unprecedented high level. This was fortunate not only because this was a heretofore unknown kind of energy but also because from the beginning it was surrounded with some uneasiness by experts as well as the general public. Nuclear energy was invisible; it had no color, taste, smell, or sound yet was deadly if let out of control. The safety measures had to address all imaginable hazard scenarios, though there had been minimal experience with actual reactor operations. Besides, both the possibilities of accidents and deliberate sabotage had to be taken into account.

In most industries, the guidelines of safety are worked out as experience accumulates from accidents. Teller and his colleagues knew that this could not be the approach for them because even one major accident might have signaled the end of this new technology. They decided to identify in advance all possible sources of accidents and set up the necessary safeguards to avoid them. Such a task seemed impossible, but Teller and his committee were ready to face the challenge. The designers of the reactors were in the process of learning what they had to do while they were already doing it. Teller and the committee subjected them to the most onerous examinations. It proved to be particularly useful to ask the designers to submit worst-case accident scenarios along with their plans for the reactors they intended to build. The designers were then asked to introduce the necessary measures in their design to exclude the possibility of those accidents.

By the time the RSC was formed in 1948, there was already some experience with nuclear reactors.[75] Three reactors had been operating in Hanford, Washington, for years, and no accident had occurred. On the other hand, there had been two accidents at Los Alamos after the war. Two people were killed and more injured, but the circumstances of these accidents occurred during the handling of bomb material rather than working with reactors. Teller recognized that reactors carried the risk of accidents that might release even a fraction of the large amounts of the accumulated radioactive material. Such material in the atmosphere would spread fast, and the population downwind from the reactor would be exposed to harmful, possibly lethal radioactivity.

One of the major safety concerns was the proper distance from which reactors could be operated in populated areas. These were not wartime emergency issues but long-term projects that would produce energy, provide fission and fusion fuels, and conduct vital research. Highly qualified engineers and researchers would have to be attracted to work there, so the distance that they had to travel daily from their homes was not a negligible consideration. The criteria of the requirements for selecting reactor locations needed to be formulated. The primary considerations of the Teller committee required that the AEC had complete control over the area surrounding the reactor; that the population within a certain vicinity of the reactor did not exceed ten thousand; and that no installations of importance for national defense operated in the region. Of course, the parameters of these criteria varied, depending on the power of the reactor.

During the course of his studies, Teller had had some experience with the possibility of accidental criticality during the Manhattan Project. Accidental criticality is when a nuclear chain reaction occurs spontaneously in the fissile material and releases radiation into the environment. The RSC interacted closely with the scientific directors and other officials of the existing reactor plants and those under design and construction. The AEC and the reactor designers and operators found out soon enough that Teller and his committee took their mandate seriously. In some cases, the laboratories operating or developing reactors were instructed to redesign them to meet the stringent safety criteria of the committee.[76] Tests were carried out to examine safety considerations. The concept of maximum credible accident was introduced, and the consequences of possible accidents had to be evaluated by the designers of the reactor because they were in the best position to do so.

In spite of precautions, accidents happened. One occurred at the Argonne National Laboratory near Chicago, and in this case the maximum credible accident had been grossly underestimated.[77] There were no fatalities, but four persons became sick from radiation. The cause was human error, but a truly safe design would have prevented the possibility of any human error causing such consequences. The accident occurred during tests for a new reactor to be used in a nuclear-propelled aircraft. The fuel was light, but the shielding would be heavy; furthermore, a crash of such an aircraft in a populated area would have been a major disaster. The program was soon terminated. Likewise, plans to construct nuclear locomotives were abandoned. Teller characterized them as "a most ingenious solution of the question how to combine minimum utility with maximum danger."[78]

Teller and the committee worked assiduously on developing nuclear power in the critical years of 1948 and 1949. Clashes of safety requirements with other considerations occurred, but Teller was not open to compromises, and he was always willing to fight for what he considered to be the right cause. Animosities developed, which were exacerbated when the Teller committee turned down the AEC proposal to increase power production of a reactor while also reducing the buffer area around it. The General Advisory Committee felt that "the Teller group had exaggerated the consequences of a reactor accident and perhaps without adequate justification had retarded reactor development."[79] Teller's RSC was dubbed the "Committee for Reactor Prevention" by some nuclear scientists and nuclear engineers.[80]

One of the interesting safety issues was protection against possible earthquakes. Ideally, reactors would be installed in areas that experience no earthquakes, but this is not possible. Thus the design of a nuclear reactor should be such that in case of a tremor, the reactor would shut down itself. Teller thought it would be especially advantageous in terms of safety to install nuclear reactors underground, but he could not get this recommendation approved. Instead, sturdy domes were built around the reactors that would contain the radioactivity in case of accidents. Superficially, the dome that was erected over a reactor that exploded in Chernobyl, Soviet Ukraine, in 1986, could be considered such a device. However, it was more of an environmental shield to reduce the access of wind and rain than a casing for containment.

With time, the authority of the RSC was slowly eroding. The safety control of reactors under AEC jurisdiction was also gradually changing. As a consequence, Teller's activities diminished in this area. However, he remained a zealous advocate of maximum safety for nuclear reactors to the end of his life. When his activities in the public arena were summarized soon after his death, it was stated: "He was ahead of his time, and perhaps ours, in seeing that nuclear power, which he thought essential, would go nowhere if its safety and security were in question."[81] As Teller's activities related to reactor safety fizzled, he was ready to throw himself into other struggles.

## THE BIG DEBATE

On August 29, 1949, the Soviet Union exploded its first nuclear device, thereby ending the American monopoly on nuclear weapons. The authorities of the United States learned about the explosion within days, but it took them longer to inform the public of it. The Soviet explosion took the American administration by surprise, although it should not have. That the American authorities learned about it quickly and reliably was due to their unqualified success and especially that of Lewis Strauss. While most Americans were complacent about the Soviets at that time, he was an exception.

Strauss played a pivotal role in developing American awareness of the Soviet nuclear explosion. When he became one of the five commissioners of the Atomic Energy Commission, he inquired with the armed forces whether there was any aerial monitoring of the atmosphere. To his astonishment, there was none. The military had not yet adjusted itself to the changed conditions that nuclear weapons brought; besides, the Soviets were not expected to be making swift progress in creating their atomic bombs. The then army chief of staff, General Dwight D. Eisenhower, however, decided to remedy the situation when Strauss brought the omission to his attention. Still, things did not move fast enough since they lacked the necessary authorizations for appropriation. So Strauss lent the initial amount of money needed to start the monitoring rolling. It commenced in early 1948, and in less than two years it paid off. On September 3, 1949, the monitoring devices observed a cloud with radioactive material and tracked it from the North Pacific to the vicinity of the British Isles.

It took two and a half weeks before the experts convinced themselves that an atomic explosion must have occurred somewhere on the Asiatic mainland. Eventually the Americans managed to assign the exact time of the Soviet explosion to August 29, 1949, in Semipalatinsk in Eastern Kazakhstan with an accuracy of less than one hour![82] Strauss later remembered: "It was hard to convince a number of people who had been skeptical of Soviet success at so early a date that some mistake had not been made by our monitoring system."[83]

President Truman made the public announcement about the end of the American nuclear monopoly on September 23, 1949. This news was a crucial ingredient in shaping the events in the debate about the development of the American hydrogen bomb in the subsequent months. Strauss speculated about the direction events might have taken in defense matters had the monitoring system not been in operation in 1949. America would not have learned about the Soviet atomic explosion and might not have developed the thermonuclear weapon, and the Soviet Union might have blackmailed America and the Free World with its military superiority.

When Strauss and Teller met for the first time, Strauss was already a commissioner. They met at the famous Temple Emanu-El on the Upper East Side of New York City, the largest synagogue in the world. Teller did not attend services at the synagogue; he was there for a different reason. In 1948 he was still involved with the loose movement to promote a world government, and as a result of that involvement he gave a speech at the synagogue on behalf of the United World Federalists. When it was over, some members of the audience approached Teller for a chat. An elderly woman was interested in his opinion of the Atomic Energy Commission, of which Teller thought highly, and the woman introduced Teller to her nephew, Lewis Strauss.[84]

They became great friends and allies in spite of their grave difference in how they related to religion. Strauss was deeply religious and projected his religious views onto politics. Referring to the Soviet Union, he pronounced that "[a] government of atheists is not likely to be dissuaded from producing the weapon on 'moral' grounds."[85] This quote figured in his memorandum of November 25, 1949, to President Truman, in which he argued for the development of the hydrogen bomb. Teller was among a group scientists who were atheists or agnostics and for whom this particular argument would not have carried much weight.

Few people were as affected and energized—although not very surprised—by the news about the Soviet atomic explosion as Teller. He knew he had to act, but he was first looking for guidance. He turned to the most relevant authority he knew from wartime experience, J. Robert Oppenheimer, who was in the best position to initiate action. Oppenheimer was not only the chairman of the General Advisory Committee, he was also in charge of other advisory bodies related to national security. Teller had no advance knowledge of the president's announcement; he was not in any official capacity to be entitled to any. He was just returning from a visit to England, where he had been involved in discussions about reactor safety with his British colleagues.

Upon his arrival, he attended a meeting at the Pentagon with a general, with von Neumann, and with scientists from Los Alamos.[86] They discussed tactical and strategic nuclear weapons. The meeting had been scheduled weeks before, but the recent developments lent it added significance. It was there that Teller heard about Truman's announcement of the news. When he called Oppenheimer and asked, "What do we do now?" the answer "Keep your shirt on" was hardly soothing to his nerves.[87] Oppenheimer's reaction had a sobering effect on Teller. He understood that he could no longer count on Oppenheimer, and this strengthened his resolve to fight his fight alone; though he soon found some allies.

The news of the Soviet nuclear explosion aroused the concern of some influential scientists on the West Coast, including physicists Ernest O. Lawrence and Luis W. Alvarez and chemist Wendell M. Latimer, dean of chemistry at Berkeley. They all agreed that developing the thermonuclear bomb would be the proper response to the Soviet threat and that the AEC should be alerted to it. Lawrence had been scheduled to travel to Washington on another matter; now he decided to extend his visit and contact the commission. For help in influencing the committee, he invited Alvarez. They knew of Teller's interest in the Super, so they arranged for a stopover in Los Alamos to meet with him. Teller was gratified by their visit, which took place in early October.[88] Lawrence and Alvarez also talked with other scientists involved with the Super at Los Alamos, including the associate director John Manley, George Gamow, and Stanislaw Ulam.[89]

There was an amusing episode during Teller's meeting with Lawrence. The Californian demonstrated a new kind of shirt to Teller, the drip-dry shirt. He showed how to wash the shirt in the evening, scrub the collar and cuffs, hang it, and have it ready to wear in the morning without the need for ironing. To Teller, this was Lawrence's way of telling him to get ready for extensive travel and lobbying in order to shore up the necessary support for the thermonuclear weapon. Lawrence's backing was a welcome counterweight to the reluctance of the Los Alamos leadership whose consent he was also eager to gain. Teller realized that the program would get a big push if Fermi or Bethe were put in charge of it, so his first self-appointed task was to visit them and try to convince them to join the program. This was all before there was any higher-up decision to go ahead with the development of the thermonuclear weapon.[90]

The Atomic Energy Commission held a meeting on October 17, 1949. The acting chairman, Commissioner Pike, sent a set of questions to the General Advisory Committee. He asked whether the United States would use a thermonuclear weapon if it could be built and what would its military worth be in relation to fission weapons?[91] The topic was picking up speed, but not the speed Teller would have liked. He was anxious to gather scientists at Los Alamos for work on the Super. On October 10, 1949, he sent a memorandum to the director of the laboratory listing all those who would be "most useful in the initial phases of super development."[92]

Teller listed a total of twenty-eight scientists in the following categories: theoretical physicists, experimental physicists, and chemists. The list started with Fermi in a category of his own. Teller gave a succinct characterization of each scientist, covering his qualifications, special usefulness to the program, and the difficulties to overcome in getting him to Los Alamos (all were men). At this point Teller appeared optimistic about the scientists' willingness to join the Super program. A few of Teller's characterizations were as follows:

> **Enrico Fermi.** One should make a strong attempt to obtain Fermi's help and give him the maximum amount of responsibility he is willing to assume. I feel sure that Fermi will want wide responsibilities in the laboratory but no overall administrative responsibilities.

> **Hans Bethe.** The same comments hold for Bethe as for Fermi. I feel sure that the respective fields in which they will want responsibility will not conflict.

> **Julian Schwinger.** The help of Schwinger is essential in case we cannot get Bethe to help. . . .

**Norman F. Ramsey.** He would be particularly valuable as an excellent experimental man who is thoroughly familiar with theory and who has been in contact with military applications.

**O. Chamberlain.** His work both here and in Chicago has earned him the most excellent reputation.

**Richard Garwin.** In Fermi's opinion, the best student he ever had.

Teller could not meet with Fermi in Chicago because Fermi was in Europe, but Teller met him at Chicago's airport upon Fermi's return on October 24. Fermi was perhaps tired or did not appreciate what Teller told him, because he did not seem too excited by the events.[93] However, this was not atypical of Fermi's behavior; he did not seem to sense the danger of the possibility of a Nazi atomic bomb when Szilard and Rabi first talked with him about it some ten years before. Still, the lack of Fermi's encouragement disappointed Teller because he knew that Fermi could have been in charge of the intensified efforts for the Super and also because Fermi was a GAC member. Teller sensed that he did not catch Fermi at his best moment, but he had to return to Los Alamos to meet with influential congressmen who were interested in weapon development.

In his letter to Maria Goeppert Mayer, Teller wrote bitterly about Fermi's reluctance to commit himself to the development of the hydrogen bomb. Fermi had told Teller, "You and I and Truman and Stalin would be happy if further great developments were impossible. So, why do we not make an agreement to refrain from such development?" At the same time, Fermi admitted that it would be impossible to verify whether or not the Russians would adhere to the agreement. This distrust was too mild for Teller's taste, and he vowed that he would never participate in such a dangerous arrangement. He viewed it as the "only one difficulty in the world: wishful thinking."[94]

Teller kept recruiting scientists for the project about which no decision had yet been made. Bethe's participation would have been especially prized by him. He was elated when Bethe first agreed and bitterly disappointed when Bethe changed his mind, something Teller ascribed to Oppenheimer's negative influence. In the background, the Los Alamos leadership was not at all enthusiastic about an all-out effort for the thermonuclear bomb. This was because the outcome would be uncertain and because they feared that the concentration of resources on this one project would jeopardize their other research. They were following Teller's activities with mixed feelings at best and with considerable resentment at worst. At this point, however, Teller seemed unstoppable.

Oppenheimer called the GAC meeting to discuss the Super for the last weekend of October 1949.[95] The first session took place on October 28, Friday afternoon; those present included Fermi, Rabi, Oliver E. Buckley, Cyril S. Smith, and Lee A. DuBridge; those absent included Seaborg, Conant, and Hartley Rowe. The GAC took a broad view of the state of affairs. George K. Kennan of the State Department gave a presentation on the world situation in general and the Soviet Union in particular. Two outside experts, Bethe and Robert Serber, gave testimonies about technical problems concerning the Super as well as other projects. The GAC meeting continued the next day when only Seaborg was absent from the GAC members. The Joint Chiefs of Staff joined them for part of the session because of the importance of the discussion to the military. However, their chairman, General Omar N. Bradley, declared that the principal advantage of the Super would be merely psychological.[96]

If we look back to the GAC meeting, we see that its outcome could not have been easily predicted. Gradually, however, the tone during the sessions shifted increasingly toward opposing a crash program to develop the thermonuclear bomb. The only dissenting voice was that of Glenn T. Seaborg, who had sent a letter to the chairman of the GAC.[97] It is impossible to know how Seaborg's opinion might have influenced the discussion, but it is probable that Oppenheimer failed to inform the GAC members of his letter. There was much made of this fact during the course of the Oppenheimer security hearing in the spring of 1954. His accusers implied that he kept silent about Seaborg's letter because it went against Oppenheimer's opinion. Seaborg's letter was read out loud in the course of the hearing.

There were two crucial sentences in Seaborg's letter that proved unambiguously his stand in the matter of the discussion. Both sentences were formulated with utmost care, and one can almost sense the anguish of their author: "Although I deplore the prospects of our country putting a tremendous effort into this [the thermonuclear bomb], I must confess that I have been unable to come to the conclusion that we should not." Then, a little later in the letter, "My present feeling could perhaps be best summarized by saying that I would have to hear some good arguments before I could take on sufficient courage to recommend not going toward such a program."[98]

The concluding session of the GAC meeting took place on Sunday, October 30. The documents produced on that occasion have remained the matter of record from this body. Three memoranda were prepared.[99] One was addressed to David E. Lilienthal, chairman of the Atomic Energy Commission. It consisted of three parts and was signed by Oppenheimer. Part I discussed the production of fissionable material, the efforts to intensify the delivery of atomic weapons for tactical purposes, and neutron production. Part II was devoted entirely to the Super and it gave primarily technical justification for the committee's stand against a high-priority program for developing it. The committee cited the need for large amounts of tritium, which would take up a lot of reactor capacity, and the absence of completed theoretical studies to ascertain the feasibility of the project. They added, however, that even if the theoretical considerations would make the project promising, further experimental work would be needed to prove that the undertaking was on the right track. This would be followed by solving formidable engineering problems, but the weapon could be produced within five years.

The consideration of the technical questions was followed by a brief analysis of more general issues. The committee mentioned the virtually limitless power of the super bomb (in contrast with the fission bomb) and that it was relatively inexpensive to produce. The bomb's limitless nature was in contrast with the limitations of the possibilities of delivery. It was also pointed out that the new bomb would have an explosive effect hundreds of times larger than the fission bombs, with severe contamination effects that could be further enhanced by easily applied technical means. The last paragraph of Part II dealt with possible consequences of deploying such a bomb, which would represent "the policy of exterminating civilian populations." Finally, there was a consideration of cost-effectiveness, concluding with the words "[I]t appears uncertain to us whether the Super will be cheaper or more expensive than the fission bomb."

Part III is the shortest of the three; it expressed the hope that "the development of these weapons can be avoided. We are all reluctant to see the United States take the initiative in precipitating this development." The GAC noted some division among the members as to "the nature of the commitment not to develop the weapon." This was expressed in two additional documents. The GAC further recommended that a public disclosure be made about the weapon without communicating technical details. The disclosure would state the desire of not developing the weapon and would explain "the scale and general nature of the destruction which its use would entail."

Attached were the additional majority and minority opinions of the GAC members. The majority opinion, signed by Conant, Rowe, Smith, DuBridge, Buckley, and Oppenheimer, recommended strongly against initiating an "all-out" effort to develop the hydrogen bomb and condemned it as a tool of genocide. It stated that

[t]he existence of such a weapon in our armory would have far-reaching effects on world opinion: reasonable people the world over would realize that the existence of a weapon of this type whose power of destruction is essentially unlimited represents a threat to the future of the human race which is intolerable. Thus we believe that the psychological effect of the weapon in our hands would be adverse to our interest.

We believe a super bomb should never be produced. Mankind would be far better off not to have a demonstration of the feasibility of such a weapon until the present climate of world opinion changes.

It is by no means certain that the weapon can be developed at all and by no means certain that the Russians will produce one within a decade. To the argument that the Russians may succeed in developing this weapon, we would reply that our undertaking it will not prove a deterrent to them.

Should they use the weapon against us, reprisals by our large stock of atomic bombs would be comparably effective to the use of a super.

*In determining not to proceed to develop the super bomb, we see a unique opportunity of providing by example some limitations on the totality of war* and thus of limiting the fear and arousing the hope of mankind.[100] (italics added)

The minority report by Fermi and Rabi was worded even more strongly and would make the Super program dependent on the stand of the Soviet government. It stated that:

. . . the use of such a weapon cannot be justified on any ethical ground which gives a human being a certain individuality and dignity even if he happens to be a resident of any enemy country. It is evident to us that this would be the view of peoples in other countries. Its use would put the United States in a bad moral position relative to the peoples of the world. Any postwar situation resulting from such a weapon would leave unresolvable enmities for generations. A desirable peace cannot come from such an inhuman application of force. The postwar problems would dwarf the problems which confront us at present. The application of this weapon with the consequent great release of radioactivity would have results unforeseeable at present, but would certainly render large areas unfit for habitation for long periods of time.

The fact that no limits exist to the destructiveness of this weapon makes its very existence and the knowledge of its construction a danger to humanity as a whole. It is necessarily an evil thing considered in any light.

For these reasons we believe it important for the president of the United States to tell the American public, and the world, that *we think it wrong on fundamental ethical principles to initiate a program* of development of such a weapon. At the same time *it would be appropriate to invite the nations of the world to join us in a solemn pledge not to proceed in the development or construction of weapons of this category*. If such a pledge were accepted even without control machinery, it appears highly probable that an advanced stage of development leading to a test by another power could be detected by available physical means. Furthermore, we have in our possession, in our stockpile of atomic bombs, the means for adequate "military" retaliation for the production or use of a "Super."[101] (italics added)

Thus the GAC unanimously rejected the development of the Super, although its opponents outside the GAC expected a less than uniform outcome. Emotions must have been running high; suffice it to mention that Lilienthal "considered the 'bloodthirsty' attitude of some scientists" without naming names. Elsewhere he had expressed apprehension that some members of the GAC might have favored "the Teller–Alvarez proposal."[102]

In hindsight, and knowing about the Soviet Union what we do today, it is obvious that the majority opinion underestimated the potentials of the Russians' nuclear weapons project. The minority opinion appeared more realistic, but it found the American stockpile of fission bombs sufficient for countering a Soviet thermonuclear attack. In this, there was no disagreement between the two opinions, except that Fermi and Rabi took a stronger stand against creating the Super. They suggested "to invite the nations of the world" for a joint pledge not to produce the Super. Fermi and Rabi appeared more determined to oppose the United States initiating such a project.[103]

In his 1976 book, Herbert York reported about a more recent conversation in which "Rabi stated it was his firm recollection that he and Fermi definitely intended *to couple American forbearance with a Soviet pledge to do the same*" (italics added).[104] Unfortunately, there was no information about what was going on in the Soviet Union. Some of the GAC people may have assumed that if the United States would not make a hydrogen bomb, the Soviet Union would not either. By then, however, an accelerated Soviet program was under way with the participation of outstanding physicists. Nonetheless, James Conant, in a correspondence with Oppenheimer, had declared that the United States would develop the Super "over my dead body."[105] He hardly would have thought that the Soviet efforts were already being carried out literally over the dead bodies of many slave workers who toiled under inhumane and unsafe conditions.

Lewis Strauss noted the tendency to consider the question of whether or not to develop the hydrogen bomb as an internal problem of the United States or at best as an internal problem of the Free World. In his words, "To a surprising degree the world has maintained its wide-eyed and childlike credulity about Soviet pronouncements. It is a state of mind that, astonishingly, survives periods of disillusionment when the actions of the Soviets expose their insincerity and untruthfulness."[106] The approach to Soviet affairs that Strauss criticized was a characteristic pattern not only on this occasion but, more generally, especially among many intellectuals in the West. As to whether the Soviet Union possessed the means and the necessary brain power, estimates varied: "[I]n general none expected the Soviets to detonate any atomic device before 1952. The majority opinion set the time substantially further in the future, while not a few believed it beyond Soviet capacity in *any* time scale likely to be of much concern to us"[107] (emphasis by Strauss).

The Soviet scientists were not aware of the happenings in America, and apparently no debates emerged among them. Nobody raised the question of whether or not it was moral to develop such a weapon. According to Vitaly Ginzburg, "Even asking such a question would have been a stupid act. We have to be realistic about life."[108] The isolation of Soviet scientists from the rest of the world was complete. In those years, they could not even have communications with their American and West European colleagues about purely scientific matters.

The Soviet Union had just battled a bloody war in which they prevented a ruthless foreign power from taking over their country. Stalin had freed enormous resources by invoking feelings of nationalism among the Soviet, and in particular Russian, people during the war that had been taboo in Soviet society prior to World War II. After the war it did not take too much convincing before the Soviet scientists felt their further work was necessary to protect their country from yet another foreign aggression.

Teller may have never understood the motivations of the Soviet scientists. In a television interview in 1999, he expressed his surprise: "I find it a little strange that Soviet scientists, who actually in their majority hated the communist government, did work strongly and effectively on the atomic and hydrogen bombs."[109] The Soviet scientists did not think about the dangers of the hydrogen bomb in more general aspects, regardless of whether it was in the hands of a democracy or a dictator. At least most of them felt that working on these weapons was their patriotic duty. Lev Landau was an exception because he understood the true nature of Stalin's dictatorship. Even he diligently fulfilled what was expected of him in the weapons programs because that was his ticket for survival.

Ginzburg's ideas contributed to the success of the Soviet hydrogen bomb, although he was excluded from direct access to the work due to his lack of security clearance (his wife was in exile ostensibly for anti-Soviet activities). He was not even allowed to know whether or not his suggestions were used in the program, and he found out about their fate only after the collapse of the Soviet Union. Today, it is ridiculous to maintain that the United States could have stopped the development of the hydrogen bomb in the Soviet Union by providing an example of refraining from it.

The physical chemist Kenneth Pitzer, research chief of the US Atomic Energy Commission at the time, stated the following after the information about the Soviet nuclear program became available: "They were going to go right ahead with that [(the hydrogen bomb)]. We got there first but not by much. It would have been quite a situation in this country if they had gotten there first and it would have been known that we had been intentionally slow about it."[110] The Soviet leadership was set on developing their thermonuclear weapons regardless of the American actions, according to Andrei Sakharov, "the father of the Soviet hydrogen bomb." He noted that American restraint from developing the Super, "would have been perceived either as a cunning and deceitful maneuver, or as evidence of stupidity or weakness."[111]

Teller must have sensed that the GAC did not provide much hope for his lobbying activities, so he instead concentrated his efforts on the legislators and the military. On November 2, 1949, he arrived in Washington, DC, to meet with Senator McMahon. The senator found their exchange pivotal in strengthening his determination to push President Truman to adopt an all-out policy toward the development of the thermonuclear bomb. At this time, Teller did not know yet the outcome of the GAC meeting, although he sensed that it would not be to his liking. All that McMahon could tell him was that the GAC recommendations made him sick.[112]

The General Advisory Committee gave advice and the Atomic Energy Commission made recommendations. The problems with underestimating the capabilities of the Soviet Union did not end with the GAC. Even President Truman "simply could not bring himself to believe that 'those Asiatics' could build something as complicated as an atomic bomb."[113] Truman asked some members of the AEC to sign a statement that they really believed the Russians had exploded a nuclear device.

Teller took an entirely different approach; rather than underestimating the Soviets, he tended to overestimate them. While the Soviet Union was a technologically backward country, it had excellent physicists—Teller could rightly testify to that from his personal experience—and it had material resources as well. It also had the ability to focus on selected tasks, and doing so was easier for a totalitarian state than for a democracy. It was an advantage in this respect that they did not have to waste time and energy on debates that a democracy could not spare. Of all these advantages, the most conspicuous was the high level and number of their theoretical physicists.

It is rather one-sided to single out Teller as the principal advocate for the development of the hydrogen bomb, as there were others among the scientists, politicians, and military. Lawrence, like Teller, did not have any formal involvement in the decision-making process, but he continued to be an important contributor to the cause of the Super. Alvarez exercised increasing influence, enhanced by his recent great services to the military. His Ground Controlled Approach (GCA) made it possible for American planes to land safely in Berlin and thus to break the Soviet blockade in 1948.[114]

On November 7, 1949, Lilienthal resigned as AEC chairman; his resignation was to be effective on February 15, 1950. During the remaining period of his chairmanship he was not fully active. Commissioner Strauss, on the contrary, displayed heightened activities. Originally, he was the only one among the five commissioners who supported the Super program. Eventually another member, Gordon Dean, changed his mind, and by the time the AEC met, three were against it (Lilienthal, Pike, and Smyth) and two for it (Strauss and Dean). Lilienthal tried to formulate a policy on a broader basis than merely deciding about the fate of a new weapons system. He stated that "the Super . . . as a weapon of mass destruction without any apparent peaceful applications, would convince the world that the United States had resigned itself to war."[115]

Teller and his colleagues at Los Alamos finally read the GAC statements on November 11. Ulam considered the GAC recommendations a mere waste of time and could not imagine that the final decisions about the Super would follow what the GAC had prescribed. In a letter to von Neumann, Ulam made the intriguing comment "that some of the opposition to the Super might have been a reaction against Teller's insistent advocacy of the new weapon."[116] Bethe, who opposed the development of the hydrogen bomb, found merit in the GAC deliberations even though he considered the presidential decision for a go-ahead inevitable. He stated in his biographical memoir about Oppenheimer, "The effort of Oppenheimer and the GAC to make the US Government pause and think about this step stands as a most important milestone."[117]

In the middle of November, Senator McMahon and his aide William Borden visited Los Alamos, as did the chairman of the Military Liaison Committee, Robert LeBaron. They shared the opinion that "the Super might well offer the nation a measure of security no other weapon system could provide."[118] Teller was one of the presenters to inform the visitors, and it was noted that he gave a "balanced appraisal of the Super."[119] He was unambiguous about his support for the program, but he stressed that its feasibility had not yet been proven. Some of his colleagues at Los Alamos viewed his presentation as a dilemma between his enthusiasm for the program and his judgment as a scientist.

As a consequence of the visit, Senator McMahon wrote a letter to President Truman on November 21. He admitted to the horrors of the hydrogen bomb but suggested that the same horrors might save the United States from enemy attack. He considered it imperative for America to become the first power possessing the new weapon and urged the president "to take the entire question to the people of the United States and the world."[120] This was an interesting suggestion because one might have thought that such a revelation would alert the Soviet Union to the American contemplation to develop the Super. Today we know that this was the appropriate approach because there was nothing new in it for the Soviet Union.

Reviewing the events of the end of 1949 and of January 1950, the picture emerges of the significant—but by no means exclusive—role that Teller played in them. He was a star witness for McMahon but hardly influenced the executive branch. Of course, the various branches interacted, and importance should be assigned to the roles of the congressional forces, the military, the Departments of State and Defense, the AEC (as well as its GAC and Military Liaison Committee), and the individual scientists.

Teller's divisive rather than uniting character made him stand out more than scientists like Ernest Lawrence and Harold Urey, who carefully avoided alienating their colleagues. Teller's lobbying for the hydrogen bomb was in stark contrast with what he wrote about the scientists' role in an article of about the same time. It carried the title "Back to the Laboratories" and stated that "it is *not* the scientist's job to determine whether a hydrogen bomb should be constructed, whether it should be used, or how it should be used. The responsibility rests with the American people and with their chosen representatives" (emphasis in the original).[121]

The Atomic Energy Commission voted three to two in November 1949 to recommend against the development of the Super. By the end of January, with Lilienthal often absent from important gatherings, the remaining two, Smyth and Pike, wavered in their conviction. A few months later, it was Smyth who called the physicist John A. Wheeler, asking him to interrupt his leave in Europe and return to Los Alamos. Smyth told Wheeler that "he shared Teller's conviction that the United

States had to react to 'Joe 1' [the American nickname for the first Soviet nuclear explosion, named after Joseph (Iosif) Stalin] . . . with a renewed effort to design thermonuclear weapons."[122]

Lilienthal's slim majority in the Atomic Energy Commission was eroding, but he continued to represent the commission at the meetings of Truman's National Security Council, which consisted of Defense Secretary Louis Johnson, State Secretary Dean Acheson, and the AEC chairman. They voted two to one to support the development of the super bomb, with Lilienthal casting the dissenting vote. Johnson had no hesitation, but Acheson gave a lot of thought to forming his opinion. He had worked closely with Lilienthal and Oppenheimer on plans for international cooperation with the Soviet Union in the field of atomic energy. Nothing had come of these plans, but Acheson learned to appreciate his partners' intelligence and preparedness.

Referring to the recommendations of the GAC, Acheson "was deeply pessimistic about the possibility of achieving any useful agreement with Stalin and the Soviets in the matter of the Super."[123] Immediately after the end of the October 1949 meeting of the GAC, Oppenheimer told Acheson that the United States by restraining itself in the matter of the hydrogen bomb would provide an example for the Soviet Union to follow. Acheson doubted that, asking, "How can you really persuade a hostile adversary to disarm 'by example'?"[124] In spite of Acheson's unambiguous stand, he was bitterly attacked by Joseph McCarthy as being soft on communism, and the senator accused the State Department of sheltering communists.[125]

It was a great victory for Teller when President Truman announced his decision on January 31, 1950, giving directions to continue research on all atomic weapons, including the hydrogen bomb. It was a victory not without reservation, since the expression "to continue research" could also be interpreted as the president's merely calling for more of the same. Teller compared the situations in 1939 and 1950. In 1939, in their first meeting with the representatives of the military, Colonel Adamson argued that new weapons do not win wars, and the scientists argued for the atomic weapons. In 1950, most of the scientists were against the new weapon.

Humankind would have been better off without the hydrogen bomb, but letting Stalin become its sole possessor would have been disastrous since he might, in fact, have tried to enslave the entire world. This is why Harold Urey warned that America should not "intentionally lose the armaments race; to do this would be to lose our liberties, and, with Patrick Henry, I value my liberties more than I do my life."[126] The same is expressed on every license plate in New Hampshire, "Live Free or Die."

In view of the general secrecy surrounding the American nuclear program, it may appear surprising that Truman's decision was made public. At least one of the reasons must have been that information about the work on the Super had been divulged by an indiscreet member of the Joint Committee on Atomic Energy on November 1, 1949. There was an increasing amount of related material in the American media in the subsequent weeks.[127]

An Austrian physicist, Hans Thirring, described the principle of the hydrogen bomb as early as 1946 in a German-language book about the history of the atomic bomb.[128] He utilized sources in the open literature, and the *Bulletin of the Atomic Scientists* printed a translation of the relevant chapter dealing with the hydrogen bomb in its March 1950 issue.[129] The journal waited some years before bringing out such a paper lest it give extra exposure to a weapon and thereby to an escalation of an international arms race. But now, with the president's decision out, there was no longer any reason to be silent about Thirring's publication.

The public did not realize that the technology to actually make the hydrogen bomb was not yet available, and it was not even known at this point whether the bomb could be made. However, there was assurance that if the hydrogen bomb could be made, it would be made in the United States.

# Chapter 6

# FATHERING THE HYDROGEN BOMB

The Dark Ages may return on the gleaming wings of Science.
—WINSTON CHURCHILL (ATTRIBUTED)[1]

Some cures are worse than the dangers they combat.
—SENECA[2]

*Amid the tension of a program that had difficulty succeeding, Teller was able to make a pivotal discovery—acting upon a suggestion by Stanislaw Ulam—that would make a thermonuclear explosion possible. Bethe found Teller's idea as surprising as the discovery of nuclear fission. After much effort, the American hydrogen bomb would be produced, and, not much later, so would its Soviet counterpart. The stage was set for the super-powers to embark on a vicious arms race. Had the United States not engaged in equipping itself with this most terrible of weapons, the Soviet Union alone would have possessed it and might have used it to blackmail the Free World. The history of the hydrogen bomb has been muddled by biases and by Teller's varying attribution of credit over the years.*

President Truman's decision about the Super indicated the beginning of a long journey; the hydrogen bomb was not yet hardware, it was merely an idea.[3] There was no assurance that Los Alamos could produce it; there was no assurance that anybody could produce it. Teller knew, however, that if the United States expected to succeed, the project would require the best scientists to participate. He was very disappointed during the deliberations at the end of 1949, when he realized that he could not count on many of them. Bethe's change of mind was especially nerve-racking for him.

When Bethe nonetheless returned to Los Alamos, he was ostensibly not working for the development of the hydrogen bomb but merely finding out whether and how it could be made.[4] He changed from this exploratory mode of operation to full engagement only after the Teller–Ulam solution to the problem had been found in the spring of 1951. It seems that Bethe never found peace with himself because of his participation in the development of the hydrogen bomb. His inner conflict may have triggered his criticism of Teller in later years. Ironically, Teller never craved anybody else's approval as much as Bethe's.

There was a difference between Fermi's and Bethe's reactions to Truman's decision. Fermi returned to Los Alamos, not full time but sufficiently to help the program a great deal, and he brought with him important associates to advance the work. Bethe, on the other hand, decided to launch one more attempt to change the administration's policy. After President Truman had announced his decision to go ahead with the hydrogen bomb, Bethe stated, "It is argued that it would be better for us to lose our lives than our liberty, and with this view I personally agree. But I believe this is not the choice facing us here; . . . in a war fought with hydrogen bombs we would lose not only many lives, but all our liberties and human values as well."[5]

Bethe argued that even if the United States did not have the hydrogen bomb and the Soviet Union did and used it, "Though it might devastate our cities and cripple our ability to conduct a long war with all modern weapons, it would not seriously affect our power for immediate retaliation . . ."[6] One wonders whether Bethe might have seriously thought that the political leaders in the United States would take upon themselves the responsibility of allowing their country to face such an eventuality. Bethe gave the answer when he admitted many years later that "Truman had no choice in the political atmosphere of the time. Had Russia developed the H-bomb and the US not, he and the scientific community that opposed it would have been considered traitors."[7]

Twelve scientists, including Bethe, signed a statement on February 4, 1950—right after President Truman's decision about the hydrogen bomb. The statement did not just oppose the thermonuclear weapon but also contained an appeal to the leaders of the United States. The statement pronounced that the hydrogen "bomb is no longer a weapon of war but a means of extermination of whole populations."[8]

The ever-practical and very hawkish John von Neumann found justification for the development of the hydrogen bomb in the inaccuracy of the then available missiles.[9] If the missiles missed their targets by a mile or more, a hydrogen bomb would still destroy what was meant to be destroyed. At that time von Neumann did not foresee the tremendous increase of accuracy of missiles. In time, his activities would be instrumental in developing a huge program to supply the American armed forces with an arsenal of efficient and accurate long-range missiles.

Bethe and his co-signers urged "that the United States . . . make a solemn declaration that we shall never use this bomb first. The circumstance which might force us to use it would be if we or our allies were attacked by *this* bomb" (emphasis in the original).[10] Then the crucial sentence came, which not only posed limitation but, most significantly, found justification for making the hydrogen bomb. This statement was not made by Teller or other proponents of the program, but by its staunch opponents: "There can be only *one justification* for our development of the hydrogen bomb and that is *to prevent its use*" (italics added).[11]

Probably many more scientists would have signed this declaration had it not been for a gag order. Those in the GAC and those involved in other ways in the program were not allowed to voice their opinions in connection with the development of the hydrogen bomb. Einstein was the head of the Emergency Committee of Atomic Scientists, and in this capacity what he said in connection with the development of the hydrogen bomb was important. He did not support it, but he declined opposing it either. When he was asked to join an appeal to call for delaying its construction, he said, "As long as competitive armament prevails, it will not be possible to halt the process in one country."[12]

In a dramatic development, on February 2, 1950, the Atomic Energy Commission learned about the spying activities of the distinguished physicist Klaus Fuchs. He was a member of the British delegation at Los Alamos and had returned to Great Britain after the war. In the summer of 1950, the inception of the Korean War gave greater impetus to weapons work. On June 25, 1950, North Korea launched an attack against South Korea across the 38th parallel, and within ten days American troops were engaged in a war whose outcome seemed very uncertain and whose escalation at times appeared most probable.

## HYDROGEN BOMB QUEST

It was somewhat ironic that during all the discussions about whether or not to develop the Super, nobody really knew how to do it. It was taken for granted that once there was a go-ahead for it, a highly focused and all-out program would solve it. It proved to be the case, eventually, but it was far from a trivial pursuit. Teller wrote to Maria Goeppert Mayer after Truman's decision: "I never felt better starting a job, and I never needed more optimism than I need this time. . . . I know objectively that we are in a situation in which any sane person must and does throw up his hands and only the crazy ones keep going."[13] One senses the simultaneous enthusiasm and despair in Teller's words.

There is an interesting side question at this point. Part of the argument for the development of the American hydrogen bomb was the anticipation that the Soviets would develop it. If that was the case, then why was not there more information about the Soviet nuclear program? Richard Rhodes reported that there was one channel through which a trickle of information may have reached Teller and Ulam in April of 1950. It concerned the crucial question of deuterium compression and was part of an intelligence report due to Arnold Kramish.[14] Yet it did not seem to have influenced Teller and the American program.

The suggestion about using a particular fuel for the thermonuclear design was identical in the American and Soviet projects, and this was before President Truman's decision. Deuterium was a gas and could be stored as a liquid only under high pressure and at very low temperature. The suggestion to use lithium deuteride was made even before the Los Alamos Laboratory was established, as early as 1942. Teller wrote about it to Oppenheimer.[15] It was also stated that lithium hydride would not be feasible, and Teller hinted that the isotopic composition of the substance might have to be changed. This may be an indication that at that point they were not necessarily thinking of the particular lithium isotope lithium-6 that later was found useful. Lithium deuteride also figured in the Soviet hydrogen bomb designs due to Vitaly Ginzburg's suggestion.[16] It was a convenient fusion material because it was a solid and easy to store. Besides, lithium itself could also participate in fusion reactions. Another similarity between the two projects was in the early suggestion of how to bring together the necessary fission and fusion fuels in the thermonuclear device. Teller's "alarm clock" idea dated back to 1946, and it meant interleaving spherical layers of fusion and fission fuel. Somewhat later, Sakharov suggested the so-called "layer cake" design to accomplish the same in the Soviet program.

Some work on the Super was going on in Los Alamos even before Truman's decision. Ulam alleged that Teller tried to downplay the significance of that work because he wanted to minimize the credit for Los

Alamos and its director. According to Ulam, "work on the 'super' had been going on efficiently and systematically."[17] One wonders whether Ulam might have exaggerated the work on the Super prior to the Truman decision when he was looking back after many years of controversy with Teller. At one point, there was a committee formed by Bradbury to organize and direct the thermonuclear project. Teller was the chair of the so-called family committee, which brought different people from different sections of Los Alamos together. Gamow and Ulam shared with him the responsibility and the mandate to be in charge of the committee.[18]

Gamow contributed a suggestion for a scheme and prepared a cartoon to popularize it; the cartoon showed the three of them—Gamow, Ulam, and Teller—accompanied by a flying Stalin with wings and a bomb under his belt, and a glorified Oppenheimer with an olive branch in his left hand and a warning index finger of his raised right hand.[19] Ulam is represented as spitting in a spittoon; Gamow as squeezing a cat's tail; and Teller as wearing an Indian fertility necklace symbolizing the womb (in Gamow's peculiar pronunciation it sounded like "vombb" according to Ulam).[20] The meaning of Ulam spitting in a spittoon is a puzzle, but its significance may have been due to a photograph in which Ulam seems to be spitting (without any visible spittoon) that Ulam gave as a present to Teller and wrote on the back: "With love to my enemy/Yours/Stan." Ulam had a lighter approach to life than did Teller with his constant worries.

Gamow was absent from the atomic bomb project during World War II, because he could not get security clearance due to his having been a commissioned officer in the Red Army back in the Soviet Union. His appointment was made so that he could teach meteorology at the Soviet equivalent of the American West Point Military Academy. Otherwise, Gamow was fiercely anticommunist and anti-Soviet. His clearance issue was resolved only after the war.

Once the president declared his decision on January 31, 1950, people were needed for Los Alamos. A number of experienced scientists soon joined Teller, including Emil Konopinski, Marshall Rosenbluth, and Lothar Nordheim. Bethe, Fermi, and von Neumann came for shorter periods, as did some of their young associates. Von Neumann interrupted his vacation in France and returned to continue his calculations in Princeton. Frederick de Hoffmann, originally an immigrant from Vienna, not only joined Teller but played the role of "reality checker," a role Teller once played himself.[21] Teller would assign much credit to de Hoffmann, maybe more than was deserved, ostensibly because it would detract from Ulam's work but not from Teller's.[22]

## STANISLAW M. ULAM (1909–1984)

Stanislaw Ulam was born in Lemberg in Austria–Hungary (Lwów in Poland between the two world wars; L'vov in the Soviet Union after World War II; today it is L'viv in the Ukraine) into a wealthy Jewish family. He received his doctorate in mathematics from the Polytechnic Institute in Lwów in 1933 and became a member of the famous Polish school of mathematics. In the mid-1930s he immigrated to the United States and worked at the Institute for Advanced Study in Princeton, at Harvard University, and at the University of Wisconsin. John von Neumann was instrumental in Ulam's coming to America and also in his joining the Manhattan Project at Los Alamos in 1944. Their interactions greatly influenced Ulam's career and the development of his research interests.

Ulam worked on the problem of implosion for the plutonium fission bomb and on the design of the thermonuclear weapon. His calculations, together with those of Enrico Fermi and Cornelius Everett, demonstrated the marginal feasibility of the so-called classical Super. In 1951, Ulam made a decisive contribution to solving the impasse in the development of the hydrogen bomb, leading to the successful Teller–Ulam approach.

Ulam started his career as a pure mathematician and made a remarkable transformation toward mathematical physics and the successful solution of practical problems. His best-known brainchild is the Monte Carlo method, which offers approximate solutions as estimates to problems for which there are no exact solutions, using sampling and probability considerations. The method has gained wide applications in highly diverse areas and many of its users are not aware of Ulam's pioneering contribution to the development of the technique.

He worked for many years full time for Los Alamos and later part time as adviser. In 1965, he joined the University of Colorado as professor of mathematics; at one time he also served as chairman of the Mathematics Department and later as professor of biomathematics at the Medical School of the University of Colorado. His autobiographical book *Adventures of a Mathematician* is an enjoyable read about his life and work, about his wide-ranging interests, and about his friends. His engaging personality, sense of humor, and love of life shine through its pages.

George Gamow's collage. From upper-left corner, clockwise:
Stalin, Oppenheimer, Gamow, Teller, and Ulam.
From George Gamow, *My World Line: An Informal Autobiography*
(New York: Viking Press, 1970), p. 154. Courtesy of Igor Gamow.

Wheeler was working in Paris when he learned about the accelerated program for the hydrogen bomb, and he struggled between what he perceived as his patriotic duty and his love for physics. The next time he met Niels Bohr, he presented his dilemma to his old mentor. Bohr was known to be hesitant and was not one to give unambiguous answers even to less controversial questions. However, on this occasion Bohr did not hesitate. He asked Wheeler, "Do you think that Europe would be free of Soviet control today had it not been for the atomic bomb of the West?" Wheeler soon joined the hydrogen bomb project.[23]

Of course, not everybody thought like Bohr and Wheeler. The Wheelers' next-door neighbors in Princeton were the Panofsky family, former refugees from Germany. Erwin Panofsky was a famous art historian, and his two sons were physicists, known by their nicknames: the bright Panofsky, who was the top student in his class, and the dumb Panofsky, who was second-to-the-top student. Erwin Panofsky did not approve of the development of the hydrogen bomb. When FBI agents came to the neighborhood to check upon Wheeler's reliability, they asked the Panofskys whether they knew of any subversive activities of the Wheelers. Erwin Panofsky's response was, "They are not subversives; they are mass murderers! We are the subversives."[24] Incidentally, the "dumb" Panofsky, Wolfgang, would become a world-renowned physicist, director of the Stanford Linear Accelerator Center, and adviser to US presidents.

Teller was engaged in at least three aspects of the thermonuclear project. One was his lobbying efforts in Washington because he was never satisfied with the way the leaders of Los Alamos were running the program. He talked with people at the Atomic Energy Commission and even more with the chairman of the Joint Congressional Committee on Atomic Energy, Senator McMahon, and the staff director of the committee, William L. Borden. When he appeared on March 3, 1950, before the committee, he lamented that the program was being delayed because there were not enough high-caliber people involved. Many had left Los Alamos at the conclusion of the war and did not want to return; others doubted the success of the project.

There was yet another factor that some scientists may have considered when they hesitated joining the nuclear program. This was at the time when Senator Joseph McCarthy began making widespread accusations of people in the State Department, in Hollywood, and elsewhere of being communists. He would be censured by his senate colleagues in 1954 following his baseless accusations against army personnel. His performance was televised, so a huge audience witnessed his outrageous activities. But his political demise was yet years away and his hysterical battle against communism was adversely influencing private citizens nationwide and officials in Washington.

In March 1950, Stephen Brunauer turned to Teller for support, as he had also been denounced by McCarthy. Teller did not hesitate to address Senator Millard Tydings of the Senate Armed Services Committee in a letter dated March 21, 1950. The letter is much more than a testimonial on behalf of Brunauer. Teller vented his frustration with recruiting scientists for his project and ascribed blame to the troubles similar to Brunauer's ordeal.

First he referred to his past association with Brunauer and vouched for his trustworthiness, loyalty, and character. He also issued a caveat that after the Klaus Fuchs case, it would be impossible to be absolutely sure of excluding anybody's involvement in potentially treasonous activities. However, he pointed out, McCarthy's accusations were made after Brunauer and his wife had been investigated and cleared by the proper authorities. Teller stressed the great importance of Brunauer's contributions to the national preparedness after the war, which he was able to do on the basis of his direct contacts with Brunauer. Then came something closely related to Teller's own project:

> There are unfortunately only too few of the scientific men who are unselfish enough to postpone their scientific work and continue to make contributions to national preparedness. If such persons are publicly accused without good evidence, great damage will be done to our whole program of preparedness. One of the reasons that keeps scientific men out of Government Laboratories, as we have seen in our recruiting program at Los Alamos, is precisely this kind of criticism to which Dr. Stephen Brunauer is now exposed. I should indeed say that this is probably one of the strongest reasons which makes adequate staffing of our war research laboratories difficult. We may lose the armament race in which Russia and this country are engaged at the present time precisely because of lack of cooperation between the war research laboratories of our Government and the best of the scientific brains that can be found in the country.[25]

If the above account had not sufficed, Teller added a weighty concluding statement, saying that "unfounded public *accusations may hurt* our position in the armament race *as greatly as actual disloyalty*" (italics added). This statement is tantamount to accusing Senator McCarthy and his helpers with treasonous activities. This was a brave, almost reckless action on Teller's part at this very early point of the McCarthy era.

Szilard described how worried he was in the fall of 1949, when he understood the state of affairs about the hydrogen bomb. All the current ideas on the Super came from a single person, Edward Teller, who was not even working on this project full time. After he had learned about the Russian nuclear explosion, Szilard found it "so incongruous that there should be only one first-class mind left working on the problems of the bomb," he informed a White House official of his worries. The administration representative then responded with shocking advice for Szilard: "For God's sake, keep the name of this man secret. If the Russians find out who he is, they can blacken his name in such a way that it will not be in the power of the president to keep him at work in his job."[26]

Szilard talked about this episode in a speech in Los Angeles in 1954. Fifty years later, Harvard biochemist Matthew Meselson still remembered Szilard quoting the official saying something like "Don't even tell, because if the communists find out they'll accuse Teller of being a communist—they will have a way of doing this—and he will not be able to work on it anymore."[27] Szilard's story, corroborated by Meselson, captures the atmosphere of the early 1950s in the United States.

Security consciousness was in the air, particularly regarding the projects under the jurisdiction of the Atomic Energy Commission. Instructive case studies were discussed from the AEC personnel hearings conducted at Berkeley in 1948 and 1949.[28] During the investigation of future Nobel laureate Melvin Calvin, his association with Michael Polanyi came up, since Calvin had spent two years as a postdoctoral fellow with Polanyi in Manchester, England.[29] Calvin had to explain that rather than sympathizing with communism, Polanyi represented its antithesis.

Teller had difficulties not only with recruiting, but also with the organizational structure within whose framework his program was operating. There was no separate division assigned to the hydrogen bomb project at Los Alamos. Teller understood that he did not make a very good administrator, yet he resented when other people were put in charge of projects in which he was deeply involved. He developed an especially poor working relationship with Marshall Holloway, with whom he had a clash of personalities. According to Carson Mark, Holloway was "a slightly conservative guy so if he thought it would take a week from start to finish he was inclined to say he needed two weeks, . . . so that would lead him into fights with Edward."[30] When Holloway was given general responsibility for the hydrogen bomb work, Teller resigned. This happened in September 1951, when the principal questions of the scientific part and design for the first big test had already been resolved. In the meantime, Teller tried to increase awareness in the thermonuclear program at Los Alamos. He gave two staff lectures in early 1950.

The initial concept of the Super was based on the idea that the extreme high temperature from a fission explosion would ignite the thermonuclear reaction in the fusion fuel, and there the next task would be maintaining the burning of this fuel. Both Teller and Ulam questioned earlier calculations and estimates. They had to work together, but they were increasingly at odds with each other. According to Ulam's calculations, the hydrogen bomb would not work if they stuck with the design they were working on at the time. Ulam was a group leader with the sole member Cornelius Everett in his group. According to Ulam, their "calculation showed . . . enormous practical difficulties and threw grave doubts on the prospects of Edward's original approach to the initial ignition conditions of the 'super.'"[31]

Teller was skeptical of Ulam's calculations, but Ulam proved to be correct. As doubts about the feasibility of the original design strengthened, an idea emerged in Teller's mind about a second weapons laboratory. He did not leave anything to chance, arguing that a second laboratory would be desirable even if the first tests of the thermonuclear design were successful. In such a case many new projects should be initiated. Teller's distrust toward Ulam was dampened by the fact that von Neumann's computations were invariably consistent with Ulam's findings, and Teller trusted von Neumann without reservation. In March 1950, Ulam and Everett estimated that "the model . . . is a fizzle."[32]

Ulam, Fermi, and von Neumann had separate discussions in Princeton about the model and decided that some of its conditions had to be changed. When Ulam communicated this news to Teller, Ulam observed that Teller "was pale with fury."[33] When Teller calmed down, his reaction was "an even more determined and fiery assault, involving a further marshalling of the nation's scientific leadership."[34] The deteriorating international situation lent further urgency to Teller's project. There was no doubt that von Neumann continued his support of the development of the hydrogen bomb, and he carried out a huge number of calculations on his ever-improving computers, the ENIAC and the MANIAC. Von Neumann estimated that one of their calculations required more multiplications than the total number of multiplications performed by humankind up to that point. The calculations reinforced the doubts about the original design of the hydrogen bomb. It was a frustrating period.

Had Teller been sure that the hydrogen bomb would be impossible to build, he would have thought nobody could build it. However, it was a tormenting thought that while the Americans might not be able to produce it, the Soviets might. It is to his credit that under such impossible tension he was still capable of thinking about new solutions. His thinking might have been paralyzed under the pressure, but it seems that the difficulty and competition drove him all the more. Luis Alvarez was amazed by Teller's determination to carry on with the work: "Edward Teller told everyone for years that he was going to make a hydrogen bomb. His Super proved unworkable, but since he had committed himself he couldn't walk away. . . . Public commitment is often an essential driving force in invention."[35] The word "frustration" still best characterized the atmosphere of this period. When much work but little real progress was reported at the Sep-

tember 10, 1950, GAC meeting, Oppenheimer expressed "frustrated gratitude" to the participants of the project.[36]

For the time being, the initial, or "classical" Super remained the model
for the hydrogen bomb, although it seemed less and less promising. Considering its difficulties, people tend to fault Teller for having been too optimistic and thereby misleading the authorities. The classical Super was a
long cylinder filled with liquid deuterium; the idea was that it would be
ignited at one end by having a fission bomb bringing it to a high temperature. The question was whether there would be sufficient energy to make
the burning wave propagate down the cylinder. That seemed to be a relatively simple thing to calculate, but it was still complicated because there
were competing effects and the outcome of the competition decided
whether it would work or not. The higher the temperature, the faster
would be the burning and the fusion of deuterium to deuterium.

At that time it was known that the reaction involving deuterium, the
deuterium plus deuterium reaction, had two channels: one, to make
helium-3 plus a neutron, and the other, to make tritium plus a proton. The
tritium then also reacts and its reaction has about one hundred times
higher cross-section than the deuterium plus deuterium reaction. Tritium
would burn instantly and would add considerably to the energy release.
On the other hand, there is also a high-energy neutron in the tritium reaction, which is likely to escape and not contribute to the heating. All these
processes required careful calculations.

The scientists knew what they wanted to accomplish; they just did not
know how to do it. The only example they could learn from at the time was
the thermonuclear reactions in the sun, but the example of the sun was of
limited value to them. Under terrestrial conditions, higher temperatures are
needed than even the temperature in the center of the sun, which is estimated to be sixteen million degrees. There the gravitation due to the
tremendous mass of the sun compresses the material to an extreme large
density. The process of thermonuclear reaction in the sun is relatively slow,
fortunately for us, meaning that the sun is expected to serve as our energy
source for ten billion years. At the much higher temperatures necessary for
maintaining the thermonuclear reactions under terrestrial conditions, much
of the energy is in the form of electromagnetic radiation, which may escape,
causing the reaction to die. This was the main problem with the classical
Super, and this problem was later solved by the Teller–Ulam approach.

One of Fermi's former graduate students, Richard Garwin, by now a
young faculty member at the University of Chicago, was a visitor at Los
Alamos during both summers of 1950 and 1951. He told me in 2008

about the way they operated during this period.[37] The first summer, Garwin shared an office with Enrico Fermi; they had two facing desks. Ulam would come in every morning, sit down in a chair, and Fermi and Ulam would go over the spreadsheet calculations that had been done by an assistant, Miriam Caldwell, the night before. They would plot up the results to see what the temperatures were and to see whether there were any promising data. Then Fermi would take out a clean spreadsheet and make another case to be calculated, varying the diameter and perhaps the temperature as well, and Fermi would fill out the first couple of rows in the spreadsheet. He had converted the higher-order differential equations into first-order differential equations. All that was required was additions and multiplications, and table-look-up, and sometimes raising things to some power in order to determine cross-sections and pressures. Fermi would do the calculations for a couple of lines just to make sure that it could be done, and then the calculations would be turned over to Miriam Caldwell. She would work on the problem and return with the results of her calculations the next day.

If people like Fermi and Ulam were willing to spend so much time on Teller's classical Super, then accusing him of "optimism" may have been unjustified. In hindsight, though, we know it did not work, although it may be better to say that its impossibility has never been fully demonstrated. Eventually, a much better and more straightforward solution arose. However, in the meantime, Fermi kept working on the original idea, trying to contribute to the understanding of the process. This was probably the last time people did calculations by hand for such work. Fermi used the slide rule for multiplications and logarithms, and an electric desk calculator for the additions. Fermi was very good at it; he had done it for a long time.

No set of parameters had been proposed that would make the classical Super work, but there was no theorem precluding the possibility that it might work. Teller maintained to the end that it might have worked. Back at Los Alamos, he expressed his ambivalence about the place by saying that Los Alamos would be able to survive a failure of the tests, but should the tests be successful and show the viability of the Super, "the laboratory might not be strong enough to exploit the triumph."[38]

Fermi managed to stay out of personal controversies and he remained on good terms with Teller until he died in November 1954 at the age of fifty-three. There were increasingly visible fault lines between Teller, Wheeler, and von Neumann, in one group, and the leadership of Los Alamos, which included Bradbury, Manley, and Holloway, in the other. Teller went for another visit to Washington in February 1951 and met with

Strauss, among others. Strauss was not a member of the Atomic Energy Commission at this time, but he remained a strong force and a staunch supporter of the hydrogen bomb.

Since there was little hope of getting unambiguous answers to the questions about the classical Super from theory, the scientists focused on the tests planned for the spring of 1951. The series was labeled "Greenhouse." Los Alamos continued devising these tests to see how to produce thermonuclear reactions even if they would not yet be suitable for weapons design. The best-known test of the series was called "George," and it was executed on May 8, 1951. It was a success because it showed the presence of a thermonuclear reaction for the first time.

The test was performed at the Enewetak (or Eniwetok) Atoll in the Marshall Islands under idyllic conditions—the contrast between what nature offered and what man was doing to it could not have been more striking. There was no straightforward path from this test to the hydrogen bomb. Still, the leaders of Los Alamos were content with it as an important step. It provided food for thought for the coming months. One problem was that it used a large amount of tritium that could have rendered it impractical for a bomb. Despite this, Teller was elated by the success of the test; he was happy to pay the five dollars to Lawrence he lost in a bet. Lawrence also came to watch and was more optimistic about the expected outcome than Teller, hence the bet.[39]

By the time the George test occurred, it had become clear that the concept of the classical Super was obsolete and had been overtaken by a brand-new idea, if not a new design. As we have seen, one of the problems with the classical Super was that a lot of energy during the fission explosion would be dissipated by radiation. The higher the temperature, the more probable fusion becomes, but at the same time more of the energy leaves the system. However, as Teller quickly realized, the particles carry energy in proportion to their number, but radiation carries energy in proportion to its volume. If the number of particles could be increased with a simultaneous decrease of the volume, more energy would stay with the particles. So the question was whether the mass of deuterium that they planned to use would be possible to compress. Since they knew that even heavy metals could be compressed in the implosion process, then surely liquid deuterium should be possible to compress. If the deuterium mass were to be strongly compressed, a thermonuclear reaction would be possible. This is how Teller remembered thinking about these events.[40] According to others, though, when the question of compression had come up in discussions Teller always rejected it.

One day, Ulam suggested a practical and feasible solution to the problem. He suggested that instead of the heat produced by the primary fission bomb, they should use the energy in the expanding mass of the bomb's core, in other words, use compression by shock wave. Despite the importance of Ulam's idea, he writes about it in a rather low-key manner in his autobiography: "[P]erhaps the change came with a proposal I contributed. I thought of a way to modify the whole approach by injecting a repetition of certain arrangements."[41] Ulam's wife, however, remembered a dramatic scene. On the day of the discovery, she found her husband in their living room with "a very strange expression on his face." He told her, "I found a way to make it work. . . . It is a totally different scheme, and it will change the course of history."[42]

Whether it happened by accident or by design—quite possibly the latter, if we consider the relationship between Ulam and Teller—Teller was the third person Ulam told about his idea. The first was Carson Mark, who was in charge of the theoretical division at the time, and the second was Norris Bradbury. It was not until the following morning that Ulam told Teller about his brainchild, and this is how Ulam remembered Teller's reaction: "At once Edward took up my suggestions, hesitantly at first, but enthusiastically after a few hours. He had seen not only the novel elements, but had found a parallel version, an alternative to what I had said, perhaps more convenient and generalized."[43] Teller told Ulam that he had thought of something that would work even better. They should use the X-rays emitted by the primary fission bomb to compress the deuterium; that is, the compression should happen through radiation.

Teller had the impression that Ulam refused to listen to him. Everybody agreed that compressing the deuterium by radiation was better because the compression had to be symmetrical, and continued symmetrical compression could be better achieved through radiation. In any case, the impasse was resolved by Ulam even if Teller did not seem to welcome it with sufficient enthusiasm. In short, Ulam saved the project at that moment. In fact, Teller remembered years later that he himself had also thought about the possibility of compression before Ulam came to him with his ideas.[44] It was characteristic of the deteriorating relationship between Ulam and Teller how cynically Ulam referred to Teller in a letter he sent to von Neumann, in which he reported the new idea for the Super: "Edward is full of enthusiasm about these possibilities; this is perhaps an indication they will not work."[45]

Peter Lax was a young scientist working at Los Alamos in 1950–1951. He observed both Ulam and Teller, and he used a strong word in characterizing their relationship: "They hated each other; certainly, Ulam hated Teller. There was nothing subtle about it."[46] According to Lax, Ulam had an idea about the technical aspects that may have had its origin in the plutonium bomb in which the plutonium was compressed by shock waves caused by explosion. Ulam thought that shock waves could compress similarly the fuel of the hydrogen bomb. Lax, who had been an expert of the mathematics of shock waves, thought that it would be tricky since the shock waves would have to be shaped in a very special way.

Teller's idea, as noted, was to compress the material through radiation, for which it was also easier to carry out the necessary calculations. Lax added, "Teller and his associates made the calculations and showed that radiation would compress the material in the desired extent. Ulam never made the calculations about the shock waves. Ulam was not a calculating person." Lax had no doubt that Ulam's idea about the compression came first, before Teller's idea, but "Teller then went on and pushed it and pushed it whereas Ulam threw out the idea and did not do anything with it."[47]

It seems that there were at least some elements of the implosion technique of the plutonium-based fission bomb that reappeared in the new approach. This makes one wonder why nobody had thought of the new solution before. It is not uncommon, though, to see a similar approach being reinvented again and again in different applications as if it had been a completely new idea. As for the question of priority, it makes it even more difficult to assign it when the new approach may have been around "in the air" for some time.

Teller may have honestly though erroneously underestimated the value of Ulam's initial suggestions, since Teller likely believed that sooner or later he would have arrived at them himself. The fact remains, however, that it was Ulam, and not Teller, who came up with the initial idea that then sparked Teller's further innovative thinking. Teller may have also thought that the ideas had been there for some time in his head when Ulam came out with his own idea. However, as long as a new idea is not recorded or communicated so that others can learn about it, it does not count from the point of view of assigning priorities. Truly great discoveries are especially apt to appear in hindsight as trivial achievements that many others could have made.

Years later, the Danish–American physics professor J. Rud Nielsen asked Niels Bohr about how great Teller's contribution to the hydrogen bomb had been. Bohr told him, "Old physicists who have turned administrators might not think of this solution. However, if you had asked a good class of physics students, two or three of them would have suggested this solution. Anyway, the Russians did the very same thing."[48] This sounded as if he were belittling Teller's achievement. In the same conversation, Bohr was highly critical of how Oppenheimer had been treated. One cannot escape the thought that the two comments were not unrelated. Richard Garwin also said that others could have come to the same idea.[49] But, again, although they could have, they did not, whereas Teller did.

Richard Rhodes discussed this question with Carson Mark, and his comment is no less puzzling. To Mark, anyone who might have heard of Ulam's idea of compressing the fusion fuel by hydrodynamic shock and who was familiar with previous tests of the thermonuclear program could not have avoided the better idea of radiation implosion.[50] According to Richard Garwin, Mark said that anyone beginning to calculate the compression by nuclear shock would immediately find that the radiation got there first and was effective in itself.[51] Mark did not say, however, why he himself did not realize this. One would have thought he was in the right position for doing so as the head of the theoretical division at Los Alamos, and Ulam had told him about his idea before he told it to Teller the next day. Teller's underestimating Ulam's contribution and the others' underestimating Teller's contribution arise from the same unwillingness to recognize what it takes to come first to an idea and dare to enunciate it, even if it appears trivial.

Teller and Ulam first wrote a sketch of their proposal and, after some further changes, they wrote a joint report, dated March 9, 1951. It was titled "On Heterocatalytic Detonations I: Hydrodynamic Lenses and Radiation Mirrors." It is usually referred to as the "Teller–Ulam design." It is still classified, and there is some skepticism whether its declassification will ever resolve the question of proper assignment of credit between the two of them. However, if each presented his own contribution in the joint report, there is hope that some aspect of the controversy will one day be resolved regardless of whether the report was compiled jointly or by Teller alone. However, it is highly probable that Ulam's and Teller's contributions are not delineated in their joint report, which was written, in fact, by Teller alone.[52]

Soon, in the spring of 1951, there was a second report, which described de Hoffmann's detailed mathematical analysis. It was composed at Teller's suggestion and placed a second fission component—a subcritical amount of uranium-235—at the core of the thermonuclear material. The analysis fully confirmed what Teller had hoped to achieve. For all practical purposes, it was a joint work between Teller and de Hoffmann, but de Hoffmann put Teller's name on it as the sole author.[53] Teller never forgave himself that he did not decline this gesture and insist on adding de Hoffmann's name to the report.

Teller, however, was as busy as ever in this period. He continued his lobbying in Washington, causing headaches for the AEC chairman Gordon Dean. It was an impossible situation in which a member of the Los Alamos Laboratory and the leadership of the laboratory represented opposite goals before their superiors. In April 1951, Teller wrote a memorandum proposing to open a new laboratory to which the frontlines at Los Alamos hardened further. Otherwise, the spring of 1951 was a triumphal period for Teller, not that he gave himself over to elation. There were two pivotal events both for him and for the entire program. One was the George test, and the other was the emergence of the Teller–Ulam design.

A crucial meeting of the Atomic Energy Commission took place at the Institute for Advanced Study in Princeton in Oppenheimer's office. The AEC commissioners, GAC members, Los Alamos leaders and associates, and other experts attended the meeting on June 16 and 17, 1951. The first day was devoted to presentations about and evaluation of the George test. On the second day came the introduction of the new Teller–Ulam design, which won over even the staunchest opponents of the hydrogen bomb. There has been some controversy about how the June 16–17 meeting proceeded, but there was no controversy about its outcome. Even Oppenheimer called the new approach technically so sweet that it had to be produced. He also said that had they known in 1949 what they knew in 1951, the GAC would have supported the development of the hydrogen bomb.

This is how Oppenheimer looked back at the debate over whether to develop the hydrogen bomb and about the change in his opinion in 1951 when the Teller–Ulam approach emerged. The following quote is from his responses at the questioning by the board chairman during his security hearing in the spring of 1954 (see chapter 8):

> . . . my feeling about development became quite different when the practicabilities became clear. When I saw how to do it, it was clear to me that one had to at least make the thing. Then the only problem was what would one do about them when one had them. The program we had in 1949 was a tortured thing that you could well argue did not make a great deal of technical sense. It was therefore possible to argue also that you did not want it even if you could have it. *The program in 1951 was technically so sweet that you could not argue about that.* It was purely the military, the political and the humane problem of what you were going to do about it once you had it.[54] (italics added)

Here Oppenheimer reduced his 1949 opposition to the development of the hydrogen bomb to technical questions. Where were then the strong moral scruples that received so much emphasis in both the majority and the minority opinions of the GAC at the end of October 1949? It is true that even at this later point, Oppenheimer debated the question as to whether to make the Super and what its possible use might be.

Teller was very impressed by Oppenheimer's support of the new approach at the June 1951 meeting, and he remarked about it at the Oppenheimer hearing, later in 1954:

> In June 1951, after our first experimental test, there was a meeting of the General Advisory Committee and Atomic Energy Commission personnel and some consultants in Princeton at the Institute for Advanced Study. The meeting was chaired by Dr. Oppenheimer. Frankly I went to that meeting with very considerable misgivings, because I expected that the General Advisory Committee, and particularly Dr. Oppenheimer, would further oppose the development. By that time we had evolved something which amounted to a new approach, and after listening to the evidence of both the test and the theoretical investigations on that new approach, Dr. Oppenheimer warmly supported this new approach, and I understand that he made a statement to the effect that if anything of this kind had been suggested right away he never would have opposed it.[55]

In any case, the June 1951 meeting could have been a resounding victory and vindication for Teller. However, by then he had alienated so many of his colleagues, especially at Los Alamos, that his victory resulted neither in vindication or the elimination of all his frustrations.

Switching from the classical Super to the new Super was not as easy as one might have thought, looking back. New tests for the classical approach had been in the making. The group knew it was not the ultimately useful approach, but something useful might have been learned from these tests. In the cautious approach by the Los Alamos leadership, inertia seemed to take center stage. Now that Teller was embarking on a new plan, the Los Alamos leaders appeared to adhere to the classical version.

Nonetheless, cautious plans were made to complete the feasibility studies for the new design as well, with a target date of October 1951, and a test date in the spring of 1953. Things soon started to accelerate beyond this schedule. Richard Garwin, spending his second summer at Los Alamos, designed the thermonuclear device according to the new proposal.[56] Even Bethe approved the design that would become the "Mike" test on November 1, 1952 (October 31, Washington time).

With the forthcoming test, there was renewed concern over whether the thermonuclear blast might ignite the atmosphere and oceans. Teller asked his old friend Gregory Breit, who was at Yale University at the time, to assess such a possibility. Breit gave an account titled "Notes for Talk on Atmospheric Ignition" at an Oak Ridge meeting not long before the "Mike" test. The report, which remained classified until about 1975, gave assurance that there would be no such consequence of the explosion. Breit augmented the theoretical considerations with some experimental work. It is especially important that even though Teller by then had dissociated himself from the preparations for the "Mike" test at Los Alamos, he cared about its possible consequences and initiated Breit's investigation.[57]

Unfortunately, the Teller–Ulam controversy has spoiled all attempts to unambiguously assess the events around the development of the American hydrogen bomb. The impression is that Teller's later descriptions, as we will see below, alternated between giving more credit or less credit to various individuals, especially to Ulam, in connection with the hydrogen bomb. Thus he further confused the story. According to Ulam's wife Françoise, when it became clear that Ulam had saved the project, "from then on Teller pushed Stan aside and refused to deal with him any longer. He never met or talked with Stan meaningfully again."[58] Gamow, who was in Los Alamos at the time and could not be considered unfriendly toward Teller, thought that "Teller behaved badly in hogging credit for the hydrogen bomb when the key intellectual trick came from . . . Ulam."[59]

When the blast effect of the hydrogen bomb was investigated, it turned out to be far from unlimited: the conclusion was that the equivalent of ten million tons of TNT would be typical. But even a bomb of this size would destroy everything within several miles in every direction. The question might be asked whether there would be a chance to blow away a huge chunk of the atmosphere because the incredibly high temperature might provide the air molecules sufficient kinetic energy, that is, velocity, to leave the earth's gravitational attraction. Although the local blast would be limited by finite atmospheric depth, this would not mean that the air would reach escape velocity. It would mean only that the five-mile depth of air would be blown up and out, descending at an altitude. The radioactivity would be blown more into space than on the earth. Thus, the size of the bomb could be further increased. The Soviets had the dubious title of record in this respect with an explosion of the equivalent of tens of millions of tons of TNT. However, the most cost-effective size would be established at one million tons of TNT. What was especially frightening was that the hydrogen bomb proved to be relatively inexpensive to produce. This might prompt those who could make it to produce it in large numbers.

## CAN WE KNOW THE PAST?

One of Teller's favorite sayings was "Although we cannot change the past, we can know it. We can change the future, but we cannot know it."[60] He also stated, though less frequently, "The future is, of course, unknown. By contrast, the past is wrongly known. History, most often, is recorded with a great amount of undeniable and systematic distortion." How the story of the hydrogen bomb was reflected in Teller's writings illustrates the difficulties of conveying and knowing the past.

Two journalists, James R. Shepley and Clay Blair Jr., published a small book about the hydrogen bomb in late 1954.[61] A foretaste of bias, exaggerated praise, and unfounded accusations were already in a prior article in *Time* magazine by the same authors.[62] This article upset Teller, but he merely sent off a mildly worded letter of protest to the editor.[63] When Shepley and Blair were preparing their book, they interviewed Teller.

The Shepley–Blair book is an easy read, but it is often simplistic and wrong. Some of the book's erroneous statements might even sound hilarious. In addition to important misinterpretations, there are many trivial errors. The narrative is soaked in admiration for Teller and animosity toward Oppenheimer. Scientists are especially sensitive to assigning credit, and the book grossly failed at this. Not only did the authors exaggerate Teller's role at the expense of others, including colleagues, the whole Los Alamos laboratory was treated unjustly whereas Livermore was praised unjustifiably in connection with the development of the hydrogen bomb. A scientist called the book "a valentine to Teller with Oppy's [Oppenheimer's] blood on it."[64]

A few of the statements in the book are mentioned here to illustrate what may have caused Teller's discomfort. Shepley and Blair reduce the big controversy over whether or not to develop the hydrogen bomb to a struggle between two theoretical physicists, Oppenheimer and Teller.[65] They use powerful contrasts of doubtful validity. They write, referring to Oppenheimer, "No scientist in the history of the United States was ever in a more powerful political position or in a position to influence more directly governmental policy on a high level." In contrast, they said of Teller, "No scientist could have had less political influence than Edward Teller."[66] The authors allege that "Teller was denied a specific job in connection with the development of the atomic bomb," and they ascribe this to the fact that "Oppenheimer did not like him personally—a fact that was perhaps traceable to their differing political views."[67]

Otherwise, the authors describe events and attitudes in a similar way, as Teller would many years later, such as underplaying Stanislaw Ulam's contribution and blaming Norris Bradbury for diminishing Teller's leadership role. Referring to the June 1951 GAC conference in Princeton, however, even Shepley and Blair find nothing objectionable. The authors hint that Bradbury forced Teller out of Los Alamos.[68] Even Shepley's and Blair's obvious bias toward Teller could not make them refrain from some slight criticism of their hero. They noticed that he was not given to self-criticism or to admitting his errors in any way: "Teller was never a man to look up into the western night and beat his breast in anguished cries of *mea culpa*." They also noted the rigidity of his political views: "The younger scientists at Livermore found him a man of fixed political views."[69] Some of the Shepley–Blair statements gained notoriety. They referred, for example, to Oppenheimer's opposition to the hydrogen bomb in the following way: "It is not criminal to be wrong about the weapons of the atomic age, only fatal."[70] This was stated in a book that appeared a few months after the AEC investigation, which had concluded that Oppenheimer must be stripped of his security clearance.

The Shepley–Blair book was so blatantly erroneous that when Teller was introduced to Senator John F. Kennedy in 1958, and the politician praised him, referring to the Shepley–Blair book, Teller took it as an insult. Kennedy was very sophisticated, and the praise was a slightly veiled criticism.[71] This was their first but not their last unpleasant encounter. The book was so biased in Teller's favor that he felt compelled to publish an article in *Science* with the title "The Work of Many People."[72] He liked the title so much that he used it repeatedly in subsequent publications dealing with the history of the hydrogen bomb.

He began his article by stating that collective efforts in modern scientific and technical developments are often presented to the public in highly simplified ways. An example is when a brilliant idea is identified with the name of a single individual. This is an obvious reference to the Shepley–Blair book, although the book is not mentioned explicitly. Moreover, the nine-page article carried not a single reference—a conspicuous feature for such a prestigious publication as *Science*. Teller went out of his way to assign credit—and much of it—to many people. Mao Tse-tung's (Mao Zedong's) saying comes to mind: When your sword is bent, in order to straighten it, you have to bend it first in the other direction. Teller must have felt he had a lot to compensate for.

James Franck was one of the people Teller had sent his article to for comments before publication. Franck noted the "exaggerated modesty" with which Teller described his own contributions and warned that such an approach "might even raise doubts as to whether the credit you give to others is a fully honest evaluation of the situation."[73] According to Teller's later hints, he was willing to even tell white lies to compensate for the damage done by the Shepley–Blair book.[74]

Describing the situation after the war, Teller profusely praised Bradbury's leadership and the dedicated contributions by de Hoffmann, Ulam, and Everett. A conspicuous omission was the failure to mention Richard Garwin's work on the design of the thermonuclear device based on Teller's successful new idea. One might ascribe this omission to Teller's explicit focus on the theoretical considerations and his exclusion of political decisions and actual design. However, he did mention that "the most complex kind of apparatus was being built in order to observe the results of the test."[75] Building is one step further from theory than design, so his omission of Garwin in this article remains a puzzle.

Mention was made of "an imaginative suggestion by Ulam" along with "a fine calculation by de Hoffmann," but leaving no ambiguity, it was reaffirmed that "I had made the suggestion that led to his calculation."[76] As for de Hoffmann's calculations, he said that "I expected that we would jointly sign the report containing the results." De Hoffman, however, "argued that the suggestion counted for everything and the execution for nothing."[77] De Hoffmann's approach was thus very different from how Teller treated Ulam's "suggestion."

Significantly, from the point of view of later accusations, Teller was laconic about the Princeton meeting in June 1951, saying that "everyone clearly recognized that with a little luck, only a great deal of work stood between us and final success."[78] There was not even a hint of anything unpleasant that might have colored his experience at Princeton. The article is a glowing recognition of the abilities and dedication of Los Alamos. We have to remember, this was published in 1955. Only those who had inside knowledge with the true history could notice that Teller was not faithful to the facts in the details of his story. Others, however sophisticated, had to take it at face value. Warren Weaver, who was instrumental in launching molecular biology, wrote a nice letter to Teller upon the appearance of his account in *Science*. Weaver read the article with "pleasure, intense interest, and great appreciation." He added that Teller made "all the small-minded politicians look very small indeed."[79]

Teller's subsequent writings and statements on the same story showed marked changes, however. We single out here for comparison what he had to say in 1962, 1979, 1986, and in his *Memoirs* in order to have some perspective. The 1962 book *The Legacy of Hiroshima* by Teller and Allen Brown is a narrative by Teller.[80] The first of a total of four parts carries the title "The Work of Many People," and it covers broadly the efforts of creating nuclear weapons. Its third chapter, "The Hydrogen Bomb," is of particular interest. There are discernible differences as compared with the 1955 *Science* article.

The most conspicuous change in the 1962 description occurs in Teller's introduction of the new Super on February 1, 1951. Here the description omitted Ulam's name. Teller noted merely that others could have come up with the same idea, and they could have come up with it even sooner. In another departure from the 1955 *Science* paper he lamented that during the spring of 1951, he discussed the new approach with a number of responsible people, such as Norris Bradbury, Carson Mark, and Gordon Dean, but nobody listened. Here as elsewhere Teller explained that Dean's distraction was due to the fact that Teller's fly was open, thanks to a faulty zipper. One of the most significant departures from the *Science* paper is the description of the June 1951 Princeton GAC meeting. Teller now complained that the initial speakers neglected to even mention his new idea. He admitted, though, that when he finally got the floor and described his new idea, the response was "enthusiastic and unanimous."[81] In the 1962 version, he gave extra credit for reliable calculations on the new concept to Marshall Rosenbluth, Conrad Longmire, and Lothar Nordheim, and cited Marshall Holloway as the one who led the preparations for the forthcoming test. Garwin was still missing from the roster.

It might then be surprising to read about Garwin's significant contribution in Teller's memorandum seventeen years later, in 1979. We have already described the special circumstances under which this memorandum was compiled (chapter 4). Here we review some of its assertions concerning the history of the hydrogen bomb with attention to possible discrepancies with previous and subsequent descriptions. With respect to Oppenheimer's attitude toward thermonuclear work at Los Alamos, Teller noted that "not long after we arrived Oppenheimer discouraged further efforts on the hydrogen bomb." However, this discouragement was not unqualified, as "Oppenheimer did not object to my putting a considerable portion of my time into hydrogen bomb work, and I think it was understood that two or three other people might help me with it."[82] Teller reiterated Oppenheimer's conflicting decisions; first, after the July 16 test, giving more attention to the hydrogen bomb; then, after Hiroshima, seeing no future in the work on the hydrogen bomb. Then, again, Teller mentioned Ulam's calculations, whose value was in demonstrating that the classical Super would not work and that eventually von Neumann's computations came to the same conclusion.

Teller made it explicit that by January 15, 1951, his forty-third birthday, he had already come up with his new idea for a workable hydrogen bomb. It may have come to him some time during the preceding December. He even gave a name to the new design, the Equilibrium Super. According to the memorandum, Ulam came to his office sometime in February 1951 with the suggestion to compress the deuterium fuel. Teller's immediate response was in the affirmative, except, he said, it would not work the way Ulam had imagined it, by mechanical compression. Teller's suggestion was compression by radiation. Teller put everything down on paper at that point, and they both signed it. Teller then asked de Hoffmann to perform the necessary detailed calculations. De Hoffmann prepared the report, and in 1979 Teller said that both de Hoffmann and he signed it.[83] Prior and subsequent descriptions by Teller stressed that he was the sole signer of the document at de Hoffmann's insistence.

The 1979 testimony gives some justification for Teller's rather peculiar approach to the question of priority in connection with the development of the hydrogen bomb. He stated that he would have had no problem in sharing the authorship of discovery with Ulam had Ulam not advocated the failure of one of the tests, called "George," and his prediction that the hydrogen bomb would never work. Teller was very explicit in 1979, saying that "Ulam did not have the idea, he did not write the paper, and when it came at last to the decision after [the] George shot, he declared that he did not believe it." Here, as later, Teller defined his criteria for authorship: "To me the authorship in a paper or in a report does not mean a question of priority, it means a question of responsibility." By responsibility he meant accepting and advocating the position of the paper. He stresses that Ulam never stood up for their paper and brought up the Princeton meeting as example "where Bradbury tried to suppress the mention of this new idea, I carried the ball alone" (ignoring the fact that Ulam was not present at the meeting).[84]

This is the point at which Richard Garwin's name appeared in Teller's 1979 narrative. Teller had not mentioned Garwin previously in connection with the story of the hydrogen bomb. Now he wrote that Garwin was "a student of Fermi's, very young at that time. He came to visit [Los Alamos] in the summer in 1951 . . . I told him about all these things and I asked him to put down a concrete design with dimensions." Garwin completed the design before his return to the University of Chicago at the end of the summer. Teller continued, "So that first design was made by Dick Garwin. It was then criticized forward and backward, in the end it stood up to all criticism." An important point was Bethe's approval. "Bethe had come out and looked it over, and first tried to change it in some ways then gave in and said all right that will be a reasonable way to shoot it." As if to make sure that he was understood about Garwin's role, Teller repeated, "[A]s far as I'm concerned the preparation for the hydrogen bomb was completed by Dick Garwin's design shortly after the famous meeting in Princeton."[85]

We now briefly mention Teller's book *Better a Shield Than a Sword*, published in 1986, seven years after both his heart attack and his memorandum.[86] One might have thought that Garwin would have by now taken his place in Teller's narrative about the hydrogen bomb, but this did not happen. Teller's new book reproduced the story of the hydrogen bomb already printed in the 1962 book; this is clearly stated in the acknowledgments. Closer scrutiny, however, reveals that in the chapter about the hydrogen bomb, there were some changes pertaining to less important affairs than Garwin's contribution.

For example, a whole new paragraph was inserted in which Teller mentioned that John A. Wheeler read in the Bible the sentence "Six days shalt thou labor," which showed remarkable consistency with Los Alamos having switched to a six-day workweek in their efforts to develop the hydrogen bomb.[87] An even longer paragraph was added at the end of the chapter in which Teller described his last meeting with Fermi, who read the manuscript "The Work of Many People" and urged him to publish it, although Teller did not quote Fermi directly.[88] He only said that Fermi asked him the reasons for his (Teller's) hesitation to publish the article and then he followed Fermi's advice.

There were other changes and additions in the text, one of which stressed that the work on the hydrogen bomb contributed not only to the safety of the United States but also to science. But Richard Garwin's name was not mentioned. Garwin returned in the *Memoirs* but only to the extent he is mentioned in the 1979 memorandum.[89] This was his sole appearance in the book, even though Garwin had a distinguished record in defense matters and had numerous clashes with Teller on various occasions (see chapter 11).

Incidentally, according to Teller's interpretation, Fermi's deathbed reaction to the manuscript "The Work of Many People" also underwent some transformations in time. Fermi's friend and former disciple Emilio Segrè described his conversation with his former mentor on his deathbed. Segrè reported that Fermi found Teller's behavior in the hydrogen bomb debate and in the Oppenheimer hearings "reprehensible," with one of the consequences being the split of the scientific community. Then Fermi added these words, "The best thing Teller can do now is to shut up and disappear from the public eye for a long time, in the hope that people may forget him."[90] Fermi asked Segrè to "summon" Teller to see him, adding, "What nobler deed for a dying man than to try to save a soul?"[91] Segrè relayed the summons to Teller without delay, and Teller visited Fermi. Then Segrè merely referred to the description of the visit given in Teller's book *Energy and Conflict*, published in 1976. After Fermi read "The Work of Many People," he said—according to Teller's account—"I think you should publish it."[92] In Teller's *Memoirs* Fermi is quoted as "Enrico advised me strongly and insistently to publish it."[93]

In summary, there is a series of accounts by Teller about the history of the hydrogen bomb, which include:

1955, "The Work of Many People" (article in *Science*)
1962, *The Legacy of Hiroshima* (book)
1979, Unpublished memorandum after his heart attack
1986, *Better a Shield Than a Sword* (book)
2001, *Memoirs* (book)

It is almost as if we could plot the information in a graph, as scientific data are often presented. Of course, there is no way to do this with any rigor, but what should have been a fairly constant set of attributes appears to be fast-moving variables. These variables include such names as Ulam and Garwin, and such deeds as coming to a crucial idea or making an important design. The 1979 memorandum can be looked at with confidence as far as the credit due to Garwin is concerned, and it is consistent with Garwin's testimonial (see below). Its validity with respect to Ulam's contribution, however, is questionable. It might even be looked at as another attempt to diminish Ulam's contribution at this critical hour, which could have been Teller's last for all he knew, given the precarious state of his health after his heart attack.

Looking back at Teller's frustration during 1950 and at his arrival upon the solution, we might ask—with the benefit of hindsight—why didn't he commit his new solution to a document? It may well be that the solution was being formed in his head, but it was only after Ulam had told him about his solution that Teller brought himself to the exact formulation of what he may have been contemplating. In this sense, even if it was merely a triggering effect, Ulam made an important contribution. That Teller could instantly snap his improved solution at Ulam in the same conversation that Ulam told him about his idea indicates that the right approach might have been brewing in Teller's thoughts for some time. Teller made a point of his telling his new idea to Lee DuBridge, then president of the American Physical Society at the time of the society's Christmas meeting in 1950 (apparently, DuBridge never told anyone about their exchange).

Teller's changing recognition of Garwin's contribution still remains a puzzle that might lend itself to different interpretations. He may have focused his attention on the contributions of the people actually working in Los Alamos in his 1955 and 1962 narratives. Then came his heart attack, after which he reviewed this most important component of his oeuvre and realized the omission. If this were the case, it should be a puzzle why Garwin was, again, omitted in Teller's subsequent accounts of the hydrogen bomb story. Garwin clashing with Teller over the years on questions of disarmament might have contributed to Teller's not giving him proper credit, both before and later. Also, Garwin was a powerful opponent of the Strategic Defense Initiative (SDI), something Teller strongly believed in. But then, in preparing his *Memoirs*, Teller did give Garwin credit. Garwin's role in the SDI controversy was no longer looming so heavily over Teller, whereas, again, his value in overshadowing others was enhanced.

The question also arises, why didn't Garwin protest having almost been erased from the history of the hydrogen bomb? There could be two reasons: one, it was his nature to be more interested in having the job done and moving on than receiving credit for it. The other reason might have been the fact that he didn't feel comfortable with having made such an important contribution to the weapon that, in his later years, he was so much opposed to. Garwin, however, did not hide his interest in the history of the hydrogen bomb or his contribution to its development. It is curious that Wheeler, in his fairly detailed description of the scientists participating in the hydrogen bomb project, made no mention of Garwin in his autobiographical volume. He may have been one of the most knowledgeable among those scientists who were still around. He could have set the record straight, though for some reason he did not.

As for Ulam's contribution, Teller seems to have diminished Ulam's role in his writings. According to Garwin, Ulam's basic idea was to have an auxiliary bomb compress a large amount of material without limiting the applicability of this idea to thermonuclear fuels.[94] The nuclear explosion would compress the fuel via the shock wave it generated. Teller saw at once why it would not work: even the increase in the number of particles would not suffice to increase the reaction rate to the desired level. He suggested instead utilizing the X-rays generated in the nuclear explosion that constitute most of the energy in such an explosion.

Garwin witnessed the frustrating time of the classical Super when he spent the summer of 1950 in Los Alamos, and he noticed the invigorating change brought by Teller's new idea when he returned there the following year. It was at Fermi's suggestion that Garwin be employed as a consultant at Los Alamos. During his first summer, he went to the classified reports lab and read all the weekly progress reports generated during the war and the postwar period, for which he had security clearance. Fermi took Garwin under his wing in Los Alamos, introducing him to the many scientists who came to see him. Fermi also took Garwin to various laboratories to get to know people and to get acquainted with various experiments.

When Garwin returned to Los Alamos in May 1951 for his second summer, he asked Teller what had happened since he had left. Teller showed him the paper he had co-authored with Ulam. He then asked Garwin to devise an experiment that could prove whether the new idea would or would not work. What Garwin designed would eventually become the "Mike" test, and it succeeded spectacularly. Garwin excelled in bringing together different ideas along with theory, experiment, and technology. He did not generate new ideas, but there was no need for new ideas at this point. What he did was decide between various approaches for which he was well experienced in spite of his young age. He was probably uniquely qualified for the task Teller had asked him to do. Teller must have sensed Garwin's intelligence and special abilities, and although the two did not work together at Chicago, Teller must have observed Fermi's outstanding graduate student turned young faculty member. As noted, Teller characterized members of the scientific community in his memorandum to Bradbury in October 1949 (see chapter 5). It may also be that Teller saw an advantage to asking a consultant rather than a permanent member of Los Alamos to design the experiment.

Although the possibility of using lithium deuteride was considered, liquid deuterium and liquid hydrogen were used for this test. Garwin had worked with liquid hydrogen in Chicago with his particle physics experiments. The whole device weighed some seventy tons; it contained a couple of cubic meters of liquid deuterium, which was kept cold by liquid hydrogen. There was a nuclear explosive at one end of the device, which had to be kept warm because a cold explosive could not be detonated. It was not a bomb, as such, but the device could have been used for bombing because it could have been carried by airplanes, flown over the Soviet Union, and destroyed targets there. Actually, the Atomic Energy Commission had a few of these devices built. It looked more like an industrial plant than a bomb.

It may seem curious that such an important job as that of making the design for the "Mike" test was given to a visitor—a consultant who was there for summers only. Had Garwin been less proficient, his involvement might have been considered a slight to the permanent Los Alamos people, but Garwin proved to be the best person for this job. It may also be surprising that, given the boost Teller's new idea gave to the development of the hydrogen bomb and the expansion of the program, Garwin was not offered a permanent job at Los Alamos. Garwin remembers that he "offended a lot of people there, because [he] would tell them exactly what [he] thought about this or that."[95] It may be that this is why the Los Alamos people told IBM when the company offered Garwin a job in 1952 that, as Garwin said, "they would be sorry if they hired me, but even more sorry if they did not."[96]

It is interesting that there was not much interaction between Teller and Garwin on a personal level; they never talked about private matters even though they had, to some degree, similar Jewish–Hungarian roots. Garwin was not interested in politics and he considered the hydrogen bomb merely an exciting technical puzzle. To the question of whether he regretted his role in creating the hydrogen bomb, this is what he said in 2004: "No. I wish hydrogen bombs were not possible. I wish fission bombs were not possible. I do believe that we are lucky not to already have had a terrorist nuclear explosion in one of our cities. I confidently believe that we will have one within the next few years."[97]

Bethe himself gave a history of the hydrogen bomb, which he wrote in 1954 but could not publish until 1982 after it had been declassified.[98] Bethe was induced to write the article by two books, the Shepley–Blair book and Peter Goodchild's book about Oppenheimer, which came out in 1980.[99] He was unhappy with both books for the way they presented the history of the hydrogen bomb. Thus, he amended the article written in 1954 with some comments about the second book when he revised his article for publication in 1982. Bethe admitted that he wrote the original article "in some anger," with obvious reference to the Oppenheimer hearing and its outcome in 1954.

Bethe records the frustration of those involved with the hydrogen bomb project up to the spring of 1951, as none of the methods previously suggested seemed to lead to a successful bomb design. The breakthrough came with the joint paper by Teller and Ulam in March 1951 and with the report by Teller (and de Hoffmann) in April. Bethe gave tremendous praise to Teller for his new concept, which is detailed below.

Bethe said what Oppenheimer had also stated concerning Teller's new Super, "If this had been the technical proposal in 1949, they ["they" being the GAC] would never have opposed the development."[100] Here again it seems that in the stand taken by the GAC in 1949 the technical and moral considerations were merged. Yet Bethe admitted that the fear of Stalin's

Soviet Union possibly developing the hydrogen bomb had to be a strong motivation for the Americans to develop it. It was not just a compelling argument but "the *only* valid argument" (italics in original).[101] He further expressed his doubts as to whether the Russians were really engaged in developing their hydrogen bomb as early as 1949. In 1954, and even in 1982, he could not yet know the status of the Soviet hydrogen bomb. Today we know that by the time the debates about the accelerated program were taking place in the United States in 1949, the Soviets had already been very involved in their program.

Bethe's article referred to the June 1951 conference of GAC in Princeton. He dealt with it in the section he added in 1982 to correct some mistakes in Goodchild's book. Goodchild repeated what Teller had stated on various occasions: that he had difficulties finding an opportunity to present his new concept at this meeting. According to Bethe, "The whole meeting was held in order to discuss Teller's new concept for the design of an H-bomb. . . . Teller himself gave the main presentation, followed by me [Bethe] and the three others."[102] In a letter published in *Science* in 1982, Teller reiterated his complaints about the June 1951 GAC meeting.[103] In an immediate response to Teller's letter, Bethe repeated what he had said in his *Los Alamos Science* article, but with a subtle difference. Concerning the presentation of Teller's method, Bethe wrote, "[T]his was done *after* reports on the most recent test series had been given (italics added)."[104]

The qualifier in Bethe's last statement is important because an independent and objective description of the meeting is consistent with Teller's account. Hewlett and Duncan include a section on the Princeton meeting in their book *Atomic Shield, 1947/1952*, constituting volume II in *A History of the United States Atomic Energy Commission.*[105] According to them, the Princeton meeting followed the successful Greenhouse test, which was the first explosion in which there was a thermonuclear component.

The conference at Princeton was not merely a GAC event, rather, the commissioners of the Atomic Energy Commission attended, as well as some of their staff members interested in weapon development. There was a contingent from Los Alamos headed by the director, Norris Bradbury, and a few additional experts, like Hans Bethe. Although Teller at this time was still a member of the Los Alamos Laboratory, Bradbury did not include him among the Los Alamos speakers. This was good for his independence but bad because his presentation was not scheduled in the official program.

Already in the preparations for the meeting there were some bad omens at Los Alamos: the prevailing mood was that the evaluation of and follow-up work to the Greenhouse tests would keep them busy for a while, and they were apprehensive of any new development that would disturb their well-thought-out schedule. This is puzzling, because by then Teller's new idea described in the joint Teller–Ulam report and the more detailed one prepared by de Hoffmann had become available and known. It meant a drastic change, compared with the classical Super approach. The Greenhouse tests were successful and important in principle for the thermonuclear weapon, but only the new concept carried the promise of a real solution.

In spite of the new developments, the Los Alamos leaders still had testing of the classical Super in their long-ranging plans, in addition to testing a device based on the new concept. It seems that Bradbury and others miscalculated, and personal animosity toward Teller may have interfered with their judgment. They may have thought of the new concept as too risky to rely on exclusively if they had not grasped its technical brilliance. At some point, Teller could no longer restrain himself, and he was finally given the opportunity to speak, causing a turnabout right on the spot at the Princeton meeting.

Of both Teller's and Bethe's accounts of the Princeton meeting, Teller's version seems the more realistic and is supported by Hewlett's and Duncan's description. Bethe's frustration over the Oppenheimer hearing must have seeped into his narrative, prepared right after the hearing, but he could have checked the story and changed his paper when he finally published it in *Los Alamos Science* in 1982; apparently, he did not. The article in *Science* reporting about Bethe's account in *Los Alamos Science* assumed that Bethe was correct, which is not surprising, given Bethe's generally well-deserved authority. In this case, alas, his bias likely took the better of him, as he himself also alluded to his frustration.

There are two other aspects of Bethe's account that most directly concern Teller, and do so in a conspicuous way. One concerns the nature of Teller's discovery and the other refers to Teller's creativity. Starting with the first, we note that while Bethe mentioned Teller explicitly, his evaluation obviously included Ulam's contribution as well. However, rarely has a greater authority in his own field praised more eloquently the discovery of a colleague than Bethe did here with respect to Teller:

> . . . the new concept . . . was entirely unexpected from the previous development. . . . The new concept was to me, who had been rather closely associated with the program, about as surprising as the discovery of fission had been to physicists in 1939 . . . the new concept had created an entirely new technical situation. Such miracles incidentally do happen occasionally in scientific history but it would be folly to count on their

occurrence. One of the dangerous consequences of the H-bomb history may well be that government administrators, and perhaps some scientists, too, will imagine that similar miracles should be expected in other developments. . . . Everybody recognizes that Teller more than anyone else contributed ideas at every stage of the H-bomb program.[106]

Bethe recorded his opinion at least one more time that for the development of the hydrogen bomb "the crucial invention was made in 1951, by Teller." Significantly, he made this statement in 1968 in his article remembering Robert Oppenheimer.[107]

Returning to Bethe's 1982/1954 account, we find that he mentioned something disapprovingly, namely that "[n]ine out of ten of Teller's ideas are useless. He needs men with more judgment, even if they be less gifted, to select the tenth idea which often is the stroke of genius." This is a curious caveat, as if Bethe tried pushing Teller off the pedestal Bethe had just erected for him. Scientists know that truly original ideas seldom occur even to the greatest minds. If one has such an idea it is not very relevant whether he had had many additional ones that were not as great.

Linus Pauling generated many original ideas. When he was asked about this, he said he had many more, but most of them were worthless so he threw them away and kept only the good ones. The Nobel laureate Marshall Nirenberg, who made the first step in cracking the genetic code, used a graphic comparison to explain the process. When one throws darts at a target on the wall, the darts hit not only the target but the entire wall; nonetheless, some may hit the bull's eye, and that is what counts. Pauling himself sifted through his many ideas while Nirenberg leaves the task to others after he is gone: he records his ideas in his notebooks that he has donated to the National Library of Medicine.[108]

As Bethe noted, Teller let his peers evaluate his ideas immediately, and Teller did not mind that, as he himself had enjoyed performing such a task for George Gamow. There was a difference between Gamow and Teller in that Gamow could easily give up an idea that Teller judged impractical, while Teller pursued "his ideas with great insistence and [which made] him act at times like a prima donna," in Rudolf Peierls's words.[109] Also, there is a difference between an idea in fundamental research, which may be impractical without much consequence, and in applications, where a diversion may be very costly, and therefore immediate scrutiny and criticism are of greater importance.

In conclusion, it is worthwhile to return to the political aspects of the 1949 debates about the hydrogen bomb program in the United States. Lately, there has been a revisionist approach in evaluating these debates. Reading the actual documents containing the GAC recommendations, one can form the opinion unambiguously that the GAC opposed the development of the hydrogen bomb. According to the revisionist approach, the GAC opposed the development of the classical Super but did not oppose the development of the hydrogen bomb as such. It is contended that it was Teller and his concept of the classical Super that delayed the development of the hydrogen bomb. One might then ask: where were the other scientists? Why didn't they come up with a more credible scheme?

Furthermore, even when the Teller–Ulam design became available, the Los Alamos leadership was in no hurry to abandon the tests for the classical Super. So it was not only Teller who was sticking to the classical Super, and it was not even he who was sticking to it the longest. In fairness though, if Bethe considered Teller's new concept a miracle, lesser scientists could have hardly been expected to discover it. The scientists clearly felt ambivalent about this most horrible weapon, first adamantly opposing it but then endorsing it enthusiastically, when a "technically sweet" solution was proposed for its creation.

The accusation that Teller caused the delay in developing the American hydrogen bomb might have its roots in another consideration. It has been alluded to above (chapter 5) that in April 1946 there was a meeting at Los Alamos reviewing the state of affairs of the thermonuclear bomb. The document about the meeting is still classified, but the idea of radiation implosion may have been floating around at the time. It may have originated from Teller; it may have originated from others. In any case, if it was mentioned, it did not take. Again, it was not only Teller who overlooked it, if it had been mentioned, but all the others who were involved with the project. Former Livermore Laboratory director John Nuckolls remembered that they asked Teller, "What took you so long?"[110] This question seems to remain unanswered.

There are signs indicating that the idea of radiation implosion might have been around prior to 1951 and that Teller might have heard about it.[111] This would subtract nothing from the importance and uniqueness of his discovery in 1951. In a letter to Maria Goeppert Mayer in mid-winter 1949, he wrote about his consideration of the hydrogen star. He correlated the possible temperature in its center with its size and found the temperature estimate realistic "for a star which has stopped growing because radiation pressure counterbalances gravity."[112] This shows that the idea of radiation pressure must have been on Teller's mind at least as early as 1949.

Another sign is merely a conjecture and cannot be considered as evidence. Livermore theoretical physicist Neal Snyderman used to have frequent discussions with Teller on a broad spectrum of modern physics topics during the 1990s. On one occasion Snyderman asked Teller whether his discussions with Gamow at Los Alamos in the late 1940s might have influenced his thinking about the hydrogen bomb. Due to Gamow's intimate knowledge of the processes of energy production in the stars, this might have been another opportunity for the idea of radiation implosion to come out. When Teller reacted to Snyderman's question, that was the only time Snyderman had seen him so upset. Teller violently denied that it could have been the case. His reaction was so unusual after so many amicable conversations that Snyderman could not help but conclude that he might have touched on a sensitive topic.[113] Teller may have indeed delayed his own discovery of the new Super if he had remained enamored for too long with his classical Super.

In some of my own encounters with participants of the American nuclear weapons program I felt that I might have touched on a sensitive point by inquiring about the 1946 meeting on the Super. This was surprising to me, since I was not asking about the technical sides of the issue. I briefly mention these encounters here. The first refers to an e-mail exchange with Gregory Canavan, who later played an important role with Teller in the SDI program (chapter 11).

Canavan participated in the group interview with Teller at Los Alamos in 1993 in which they talked about the April 1946 meeting at Los Alamos alluded to above. According to the transcripts of the 1993 discussion, Canavan asked penetrating questions about the slowness of the emergence of the radiation implosion concept. I found his questions interesting from the point of view of understanding the nature of scientific discoveries. Often ideas that later seem straightforward appear as shocking when somebody pronounces them for the first time. Also, sometimes ideas that are eventually found revolutionary may have been floating around without many knowledgeable people recognizing their utility.

The concept of implosion had already been employed in the plutonium bomb, yet even for Teller an additional spark was apparently needed to use this concept in terms of the hydrogen bomb. And this spark, by all indication, was Ulam's original proposal. I raised this question in an e-mail message to Canavan. I added that if the 1946 meeting discussed radiation implosion, Klaus Fuchs, who was present at the 1946 meeting, might have informed the Soviets about it. In that case the solution for the Soviet hydrogen bomb might have not been their genuine invention. Canavan found my note interesting and my questions "very perceptive."[114] He promised a response, and a few days later he communicated that he had compiled his response, but parts of it had to be reviewed before he could send it to me. His response never came.

Another encounter that touched on the 1946 Los Alamos meeting was with former director of Los Alamos Siegfried Hecker. It was an amicable conversation, which I recorded in his office at Stanford University. He lost his calm demeanor only once and only for a moment when I posed an innocent question to him about the 1946 meeting on the Super at Los Alamos, and he told me that I was "pushing the envelope." I still have no idea what he meant, but I must have touched on a sensitive point.[115]

Another idea whose inception would be interesting to further scrutinize is the application of lithium-6 deuteride as fusion fuel. This idea had been around long before Fuchs's departure from the United States, so he might also have conveyed this idea to the Soviets. Above we mentioned Ginzburg's suggestion to use lithium-6 deuteride for the Soviet program. There is no doubt as to the originality of Ginzburg's suggestion because—lacking security clearance—he could not have had access to the secrets of espionage, whereas the other Soviet scientists in the program might have had such access. Ginzburg did not even learn whether lithium-6 deuteride was ever used in the Soviet program until the collapse of the Soviet Union. Then he learned that it was, but it is not known whether it was based on his suggestion or on prior intelligence.

The Americans did not use lithium in their thermonuclear devices until after the "Mike" test, that is, a little more than ten years after Teller's letter to Oppenheimer in 1942 (see above). The idea of using lithium in fusion fuel had been around from the early 1940s. Nuckolls gives a compelling example. In 1947 he was a seventeen-year-old high school student living on a farm in Illinois. He read in the newspapers that the great wave of the future was fusion to help humankind to prevent wars and provide unlimited energy. Nuckolls prepared an entry for the Westinghouse Science Talent Search in which he described a scheme for a hydrogen bomb. He used his father's nuclear physics books, the Smyth Report, *Atomic Energy for Military Purposes*, and some magazine articles in *Scientific American*. According to his description, there was a lithium-6 container of a cylindrical shape that was filled with deuterium. He then put an atomic bomb in the middle to heat up the system to ignition temperature. Nuckolls submitted his project to his physics teacher, but he never heard about the fate of his submission. The point is that the idea of using lithium, even lithium deuteride, was floating around in the 1940s, much before it would have been used in the actual hydrogen bomb program.[116]

As for the classical Super, Teller did not consider it a failure, according to his 1979 memorandum. In the conclusion of the memorandum, he said that eventually, using better computers, it was shown in Livermore that even the classical Super would have worked. In the early 1970s, Teller's young protégé, Lowell Wood, used the state-of-the-art computational facilities of the Livermore Laboratory and indicated that not only would the classical Super have worked, it could have produced heretofore unheard of strong blasts in the gigaton-TNT-equivalent range. Luckily, no device was ever built, let alone tested, to substantiate this claim.[117]

The hydrogen bomb was (and continues to be) the most horrible weapon, a shameful development in the history of humankind. Some say that its development has proved unnecessary because in spite of its tremendous expense, it was never used. It must be emphasized, however, that even though it was never employed, both superpowers did possess it. Had Stalin (and his successors) unilaterally become the sole possessor of the hydrogen bomb, might he have not blackmailed the Free World by using it as a threat?

Teller has been accused of having forced his agenda onto the United States. However, it would almost be ascribing him supernatural powers to suppose that he could have single-handedly imposed his will upon the Truman administration and President Truman. With a proper system of interactions between government and science, including a reliable flow of information, Teller's lobbying might have been superfluous and deemed out of place. However, he remembered the situation, merely one decade before, when a small group of immigrants took it upon themselves to warn the US president of the possible danger of an atomic bomb of heretofore unimaginable power possibly coming into the hands of Nazi Germany. Nobody has accused Leo Szilard of exaggerating the dangers, although he has been accused of inefficiency in enabling the development of the atomic bomb within a speedy time frame. Having experienced the time of the Nazis, Teller felt the torment of facing a yet stronger adversary of the United States. He feared once again the possibility of the enemy acquiring a thousand-times-more powerful weapon than the atomic bomb while the democracies stood idly by.

# Chapter 7

# FROM WORRIER TO WARRIOR

> The accomplishment I am most proud of is not the hydrogen
> bomb; it's the establishment of Livermore.
>
> —EDWARD TELLER[1]

*After his triumph with the new Super, Teller's next major goal was the establishment of the second weapons laboratory. Teller's relentless struggle for the lab resembled his battle to obtain a presidential decision to develop the hydrogen bomb. He was moving among the politicians and military leaders with increasing ease. The Livermore Laboratory was established, and its program, again, not without controversy, was taking shape to Teller's liking.*

Because the Los Alamos Laboratory was established for a specific wartime project, it might well have been dissolved at the end of World War II. However, there was no ceasing of war preparations with the defeat of Japan. The hot war was giving way to a cold war with the Soviet Union, so Los Alamos would continue. One might have thought that with the hydrogen bomb program having been elevated to the level of national priority, the Los Alamos Laboratory and its leadership would have felt even more secure. However, the hydrogen bomb program had no assurance of success despite its absorbing a large proportion of the resources of the laboratory, both in materials and in human efforts. Perhaps, even worse, its main proponent, Edward Teller, was increasingly a thorn in the side of the Los Alamos leadership. He was not a good organizer, but he was a strong critic of those who were in charge.

Teller had predicted that Los Alamos might be capable of dealing with failure but not with success. He meant by this that as long as the work on the Super was being done in the dark, the existing arrangements would be fine; however, its inadequacy would be obvious once the way to design the hydrogen bomb was found. When the Teller–Ulam solution became known, this inadequacy surfaced and Teller found ample basis for proposing the establishment of a second weapons laboratory. There were attempts to find a solution within the existing framework, and had there been a different chemistry between the Los Alamos leaders and Teller, a solution might have been found—but this was not to be. Teller's Los Alamos colleagues were watching with distrust and apprehension as Teller schemed for a second laboratory. Looking for support, Teller once again went outside the circle of scientists and the Los Alamos Laboratory, and took his point directly to generals and politicians. Ultimately, he got what he wanted.

## ROAD TO LIVERMORE

As early as April 1951, Teller proposed to Gordon Dean, chairman of the Atomic Energy Commission, the establishment of a second weapons laboratory.[2] Teller came to Washington and spent a long session with Dean on April 4. The chairman found the usually intense Teller very objective in his argument for a second laboratory and thought that the difficulties Teller described regarding the Super were not insurmountable. Soon, on April 16, Dean listened to the position of the Los Alamos leadership. When Teller returned to Los Alamos, he put everything in writing for Dean. He argued that Bradbury was not moving fast enough and he made factual proposals. For a location, he suggested Boulder, Colorado, and recommended hiring 50 senior scientists, 82 junior scientists, and 228 assistants. A theoretical group could be in place and working by the fall of 1951, and routine operations could commence in the summer of 1952. Teller suggested that Frederick de Hoffmann be manager of the operation.[3]

Teller was still a member of Los Alamos, and when he finally resigned from the laboratory (after some not quite final resignations), he was free to take up more active lobbying. On September 18, 1951, there was another Soviet nuclear test that was announced by the United States on October 3. This was yet one more reason to speed up the development of the American hydrogen bomb. In October 1951, the General Advisory Committee discussed the question of a second laboratory but took a stand against it.

Back in 1946, Teller's departure from Los Alamos and his return to the University of Chicago was more than another change of location. Los Alamos was a watershed in his development. He arrived there as a popular figure who went out of his way to please his peers and came away as someone who was despised by some of his former friends. In the beginning, his political leanings extended only as far as anti-Nazism, but he came away as an anticommunist. He went there as a foot soldier and came away a prima donna. His relationship with his wartime director also changed over time. In 1949, Oppenheimer thought he could easily brush off Teller when Teller asked what they should do once the Soviets had exploded their nuclear device. By 1951, Teller had become a man to reckon with.

Teller visited Oppenheimer early in November 1951 and told him "with an intensity few others could muster" that the GAC was wrong to deny support.[4] He asked for the chance to speak with the GAC members when they got together in December, and Oppenheimer promised him such an opportunity. Teller was finding his voice even with Oppenheimer; he appeared more determined and self-confident—qualities that may have come from his success with the new Super. But he also sensed that Oppenheimer was diminishing in importance. Still, Oppenheimer was not done yet, and giving Teller an opportunity to speak to the GAC also meant ample opportunity for counterarguments.

Teller was preparing for the meeting with grave expectations. In his *Memoirs* he described the dream he had during his overnight train ride from Chicago to Washington, where he was to address the GAC meeting on December 13, 1951.[5] In the dream, he was defending a position alone in a trench; he had a rifle with eight bullets, but nine men were attacking him. He was frightened, and the dream reminded him of his dreams as a child in Budapest. Then, the fright was replaced by elation when he realized that he had a target for each of his bullets, so none of the bullets would go unused. He was even more elated by the realization that he was no longer afraid of firing off his bullets, regardless of the outcome of his fight. This dream symbolized the transformation from someone who had been plagued by his worries into someone who was a fighter. Of course, he was not timid in his quest for the hydrogen bomb, but he spelled out his confidence in his description of his dream almost half a century later. It is also telling that he equated the GAC members with the men who were coming against him in the trench and against whom he had to defend the trench. When he got off the train, Teller had to walk to the AEC office because there had been a heavy snowfall and there was no available transportation. He made a rare reference to his missing foot and unsteady balance, which, according to him, did not bother him but apparently bothered him enough for him to remember this long walk almost a half century later.

The meeting went well, and he felt satisfied with the way he made his arguments. His main points there could be summarized as follows:

- The Los Alamos people were fine experts, but they liked to set themselves limited goals, and for thermonuclear research such a timid approach meant that many possibilities might be left unexplored.
- The rigid organizational setup discouraged some scientists from joining Los Alamos, scientists who otherwise would have been interested in thermonuclear research (an obvious example might have been Teller himself).
- Not all thermonuclear research should be shifted to the second laboratory, but it should be its main priority; other weapons development and basic research should also find a place in the second laboratory.
- The second laboratory should be a small institution, up to three hundred people, not more.[6]

Teller did not expect to change the minds of those GAC members who had opposed his suggestion—and that did not happen—but his main goal was to face his opponents firmly. His meeting with the GAC did not go down well with the Los Alamos leadership. Obviously, the main reason for the need for a second laboratory appeared to be the inadequacy of Los Alamos. Bradbury went on record about "the 'rather thinly veiled criticism' that progress on weapon research and development at Los Alamos was not adequate to the national need."[7] Bradbury labeled Teller's arguments emotional and intuitive, and he found the AEC's attitude damaging for the morale of Los Alamos in that it delayed approvals for construction of buildings they badly needed, and so on. Bradbury also wanted to counter the perception that Los Alamos failed to attract the needed personnel. He mentioned the example of the research unit for thermonuclear work organized by John A. Wheeler at Princeton.

A decade later, Teller explained the reasoning behind his suggestion for the second laboratory:

It was an open secret, among scientists and government officials, that I did not agree with Norris Bradbury's administration of the thermonuclear program at Los Alamos. . . . we differed sharply on the most effective ways to produce a hydrogen bomb at the earliest possible date. . . . The dissention with Bradbury crystallized in my mind the urgent need for more than one nuclear weapons laboratory.

. . . [S]cience thrives on friendly competition, on the fostering of different points of view, and on the exchange of ideas developed in different surroundings. . . . a single group of scientists working together can easily become fascinated by special aspects of a development—to the neglect of other hopeful approaches. . . . I began to doubt that one laboratory would be physically capable of handling all the work that had to be done.[8]

Of course, Teller's strongest statement in this debate was his departure from Los Alamos. This happened at a time when his new design had been accepted and was being put into action. The general problem of the thermonuclear reaction had been solved, but the Los Alamos theoreticians had to complete the work. Teller left in the summer of 1951, but his family stayed behind and went to Chicago once he found a place for them to live. In Chicago, he carried out calculations of the blast effect of the hydrogen bomb. He found that as the size of the hydrogen bomb increased, the blast effects had limitations in their size. This limitation was in contradiction with the GAC majority and minority reports in October 1949, which warned of the limitless destructiveness of the thermonuclear reaction.

During all this time he continued his fight for the second laboratory—relentlessly. Again he departed from what he had advocated: that scientists should limit their activities to science.[9] He felt he had to choose between staying idle and worrying, or aggressive lobbying; and he chose the latter. He eventually found an ally in Lewis Strauss, with whom he shared a concern as "ethnic Jews"—another rare reference—about the horrors of World War II.[10] Willard Libby was also an ally. Libby, a member of the GAC, was still in Chicago and supported the idea of a second laboratory—the only GAC member at the time to do so. Likewise, at that time there was only one commissioner of the five, Thomas Murray, who supported Teller's suggestion.

The GAC was trying hard to find a compromise because it was obvious to the committee that Teller should be pacified. They were trying to find a solution that "would make the best use of Teller's abilities."[11] They contemplated, for example, organizing a mission-oriented new division at Los Alamos that would concentrate on thermonuclear work. This might have been an acceptable solution for Teller in 1950, but by this point it was too little too late. On the other hand, having a new division at Los Alamos would be a hard sell to Bradbury, who had successfully resisted giving such an elevated status to thermonuclear research—that is, to Teller. So the compromise suggestion would have been as difficult for Teller as it was for Bradbury.

One of the alternative approaches under consideration was to outsource more work from Los Alamos to other locations and thus free Los Alamos's capacity for thermonuclear work. The GAC could not ignore Teller's criticism since the hydrogen bomb was now on the national agenda, and the GAC was trying everything to find a solution within the existing framework and without setting up a second laboratory. In the meantime, the importance of nuclear weapons in the defense plans of the United States was becoming increasingly recognized. Of course, nobody could determine how much was enough, but the mood of the military was that if there was an error in their estimates, they would rather err in the direction of too much than in the direction of too little.

Once again, time was on Teller's side as he gained influential allies. As further proof of his influence, the US Air Force expressed interest in involving Teller in its strategic planning. There was a personal connection in Theodore von Kármán, who had long been providing scientific background for the American Air Force, and his contribution was much appreciated. Von Kármán recognized early on the importance of nuclear weapons and wanted information from the AEC about the size and weight of projected nuclear bombs. He wanted the air force to be ready to deliver them. But the AEC declined to give him such information (it's possible that they did not have the information; regardless, they just brushed him off).

This was not Von Kármán's first unsatisfactory experience with officials of the nuclear program. At the time of the Manhattan Project, he went to see General Groves and asked him for advice about nuclear bombs, as he was already involved in long-range planning for the postwar period. Groves was rude to him and requested that the Pentagon send somebody whose English he could understand. By then, Groves must have been used to foreign accents, including Hungarian, so his attitude was unwarranted. Von Kármán ascribed Groves's behavior to interservice rivalry. For revenge, he liked to tell the following story: "General Groves and Robert Oppenheimer are in an atomic shelter watching the first A-bomb explosion. 'What did you see?' a reporter asked. 'I saw the end of the world,' Oppenheimer replied. 'And I saw a third star,' said the two-star general."[12]

Von Kármán had suggested setting up the Scientific Advisory Board (SAB) of the air force, and it was initially thought to be temporary, but the board proved to be so useful that it became permanent. Teller was invited to join in 1951 and he remained on the board until the early 1990s. The air force was charged with delivering the nuclear weapons and was counting on large bombs, as it was not thought that nuclear warheads would become so small as to be deliverable by missiles. As a consequence of lacking AEC cooperation, von Kármán suggested that the air force set up a Nuclear Weapons Panel to investigate questions related to nuclear weaponry. John von Neumann became chairman, and familiar names figured among its members: Teller; Hans Bethe; Norris Bradbury; George Kistiakowsky of Harvard University; David Griggs, chief scientist of the air force; and others. The confluence of Hungarian scientists caused author Richard Rhodes to remark that "sometimes the arms race looks like a Hungarian scheme for fighting Russia by proxy."[13]

Von Kármán noted in his autobiography[14] that the panel was to play a significant role, but initially its members were divided emotionally as well as technically. At the root of the division was the fact that Teller and Griggs advocated strategic bombing whereas Bethe and Bradbury were for tactical bombing. Teller and the future Livermore Laboratory pushed for more powerful hydrogen bombs with radical solutions, whereas the Los Alamos people stressed improvements of existing designs. When the Oppenheimer case developed, the participants were similarly divided between those who opposed Oppenheimer and those who supported him. Von Kármán praised von Neumann's skill in operating in a harmonious way with others; it was something Teller might have been incapable of doing.

There were other policy questions for which opinions diverged. For example, the air force wanted to have intercontinental ballistic missiles developed as early as 1949. However, Vannevar Bush, the wartime supreme leader of defense-related scientific research and development, did not believe that it was possible to control long-range rockets. He concluded—incorrectly, as it was later proved—that they would not be useful for a long time.[15] He had also been wrong in his estimate of how soon the Soviet Union would have nuclear bombs. As has already been noted, von Neumann was also initially skeptical of the accuracy of missiles, hence his enthusiasm for the hydrogen bomb.

The interest of the air force in nuclear weapons was natural, because it was this branch of the service, more than any other, that would be charged with their delivery. The next would be the navy, with the land forces bringing up the rear. Teller had met David Griggs at Los Alamos, when Griggs worked at University of California at Los Angeles. He was one of the first members of the RAND Corporation in Los Angeles, which did scientific research for the government and for private industry. Griggs was brought into the air force as a civilian and was made air force scientist, which was a rotating position. Griggs helped Teller make contact with top air force people at the SAB meetings, among them General Jimmy Doolittle, who became another influential supporter of the second laboratory.

Just as in his quest for the hydrogen bomb, Teller could count on Lawrence, who was anxious to get into defense. The Berkeley people identified three principal reasons for establishing a second laboratory, which they wanted to build in Livermore, California:

- Lawrence, Alvarez, and Latimer wanted more active participation in response to the Soviet nuclear threat;
- Teller's conflict with the Los Alamos leadership;
- There was a group of young scientists at Berkeley that participated in the George experiment of Operation Greenhouse; they were familiar with the thermonuclear weapons design and were willing to continue this line of work.[16]

There was more than just a slight interest at Berkeley, because they needed projects to work on. Earlier, Teller had told Lawrence that for future thermonuclear devices they would need large amounts of tritium and deuterium if they were to develop the fusion bomb. In order to produce these substances, large excess amounts of neutrons would be needed.[17] Initially, they were afraid of a shortage of uranium, but the situation changed when large uranium ore deposits were discovered in Colorado and Canada. Further change came about with the new Super, which would necessitate smaller amounts of tritium and deuterium. So the neutron factory project they were building up at the old navy training ground at Livermore was becoming superfluous. The second weapons laboratory advocated by Teller seemed to be a good undertaking to absorb those young scientists who were anxious to stay employed in defense-related work. This was an important consideration and countered the complaints of the Los Alamos leadership that had expressed fear that a second laboratory might raid Los Alamos for talent.

At one point, Lawrence invited Teller to Berkeley and introduced him to the Livermore site on February 2, 1952. Livermore is about a half hour's drive east of the San Francisco Bay area. At that time, it was a small farm town of forty-five hundred inhabitants and was known for wine, roses, and cattle—not yet for bombs. Some later disadvantages—"of being situated among environmentally sensitive and politically active neighbors"—were

## ERNEST O. LAWRENCE (1901–1958)

Ernest Lawrence was born into a family of high school teachers of Norwegian extraction in Canton, South Dakota. He received a BS degree from the University of South Dakota, an MA degree from the University of Minnesota, and a PhD degree from Yale University in 1925, the latter two in physics. He started his career at Yale but soon moved to the University of California at Berkeley, where he founded the Radiation Laboratory in 1936. Lawrence invented the cyclotron, a circular accelerator that he used for producing artificial radioactive elements and for further discoveries. Lawrence was both a scientist and a politician who knew how to get things done.

He was considered less as a deep thinker and more as a practical man who was very good at securing huge funds to construct and build new machines. He was awarded the Nobel Prize in Physics in 1939 for the cyclotron and for producing artificial radioactive elements. Lawrence's Nobel Prize was a first for hardware in physics; the first for work done entirely on a University of California campus; and the first awarded to a professor from a public university in the United States. More Nobel laureates came out of Lawrence's laboratory, such as Luis Alvarez, Melvin Calvin, Edwin McMillan, and Glenn Seaborg, and they proudly referred to themselves as members of Lawrence's team.

Lawrence and his associates actively participated in war research during World War II for a host of projects. They used electromagnetic separation to enrich and separate the uranium-235 isotope, which became the fission material for the atomic bomb that was exploded over Hiroshima. When Oppenheimer had difficulties in getting security clearance in 1943, Lawrence vouched for his fellow Berkeley professor.

After the war, Lawrence continued amassing instruments, conducting experiments, and building his laboratory. He supported Teller's efforts in the hydrogen bomb program. He joined forces with Teller for the establishment of the second weapons laboratory at Livermore. At the end of his life he was a member of the American delegation, negotiating with the Soviet Union about a proposed treaty of a test ban for nuclear weapons. He suffered from chronic colitis and was rushed back to California for emergency treatment from Geneva, but he could not be saved.

Politically, Lawrence was considered a right-wing ultra-conservative. He supported Teller, but they were not close on a personal level. In his *Memoirs*, Teller referred to their shared quality of being willing to risk their personal reputations by pointing out what needed be done. However, this was more characteristic of Teller than Lawrence. Lawrence achieved his goals without taking such risks.

not yet manifest.[18] Lawrence acquired a one-square-mile site in 1950 for the AEC; it was a former naval air station from World War II. At Alvarez's suggestion, a materials testing accelerator was built on this site. Some future well-known physicists got involved in this project, such as, for example, Wolfgang Panofsky.[19] Lawrence asked Teller whether he would be willing to leave the University of Chicago and move to Livermore. Teller said yes, if the new laboratory provided thermonuclear work.[20]

Teller could count on Lawrence as an important ally. Lawrence was known to have a dominating personality and even more extreme political views than Teller, but Teller realized that he could not be too choosy in picking his supporters. Another significant ally was, once again, Senator McMahon, who convened a closed hearing of the AEC members before his committee on February 21, 1952. McMahon did not hide his disappointment in the AEC's lack of action concerning the second laboratory. When pressing for information, one of the commissioners stressed that everybody who could participate in the nuclear weapons program was already in Los Alamos, including one of its frequent visitors, Edward Teller. This was not quite true, because Teller was not willing to continue at Los Alamos. This was only a hearing, and a congressional committee could not decide the fate of the second laboratory, but it could create pressure, and it did.[21]

In the meantime, General Doolittle and David Griggs were busy paving Teller's way to influential people at the air force. In mid-February 1952, they arranged a meeting for Teller with the secretary of the air force, Thomas K. Finletter. This was thus far Teller's highest-level contact. Teller was now lobbying extensively with top brass, and his power of persuasion worked: From a cold and slow start, the meeting with Finletter became a fruitful encounter. Teller explained to the secretary the importance of competition and implied that the Los Alamos leadership was afraid of such competition. Finletter decided to check it out and visited Los Alamos right away. He was so much sold on Teller's ideas that he was willing to establish a second laboratory even if the air force had to do it alone.

The next level was secretary of defense Robert A. Lovett, who showed interest even in the technical details of the hydrogen bomb. Although he had been against the second laboratory, he eventually changed his mind. The meeting with Teller must have played a role in this, and Lovett wielded great authority. Teller's next meeting was with secretary of state Dean Acheson and was attended by deputy secretary of defense William Foster and AEC chairman Gordon Dean. Dean was not at all friendly with Teller and still did not support the idea of the second laboratory. Again, the similarity of the situation with the development of the hydrogen bomb controversy was striking, including the long-held opposition of the AEC. However, the opponents of the second laboratory were losing ground. It was part of rear-end defense in which Bradbury visited Berkeley in May to work out the division of work between Los Alamos and Livermore even before there was a decision to set up a second laboratory. Bradbury wanted to avoid Los Alamos becoming a recruiting ground and a second-class supply house for Livermore.[22]

Teller's meetings and the will of the political and military establishment—especially that of the air force—made it increasingly inevitable that a second laboratory would be created. Finally, the June 1952 meeting of the AEC resulted in a decision to establish a second laboratory. The next pressing matter was the choice of its location; there was no formal decision in spite of Lawrence's groundwork. Teller may have hoped that the site would be in the vicinity of Chicago, a reasonable wish, given his role in the process and his professorship at the University of Chicago. In the vicinity were Teller's colleagues and students, as well as the Argonne National Laboratory.

Teller wrote that Fermi disapproved of the establishment of a second laboratory, which discouraged Teller from planning the location right under Fermi's nose.[23] This does not sound very convincing—that Teller would just give up the site most convenient for him in deference to Fermi's feelings. It might have been more realistic to consider that Teller would not find sufficient support from among his fellow professors and other scientists in his Chicago base. Also, just as it might have been lucky to have Argonne for a neighbor, it might have also become an attractive alternative to those who were willing to enter employment at a government laboratory.

Lawrence's interests and determination were much more crucial factors in choosing the location than were Fermi's sentiments. At some point, Lawrence dispatched Herbert York to see Teller and convey to him the message that Livermore would be a good location; that the director of the second laboratory should be reporting to Lawrence; and that York should be the director. This was straight talk, and Teller agreed with all three points on the spot. Looking back, he added that it was a nice gesture on Lawrence's part to inform him of his choice of director in advance.[24]

Teller's attitude toward Lawrence's conditions further demonstrates not only Lawrence's authority but also how much Teller must have felt he needed Lawrence's support. It is hard to believe that Teller had not entertained the idea that he would head the new laboratory. This is not to say that he would have been a good manager and that York was not qualified for the job. From Teller's point of view the problem was likely not so much whether he should have been considered for directorship but whether he should have been consulted about the appointment. Yet he made no complaint, not even in hindsight.

The insider Alvarez noted that Lawrence "picked Herbert York to administer the new laboratory, with Edward Teller as de facto director."[25] For York and his colleagues, the Super meant more than just employment. They were fascinated by the challenge of working on a heretofore unprecedented project and in the company of such prominent scientists as Teller, Bethe, von Neumann, Wheeler, Gamow, and others. The young scientists traveled to the Enewetak Atoll for the George shot; did measurements using diagnostic equipment; evaluated their data; participated in the discussion; and did some follow-up studies. Lacking further tasks, the group was facing disbandment, so they had every reason to view the establishment of the second laboratory at Livermore as a godsend.[26] York himself was enthusiastic about the challenges represented by a second laboratory and "found the whole affair heady and exciting," but he issued an apology a quarter century later, by which time York would have become an opponent of thermonuclear weaponry in whose creation he had earlier participated with dedication.[27]

Lawrence dispatched York to see Teller in Chicago and also to Los Alamos and Washington to discuss the matter of the second laboratory. Lawrence did not want to leave anything to chance. Teller and Lawrence worked for the project separately rather than in a coordinated way, and when the authorities finally decided to grant Teller's request, Lawrence was to have the upper hand. There was a point when Lawrence would not have minded if Teller had dropped out of the project.[28] This attitude sounded like the old adage: "The Moor has done his duty, the Moor can go,"[29] and it almost came to that.

## LENGTHENED SHADOWS

The maxim "An institution is the lengthened shadow of one man"[30] appeared true for Livermore, except that the shadow changed from being Lawrence's to Teller's. From the start, there were differences in how Teller and Lawrence envisioned the operations of the new laboratory.[31] Lawrence thought that if they assembled a group of bright young men, they would find their way toward great work even if only vague goals were formulated for them. Teller did not want to leave matters to their own device and was suspicious of the AEC's intentions when it finally agreed to set up the Livermore

Laboratory. For him, the model to follow was the Los Alamos of 1943, which had well-defined goals that were pursued relentlessly. There is some inconsistency here, because Teller mostly stayed away from the principal goals of Los Alamos during the war, but he did not necessarily consider his own attitude at Los Alamos to be the example to emulate.

Lawrence charged York with drawing up the plans for the laboratory and gave him only the most general instructions, which, regardless, York followed rigorously. Livermore was to be a branch of the Berkeley laboratory, the University of California Radiation Laboratory. This was a great advantage from the point of view of the Atomic Energy Commission because the second laboratory could be easily integrated into an existing system. When York showed his plan to Lawrence, he accepted it without criticism. Alas, this plan called for merely an assisting role in the thermonuclear weapons program by conducting diagnostic experiments during tests and related research, and nothing was spelled out about its direct participation in weapons development.[32]

In Teller's *Memoirs*, York's original plan called for "work on weapons tests, on computing, on materials testing accelerator (MTA), on theoretical physics, and on controlled fusion." This plan left hardly anything out, except developing new weapons, for which Teller had launched the whole idea of a second laboratory in the first place.[33] The inclusion of controlled fusion was a strange move on York's (and Lawrence's) part, because it was very removed from weapons design and even from realistic goals. It is true that controlled fusion would have been important for generating inexpensive energy and little radioactivity as compared to fission reactors. But even in 1989, when there were reports from Utah about its accomplishment, Teller—wisely, as it happened—did not join in the euphoria. Instead, he stated that it would happen at some point, but it was yet too early. (We note, though, that York did not mean cold fusion in his plans for the laboratory but set limited goals.) At the beginning of the twenty-first century controlled fusion still remains only a dream among scientists. It would have been utopist to push for it in a national laboratory.

Teller ascribed York's plan to his young age and inexperience and to his desire of avoiding confrontation with Bradbury. In his *Memoirs*, Teller wrote about this as if York could have strayed from the original goal without Lawrence's tacit consent. Then he said that the matter was resolved quickly by Lawrence, who decided to give emphasis to weapons design and let Livermore work on controlled fusion on the side.[34] This is how Teller viewed the problem half a century later.

In the actual situation, however, Teller's dissatisfaction was so strong that in early July 1952 he told Lawrence, York, and Gordon Dean that he would have nothing further to do with Livermore. This happened in Berkeley at the very reception that was held to celebrate the opening of the new laboratory. Again, there were some who would have been only too glad to go ahead without him; York remembered Lawrence telling him so, but others felt Teller had to be committed to the new laboratory.[35] Thus, they revised the plans and made them more concrete. Thermonuclear weapons development was included and spelled out.

The way Lawrence was playing on Teller seems like a dirty trick in hindsight, but when it did not work, he changed his methods and went ahead as if nothing had happened. Lawrence's ambiguity was manifested in his way of naming the director of the laboratory. Initially, he communicated to Teller that York would be the director, but he advised York not to use the title of director and not to use titles for anybody among the scientific personnel. He did not allow recruiting through advertising, either.[36] He explained that there was no need for such things among scientists. Then, one day, Lawrence told York that out at Livermore, he should start calling himself director; McMillan, Seaborg, Alvarez, and Teller were told to assume the titles of associate director (of the Radiation Laboratory); and Lawrence informed the Atomic Energy Commission of his decision.[37]

This was typical of Lawrence: he did everything in a casual way, he did not rely on a search committee, and he did not even raise York's salary when he appointed him to be in charge.[38] Lawrence decided, and nobody questioned his decisions. A Scientific Steering Committee was set up with Teller, Harold Brown, John S. Foster Jr., and others. Brown's special interests included optimizing cost, yield, and weight of hydrogen bombs, and Foster was designing ever-smaller nuclear weapons to be shot from guns for fending off even tank attacks. Teller did not have any formal distinction over the others, except one in a strange way: he had veto power over the decisions of the steering committee. Teller became Livermore's éminence grise.[39] Although at the very first moment there was high tension surrounding him, things eventually smoothed out and Livermore operated without any of the strain around Teller that had so disturbed him at Los Alamos.[40]

Computerization—von Neumann's dream and direction—was important at Livermore from the start, and it was among Teller's first priorities. He wanted state-of-the-art computational facilities, and as soon as they installed their first modern computer, he was already making plans for the next, more powerful one. Computers saved time, personnel, and were fully justified in weapons design. It should be noted, however, that weapons design had not been the only task for Livermore, or for Los Alamos, for that matter. For example, Teller encouraged the utilization of the out-standing computational facilities at Livermore for investigating and modeling the global atmosphere and weather forecasting.[41] The Atomic Energy Commission noticed Teller's dedication to computerization; when they created a Computer Council, they appointed Teller as chairman.[42]

In 2000, in the wake of the devastating fire in the Los Alamos region, Teller turned to the secretary of energy, calling for intensified efforts to improve weather-predicting capabilities through modern computational techniques. He criticized the lack of long-range weather-predicting capabilities, referring to the original proposal by John von Neumann, whose initiative Teller followed up on at Livermore when he approved "the first program aimed at digital computer-based climate and weather prediction." He deemed the results to be "one of the larger societal benefits realized thus far from R&D conducted at the National Labs."[43]

When things settled around the setup of Livermore, the Tellers once again faced relocation. As had happened before, they bought a house, this time in Chicago, in which they would never live. They did not, however, depart from Chicago for good. Teller took only a one-year leave of absence from the University of Chicago, where he had dear friends, such as Fermi,

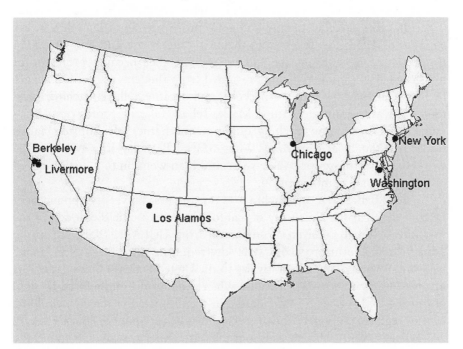

Edward Teller's principal locations in the United States.

James Franck, and Maria Goeppert Mayer. The change was another big step, because it might have signified a final departure from academia, although it did not.

Teller's moving from Chicago to Livermore meant not only moving from the Midwest to the West Coast and moving from academia to another government laboratory; it also meant moving into Lawrence's circles that many of his peers detested. Independent of each other, von Neumann and Fermi warned him not to go. Teller was especially impressed by Fermi, who said that Lawrence and his circle were less scientists and more political plotters. Within some years, many would be saying the same about Teller. Incidentally, Fermi and von Neumann were far from united in their political views, but their warnings did not stem from their politics. Rather, they were genuinely fond of Teller and appreciated him for what he was and felt this was a wrong move for him. When Teller nevertheless decided to go, he allegedly said, "I am leaving the appeasers to join the fascists." But when he was asked about the validity of this statement, he denied it.[44]

The Teller family moved to Diablo, California, on July 14, 1952. Mici and their two children loved the place and much preferred it to windy and cold Chicago. When the one-year leave from the University of Chicago ended, she was very much in favor of staying. Within that time Lawrence made an offer to Teller to stay and get a professorship at Berkeley. In the summer of 1953, the Tellers sold their Chicago house and bought one in the Berkeley hills. From the fall of 1953, Teller became professor of physics at Berkeley, after having resigned from the University of Chicago. In 1955, Hans Mark came to Berkeley and became Teller's teaching assistant. He was the son of Herman Mark, Teller's favorite professor at Karlsruhe. Teller was now teaching quantum mechanics, the subject Hans's father had taught at Karlsruhe. When Mark's Berkeley stint was over, he worked at Livermore for twelve years and then went on to a distinguished career in science administration.

Teller's Berkeley professorship was not the first university appointment he considered at the University of California; however, he didn't accept the first offer. He had received an invitation from UCLA in 1950, but in the same year California passed a state law requiring all professors in public institutions to sign an oath of loyalty to the United States. Teller's name was involved in the controversy because when some professors refused to sign, one of the UCLA regents announced that the dismissed professors would be easy to replace, and referred to Teller, who was just about to join UCLA.[45]

Teller may have not cared about the oath either way, but he found the dismissal of professors an injustice and decided not to accept the invitation. However, he wanted to remain on good terms with the administration of UCLA and the University of California in general, so he related his feelings only to President Sproul of the University of California (which comprised the entire system), who assured Teller that the controversy was temporary. Teller reported to Goeppert Mayer in September 1950 that his visits went well with one exception, Ernest O. Lawrence. In Teller's words, "Since the days of the Nazis I have seen no such thing. I have talked sufficiently gently and generally so that Lawrence did not attack me personally. But he did use threats and he was quite unwilling to listen . . . I felt somewhat sick when I left his office."[46]

By the summer of 1952, when Lawrence asked Teller to join Berkeley, the California Supreme Court had found the loyalty oath unconstitutional, and the dismissed professors were asked to return, though not all of them did. Today, there is still a loyalty oath for state employees in California, including the professors of the University of California. In the early 1950s, it was not so much the oath but the heightened political atmosphere that made the oath a symbol of governmental interference in the lives of individuals that stirred the controversy.

Teller's stand on the loyalty oath question was not some aberration from his political views, because in many respects he was liberal—to use an ambiguous political label. However, he was liberal when he stood up for the rights of people of very different political views to teach at the same university where he taught, or when he did not disapprove of his children making friends with children of very leftist families, to give but two examples. In 1970 he wrote a letter to the acting chancellor of the University of California, Herbert York, the former first director of the Livermore Laboratory and Lawrence protégé. At the time of the oath controversy, in the early 1950s, York expressed the conviction that communists should not be allowed to teach at the university. Teller did not support such exclusion then and he did not support it in 1970 either. He queried York whether he had changed his stand on this issue.[47] The irony in Teller's question could not have been lost on York. The university administrator had moved toward liberal views at the time that Teller was being considered as a target by anti–Vietnam War student demonstrations.

When the Tellers arrived in California in July 1952, Lawrence created temporary working conditions for Teller at Berkeley. Then, the people of the new laboratory moved from Berkeley to Livermore on September 2, 1952. Looking at the time line, everything happened very fast; from the end of August 1949, when the Soviets exploded their first nuclear device; to Truman's decision to develop the American hydrogen bomb on January

31, 1950; to the Teller–Ulam design of the new Super in the spring of 1951; to moving into the newly established second weapons laboratory at Livermore on September 1952; to the "Mike" shot of the first genuine thermonuclear device on November 1, 1952. This all happened during a short period of time but involved momentous decisions, large amounts of money, and the participation of many people.

As for Livermore, about 150 associates moved there right away. There were hardly any telephone lines, fewer desks than needed, and too few desk lamps. The conditions were primitive: leaking roofs and antiquated plumbing in the wooden frame buildings, but there was a lot of enthusiasm and excitement. Teller had given up his prestigious University of Chicago professorship in the company of Fermi and others for this. He might have regretted the move, but we cannot be sure whether he felt disappointment or elation.

Whereas the quest for the hydrogen bomb could be considered in the immediate interest of national defense, the second laboratory could hardly be considered so urgent. For Teller, this was a course from which there was no return; it was his choice. For Lawrence, this was a much less momentous decision; it was merely an extension of his sphere of influence. And for the many ambitious and bright young scientists, this was an opportunity, so they flocked to Livermore and made it into an important center, though it did not happen overnight. The fear proved unfounded that Los Alamos's manpower would be depleted by Livermore raiding it; there was no need for that.

Teller enjoyed the early-Los Alamos-like conditions at Livermore. He even presided over the orientation sessions, introducing the new arrivals to Livermore, just as he used to do at Los Alamos. Some numbers well characterize the rapid initial growth of the new laboratory.[48] Already at the end of the first fiscal year, 1953, there were 698 people working at Livermore with an operating budget of $3.5 million; in 1958, there were over three thousand people and $55 million; 1963, over five thousand people and $127 million.

Livermore became a vital source of important components of the United States defense, but initially there were some flops. Soon after operations had begun, there was a test in the Nevada desert in 1953 that failed. Another test soon followed, the lab's first thermonuclear device in the Bikini Atoll in 1954, and it failed too. The code name was Koon, and the test used a fission device from Los Alamos to initiate the thermonuclear reaction. This was the only part of the test that performed successfully; nothing else did. It was after such failed events that "some Los Alamos scientists filled the air with horse laughs."[49] Following the failed Bikini test, von Neumann remarked to Ulam, "There will be dancing in the streets of Los Alamos tonight."[50]

As Teller described the aftermath of the first failures, he quoted from a conversation between Lawrence and himself. Lawrence was not rancorous, but the conversation leaves no doubt as to who was in charge and who was the subordinate. Teller asked Lawrence to call off the next shot, as it would fail too. Lawrence's response was that he would leave it to York and Brown to call it off and told—in fact, ordered—Teller to fly out into the Pacific immediately to assist them. But Lawrence could also be magnanimous; he had a way to turn around the meaning of Livermore's failures when he chided his people at Berkeley. He told them that their solid successes indicated that they were not taking chances, whereas in order to be moving ahead fast, one needed to take chances and that meant occasional failures along the way.[51]

There was general agreement that the next test should be canceled, and it was. It is curious, then, that two prominent members of the Livermore Laboratory would return to the question of this test that never happened and three and a half decades later come to the conclusion that this test was sufficiently different from the previous one and that "it would probably have succeeded." They added that Teller learned his lesson and never again would have refrained from "technical endeavors of all types."[52] The reality was, however, that the beginning was not glorious, and it did not take long for the earlier opponents of the second lab to again question its justification. There was little time and interest for such controversy, however, because in less than a week after the failed Bikini test, the Oppenheimer hearing began in Washington on April 12, 1954.

Teller was as relentless and stubborn in making Livermore a success as he had been in making its creation possible. One of von Kármán's former students, Mark Mills, was made chief theorist, and he was instrumental in improving the designs so that future failures could be avoided. Indeed, Livermore badly needed successful tests and they had them. It was at Mills's suggestion that they established the system of "pre-mortem" (rather than post-mortem); that is, they reviewed every device for the forthcoming test by experts who were not participants in making that particular device.[53] This approach was reminiscent of Teller's technique of ensuring the safety of reactors and learning about all potential failures in advance rather than analyzing them after they had occurred. Soon, successful tests were performed in Nevada in 1955 and in the Pacific in 1956.

For some time, until about the 1970s, any new scientists, especially those to be hired for senior positions at Livermore, were also interviewed by Teller. He wanted to ensure unconditional dedication to weapons development among the new hires in addition to their professional excellence. John C. Browne, who would later in his career serve as director of the Los Alamos Laboratory, was looking for a job upon having received his PhD in physics from Duke University. During his interview at Livermore Teller told him that before giving him a job, "first I have to ask you a question. If while you were doing your scientific research I called and said I absolutely need you to tell me the answer to this question relating to weapons, would you work on such a problem?" When Browne started giving him an elaborate answer, Teller insisted on responding with yes or no. Browne said, "Yes."[54]

The Nebula Prize–winning science fiction writer Gregory Benford had a similar experience. He had earned his doctorate in physics at the University of California at San Diego and was being interviewed at Livermore. Just when he thought it was over, he was ushered into Teller's office with a friendly admonition not to be nervous—advice that only increased his anxiety. Teller grilled him for an hour about solid-state plasmas—his thesis research—before asking whether Benford would be willing to work "on whatever comes up." This gave added emphasis to the question he had already been asked regarding whether he would be willing to work on weapons. Benford agreed and was hired.[55]

Not all of Livermore's recruiters were so straightforward. When one of the future celebrities of the SDI work, Peter Hagelstein, was being courted (not by Teller), he realized that the laboratory was a "bomb shop" only after arriving at Livermore.[56] Teller was involved in everything at Livermore; his insatiable curiosity served him and his associates well, because they had to be alert when he interrogated them on their projects. He continued to originate ideas as before, except his peers who used to filter them out were not around. To some, he was a man of vision; to others, he was crazed.[57]

Teller's fight for the establishment of the second weapons laboratory differed in one significant aspect from his quest for the development of the hydrogen bomb. He did not have to make any scientific contribution for achieving success. At Los Alamos, Teller may have not authored many publications but he produced scientifically important classified reports that eventually became declassified, and he had an impressive research output during his Chicago period (chapter 5).

In contrast, the year 1952 is conspicuous for its absence of Teller's scientific publications. This is not surprising in view of his other activities. It is also noteworthy that the two papers bearing the year 1951 on his list of publications dealt with the multilayer adsorption problem, the BET equation. This was a rare case when Teller reached back to a previous project and added a further contribution to it. There was a significant paper in 1953 co-authored with Los Alamos colleagues, and from 1954 his publishing record increased again with his physics papers alternating with articles that covered topics of nuclear weaponry and testing.

The scientific paper in 1953 carried the innocuous title "Equation of State Calculations by Fast Computing Machines."[58] Of its five co-authors, there were two married couples, Edward Teller and Augusta H. Teller (Mici), and Arianna W. Rosenbluth and Marshall N. Rosenbluth. Teller's affiliation is given as the University of Chicago, but a footnote indicates his current workplace as Livermore. In full, it was the Radiation Laboratory of the University of California, Livermore, California. This may have been his first paper where his association with Livermore was indicated.

The somewhat mysterious expression "equation of state," is often used in science and refers to the relationship of various characteristics of materials. In it there may be much information from basic science that can be used in industrial processes. The application of fast computers in these calculations was a noteworthy advance. Teller and Rosenbluth worked on finding a solution by which the number of calculations for arriving at the sought-after results might be reduced. They used a computer built by the fifth author—the first among the authors on the paper—Nicholas Metropolis. The two wives carried out the computational work.

The procedure described in this paper proved to be so fruitful that the fifty-year anniversary of its appearance was marked by a celebration at Los Alamos. The foundation of the technique was a new mathematical approach called the Monte Carlo Algorithm worked out by Stanislaw Ulam, John von Neumann, and others. Ulam gave it the name Monte Carlo, by which it has become famous. The original concept came to Ulam while playing solitaire during an illness. The idea is to do sampling rather than try out all possibilities in a systematic way and thereby reduce the process, which leads to a probable solution. The 1953 paper has been cited over ten thousand times, which is quite exceptional, and the mathematical procedure was selected among the ten top that had "the greatest influence on the development and practice of science and engineering in the twentieth century."[59]

By the time the "Mike" shot took place in the Pacific as part of the IVY series, on November 1, 1952, Teller had already been a member of the Livermore Laboratory. It was on the same Elugelab Island of the Enewetak Atoll that the George shot took place in the spring of 1951. Elugelab was one mile in diameter, and it disappeared as a result of the 10.5-megaton

(10,500,000 tons TNT equivalent) blast. The bomb was approximately eight hundred times larger than the Hiroshima bomb. "Mike" was the first detonation of a truly thermonuclear kind. It was Teller's brainchild, and Bradbury graciously invited him to Enewetak to witness it. Teller declined, citing the importance of his presence at the fledgling Livermore Laboratory.

For the development of the hydrogen bomb, and for Teller personally, "Mike" was a resounding success, but it was not yet a deliverable bomb. It was huge, with a lot of cryogenic equipment: the mass of otherwise gaseous deuterium, which served as the thermonuclear fuel, had to be cooled below −250°C in order to keep it as liquid, while the fission device had to be kept at ambient temperature. The whole construction looked more like a building than a bomb. The principal purpose of the IVY series was to prove the feasibility of the thermonuclear device.[60]

At the time of the classical Super, in 1949, Oppenheimer characterized technical problems of size and weight with the anticipated hydrogen bomb; he doubted whether "the miserable thing will work," and whether "it can be gotten to a target except by oxcart."[61] The "Mike" device could not have been pulled to any target by an oxcart, but Oppenheimer's 1949 comment had some resonance with a conversation in 1952 that took place between him and Teller following the "Mike" test. After the "Mike" shot, later in November, Teller visited Princeton to personally thank John Wheeler for his work on "Mike." On the occasion of this Princeton visit, he met with Oppenheimer. According to one account, Teller, Oppenheimer, and Isidor Rabi had lunch together, during which Oppenheimer turned to Teller and said, "Well, Edward, now that you have your H-bomb, why don't you use it to end the war in Korea?" The Korean War had been going on for more than two years. Teller's response was "The use of weapons is none of my business. This is a political decision and I will have no part in it."[62] When Teller told the story to his biographers, he cautioned them that Rabi would probably not remember the conversation, and he was right.[63] The story was repeated in a second biography a decade and half later.[64]

Teller describes the same meeting in his *Memoirs*, according to which Oppenheimer invited Teller for a drink in his home and raised the possibility of using the thermonuclear device to end the Korean War. Rabi's presence is not mentioned. In this narrative, Oppenheimer elaborated on how it could be done given the inconvenient size and weight of the device. He suggested that it should be positioned somewhere and the communist troops should then be induced to concentrate in its vicinity before the explosion could wipe them out.[65] Here the story resonates with Oppenheimer's famous comment on the thermonuclear device being pulled by oxcart.

That Oppenheimer might have asked Teller sarcastically about the thermonuclear device does not sound unlike him, despite the bad taste of such a question. In 1943, Oppenheimer and some other participants in the Manhattan Project, among them Conant and Teller, contemplated the use of radioactive poison to incapacitate hundreds of thousands enemy troops, but the plan was never turned into action.[66] Also, Oppenheimer might have had the use of a fission bomb in mind to end the Korean War. He was on record as having raised the question about the possibility of using nuclear bombs in Korea when he was lamenting that such a decision should depend so much on the opinion of America's allies.[67]

There is at least one more description of the episode by Teller in a letter to Los Alamos veteran Robert Bacher in the summer of 1977. Apparently Teller gave two talks before large audiences and met with a smaller group of students during a visit to Caltech. From Teller's letter we can conjecture that Bacher, himself a Caltech professor, was not present at Teller's encounters with the students but had heard about them and had inquired about them; hence Teller's response. Teller described a meeting at Princeton following the "Mike" test in November 1952 when Oppenheimer suggested to him that he should try to have the hydrogen bomb used in Korea. Teller quoted his emphatic reaction that he would have nothing to do with such an attempt.[68]

Apart from teasing Teller—most inappropriately, to be sure—it is unlikely that Oppenheimer could have had in mind using a "Mike"-like device in Korea rather than an atomic bomb, which by then the United States possessed plenty of. The implication was that he would not have a problem with dropping an atomic bomb on Korea. Oppenheimer's question—if he had really asked Teller—could not have been part of a real policy discussion, but he may have easily offended Teller. The episode could only have served as an insult for Teller.

As for the relationship between the two weapons laboratories, York noted that it was strained from the beginning and rapidly grew worse.[69] A major issue of contention was who should have the credit for the hugely successful "Mike" shot. We have seen in the previous chapter how the media, including a biased book, overemphasized Teller's role. This injustice spilled over to involve the two laboratories in that Livermore was given full credit and Los Alamos practically none. The situation was made worse by the ban of the AEC on the associates of the laboratories and others involved with the work from talking to the press until about two years after the event. The animosity did not disappear easily.

Thirty years later, during the Strategic Defense Initiative program (see chapter 11), the "Goldstone" device of Livermore was detonated in the Nevada desert on December 28, 1985, and the X-ray laser demonstrated orders of magnitude lower efficacy than hoped for. Soon after, a journalist talked with a group of elated Los Alamos scientists who told him laughingly that it would take a century or two to bring the X-ray laser to the prescribed level.[70] The controversy with Teller had its mark on the relationship between the two laboratories.

The negative examples referring to the rivalry between the two giant weapons design labs, however, should not obscure the fact that fruitful and multifaceted cooperation also formed between them. York noted that the Los Alamos leadership treated Livermore appropriately from the start and provided valuable technological assistance to Livermore.[71] Moreover, there was one conspicuous area where the two laboratories were in unison, and that was their opposition to banning nuclear tests.[72]

The "Mike" shot of the IVY series was followed by a second test, the King shot, again, by Los Alamos. It was a fission bomb, but its blast was so large, 0.5 megatons, or 500 kilotons, that for its strength, it could have been a hydrogen bomb. York argued that it had been developed at least in part as proof that fission bombs could also be effective counterparts of hydrogen bombs.[73] This very large fission bomb used a sophisticated technique of the implosion approach, so in one way it also carried Teller's marks.

The nuclear weapons development reached the next level with the CASTLE series, which tested deliverable hydrogen bombs containing the solid fuel lithium deuteride instead of liquid deuterium. Its success once again reinforced the validity of the "Mike" design. It was also on the basis of the "Mike" design that the US Air Force initiated a crash program to build intercontinental missiles that would have the capability of delivering nuclear warheads. Thus this program was not a derivative of the CASTLE series but was decided entirely based upon the "Mike" device.

The utility of the Teller–Ulam approach had long-range consequences. The Scientific Advisory Board of the Air Force and its Nuclear Weapons Panel helped turn around American defense policy and, accordingly, American foreign policy, as well. The panel came to the conclusion in 1954—after "Mike" but before yet another test named Bravo—that they should build rocket-powered ballistic missiles that would carry nuclear warheads across a quarter of the world and would deliver them with accuracy.

In this field, the Soviet Union might have been ahead; the Sputnik program would begin in 1957, but the United States was catching up, and what began in 1954 bore fruit in the years to come. The resulting intercontinental missiles included the Atlas, Titan, and Minuteman; the intermediate-range missiles included the Thor and Jupiter; and there was the submarine-launched Polaris. Teller and his Martian friend, von Neumann, joined forces in this enormous operation in which they played a small but essential part.

The Bravo shot occurred in the spring of 1954. The uninhabited atoll of Bikini (180 miles east of Enewetak Atoll) was chosen for its test. The shot's yield was much larger than expected because the advance calculations missed some nuclear reactions that took place in the lithium deuteride fuel. The blast of the Bravo shot was fifty megatons, the first enormous American hydrogen bomb and the largest-ever American bomb tested. Its fireball extended four miles in diameter.[74] American observers on a ship thirty miles away were covered by dangerous radioactive dust.

The fallout from "Mike" dissipated and disappeared into the endless ocean. In contrast, the larger blast of Bravo, combined with misjudged weather conditions, caused the radioactive fallout to land on three inhabited Marshall Island atolls east of Bikini and on a Japanese fishing boat, the *Lucky Dragon* (Fukuryu Maru), in the vicinity. After they had been exposed to the fallout, the 250 natives and 28 military personnel were evacuated. Eventually there were fatalities attributable to the fallout, followed by worldwide condemnation and protest. At one point, York proposed to the AEC testing a larger-than-20-megaton bomb. It was vetoed by President Eisenhower, who found the already existing bombs too big.[75]

York remained director of Livermore for five years, then he left for a Washington position. Teller took over as director in 1958, somewhat reluctantly, because the position did not suit him and it restricted his freedom of moving around and voicing his opinion on a broader scope of issues than a laboratory director should give. Teller talked with Lawrence, who named him director after having complied with the necessary formalities. Teller served as director with dedication, although to be a good director, he had to change a lot of his activities and commitments. The tenure of his directorship was not uneventful either. Two incidents presented additional challenges for Livermore and for Teller, above anybody else. One was Lawrence's death in 1958, which was especially hard on Teller, as his responsibilities were increased not only by the amount of additional work but because he remained the sole elder statesman for the laboratory.

Then came the news about the forthcoming test ban on nuclear explosions (see chapter 9). Accelerated efforts were necessary to complete the ongoing tests, as the deadline of the ban was approaching. The ban also created some uncertainty about the future of Livermore, so Teller had to be vigilant to not let the morale drop, let alone an exodus of gifted scientists. At the same time, he relished some pleasurable duties of his director-ship. An important community affair was the ribbon-cutting ceremony of the opening of the new access road to the laboratory, which decreased the traffic going through the town of Livermore. The ceremony took place in August 1958, and "the Lab provided a metal rod coated with radioactive table salt which, when placed near a detector by a county official, activated a circuit to detonate a black powder charge which severed the ribbon."[76]

When Teller took over the directorship, he did so with the under-standing that Mark Mills would soon take it over from him. So it was a heavy blow for Teller, in addition to losing a friend and valued colleague, when Mills was killed in a helicopter accident during a test series at Enewetak in 1958. It was only in 1960 that Teller could finally resign, and he was followed by Harold Brown, whose directorship lasted only one year. Brown was eventually elevated to the highest office among former Livermore associates, where he served as President Jimmy Carter's secre-tary of defense. Livermore was characterized by its internal breeding of its leaders, which is not very typical in the United States. On the other hand, many among its cadres started very young and their rise was fast, and many were catapulted into prestigious outside jobs. Soon after Lawrence's death, the laboratory took on his name, and today it is known as the Lawrence Livermore National Laboratory (LLNL).

Teller finally arrived home at Livermore, and his constant wanderings came to end when he settled there. When he said that Livermore was the accomplishment he was most proud of, he did not exaggerate. His vision about the need and usefulness of competition proved to be accurate, but rivalry is different from competition, and the line between the two is blurred. As late as 1977, Harold Agnew, the third director of Los Alamos, referred to their competition in the following statement: "It's like Hertz and Avis, and we think we're no. 1."[77] I heard a far less complimentary char-acterization of the relationship between the two laboratories from a Liver-more associate: "The Russians are not the real enemy; Los Alamos is."

For a decade and a half, Teller was persona non grata in Los Alamos, until Agnew invited him back to spend a summer there. Agnew was one of Fermi's students in the early 1940s in Chicago, and he participated in the construction of the atomic pile, the first nuclear reactor. He went to Los Alamos in 1943 and flew with the second plane over Hiroshima to mea-sure from the air the yield of the first atomic bomb. After the war, he com-pleted his studies with Fermi at Chicago, and after he received his PhD degree, he returned to Los Alamos in 1949. He was director there from 1970 to 1979 and was awarded the Fermi Prize in 1978.

We talked in Budapest in August 2003, just about a month before Teller died.[78] Agnew must have been atypical at Los Alamos because he thought at the time that Livermore was established that competition would be good for Los Alamos. He also remembers that "it took Livermore ten years before they got something successful in the stockpile"—because they would not build on what Los Alamos had done.[79] Agnew found it important to bring Teller back to Los Alamos, especially so the young people could meet and talk with him. Teller and Livermore also benefited because Los Alamos had already started a laser program by the time of Teller's visit, and the laser program at Livermore became very large.

Agnew asked Los Alamos physicist George A. (Jay) Keyworth to look after Teller during his stays at Los Alamos, "to become Edward's 'keeper,'" in Keyworth's words.[80] Teller and his wife used to stay at Keyworth's home during their visits. Keyworth "often arranged for him to play the piano with a small group of musicians from Santa Fe." Then Keyworth left Los Alamos and moved to Washington to become President Reagan's science adviser in 1981. Keyworth considers Teller one of his "three major mentors, along with Ronald Reagan and David Packard."

Another former Los Alamos director, Siegfried Hecker, stresses his own role in helping to normalize Teller's relationship with Los Alamos. He said that Teller liked to come there to carry out research in cooperation with his local colleagues. Hecker lately became a frequent traveler to Russia and has established a close relationship with his Russian counterparts. He does not hide his satisfaction when he says: "There was huge animosity between Arzamas-16 and Chelyabinsk-70 just like there was between Los Alamos and Livermore."[81]

These visits to Los Alamos meant a great deal to Teller. Even before the establishment of Livermore, he often complained to Goeppert Mayer of missing physics. He was too busy with other things and he regretted no longer working as a physicist and yearned to return to it. After the Oppenheimer hearing his remoteness from the modern development of his science became almost institutionalized. He found solace in a flurry of activities, and physics was hardly among them. Teller seldom did research alone, and his isolation from the physics community virtually paralyzed him in this respect. Livermore employed a lot of physicists, but there was relatively little emphasis on fundamental studies. Teller tried hard to get outstanding scientists to Livermore, but his success in this was limited, especially when compared with the stellar roster of his former colleagues at Los Alamos.

Up until this point, Livermore could claim one discovery that brought a Nobel Prize, and the awardee was Robert B. Laughlin in 1998, almost a decade and a half after he had left the laboratory. The discovery, the fractional quantum hall theory, was not part of Laughlin's principal Livermore activity but a side project. For fairness, he was encouraged to think about his basic science, he was allowed to use his computer account for its calculations, and his extensive travels related to his basic science were generously supported. Laughlin, however, did rather little work for what he had been hired to do in the first place, which was related to the X-ray laser.[82] He had meetings with Teller, as Teller was increasingly trying to return to physics in the early 1980s, but they did not seem to find a common language and Laughlin did not seem to have enough patience for Teller.

One of the first physicists who fit Teller's need for a partner for his physics discussions was the man who had recognized Laughlin's merits for Livermore a few years earlier. Richard More was summoned to Teller's office about once a week where they talked science. Teller grew accustomed to their encounters, and when More was spending a sabbatical in Paris, he often received a call from Teller at around ten in the evening when he had just returned from his exercises. Speaking physics with Teller was like a second workout for More. He characterized Teller's method of doing research in the following way:

> His method was "extreme Socratic" . . . Dr. Teller had enormous energy and persistence and had a very strong intuitive sense of the behavior of atoms, electrons and photons. He could focus his attention very narrowly on a specific process and his power of concentration was awesome. He did not like formalism or mathematical structure beyond the necessary minimum. I remember he was proud of doing arithmetic in his head and often was faster than my pocket calculator, but that arithmetic was not always accurate. I always felt he cared deeply about scientific phenomena and had little interest in who had said what, who had discovered what, who planned to do what, etc.
>
> . . . Dr. Teller had a profound optimism about the effects of science on our human world; he felt that engineers could achieve and then optimize anything that was not fundamentally impossible. He was less interested in cost. Sometimes this optimism could seem naive or even irresponsible.[83]

Toward the end of the 1980s, two other young theoreticians joined Teller in discussions, Stephen Libby (not related to Willard Libby) and Neal Snyderman.[84] They were not randomly selected; apparently Teller tried out several people and settled on them. These scientists provided Teller an opportunity to satisfy his insatiable curiosity for modern physics.

Yet another reason for their happy match might have been that at the end of Reagan's presidency and with the cold war soon ending, Teller found more free time on his hands than ever before. He was never the reading type, and he used the Socratic Method that More alluded to above very successfully. Besides, he had the opportunity to engage people, and it was considered a distinction and honor to become Teller's partner in conversation. Usually his secretary summoned one of the physicists to have lunch with him. These were not fancy lunches; Teller had no problem eating any kind of junk food brought to him and his guest from the cafeteria.

Teller asked specific questions and appreciated when the answers went beyond them. The discussions covered a broad range of topics. Teller could be childishly happy when he figured out an explanation and interpretation and would stop his partner in their discussion so he could figure it out. He still had tremendous mental capacity and recall of what he had done and learned decades before, while his short-term memory was diminishing. He often started the conversation the next day by asking the same question he had asked the previous day. So, much of the conversation would be repeated the following day, and progress was slow, but there was progress. There were cases when Teller's partner prepared for their discussion; in other cases no preparation was needed, but the discussions were always a challenge.

Teller had a spacious but not pretentious office at Livermore. The director's office was on the same floor as were the offices of other former directors. Teller's office and the corridor leading to it were decorated with photographs showing him with US presidents and other luminaries like John Paul II. Physicists appeared in one frame only: a framed yellowing newspaper page about several scientists with funny but kind characterizations, including Teller. The books on Teller's desk kept changing over time, reflecting his changing interest. There was one black-bound dissertation with its title in large gold type, *Longevity*, which had a permanent place.

Not all those who were summoned to Teller's office handled the situation as easily as More, Libby, or Snyderman. Of the dozen or so senior scientists I interviewed—most of whom were retired from Livermore—some sounded bitter about their encounters with Teller. In one case Teller invited someone to talk about a particular topic, and when the guest did not appear sufficiently versed in it, Teller sent him home to study and return when he was ready. To another, when he tried to explain to Teller his research and suggestions that were supposed to interest him, Teller said, "You're wasting my time; go away." The guest did. Some of these people still carry the verbal scars and a feeling of humiliation from their meetings with Teller at Livermore.

One of Teller's favorite topics was reactor safety. About a year before Teller's death, plasma physicist and nuclear engineer Ralph Moir was summoned to Teller's office, where he was told that Teller wanted to see him often during the next twelve weeks and talk with him about nuclear reactors. An intense period of discussions followed, which lasted almost a year and resulted in a joint paper, Teller's last scientific contribution. It was not original research; rather, it was an example of Teller's efforts to have applications for his science. Moir used to come to Teller's office once or twice a week, as well as visiting his home in Palo Alto from time to time. Moir had to describe all possible kinds of reactors, including their merits and potential problems, after which he would be bombarded by Teller's questions. The next day or week he would start with asking the same questions, and Moir never knew whether this was because Teller forgot or was testing him. In any case the conversation advanced a bit every time, just like those with the theoretical physicists. Apart from his day-to-day memory problems, Moir found Teller's mental capabilities to be excellent.[85]

One aspect of their discussions about reactors was the consideration whether to put them underground. We have seen that this was something Teller found important but had failed to get passed at the time of his involvement in reactor safety after the war (see chapter 5). Teller initially proposed putting the reactors about three hundred feet underground, which would have greatly decreased their economic feasibility. Moir convinced him to settle on a thirty-foot depth in their proposed solution, which would mean minimal additional costs. The thirty feet would protect the reactor from being destroyed by a commercial airplane flown by terrorists, but it would not protect it from a military attack. The concrete plates covering the reactor could be lifted by a crane, and the same crane could be used for a whole series of reactors. The concrete plates would be covered by dirt. The technology of their preferred reactor would be based on molten salt, an approach well known in nuclear technology.

When Teller and Moir reached an agreement about all the technical details, Teller suggested writing a joint paper. Moir prepared a draft and read it to Teller (who was by then completely blind). Numerous discussions followed, and every time they started with Moir rereading the draft, the portion about which they had mutually agreed kept growing. Teller paid special attention to the introduction, for which he essentially gave a dictation. The whole exercise proved to be a very instructive experience for Moir. When they were ready, they sent the manuscript to *Science*, then to *Nature*, and then to a series of other periodicals. They received rejection after rejection.

The importance of the paper was not so much in novel findings but in reemphasizing some points about reactor design, especially the use of thorium, molten salt, and building it underground, and in the fact that this had been done with Teller's participation. By the time the manuscript appeared in a lesser-known and highly specialized journal, Teller had been dead for more than a year.[86]

## TELLER TECH

Teller organized a graduate program whose primary purpose was to provide advanced training for Livermore associates. This coincided with his interest in giving proper emphasis to applied science in graduate education. Applied science is, of course, not a different branch of science, it is only the application of science, but its place is difficult to define between pure science and engineering. Teller long thought of creating the proper place for applied science in the university structure and found that the best way to do this would be by offering a PhD degree in applied science. Whereas his idea did not resonate with the administration at Berkeley, he found interest and opportunity at another campus of the University of California system, at Davis, some sixty miles north of Livermore. A new department of applied science was founded, and Teller became its first chairman, although he accepted the assignment for a limited time. Courses started in 1963 with the active pedagogical participation of many Livermore scientists. All the core subjects emphasized applications, and the advanced students could specialize in nuclear chemistry, physical chemistry, computer technology, materials science, or plasma physics. Livermore absorbed many of the new graduates, and the school has since thrived.

Livermore made a big step right from the beginning of its existence toward becoming the equal of the Lawrence Berkeley Laboratory, which had developed a joint program with the University of California at Berkeley. Teller himself taught two courses at Davis, "Production and Use of Nuclear Energy" and "Structure of Matter," and another course at Livermore, "Statistical Theory of Equilibrium and Transport Phenomena."[87] The educational project at Davis has been referred to as Teller Tech, a distinction fully deserved. This institution was consistent with Teller's long-held views on the unjustified preeminence of basic science over applications in academia. His views agreed with Alvin Weinberg's experience, who was the director of the Oak Ridge National Laboratory.[88]

Education was one of the areas that provoked Teller's interest in Israel.

During the years immediately after the war he showed a remarkable indifference toward the pioneering country. His meeting and friendship with Yuval Ne'eman contributed greatly to his developing close ties with Israel. They shared devotion to defense, science, and education.[89] Ne'eman had graduated from the Technion–Israel Institute of Technology, but he was a physics professor at Tel Aviv University. When he was serving as its vice president, he invited Teller to participate in a meeting of the Board of Trustees of Tel Aviv University. For some time, Teller regularly attended the annual board meetings. It was during this close interaction that the idea of establishing an engineering school at Tel Aviv University arose, and Teller was instrumental in making it happen. Ne'eman then served as the first dean of the School of Engineering at Tel Aviv University. Teller was always asked to give lectures during his visits to Israel; the lectures were always standing room only. In 1969, he gave a series of ten lectures with the title "The Rise of Modern Physics." Usually Teller drew a big crowd for the opening lecture, but then the audience dwindled. This time, it was the opposite; the audience kept increasing from lecture to lecture, and Teller had to change to ever-larger venues.

Teller's activities with the Hertz Foundation were also part of his oeuvre in education. He began his participation in the work of the foundation in 1963, after its inception six years earlier. The founder, John D. Hertz, came to the United States from the Austro–Hungarian Empire toward the end of the nineteenth century. He started, among other ventures, the Yellow Cab Company and the Hertz Rent-A-Car Company. By establishing the Fannie and John Hertz Foundation, he wanted to contribute to the technological education of young Americans in their undergraduate engineering studies.

From 1963, the emphasis shifted to granting postgraduate fellowships leading to a PhD degree in the fields of engineering and applied sciences. Teller played a pivotal role in this shift, and the fellowship was renamed Fellowships in Applied Science. He participated in the selection of the grantees. Eventually some friction developed between Teller and some of the other board members, as not everybody was pleased with his enhanced influence on the Hertz Fellowships.[90] Regardless of such controversies, Teller's impact on the foundation was rewarding. There were quite a few outstanding awardees, such as Ali Javan and Theodore Maiman, both distinguished for their pivotal discoveries in the field of lasers. Then there were Carl Wieman, Nobel laureate in 2001, and John Mather, Nobel laureate in 2006, and other luminaries in applied science and entrepreneurship. One of the pioneers of the X-ray laser, Peter Hagelstein (see chapter

11) was also a Hertz Fellow, and so were many other outstanding scientists at the Livermore Laboratory.

Teller's devotion to education was the least controversial among his areas of activities. He had a well-defined agenda in enhancing the prestige of science and the applications of science, and his achievements have constituted a lasting impact.

# Chapter 8

# DOUBLE TRAGEDY

## Teller and Oppenheimer

Edward understood power . . . he could have written [Machiavelli's] *The Prince.*

—GEORGE (JAY) KEYWORTH,
PRESIDENT REAGAN'S SCIENCE ADVISER[1]

*Edward Teller was the only major scientist who testified for denying Oppenheimer's security clearance at the Oppenheimer hearing. Given the circumstances at the time of the McCarthy era, and Oppenheimer's past, the withdrawal of his security clearance was not unexpected. When the material of the hearing was released, Teller's testimony caused irreparable damage to his relationship with his peers, while it contributed to Oppenheimer's image as a martyr. Interestingly, Teller was also under FBI surveillance during much of his career.*

It was a culmination of years of suspicion, testimonies, and false testimonies, as well as FBI and military investigations of Oppenheimer's deeds, intentions, and stands, when in 1953 he was accused of espionage for the Soviet Union. Oppenheimer had a complicated history and personality, and volumes have been published about his life.[2] Our coverage has a limited goal in that we will be interested in Oppenheimer's life to the extent that it intertwined with Teller's life and activities. We will, however, cover a considerable portion of it, because Oppenheimer—directly and indirectly—had a major impact on Teller's fate during the last six decades of his life.

Teller's testimony at Oppenheimer's security hearing in April 1954 resulted in his third exile. This was an internal exile and it was self-inflicted, unlike his first two exiles, which were caused by forces and events independent of him. The third exile separated him from the larger part of the community of American physicists, and impacted his life at more than one level. From the point of view of his scientific activities, his first two exiles proved to be beneficial. The third damaged his chances of interacting with other physicists—his major means of doing research—and diminished his scientific output after 1954.

Parallel to his withdrawing from a larger part of the world of science, Teller immersed himself increasingly in the world of politicians and military leaders. From this point on, lacking the friendly circle of his peers, he revealed his negative traits more readily.

## PARALLEL LIVES

J. (Julius) Robert Oppenheimer was born in New York City in 1904 into a nonreligious Jewish family. His father was an immigrant from Germany; his mother was American-born, from Baltimore. They lived in a good neighborhood in New York City, the Upper West Side. They spoke English, not Yiddish, and though they were eager to assimilate, they kept their Jewish identity. The anti-Semitism in American society toward the end of the nineteenth century and the first half of the twentieth century manifested itself in discrimination.[3] That was felt more by the aspiring Jewish upper classes—to which the Oppenheimers belonged—than the masses that were struggling for economic survival. Such anti-Semitic discrimination was common in places of employment, housing, hotels, and resorts, as well as in schools. It continued through the 1940s—not only for Jewish pupils but for Jewish professors as well.

In view of the Jewish presence in American scientific life today, it is hard to imagine the situation a few generations ago. It was difficult for aspiring Jewish youth to get into graduate schools, especially at good universities, and/or receive teaching and research assistantships. Even benevolent professors might have discouraged Jewish students from continuing their education because of their bleak prospects for employment.[4] At the time Oppenheimer was of high school age in America and Teller in Hungary, Isidor I. Rabi had just gotten his bachelor's degree in chemistry at Cornell University and could not get a job in chemical companies or a faculty position at colleges or universities.[5] In 1929, he became a lecturer at

Columbia University, and when he was promoted, he was the first Jewish professor of physics at Columbia University.

In the second half of the nineteenth century, the discrimination in education stimulated the American Jewish community to search for solutions. The outstanding school of the Ethical Culture Society, which Oppenheimer would eventually attend, was one of the few highly esteemed establishments to counter discrimination. Ethical Culture was a religion to some who did not care for Judaism; who did not want to be pronounced atheist; and who wanted to feel comfortable in following high-minded moral principles. The Ethical Culture School in New York City was "an ideal place for the precocious, dazzling, but insecure young Robert."[6] There was a similarity between Oppenheimer's and Teller's childhoods, in that both children had few if any friends and found it difficult to relate to others. Eventually, though, both became more popular among their peers. Although both cultivated their popularity among their peers, both were given to traits that could easily damage their relationships.

Just like Teller, Oppenheimer also went to study in Germany, where he also obtained his doctorate. Neither of the two had a straightforward route to his doctoral studies. Oppenheimer first joined the Cambridge physicists but did not feel comfortable among them. He was not much given to experimental work and may have felt intimidated by the stellar group around Rutherford. When the visiting Max Born invited him to Göttingen, he was happy to accept the invitation, and their interaction developed into Oppenheimer's doctoral work. Beyond the similarity in Oppenheimer's and Teller's gravitating to Germany, there was also an important difference. For Oppenheimer it was a trendy choice to go to Europe, whereas for Teller it was a necessity to depart from Hungary. Oppenheimer spent years in Europe with a secure American background to which he would return. As for Teller, he hoped to establish himself in Germany for good, since returning to his home base was out of the question. Oppenheimer did not have to dedicate all his time and energy to physics, so he studied a wide range of humanistic interests including the Sanskrit language. For Teller, his dedication was completely to physics except for light diversions such as playing the piano and table tennis with his colleagues.

There was a difference in their approach to their Jewish roots as well. There is no indication of Teller's feeling uncomfortable about his Jewishness, and he never forgot the anti-Semitism he experienced during his high school years back in Budapest. Though he did not go to a Jewish school there, his circle of friends was almost exclusively Jewish. Oppenheimer never felt comfortable with his Jewishness. According to Rabi, "Oppenheimer was Jewish, but he wished he weren't and tried to pretend that he wasn't."[7] Institutions of higher learning, such as Harvard University, imposed restrictions on the number of Jewish students accepted, but given Oppenheimer's outstanding academic record and his family's financial resources, his road to Harvard seemed straightforward. Yet he felt isolated there because of the quota system.

Both Oppenheimer and Teller showed little interest in politics during their university years and even afterward for some time. Oppenheimer's family wealth isolated him from the perils of the Great Depression, whereas Teller had to rely on much more limited sources to set up his first emigration. While he felt like an insider in German culture, he was an outsider in German politics. Oppenheimer developed a strong left-leaning interest in politics from the mid-1930s, focusing on issues like the support of Republican Spain. He was also concerned about the situation in Nazi Germany as proven by his words: "I had had a continuing, smoldering fury about the treatment of Jews in Germany. I had relatives there."[8] Thus his feeling uncomfortable about being Jewish did not make him blind in this respect. Further, he became disappointed in the economic situation in America through his observation of other people's difficulties. He took more notice of politics when he saw what the Depression was doing to his students "and through them, [he] began to understand how deeply political and economic events could affect men's lives."[9] Through friendships and personal interest, Oppenheimer became active in communist front organizations that were groups closely associated with the communist movement but formally not affiliated with the Communist Party. There is no hard evidence that Oppenheimer ever became a card-carrying member of the party.

Teller was eleven years old during the 133-day-long communist dictatorship in Hungary, which caused him discomfort but hardly any lasting impression. In Germany, he was exposed to the communist and Nazi movements early on, but only as an observer. He did not need to delve into the Nazi ideology to reject it, and we do not know how much he learned about Marxism at the time; probably not much beyond a superficial knowledge. He witnessed demonstrations and clashes in the streets from a safe distance and was not much concerned by them. His friendship with Laszlo Tisza played much more of an important role in forming his opinion on communism. Through Tisza, Teller could witness the dedication of a communist sympathizer as well as his disillusionment with the Soviet Union in a few years' time. Teller became an anticommunist as a result of a long process, and Arthur Koestler's *Darkness at Noon* completed his political education (see chapter 4).

Oppenheimer had received disappointing information about the Soviet Union when two of his physicist friends visited him in the summer of 1938.[10] Victor Weisskopf and George Placzek had spent some time in the Soviet Union and "tried to convince him of the reality of Soviet life by describing the lack of freedom and the persecutions under Stalin that [they] had seen there."[11] Weisskopf's impression was that Oppenheimer still believed in communism. Tisza's impact on Teller was stronger than Weisskopf's and Placzek's on Oppenheimer, and not only because of their closer friendship but also because Teller could witness Tisza's transformation. Weisskopf and Placzek could not quite crush Oppenheimer's optimistic viewpoint of the great socialist experiment.

In light of Oppenheimer's later troubles due to his leftist interest and affiliations in his youth, it is important to point out that communist ideology has been more tolerated in the European democracies than in the United States. It is also crucial to distinguish between the communist ideology and Soviet communism, which distorted the original ideals in the way that the Inquisition distorted original Christianity. The distortion of communism reached such a degree that Soviet communism could be paired with Nazism with good reason.

In the 1930s, many intellectuals in the West had an open mind to communism and a number of them became party members. It was considered a possible approach to correct the economic wrongs of capitalism at the time of the Great Depression. In addition, the communists were consistently against the Nazis from the beginning up to the German–Soviet Non-Aggression Pact in August 1939. The pact, which was signed just before Germany (and the Soviet Union) attacked Poland and before World War II had begun, lasted for almost two years. It caused a lot of confusion among Western communists, many of whom became disillusioned.

The situation changed on June 22, 1941, with the German attack on the Soviet Union. In the subsequent four years, the Soviet Union was an ally of the Western powers and sacrificed the most in its struggle against Germany. This made the Soviet Union and even Stalin a hero for many, which did not necessarily end when the war and the grand alliance was over and the appreciation among some turned into blind devotion. In view of this, there was some basis for fears that being a communist might be equivalent to unconditional adherence to Moscow. This fear may have been justified in some cases, but to generalize this would be similar to assume that all Catholics follow the orders from the Vatican regardless of their own feelings as well as the beliefs of their country. In 1960, such a fear found expression when John F. Kennedy, a Catholic, was running for president.

In any case, the success of communist propaganda at that time among liberal intellectuals in the West is noteworthy especially in view of the persecution of liberals in the Soviet Union. Those in the West, who viewed the Soviet system with a clear eye and noted the realities of the Soviet system, were not misled. While, for example, Leo Szilard was devoted to peaceful coexistence and advocated overtures to the Soviet Union, he knew that he could not have survived in the Soviet system.

Oppenheimer graduated from Harvard in 1925 and left soon for Europe. What Leipzig was for Teller, Göttingen was for Oppenheimer, except that the change after Harvard was less drastic than the change after Budapest had been. Similar to Teller, the first projects of Oppenheimer were in molecular physics, and he made contributions of lasting significance in this field. The collaboration between Born and Oppenheimer resulted in, among others, the famous Born–Oppenheimer approximation. Every student of chemical physics and physical chemistry today learns about it. It is based on the fact that, generally, the relatively heavy nuclei in a molecule move much more slowly than the electrons associated with them, hence, the electrons may be assumed to move about a fixed nuclear arrangement. This approximation greatly simplifies the description of molecular structures.

The significance of the Born–Oppenheimer approximation and Oppenheimer's later contributions have been recognized, yet for Oppenheimer it must have been painful that he joined the scientific revolution of the twentieth century only in the second layer, after the true creators of quantum mechanics. He, and others, like Teller, had to be content with picking up the pieces left by the likes of Heisenberg, Born, Dirac, and Schrödinger.

Oppenheimer received his doctorate at twenty-three years of age from the University of Göttingen in 1927. Teller completed his doctorate in Leipzig in 1930 when he was twenty-two. Their careers were advancing more or less along parallel lines. It must be noted, though, that whereas in Germany a second doctorate—the habilitation—was (and is) needed before one could give lectures and become a professor, in the United States the PhD degree was (and is) the "terminal degree." Thus, Oppenheimer could go after an independent professorial position in the United States right away, whereas Teller had to be content with an assistant's position in Germany for the next few years.

Oppenheimer was in great demand back home, where the lack of theoretical physicists was being acutely felt, so he split his time between Caltech and Harvard in 1927–1928. Then, he returned to Europe and toured the centers of physics in 1928–1929. His stay with Wolfgang Pauli in Zurich was especially influential on him. Much has been written about his physical and psychic instabilities as well as his exceptionally quick mind. Even when he was wrong, he could move on to the next topic so fast that his partners in conversation were unable to correct him. Ehrenfest introduced Oppenheimer to Pauli with these words in his letter of recommendation: "I have here a remarkable and intelligent American but cannot handle him. He is too clever for me. Couldn't you take him over and spank him morally and intellectually into shape?"[12]

Upon Oppenheimer's return to the United States, he had to choose whether to accept an appointment in the Northeast or take up the challenge of moving to California. Given his background and general disposition, he might have been expected to stay in the Northeast; instead, he went to California in 1929 and split his time between the University of California at Berkeley and the California Institute of Technology (Caltech) in Pasadena. In contrast, Teller simply went where he felt he was needed.

By all accounts, Oppenheimer was an arrogant young man, which may have been an overcompensation for his insecurity. Even the mild-mannered Max Born had to take measures to curb Oppenheimer's aggressive interruptions of seminar speakers, and not only of visitors, but of Born himself. To the German professor, Oppenheimer "was a man of great talent, and he was conscious of his superiority in a way which was embarrassing and led to trouble."[13] The members of Born's group threatened a boycott of the seminars unless Born stopped Oppenheimer's interruptions. Born devised a scheme to let Oppenheimer learn of these complaints without directly confronting him, and when this happened, the interruptions ceased.[14]

A much later characterization of Oppenheimer noted his negative traits, which he retained even at the peak of his career. In the words of fellow physicist and science historian Abraham Pais: "vast insecurities lay forever barely hidden behind his charismatic exterior, whence an arrogance and occasional cruelty befitting neither his age nor his stature."[15] Teller also interrupted the speakers, even his professors, during lecturing, to clarify a point, as we have seen in Herman Mark's description from their Karlsruhe days (chapter 2). Apparently, Teller managed to do this in a nonoffensive way, because no complaints are recorded.

From the mid-1930s, a dramatic change took place in the relative influence of the United States and of Germany in science. The center of gravity of scientific progress, of physics, in particular, shifted to America, and the language of science changed from German to English. The changes had two principal components. One was the American aspiration for excellence and the other was the driving desire to combat the Nazis. This was most of all a shift in the status of theoretical physics. Experimental physics in America had enjoyed more interest and respect than theoretical physics, but now theoretical physics was fast catching up. The other principal component was the sudden influx into America of physicists fleeing Nazism in Europe. Teller was part of that influx.

Oppenheimer greatly contributed to bringing up the level of theoretical physics in the United States. In Hans Bethe's words, "More than any other man, he was responsible for raising American theoretical physics from a provincial adjunct of Europe to world leadership."[16] Among Oppenheimer's students and associates were at least three future American Nobel laureates, Julian Schwinger, Willis Lamb, and Carl Anderson. Of course, building up a great school takes more than an outstanding scientist, however charismatic he may be. Oppenheimer also represented taste in science and style, and he mobilized streams of eminent and impressive visitors and carried out a stimulating seminar program. One of the invitees was Edward Teller.

Oppenheimer became a popular teacher in California for sophisticated and aspiring students. This segment of the student population was one of Teller's two principal constituencies (the other was the uninitiated). Unfortunately, Oppenheimer was not beneath being sarcastic at his students' expense. He could be charming and arrogant, endearing and contemptuous. He was not very good with those who were intellectually inferior to him and had no patience for undergraduate teaching, let alone science popularization.

In contrast, Teller with his deep and resonating throaty voice was an excellent lecturer at the popular level as well. He had the ability to explain complex concepts in an accessible way. During the Manhattan Project it happened that some military men, including General Groves, wondered whether U-238 and U-235 might become spontaneously separated. Such a separation might cause an unwanted explosion. Teller explained to them that it was as likely as "that all the oxygen molecules in this room get under the table, and we asphyxiate."[17] Once Groves understood how impossible this was, he no longer considered spontaneous separation of the uranium isotopes a possibility.

In research, Teller preferred working with people one-on-one. After World War II, he continued an academic career at the University of Chicago for some years. Oppenheimer returned for a very brief period to the California Institute of Technology and resumed teaching in November 1945, but this did not last long. There was to be a greater demand on him at the national level, and he soon moved to the Northeast to embark on a different career.

We have already alluded to Teller having been a digger rather than a driller—referring to the division of research scientists into classes of drillers and diggers. Oppenheimer also belonged to the diggers. He made important contributions to physics, including astrophysics, and his contributions were outstanding but diverse. According to some evaluations, had he lived longer, he might have been considered for the Nobel Prize for his works on neutron stars and black holes (the term was introduced much later). As noted earlier, there have been precedents when the Nobel Prize was awarded not for an obviously seminal discovery but rather for a lifetime of achievements. Even in such a case a particular discovery or improvement is singled out, since the Nobel Prize is not meant for lifetime performances. Oppenheimer might have been such a case, and so might Teller have been.

Once Oppenheimer became interested in politics, he was rather active in various left-wing organizations. The label "fellow traveler" was applied to him as it usually applied to nonparty members who were associated with the communist movement. Barton J. Bernstein gave a penetrating analysis of the significance of the level of Oppenheimer's involvement in the communist movement and the possible importance of the distinction between card-carrying membership and belonging to the Communist Party without such formalization.[18]

Oppenheimer gradually decreased his political involvement from the late 1930s onward, but he did not break with his communist friends even after the detestable Soviet–Nazi Pact in 1939. This was important because during the interval after the German–Soviet Pact—but before the Nazi attack on the Soviet Union—the Communist Party followed an isolationist policy in the United States, whereas Oppenheimer was for intervention. After the Soviet Union had become a belligerent of Germany, the Communist Party resumed its vigorous anti-Nazi policy and advocated support of the new victim of German aggression.

When Oppenheimer joined the Manhattan Project, he felt the need to overcompensate for his prior involvements in communist causes and so was eager to prove his loyalty before the authorities. The army people seem to have taken advantage of his perceived political naïveté or even guilty feelings, and General Groves told him that if he had to choose between friends and country, he should side with his country. Had Oppenheimer retained his wit and earlier posture, he might have responded to the general that there was no need to make such a choice; such a necessity did not exist. Instead, as future events showed, he made the artificial choice of country over friends.

When Groves met Oppenheimer in the fall of 1942, there was good chemistry between the two, but the relationship was more complicated than just a mutual feeling of comfort. Oppenheimer did not choose Groves; Groves chose Oppenheimer. The task was bringing together the last phase of the atomic bomb project into a single location. Oppenheimer's ranch, not far from where the Los Alamos Laboratory would be built, played a role in the selection of the location. Of all the different segments of the Manhattan Project, Los Alamos was the only one for which Groves had the opportunity to name the director, and he never regretted that it was Oppenheimer.

The choice was not obvious. It was not at all straightforward that a theoretician without prior managerial experience would be the right person to deal with a plethora of practical problems, but Groves had a gut feeling that Oppenheimer was the man for the job and he went after him. In doing so, Groves had to overrule the reservations of the security people. Groves and the project needed Oppenheimer and that was what decided his appointment.

The general realized that ideological and political differences could be ignored in such strained situations. The openly communist scientist J. Desmond Bernal's services were accepted and appreciated in Great Britain during World War II. Stalin, when he felt he was being driven to the wall, called upon Russian nationalism and included the Eastern Orthodox Church in his appeal for the defense of the Fatherland soon after Germany's attack on the Soviet Union. There was the famous Red Square parade on the anniversary of the Russian revolution, on November 7, 1941, when the Germans were already at the outskirts of Moscow, and the troops went directly into battle from the parade. Stalin gave a rousing speech in which he, himself a non-Russian, appealed to the nationalism of the troops and the Russian people in general. He declared, "May the courageous image of our forbears inspire you in this War."[19] Stalin also allowed the Leningrad metropolitan of the Russian Orthodox Church "to stir the deep patriotism of his audience." However, "[o]nce the Germans were defeated, the Soviet authorities once again erased Orthodoxy from Russian history."[20] We have seen that Hitler did not indulge in such compromises with respect to Jewish scientists when Max Planck appealed to him in defense of science in Germany (chapter 2). Hermann Göring was more pragmatic when he retained a Jewish cancer researcher and declared, "I decide who is a Jew."[21]

When he needed Oppenheimer in 1942, General Groves had no qualms about instructing the security people to issue clearance for Oppenheimer in spite of his past. He reinforced this in his 1962 book: "I have never felt that it was a mistake to have selected and cleared Oppenheimer for his wartime post."[22] He wrote this well after Oppenheimer's security clearance had been withdrawn. However, when Groves was asked during the security hearing in 1954 whether he would grant Oppenheimer security clearance under the *current* rules, he felt obliged to say that he would not.[23] There is an opinion that one of Groves's motivations in choosing Oppenheimer was his hold over him because of Oppenheimer's political past, which could always be used to keep him under Groves's thumb.[24]

In principle, Oppenheimer's past leftist associations should not have been a major consideration in gaining security clearance in 1942–1943 for the very reason that the Soviet Union was an ally of the United States. However, the alliance was never considered to be an easy one. There were forces within the United States that considered the Soviet Union to be the enemy in the long run and more dangerous than Nazi Germany. Regardless, assuming that left-leaning scientists were a threat to the security of the United States was a simplistic approach in any case. The atomic project was a latecomer as far as war-related scientific activities were concerned. And Los Alamos included a number of leftist scientists and even former Communist Party members who had not been tapped for other war projects, like radar research, and some were immigrants who had recently acquired US citizenship.

Teller and Oppenheimer initially developed a cordial relationship based on respect for each other's intellects. Then, in the fall of 1942, Teller and Oppenheimer traveled together to New York, and an incident happened. By then the army had overtaken the Manhattan Project and General Groves had been appointed its commander. Oppenheimer complained to Teller about working with Groves. He said something according to Teller's *Memoirs* that disturbed Teller greatly: "No matter what Groves demands now, we have to cooperate. But the time is coming when we will have to do things differently and resist the military."[25] Teller dated his alienation from Oppenheimer to that moment because he found such an attitude toward their own authorities unacceptable. Teller, as we have seen, always respected authority and nearly always submitted to it. Whether the episode happened exactly the way Teller described it is impossible to verify, but it sounded rather characteristic of Oppenheimer's ambivalent attitude.

It was in Berkeley in January or February 1943 that the Oppenheimers entertained at their home their friends, Haakon and Barbara Chevalier, who happened to be a leftist couple. A brief exchange between Oppenheimer and Chevalier took place, which was to have repercussions on their lives. There have been different versions of the conversation, the gist of which was that Chevalier asked Oppenheimer about the possibility of transmitting information on war-related projects to the Soviets. There seems to be no doubt that Oppenheimer rejected the idea. However, he waited eight months before he told the security people about the conversation without yet revealing Chevalier's name.

On the other hand, he went further by telling the security officers that a total of three people had been approached about giving information to the Soviets, but he kept refusing to name the person who had approached them. He never revealed the other names, and such approaches might have never been made. When this topic came up in the Oppenheimer hearing of 1954, and he was asked why he exaggerated what Chevalier was doing, Oppenheimer's response was shocking: "Because I was an idiot." Alternatively, the fact that he later changed his testimony might have been an effort to protect the other people, among whom might have been his brother Frank. This will never be known. While Oppenheimer refused to reveal Chevalier's identity, he told Groves that if the general were to order him to do so, he would reveal the name. For the time being, Groves did not.

Oppenheimer moved to Los Alamos in the spring of 1943 and began directing this unprecedented operation, being fully dedicated to its cause. Although he had Groves's vote of confidence, he never managed to gain that of the security people. He was watched and followed everywhere he went, his telephone was tapped, and his every move was recorded and analyzed. The security people supposed all along that Oppenheimer might be revealing secrets of atomic information to the Soviets. Even while the work in Los Alamos was going on, they made recommendations to remove him from his position and dismiss him from government employment. However, Oppenheimer continued his dedicated work and, whenever the opportunity arose, gave additional evidence of his loyalty.

He had ample occasion to do so because the military intelligence people continued pressing him to reveal the name of the person who had approached him about the information transfer to the Soviets. Lengthy interviews ensued in the midst of the atomic bomb project when Oppenheimer's time was most precious, and there is no trace of any protest on his part against such interviews. He was being humiliated, especially when asked about possible party affiliation of his associates at Los Alamos, and he did not refuse the suggestion that he gather information about his colleagues.

A sad picture emerges of both Oppenheimer and the security services; for the latter especially, as we can see with the benefit of hindsight that they were actually hindering the success of the project. At the same time, they never detected the ongoing espionage by Klaus Fuchs, who was a member of the British delegation of physicists, and whose espionage was uncovered after the war. Back in Britain, the German refugee Fuchs had "never concealed the fact that he was a communist."[26]

By December 1943, General Groves felt impressed upon himself to "order" Oppenheimer to reveal the identity of the contact, that is, Chevalier's name. It was only many years later, in 1954, after the Oppenheimer security hearing that Chevalier understood why he had been blacklisted and why he had lost his job and even his citizenship, while Oppenheimer had continued his friendship with him. Later Oppenheimer said, referring to his own fabrication, "I invented a cock and bull story."[27] Alas, the Chevalier story was not the only case of such disloyalty to friends by Oppenheimer.

The most conspicuous case was that of Bernard Peters, a German Jew and former communist or communist sympathizer, who early on fought the Nazis and was incarcerated in the first German concentration camp, Dachau. His mother bribed some officials and was able to have her son transferred to a city jail. Eventually he was released and escaped to the United States. Peters did his doctoral work under Oppenheimer in Pasadena. Lacking security clearance, he could not participate in the work at Los Alamos, but he worked in Berkeley during the war. In his wartime interviews with the security people, Oppenheimer gave damaging testimony about Peters. He characterized him "as a dangerous Red and former Communist."[28] Oppenheimer held against Peters the fact that he fought physically against the Nazis, as he took it as a sign of violent temper and held against his character that he was freed from the Nazi concentration camp "by guile."[29]

Oppenheimer's encounters with the security investigations were the dark side of his life during the Manhattan Project. The bright side was being the scientific head of Los Alamos, and friend and foe alike praised his performance without reservation. He kept a watch on the larger picture and was at home with the details. He provided encouragement, and his associates worked not only for the greater goal but also for his personal recognition. Stern compared Oppenheimer with Teller; whereas Teller was "voluble and gregarious," Oppenheimer was "quiet and private"; whereas Teller was "extravagant in speech and in thought," Oppenheimer was "more precise in both."[30]

Their personal lives were also different. For Teller, his wife was the only woman in his life from the time he was a teenager. His soul mate,

Maria Goeppert Mayer, with whom he conducted at times painfully candid correspondence, was just that, a soul mate, nothing more. Oppenheimer gave the impression of a confused attitude toward himself in his youth, including his relationship with women. Even in adulthood, his interactions were complicated. He started his affair with his future wife when she was married to another man. During their marriage he felt obliged to comfort another woman from the period before his marriage.

Teller blamed Oppenheimer for having prevented him from signing the Franck petition calling for a demonstration of the atomic bomb before dropping it (chapter 4). Oppenheimer's friends and followers might have expected him to oppose deployment without warning, but in the final account he did not. When he talked Teller out of signing the Franck petition and further asked him not to circulate it among the Los Alamos scientists, Oppenheimer could not let Teller know that he had been more knowledgeable about the events than Teller might have supposed.

In the immediate aftermath of the bombing of Japan and the end of World War II, on August 30, 1945, the Association of Los Alamos Scientists was set up. Curiously, this was not the first attempt to establish such an organization, but twice Oppenheimer had managed to talk his colleagues out of organizing themselves. The association elected its executive committee, which consisted of four scientists, Hans Bethe, Jerrold Zacharias, Frank Oppenheimer (Robert's brother), and Edward Teller.[31] In an official statement they warned against an arms race and called for international control of nuclear energy. Their hope that Robert Oppenheimer would help them reach high government circles was not realized.[32] The others did not know until 1949 that Frank Oppenheimer was a former Communist Party member. Teller made no mention of this association and its activities in his *Memoirs*.

The association issued a statement expressing concern over the fate of nuclear weapons and urged international control of atomic energy. Association members expected Robert Oppenheimer to be their spokesman in Washington and were disappointed when they observed his changing attitude toward their demands. In establishing the framework of handling nuclear matters in the United States, Oppenheimer sided more with the administration than with the scientists. The scientists' feeling of being abandoned by him was strengthened when Oppenheimer resigned from his directorship of Los Alamos, after he had been pleading with them to stay on in the laboratory. Teller was among those who felt betrayed by him.

Nonetheless, the executive committee of the association convened a meeting for Oppenheimer to address his former colleagues on November 2,

1945.[33] His speech was full of weighty issues. One cannot help but still see the validity of his thoughts in terms of today's problems. Here we mention one example, namely, the difficulties in trying to reconcile the preeminence of democracy in one part of the world with strong negative forces in other parts.

Oppenheimer recognized the value of democracy for which Americans were willing to die, but he reminded his audience that there were other parts of the world in which there was no democracy, yet a common solution needed to be found to act in a united front. He added, "[T]here is something more profound than that [that is, democracy]; namely, the common bond with other men everywhere."[34] He discussed these and other issues because scientists could no longer withdraw completely from the consequences of their discoveries. He warned his colleagues to stay honest in giving technical information and "in distinguishing between what we know to be true from what we hope may be true." He called upon scientists to maintain their belief "in the value of science, in the good that it can be to the world to know about reality, about nature, to attain a gradually greater and greater control of nature, to learn, to teach, to understand."[35]

Oppenheimer's farewell speech at Los Alamos gave a sense of special importance to his five hundred listeners. It was a beautiful expression of his formidable intellect and it illustrated his ability to rise above everyday problems and be eloquent and lofty. He was a great master of words, but, at least to Stanislaw Ulam, "He was more intelligent, receptive, and brilliantly critical than deeply original. Also he was often caught in his own web, a web not of politics but of phrasing. Perhaps he exaggerated his role when he saw himself as 'Prince of Darkness, the destroyer of Universes.'"[36]

When Oppenheimer resigned from the directorship of Los Alamos, he was considered to be a national hero in addition to being a great expert, and he must have felt ready for a higher calling. In addition to chairing the AEC GAC, Oppenheimer served as chair of six more government committees in the postwar era.[37] He was called to testify before Congress in the debate about the military versus civilian control over nuclear matters. It was a typical Oppenheimer performance: the congressmen had the impression that he supported the bill and the physicists thought that he was against it.

By then, his politics and his attitude toward the Soviet Union had changed. He urged the American representative in the United Nations Atomic Energy Commission to discontinue negotiations with the Russians about world control of atomic energy. He argued that the Soviet Union did not want an agreement but would use the negotiations for propaganda purposes. This was in 1947, at the time when the extended security investigations of Oppenheimer were intensifying. His FBI files were scrutinized by each member of the AEC, who were appalled by the derogatory information about his past associations and politics and in particular about the Chevalier incident. Yet they did not find the documents sufficiently damaging to remove him from his position of the chairmanship of the GAC.[38]

Still, the number of people Oppenheimer alienated with his arrogant behavior was growing. One of them was Admiral Lewis Strauss. All this was paralleled by the various investigations and press releases about Oppenheimer's leftist past and his disloyal testimonies about his friends and associates to the security organs. Even his physicist colleagues condemned him for his role in the Peters case. He was compared to a magnet with two poles. The positive was "the charmer, the persuader, at times almost the hypnotizer." The negative was "the humiliator, the witherer, the arrogant, impatient condescender of intellects lesser than his own."[39]

Wolfgang Panofsky knew both Teller and Oppenheimer and in 2004 he said that Oppenheimer

> was a strange person. I knew him way back when he was in Caltech and he used to go back and forth between Berkeley and Caltech. I used to go and listen to his lectures and understood nothing. He always had a slightly mystical aura around himself, so I never understood him. But I didn't understand Teller either. . . . Teller was very doctrinal. When we were on committees together usually he [Teller] was rather unwilling to listen and participate in an informed debate, he was not very good in that. . . . He didn't pay any attention to the length of time and he just didn't stop. . . . But they were both very good physicists. I admired Teller more as a physicist than I do Oppenheimer. Oppenheimer was an interpreter but he hadn't done very much physics. He introduced European physics to America but his own contributions were really not that profound. I don't pretend in any deep sense to understand either of them. I am too much of a down-to-earth character.[40]

Both Oppenheimer and Teller possessed a considerable amount of vanity in that both liked to be associated with high officials and both liked to let such associations be known by their peers. Oppenheimer had access to high government officials, nationally known figures, whom he referred to by their first names. He appeared to be ubiquitous in decision-making positions, which might have represented some conflicts of interest.[41] For example, he advised the Pentagon that the hydrogen bomb was not feasible technically and hence the Pentagon did not claim a military need for it. Then he told the AEC that there was no request from the Pentagon for the hydrogen bomb.[42]

According to Ulam, vanity might have played a role in Oppenheimer's initial opposition to the hydrogen bomb in addition to moral, philosophical, and humanitarian considerations. He struck Ulam "as someone who, having been instrumental in starting a revolution (and the advent of nuclear energy does merit this appellation), does not contemplate with pleasure still bigger revolutions to come."[43]

Teller also liked to appear important and influential. The Nobel laureate physicist Donald Glaser was sitting next to him during a flight, and they had an amicable conversation. As soon as they deplaned and there was an audience, Teller started speaking loudly and putting on a show.[44] Gamow was Teller's friend and he remained a friend even when most had turned away from him. Yet Gamow was aware of Teller's vanity and his fondness for appearing very important. When Gamow described the fictional professor in his book about the atoms, it was impossible not to recognize Teller in Gamow's Dr. Tallerkin. In the story, Tallerkin arrives in a hurry at a meeting where he is to speak to a lay audience. He is "an impressive-looking man with burning eyes and overhanging bushy eyebrows." He starts his presentation in Hungarian, but soon realizes his mistake and switches to English:

> Ladies and Gentlemen! I must be short because I am very busy. This morning I attended several conferences in The Pentagon and The White House; this afternoon I have to be present at the underground test explosion at French Flats in Nevada, and in the evening I have to deliver a speech at a banquet at the Vandenberg Air Force Base in California.[45]

Both Teller and Oppenheimer could be charmers and excellent debaters. Once Oppenheimer faced an initially unfriendly House Un-American Activities Committee (HUAC), which included California congressman Richard M. Nixon. At the end of the hearings, however, all congratulated him and thanked him for his performance. Apparently, Nixon was so favorably impressed by Oppenheimer's performance while he was vice president that he saved Oppenheimer from a McCarthy investigation.[46] Teller held debates in the media, and his partners not only admitted defeat but found it impossible to argue with him successfully. Both Teller and Oppenheimer let themselves be victims of interservice rivalries. Part of this was due to the fact that the army wanted smaller fission bombs and the air force preferred the big hydrogen bombs. (The story of Admiral Bogrov in *Darkness at Noon* comes to mind; see chapter 4.) Early on Oppenheimer angered the air force, whereas Teller found in it a staunch ally in his advocacy of the hydrogen bomb and the second weapons laboratory.

Oppenheimer's behavior often lacked consistency. An example was his changing of attitude toward the development of the hydrogen bomb. Initially he was interested in it, and, as Alvarez remembered, "[the] most enthusiastic person I had ever met on the program of the super weapon was Dr. Oppenheimer."[47] After the Trinity test Oppenheimer wanted to accelerate the work on the Super, but after the Japanese surrender he had the work on it stopped, and he became a vocal opponent of its development. When the Teller–Ulam approach provided the technical solution to making the hydrogen bomb, he switched to supporting it without reservation.

In 1952, Oppenheimer was not reappointed to his position in the GAC, although he continued as a consultant to the AEC. He still had other positions, but his great authority started diminishing. The Department of Defense, for example, abolished its entire Research and Development Board in 1953 in order to get rid of Oppenheimer as their consultant.[48] Soon, in the summer of 1953, Oppenheimer's term as consultant to the AEC would have expired. Had it not been extended, life would have been simpler for Oppenheimer and Teller, but the appointment was extended through June 1954.

## THE CASE AND THE HEARING

William Borden, former chief of staff of the Joint Committee on Atomic Energy of the US Congress, wrote a letter to J. Edgar Hoover, director of the Federal Bureau of Investigation (FBI), on November 7, 1953.[49] The letter contained an accusation that "more probably than not J. Robert Oppenheimer is an agent of the Soviet Union." He stated that Oppenheimer was a communist; his friends and associates, family members, even his mistress, were communist; he belonged to communist organizations; and "he was in frequent contact with Soviet espionage agents." The accusations were often vague in spite of the strong words Borden used. It was stated, for example, that there was evidence indicating that Oppenheimer either stopped giving money to the Communist Party at a certain point or he made his contributions through unknown channels. Oppenheimer's opposition to the development of the hydrogen bomb was also interpreted as his being an agent of the Soviet Union.

The letter hardly contained new facts, but toward the end of 1953 the accusations took on a new life because the atmosphere in the United States was such that they could not be ignored. Besides, he was no longer irreplaceable for projects of national importance; there was no atomic bomb project with a General Groves as its head that would brush off such accusations. This was the time of McCarthyism, when even Truman, the former president, was accused of treasonous actions.

The first consequence of the Borden letter was that President Eisenhower ordered a "blank wall" between Oppenheimer and any classified information. At this point, the AEC could have just let Oppenheimer's consultancy lapse when it expired in the summer of 1954. They offered Oppenheimer the possibility to withdraw quietly, but he wanted to clear his name. Thus the AEC set up a three-man Personal Security Board (PSB) to conduct a hearing. Heavy charges were leveled against Oppenheimer according to the standards of the day. Twenty-three of them dealt with his past associations with communists and communist causes, all prior to 1947. The twenty-fourth related to his opposition to the development of the hydrogen bomb. It was different from the other charges in that it concerned Oppenheimer's *opinion* rather than his actions.

It was stressed throughout that his hearing was not a trial, but this turned out to be disadvantageous to him. His counsel, lacking the appropriate security clearance, could not be present at certain portions of the procedure. The board, playing the role of judge, was relying on additional material in its judgment, not only what was presented during the hearing, whereas at a trial all evidence would have had to be presented. In a real trial the defendant has the right to face his accusers, whereas at the hearing accusations could be introduced from sources not present, sometimes not even properly identified. There was no impartial jury. The government representative, essentially the prosecutor, had much broader latitude of action than Oppenheimer's counsel.

Lewis Strauss as chairman of the AEC did his utmost to have Oppenheimer's security clearance withdrawn. At the same time, he wanted to avoid the appearance of a McCarthy-like procedure. For the principal government representative he selected Roger Robb, an aggressive private lawyer with considerable courtroom experience. Even Teller referred to him in an interview thirty-five years later as "attorney for the prosecution."[50] Carl A. Rolander, the AEC's deputy director of security, helped Robb to prepare the case. Oppenheimer asked Lloyd K. Garrison to be his leading counsel, and he was aided by Herbert S. Marks. Both were knowledgeable and socially conscientious lawyers, but they lacked courtroom experience and security clearance, which prevented them from getting acquainted with some of the crucial documents in the case. They were further assisted by Samuel J. Silverman and Allan B. Ecker, both members of Garrison's law firm. The three members of the board were handpicked by Strauss.

## LEWIS L. STRAUSS (1896–1974)

Lewis Strauss was born in Charleston, West Virginia, the son of a shoe merchant. He completed only a high school education. Even though he earned a scholarship to college, family problems prevented him from using it.

He was interested in science and especially in physics. He worked as a traveling salesmen of shoes, and once he'd saved enough money to quit, he became an assistant to Herbert Hoover, the future president of the United States. Hoover was charged with managing the food production in America at that time, which was a serious responsibility because there was a shortage of food in much of Europe, and soon food was indeed shipped to Europe.

After having worked for Hoover for a few years, Strauss entered private business and became a wealthy man from his investment banking. In World War II, Strauss served in the US Navy and rose to the rank of rear admiral. After the war he became involved in national politics, in nuclear matters in particular. He made many friends and as many enemies during his career. He became Teller's political ally and supported him in his quest for the hydrogen bomb as well as for a second weapons laboratory. They had a shared hostility toward J. Robert Oppenheimer.

At a congressional hearing in 1949, Oppenheimer humiliated Strauss. The topic was the export of radioisotopes to friendly countries. Strauss was very much against such export, citing their possible application in matters of atomic energy, and Oppenheimer publicly ridiculed his opposition. He compared the usefulness of radioisotopes for such a purpose to the usefulness of a shovel or a bottle of beer. Insiders understood that Oppenheimer was making a fool of Strauss, and Strauss never forgave him for it.

At the time of the Oppenheimer security hearing, Strauss was the chairman of the Atomic Energy Commission and played a prominent role in choreographing the hearing. Strauss was later ostracized by those who felt that Oppenheimer was unfairly treated. President Eisenhower appointed Strauss to be his secretary of commerce—in which position he served for a short time—but when the president sought congressional approval, the Senate did not confirm the appointment. Strauss was deeply hurt and wrote about the hearings along with many other events and decisions in which he participated in his autobiography, *Men and Decisions*.[51]

There were plenty of damaging testimonies for Oppenheimer during his trial, but most of the scientists came out in his favor. Enrico Fermi, Isidor Rabi, and Hans Bethe spoke for him as did his wartime superiors, Vannevar Bush and James B. Conant. In addition, two former AEC chairmen and three former AEC commissioners testified on his behalf. David Lilienthal compared the proceedings to the Spanish Inquisition.[52] Vannevar Bush noted that Oppenheimer was being tried, among other things, for his opinions and likened the procedure to the "Russian system."[53] He also expressed his fears that the hearing would have negative consequences in the scientific community.

In contrast, Kenneth Pitzer accused Oppenheimer not because he opposed the development of the hydrogen bomb but because he failed to "enthusiastically urge individuals to participate in the program." Pitzer thought that Oppenheimer should have disqualified himself from the position of leadership after the president made a decision contrary to Oppenheimer's advice. Oppenheimer's counsel pointed out here that Oppenheimer did offer his resignation.[54]

Teller's testimony was of the greatest interest, but opinions are divided whether or not it was decisive. There are those who believe that the board's decision would have been the same without Teller taking the stand; others consider it a turning point in the hearing.[55] In any case, Robb, the "prosecutor," did not want to leave anything to chance, and on the eve of Teller's testimony, he met with Teller and introduced him to Oppenheimer's testimony, during which time it came out that he had lied about the Chevalier incident during his wartime interviews with the security people.

Some verbatim excerpts from the actual testimony are quoted below.

*Roger Robb in the direct examination*: Dr. Teller, may I ask you, sir, at the outset, are you appearing as a witness here today because you want to be here?

*Teller*: I appear because I have been asked to and because I consider it my duty upon request to say what I think in the matter. I would have preferred not to appear.

. . .

*Robb*: Is it your intention in anything that you are about to testify to, to suggest that Dr. Oppenheimer is disloyal to the United States?

*Teller*: I do not want to suggest anything of the kind. I know Oppenheimer as an intellectually most alert and a very complicated person, and I think it would be presumptuous and wrong on my part if I would try in any way to analyze his motives. But I have always assumed, and I now assume that he is loyal to the United States. I believe this, and I shall believe it until I see very conclusive proof to the opposite.

*Robb*: Do you or do you not believe that Dr. Oppenheimer is a security risk?

*Teller*: In a great number of cases I have seen Dr. Oppenheimer act—I understood that Dr. Oppenheimer acted—in a way which for me was exceedingly hard to understand. I thoroughly disagreed with him in numerous issues and his actions frankly appeared to me confused and complicated. To this extent I feel that I would like to see the vital interests of this country in hands which I understand better, and therefore trust more.

In this very limited sense I would like to express a feeling that I would feel personally more secure if public matters would rest in other hands.[56]

This statement by Teller has been much quoted. He lamented in his *Memoirs* that this exchange came much too early in the testimony and that he went too fast.[57] He was still very much under the impact of what Robb had shown him from Oppenheimer's testimony referring to the Chevalier incident. According to Teller, this senseless testimony made him say things that sounded more general than they should have.

The FBI had interrogated Teller before, in 1952, and he was not a sympathetic witness for Oppenheimer then either. The FBI report concluded, "Teller states he would do most anything to see subject [Oppenheimer] separated from the General Advisory Committee because of his poor advice and policies regarding national preparedness and because of his delaying of the development of H-bomb."[58]

In 1977, Teller requested and received a copy of the FBI report and submitted a request to the Deputy Attorney General, Office of Privacy and Information Appeals. He wanted to clarify that his 1952 testimony to the FBI was initiated by the FBI. Further, he characterized the concluding sentence of the testimony, quoted above, "vague and meaningless when taken out of context," and requested that the "sentence should, therefore, either be deleted or an explanation should be made to the effect that the statement does appear without the full context having been displayed."[59] This letter provides direct evidence that the testimony was given and that it included the quoted statement. This means that Teller's damaging testimony at the Oppenheimer hearing on April 28, 1954, was not made on the spur of the moment; rather, he was expressing his long-held views concerning Oppenheimer's activities.

There is yet another source for Teller's views of Oppenheimer prior to his own testimony and prior to his getting acquainted with Oppenheimer's testimony at the hearing that might have influenced him. Charter Heslep of the Public Information Service visited Teller on April 22, 1954, and the two discussed the Oppenheimer case. This took place a mere six days before Teller's appearance at the hearing and Heslep had the impression that by talking to him, Teller was rehearsing what he intended to say later to Strauss in his preparation for the testimony.

Heslep prepared the summary of their conversation for Lewis Strauss, who was deeply involved in orchestrating the hearing. Only a few points from among the nine entries of Heslep's account will be mentioned here. Teller regretted that the hearing was aimed at deciding on Oppenheimer's clearance because—as he implied—it would be untenable to deny. However, once security was on the table, he wondered whether it would be possible to "deepen the charges" and document bad advising on Oppenheimer's part from the beginning of the postwar period. He did not find Oppenheimer a menace, merely "no longer valuable" to the program. Then he complained about the ubiquitous impact of the powerful "Oppie machine." Teller agreed with Heslep's expression that Oppenheimer "should be unfrocked in his own church."[60]

Let us return now to Teller's testimony at the hearing on April 28, 1954. Teller was answering questions about his work at Los Alamos during the war. He recounted the changes in Oppenheimer's opinion with respect to the work on the thermonuclear weapon. He added that Oppenheimer's negative attitude toward thermonuclear research following Japan's surrender was shared by most scientists at Los Alamos because they wanted to go home. Teller was in disagreement even then with the general opinion because he considered it important to continue their efforts with regard to the thermonuclear work. Even with no question in relation to the topic, Teller interjected a segment in his testimony lavishly praising Oppenheimer's performance as director of the Los Alamos Laboratory. However, immediately after, Teller alleged "that if we had gone to work in 1945, we could have achieved the thermonuclear bomb just about four years earlier."[61]

Teller's testimony at one point turned to the story of the Russian explosion of a nuclear device at the end of August 1949. By then Teller had returned to Los Alamos for a one-year stint, partly because Oppenheimer had encouraged him to do so. He was contributing to fission bomb development at the time. The narrative then describes the meetings and events during the fall of 1949, when, again, Teller felt Oppenheimer's opposition to developing the hydrogen bomb. His negative opinion persisted even when the president's decision to go ahead with the program was impending. At this point, Teller understood that in the future it would be futile to expect any support from Oppenheimer concerning the hydrogen bomb.

After Truman's decision, thermonuclear research accelerated. Teller chaired a committee at Los Alamos in which representatives of various divisions participated, and they made a number of proposals to further the thermonuclear program. However, Oppenheimer remained chairman of the GAC and he and the GAC did not facilitate Teller's efforts toward the hydrogen bomb. This changed in June 1951, when the new approach was presented. Teller said: "Oppenheimer warmly supported this new approach, and I understand that he made a statement to the effect that if anything of this kind had been suggested right away he never would have opposed it."[62]

Then an interesting although rarely quoted portion of the testimony followed. Teller gives his judgment of not only Oppenheimer's advisory activities and that of the whole GAC but of something that concerned the quality of advice rather than a question of security:

> *Robb:* In your opinion, if Dr. Oppenheimer should go fishing for the rest of his life, what would be the effect upon the atomic energy and the thermonuclear program?
>
> *Teller:* You mean from now on?
>
> *Robb:* Yes, sir.
>
> *Teller:* May I say this depends entirely on the question of whether his work would be similar to the one during the war or similar to the one after the war.
>
> *Robb:* Assume that it was similar to the work after the war.
>
> *Teller:* In that case I should like to say two things. One is that after the war Dr. Oppenheimer served on committees rather than actually participating in the work. I am afraid this might not be a correct evaluation of the work of committees in general, but within the AEC, I should say that committees could go fishing without affecting the work of those who are actively engaged in the work.

In particular, however, the general recommendations that I know have come from Oppenheimer were more frequently, and I mean not only and not even particularly the thermonuclear case, but other cases, more frequently a hindrance than a help, and therefore, if I look into the continuation of this and assume that it will come in the same way, I think that further work of Dr. Oppenheimer on committees would not be helpful.[63]

Here follows a rather long list of Oppenheimer's bad advice, more from hearsay than from Teller's personal experience. At the conclusion of the direct examination, there was hardly any cross-examination. The only question put to Teller concerned Oppenheimer's and the GAC's attitude toward establishing the second weapons laboratory. It came out that Oppenheimer and the GAC opposed the establishment of a clone of Los Alamos, but they did not oppose the one that finally came to life. At this point, the chairman of the Personnel Security Board took over the questioning and, eventually, returned to the principal question of the hearing:

*Gordon Gray*: Do you feel that it would endanger the common defense and security to grant clearance to Dr. Oppenheimer?

*Teller*: I believe, and that is merely a question of belief and there is no expertness, no real information behind it, that Dr. Oppenheimer's character is such that he would not knowingly and willingly do anything that is designed to endanger the safety of this country. To the extent, therefore, that your question is directed toward intent, I would say I do not see any reason to deny clearance.

If it is a question of wisdom and judgment, as demonstrated by actions since 1945, then I would say one would be wiser not to grant clearance.[64]

Teller notes in his *Memoirs* that here he had a second chance and in retrospect he wished that at this late stage of his testimony he had clarified his initial "ambiguous" remarks.[65] Instead, he amplified what he had said and went further. The question was about security clearance, and the first part of Teller's response referred to Oppenheimer's intent. The second part referred to Oppenheimer's wisdom and judgment, which he was not asked about. Teller acted as if he wanted to make sure that his words carried sufficient weight. In this light his preceding positive comment could serve only to enhance the negative effect of his conclusion.

In a brief second cross-examination, Teller was asked whether Oppenheimer's access to classified information might or might not represent danger for national security:

*Samuel Silverman:* I would like you, Dr. Teller, to distinguish between the desirability of this country's or the Government's accepting Dr. Oppenheimer's advice and the danger, if there be any, in Dr. Oppenheimer's having access to restricted data. As to this latter, as to the danger in Dr. Oppenheimer's having access to restricted data without regard to the wisdom of his advice, do you think there is any danger to the national security in his having access to restricted data?

*Teller:* In other words, I am now supposed to assume that Dr. Oppenheimer will have access to security information?

*Silverman:* Yes.

*Teller:* But will refrain from all advice in these matters which is to my mind a very hypothetical question indeed. May I answer such a hypothetical question by saying that the very limited knowledge which I have on these matters and which are based on feelings, emotions, and prejudices, I believe there is no danger.[66]

Of course, this exchange of hypotheses could not have had any effect on the outcome of the hearing either way. As Teller was leaving the witness chair, he went to shake Oppenheimer's hand and told him, "I'm sorry." Oppenheimer answered, "After what you've just said, I don't know what you mean."[67]

At that point, there may have been only one person in the room of the hearing who correctly guessed the significance of Teller's testimony. He was Ward V. Evans, who was judged by some to be the least significant of the three-man board. He had put this question to Teller: "Do you think the action of a committee like this, no matter what it may be, will be the source of great discussion in the National Academy and among scientific men in general?"[68] This question resonated with Vannevar Bush's comment about the consequences of the Oppenheimer hearing in scientific circles. Evans was to be the sole member of the Gray Board to vote for reinstating Oppenheimer's security clearance. Evans did not ask Teller this question by accident. He noted that the "scientific backbone of our nation" came out in the defense of Oppenheimer.[69]

The board unanimously found Oppenheimer to be a loyal citizen, but decided with a two to one majority to suggest denying Oppenheimer's clearance. This was followed by the general manager's recommendation for the Atomic Energy Commission to do the same. Finally, the AEC found Oppenheimer loyal, with one dissension, and a security risk, also with one dissension. When the decision became public, Albert Einstein expressed his respect and admiration for Oppenheimer in a public statement.[70]

The decision of the AEC was made just a few hours before Oppenheimer's security clearance would have expired anyway. In the process of the hearing he was judged by friends and foes alike to be his own worst enemy. Gone was the brilliance of his articulation that had won over the hostile members of the House Un-American Activities Committee five years earlier. One of his former students made a curious assessment of Oppenheimer: "In order to be an insider he then cut away more and more of his independence. In the end, they destroyed him."[71]

In subsequent years, there has been much speculation about Teller's motivations in his testimony. That he testified must have come from his feeling of duty and that he testified against Oppenheimer must have come from his conscience. Yet one cannot avoid the impression that it was partly revenge and partly to please people who were in some sense his superiors. They included the air force, which was an enthusiastic supporter of his initiatives, and Lewis Strauss, the powerful chairman of the AEC.

Teller was not the only important scientist who held a negative opinion about Oppenheimer. However, the super-conservative Ernest Lawrence withdrew from testifying at the last moment, quoting severe illness, which was genuine. His incapacitation came in handy for him because he did not find it beneficial to testify and thereby invite the wrath of the scientific community, which he had realistically anticipated. That this was the case is shown by his action in connection with Alvarez's testimony. Alvarez was Lawrence's subordinate, and Lawrence ordered him to withdraw from testifying. Lawrence explained to Alvarez that their Berkeley laboratory might suffer if they testified against Oppenheimer. After some hesitation Alvarez testified nonetheless; however, he did so cautiously. He described his disagreement with Oppenheimer concerning the development of the hydrogen bomb, but he made it clear that it was a difference in opinion rather than anything related to security clearance. To Alvarez, Oppenheimer "showed poor judgment."[72] Subsequently, Alvarez would never be accorded the excommunicating behavior by his peers that Teller would.

John von Neumann's testimony was also noteworthy. He was no less of a hardliner than Teller; had a similar background to Teller's; knew Oppenheimer well given the years they spent together at the Institute for Advanced Study; and was no Oppenheimer fan. Von Neumann stated unambiguously (as Teller did) that he had no doubt about Oppenheimer's loyalty; but he also expressed forcefully that he had every confidence about Oppenheimer's discretion in the handling of classified information and that Oppenheimer was not a security risk.[73] In addition, von Neumann made an interesting point that the atomic bomb project presented a whole new situation for the scientists, and getting adjusted to it took some time and for different people it took different lengths of time.

Von Neumann brought a human element into the discussion that at different points seemed as if machines rather than people were the subject of the hearing. He must have annoyed the government representative because at the end of von Neumann's testimony, Robb posed this question to him, "Doctor, you have never had any training as a psychiatrist, have you?" Of course, von Neumann did not.[74] In 1955, von Neumann was appointed to be commissioner of the AEC, which was considered an attempt to diminish the split in the scientific community because von Neumann "was universally respected, by the friends of Oppenheimer as well as those of Teller."[75] If this was indeed the goal of the AEC, it did not succeed.

The AEC board had no doubt about Oppenheimer's loyalty to the United States after its rigorous and comprehensive probing. Nonetheless, whether Oppenheimer supplied secret information to the Soviets has remained a much-discussed issue without any solid evidence.[76] In an unprecedented statement, the FBI was quoted in 1995 as saying that it had "classified information available that argues against conclusions" based on Soviet/Russian allegations that Niels Bohr, Enrico Fermi, Robert Oppenheimer, and Leo Szilard might have participated in espionage.[77] In other words, they did not.

## AFTERMATH

The transcripts of the Oppenheimer hearing were made public in spite of prior assurances to Teller that they would not be. Due to some administrative mix-up, the Atomic Energy Commission felt it incumbent upon itself to make the transcripts available to the public. The physicist community singled out Teller as the one who wronged Oppenheimer—who, in turn, was reaching martyr status. Teller counted on the minutes remaining classified; otherwise, it would have been reckless on his part to ignore the dangers that his testimony carried. Szilard had realized the risks that Teller might be taking by such a testimony. He tried to reach his friend on the eve of the testimony but did not find him.[78] It was before the transcripts became public that John Wheeler wrote to Teller: "I am so glad not to have had to testify in the Oppie [Oppenheimer] matter, and just want to say how much I admire your courage and honesty and clarity in speaking as you did. I have yet to see the complete testimony."[79]

Soon after the Oppenheimer hearing, Teller was in Los Alamos for two weeks and while there, Robert Christy refused his handshake. Christy was Oppenheimer's doctoral student at Berkeley and a Los Alamos veteran. He and his family and the Tellers had shared a house for a year in Chicago. In 1994, Christy stated that he did not feel it right for an honorable physicist to testify against Oppenheimer. "I happened to see Edward Teller . . . he approached me with his hand out to shake my hand. And I very deliberately refused to shake his hand . . . it was recognized by everyone for what it was."[80] Standing nearby, Rabi witnessed the scene and told Teller that he would not shake his hand either. To make sure Teller understood the reason, Rabi congratulated him on the "brilliance" of his testimony and his "extremely clever way" in which he expressed his opinion that Oppenheimer was a security risk.[81] Other scientists, who had no opportunity to demonstrate their condemnation publicly, emulated Rabi's and Christy's attitude. For example, Murray Gell-Mann "resolved never to speak to Teller again."[82]

Teller gave a vivid account to Maria Goeppert Mayer about at least some of his trials during his visit at Los Alamos. He "felt like Daniel in the lions' den." He found that of all the negative attitudes of his colleagues, "the worst of them is Rabi. He never was my friend but now he is terrible. Tomorrow I have to give a report. He will be there to heckle me." He described to Maria his dream about a raven (and hinted to Maria to translate "raven" into German, which would be *Rabe*, an unmistakable reference to I. I. Rabi). Then something unexpected happened—in Teller's dream only—the raven smiled at Teller and he slept "quite well."[83]

Incidentally, Teller saw a silver lining in the aftermath of the Oppenheimer hearing, when he said, "I am finding out who are my friends."[84] He assessed it perhaps very realistically in a letter to James Franck that he had very few friends.[85] Franck was certainly one of those few, but this did not spare Teller from Franck's criticism: "I too cannot agree with the philosophy that a government adviser should be regarded as a security risk because his judgment was wrong." Franck, of course, did not exclude the possibility that the judgment of an adviser was merely "unpopular with the Administration."[86]

For years, Teller could never be sure whether his company was welcome among his colleagues anywhere he went and whether or not his hand would be accepted for a handshake. On the other hand, those, including Gell-Mann, who participated in various defense-related projects, could not avoid coming in contact with Teller in subsequent years. The members of a loose and informal organization called "Jason" included Gell-Mann and others, whereas Teller, along with Bethe and Wheeler, was its senior consultant. Jason was started with the idea of physicists getting together under good conditions rather than working individually; and part of the rationale was that together they could wield more influence on governmental strategic policy.[87] The organization has fulfilled the expectations.

The impact of the Oppenheimer hearing has not diminished. The impression is that the animosity against Teller stemmed primarily from the role he played in his testimony. Others maintain that he had by then so much alienated most of his peers that his testimony added little to the ill feelings against him. According to a Los Alamos story, when a former leader of one of the weapons programs asked Ben Diven, a veteran of Los Alamos, "whether the Los Alamos dislike of Edward Teller that was so pervasive during the late 1950s and all during the 1960s was due to Edward's testimony at the Oppenheimer security hearings, Diven said, 'No, there are lots of reasons to hate Teller.'"[88] John A. Wheeler, who could not be considered biased against Teller, said, "During the war and later, during the hydrogen bomb project, he made many enemies with his impatience and arrogant behavior. His campaigning for the new weapons laboratory did not help either. By the time he testified at the Oppenheimer hearings, almost all scientists disliked him."[89]

In contrast, Harold Agnew, the third director of Los Alamos, does not agree with the assessment that by the time of the Oppenheimer case, Teller was unpopular. He thinks that it was only the Oppenheimer affair that brought about the wrath. And he does not think that Teller's testimony was all that bad. Teller was singled out for criticism because he was the only important scientist who testified against Oppenheimer; others, who agreed with Teller, like Lawrence and Seaborg—according to Agnew—kept quiet. Agnew is obviously fond of Teller, but he felt that "[f]or completeness, I may add that Edward can sometimes make you very angry; he just can."[90]

Richard Garwin thinks that Teller attacked Oppenheimer in such a manner as to destroy him because he got in his way. Garwin contrasts Teller with Leo Szilard: "Szilard . . . would realize which people stood in his way . . . but never did anything mean or negative to them. Teller extended his no-holds-barred physics style also to his personal relationships. If people were in his way, he might try to destroy them as happened with Oppenheimer." Garwin calls Teller's treatment of Oppenheimer "unpardonable,"[91] and is amused by Teller's saying that Oppenheimer was a complex character as if he, Teller, was a simple character.[92]

Teller reinforced his observation regarding Oppenheimer's complexity after he had viewed the BBC drama of Oppenheimer's life presented in a seven-hour program. According to Teller, Shakespeare used more than seven hours to present the life story of Henry VI, who "was not nearly as unique, ingenious or self-contradictory a character as Oppenheimer."[93] Unfortunately, Teller's article in 1983 does not help us to understand his attitude toward Oppenheimer. He said that he went to the hearing to testify for Oppenheimer's clearance and changed his mind *minutes* before his testimony after Robb had introduced him to Oppenheimer's infamous Chevalier story. Furthermore, Teller says that he "was convinced then and continue[s] to believe now that the hearing should never have occurred."[94]

Freeman Dyson blames Teller only for having gotten into the trap of politics, which had its roots in the rivalry between the air force and the army, which "were more hostile to each other than either of them was to the Russians."[95] Dyson thinks that once Teller got into the trap, he told the truth, meaning also that the others, to "some extent," did not.[96] Dyson elaborates:

> . . . if you look at what he actually said at the hearings, it was all true and he was perfectly sincere. He believed what he was saying and what he was saying was not actually very extreme; he never said that Oppenheimer was a spy, he never said that he was disloyal; he just said that he was complicated and unreliable, which is true. I think that everybody who examined Oppenheimer would agree with that. He said very strange things and often told things that were not true, for no reason one could understand. So when you were dealing with Oppenheimer you felt that this is somebody you don't really feel at ease with and what Teller said was that he would prefer the security of the country in other hands. That, I think, was a correct statement. So I don't blame Teller for saying what he said; I just blame him for getting in it in the first place.[97]

The Oppenheimer hearing, often called a trial, and with good reason, has been a subject of controversy for over half a century and has not gone away. Two other similar trials come to mind, and it is yet a question as to whether the Oppenheimer hearing will become comparable to them in the longevity of its impact. The cases are far from analogous, but there are common features among them.

In 1894, the French artillery captain Alfred Dreyfus was wrongly accused of and tried and sentenced for espionage on behalf of Germany. A few years later, he was exonerated and celebrated as a hero. The case divided French society, and many considered Dreyfus a martyr (those who condemned him often had an anti-Semitic undertone). There are two similarities between the Oppenheimer hearing and the Dreyfus case. One is that both Dreyfus and Oppenheimer were subjected to undeserved ordeals. The other is the parallel between the disappointment of Dreyfus's hero-worshippers when they found him ordinary and the disappointment and anguish of Oppenheimer's worshippers when they learned of his disloyalty to friends and associates.

The other famous trial concerned Galileo and his scientific teachings; the parallel in this case is, again, the undeserved ordeal, which reflected badly on its instigators in the Catholic Church and not at all on Galileo. There are also more subtle similarities between the two cases referring to the brilliance and the mastery of language that made Galileo/Oppenheimer to be perceived as dangerous to the Catholic Church/the US Air Force.[98]

The frequent comparisons between the Oppenheimer trial and other historical cases give added significance to the fact that Teller was Oppenheimer's most conspicuous antagonist. Strangely, very few point to Lewis Strauss, who was out to get Oppenheimer, nor do many people accuse Gordon Gray or Roger Robb, the participants in the hearing who were his instruments. The one who stands out here is Edward Teller. He was the best-known personality among them and his testimony contained sentences that appeared cleverly formulated and easy to remember. The long-range interest in and impact of the Oppenheimer case is well demonstrated by the numerous books and plays about Oppenheimer—there has even been an opera composed recently: *Doctor Atomic*. It is impossible to remember Oppenheimer without Teller, and Teller suffers in the comparison.

In the spring of 1963, President Kennedy decided to award Oppenheimer the Fermi Award, which was handed to him by President Johnson on December 2, 1963, the anniversary of the first nuclear reactor, only days after Kennedy's assassination.[99] Teller had been given the Fermi Award the previous year and he attended the ceremony where Oppenheimer received his award. To some his attendance was considered a nice gesture; to others it was a deplorable insensitivity. Oppenheimer's Fermi Award, the highest recognition from the AEC, was an informal exoneration. He was offered the possibility of renewing his case, but he emphatically declined.

In April 1962, not long before these two awards were presented, an interesting episode occurred in connection with a taped WNBC-TV show, called *Open Mind*, in which Teller was the guest. He was famous for never letting the recordings of his interviews be edited; he preferred everything broadcast as it was, without any change or not at all. On this occasion, however, a sixty-four-second portion of the show was deleted at Teller's request. The show was moderated by Eric F. Goldman, a history professor at Princeton University, who asked Teller about Oppenheimer's security clearance during the program: "In view of the high respect in which Dr. Oppenheimer is held in many scientific circles today, and his very high reputation in a number of circles abroad, both as a scientist and a humanist, are you still opposed to granting him clearance?"[100] Teller did not respond, rather, he remained silent and when asked whether he would prefer not to discuss the matter, he again gave no response, so the moderator just went ahead with the program. Later, Teller requested the deletion.

It is of interest to see how Teller viewed the Oppenheimer case in retrospect. When his *Memoirs* appeared in 2001, he was ninety-three years old, so we might not want to give too much weight to his reminiscences. However, a comparison of his *Memoirs* with his earlier writings shows consistency of his views. Only a few characteristic aspects will be mentioned here concerning the Oppenheimer hearing and its afterlife from Teller's perspective. Above all, Teller appeared puzzled by the Personnel Security Board's decision, as if he had nothing to do with it. He expressed surprise that the government did not find the former leader of Los Alamos trustworthy enough to continue working on its affairs. He blamed President Eisenhower for the hearing taking place at all; Teller thought it should not have. Further, according to Teller, the president could have avoided what appeared to be "the unfair persecution of a public hero and the capstone of witch-hunts."[101]

Teller stressed that he disagreed with Borden's accusations. He stated that even a communist can be a loyal citizen, but he quoted Borden's letter at length—and some of Borden's points read as if they could have been written by Teller. Teller called his own testimony at places "ambivalent"[102] and "ambiguous and fumbling."[103] There is a general tendency in Teller's evaluation of the Oppenheimer hearing to give maximum weight to Oppenheimer's deplorable behavior in the Chevalier affair and to minimize the impact of his opinion about the development of the hydrogen bomb. He felt, however, that his testimony may have not given this impression, so he had to clarify it: "*What I had in mind* when I said Oppenheimer's action seemed confused was not Oppenheimer's opinion about the hydrogen bomb but his contradictions and admitted lies in his testimony about his friend" (italics added).[104] But he admitted: "My testimony reads as though I felt that Oppenheimer's judgment on the hydrogen bomb was a major objection to his security clearance."[105] Further, "I must acknowledge that my testimony was damaging."[106] He stated, "I never wanted Oppenheimer's opinion on the hydrogen bomb to count in the decision on his security clearance, nor did I believe that doing so was justified."[107]

His retrospective self-justification can be contrasted with what he wrote soon after the hearing to Maria Goeppert Mayer. Here he discusses his backbone in connection with the whole affair: "I seemed to get along fine without one. Now there seem to be some growing pains. I also wonder whether it is growing in the right direction."[108] In a 1984 interview Teller acknowledged that the Oppenheimer case, "apart from hurting a few individuals, hurt the United States in a most tragic manner, because it introduced a deep split into the scientific community from which we are suffering even today." But he again blamed President Eisenhower for what happened and added, "Lewis Strauss is accused of having caused the investigation. When, in reality, he was only carrying out orders from the President." This is an interesting assertion and turnaround, since Teller always believed in the primacy of authority. Here he blamed the highest authority in the United States for his own errors—or for making him follow orders. When the interviewer volunteered that the case caused a wound, Teller found the word insufficient and characterized it as "hatred," adding, "Wounds heal more easily than hatred."[109]

Vitaly Ginzburg did not know all the intricacies of the story with which he became familiar only decades after it had happened. He thought that Teller's unpopularity was the consequence of his testimony against Oppenheimer, and Ginzburg found this unpopularity "unfair." Ginzburg said, "I don't think that Teller was incorrect. Oppenheimer was an important physicist, but he did not understand the Soviet threat and Teller understood it."[110] His opinion is rather simplified, but he is an important witness as he lived through the Soviet side of the story. Ginzburg was a co-recipient of the 2003 Physics Nobel Prize and we had our conversation in Moscow in 2004. Earlier that year I had a conversation with another co-recipient of the 2003 Physics Nobel Prize, Alexei A. Abrikosov, who had moved years before from Moscow to the Argonne National Laboratory. We met in Lemont, Illinois. He stressed that "the Soviet scientists had better ideas than Teller about how to build the hydrogen bomb" and singled out Ginzburg's role in it.[111]

As mentioned before, Teller's testimony in the Oppenheimer hearing contributed in a major way to his third exile. Thirty-five years later, in 1990, a reporter asked him what it was like to "have been mistrusted by so many people for so long now, including some of your colleagues like Dr. Rabi. . . . How have you handled that? What is it like?" Here is how Steven Heimoff described Teller's response:

> Teller was quiet. He sat with his hands folded in his lap and I saw his eyelids flutter. There was a long period of silence, perhaps half a minute.
>
> Teller finally answered, in a voice so soft I could barely hear. "As best I could."

"Does it hurt?" I asked.

He spat out his next words with a force that struck me physically. "None of your business!" Then he paused, and added, in a child's singsong, "Perhaps it does, perhaps it doesn't." There was another long pause. "I can tell you one thing," he said, his voice dropping to a whisper, "To be very open about it, a long, long time ago, like 35 or 36 years ago, it did hurt." Another very long pause, before he continued, very slowly: "I left Hungary, lost contact with my relations. Left Germany and Europe, lost contact with many friends. I retained contact and good relations with my colleagues in the United States. Now, because of political differences, I have lost many of those. Most of those; not all. There are a few exceptions."

Then he suddenly exploded. "Of course that hurts! It was meant to hurt, and it did! It was long ago, you know? In the meantime, Livermore was built. I acquired a new set of friends, at an age when most people make no more friends. In the meantime, I had lots of additional problems. No longer a question of discussion."[112]

This was part of a long interview, but it did not last much longer after this exchange, because Teller became increasingly angry with the way the interview progressed. He chastised the interviewer and wished he had not granted the interview, which he soon ended abruptly. In a more peaceful mood, when he was asked in 2002 if there was anything he regretted in his life, Teller declared: "Concerning Oppenheimer, I should have been more careful." And he added that "as far as people getting angry with me, I wish the outcome had been a little more gentle."[113]

## "DANGEROUS MOLE"

Teller himself was not immune from the probing of the FBI.[114] He referred to himself with self-deprecating humor as a "dangerous mole," when he heard that at the age of ninety-five, the FBI was still looking for proofs of his patriotism. The material in the file throws some light on Teller's stand on some issues and is useful in a quest to understand him. Given that he had been an immigrant from areas that had become part of the Soviet empire, and given his conspicuous position in American defense, it is not surprising that Teller had an extensive FBI file. What is puzzling, though, is that the investigations of his affairs were not conducted with higher efficiency. In a letter of May 4, 1948, the El Paso office of the FBI was informed that Teller and his wife were from Chicago and were visiting Los Alamos frequently. Apparently, there was an Edward Teller in New York

City in 1941 who taught political economy, Marxism–Leninism, and the history of the Communist Party of the Soviet Union.

For years, the FBI could not ascertain that the Teller involved with Los Alamos and AEC work and Teller the communist teacher were not the same person. There were extensive reports about Teller and his wife with a lot of repetition of the meager findings. The Tellers' names appeared on some lists of communist front organizations. An "Edward Teller" was found on a membership or mailing list of the United American–Spanish Aid Committee, and Mici's name appeared on the list of the League of Women Shoppers. According to the FBI reports, the Women Shoppers were infiltrated by communists or communist sympathizers. In fairness, however, the reports also stress that all the associates and acquaintances of the Tellers advised the investigators that the Tellers were loyal to the United States and were anticommunist. One of the witness testimonies noted that Teller was much more loyal to the United States than most Americans.

Teller was interviewed in January 1949 by FBI agents and he "emphatically denied that he was identical with the communist teacher, and stated that he had absolutely *no background or qualifications* which would entitle him to teach any such subjects as those listed by the Workers' School" (italics added). We might note that a rabid anticommunist, such as Teller was believed to be, might have had a good knowledge of the subjects, but it turns out, he did not. Further, he advised the agents that he never belonged to the organization of the United American–Spanish Aid Committee, but, he added that "his sympathies at that time were with the Spanish Loyalists, as he was not in favor of the fascist control extending into Spain through the interference of Germany and Italy in the Spanish Civil War." He added that he might have heard about the organization from Harold Urey, but he did not remember ever contributing anything to aid Spain, but he "might have done so due to his sympathies." This was a brave statement at the time, and there was more to come.

Teller told the FBI that he was concerned about the international situation with Russia and the tension and danger that he felt had led him to interrupt his academic life in Chicago and made him return to Los Alamos. However, the FBI report continued

> He stated that he felt in returning to Los Alamos he did not want to be in the position where he felt he could not read anything about Russia or Communism or Communist publications as he wanted to know as much about the developments in Communism and what was going on in Russia. Further, he did not want to be placed in the position where he could not asso-

ciate with his friends who had definite political views or if they were sympathetic towards Communism, as he did not discuss his work and Atomic Energy with anyone who was not cleared for such discussion.

In February 1950, Teller and his wife were interviewed by the FBI on separate occasions about their interactions with Klaus Fuchs. Teller was also asked about Theodore Alvin Hall, on whose behalf he had written a letter of recommendation. Hall was then "identified as a Soviet Espionage Agent while at the Los Alamos Project." Hall became a graduate student at Chicago, and Teller even visited him at his home, where he noticed communist literature. When Teller asked his host about it, Hall's response was that one had to look into both sides of all questions. When the FBI asked someone else about Hall, who had been Teller's student, the person remembered Teller's mentioning Hall's pro-communist remarks.

The FBI informed Teller that his name appeared on a list given to Czechoslovakian agents by the Soviet authorities. The list included names of US personnel engaged in nuclear research from which they should acquire information about the atomic bomb. In response,

> Teller stated that he was constantly concerned about the possibility that attempts might be made to obtain secret information from him and that his concern was increased by the fact that members of his family were still in Hungary . . . he requested that any information indicating [that] any of his associates were friendly towards Russia be supplied [to] him so that he could avoid any scheme which would have as its objective the obtaining of secret data from him.

Teller informed the FBI about a letter received from a relative in Hungary in 1947 that might have been friendly toward Russia and communism. During the January 1949 interview, Teller volunteered information about Laszlo Tisza. He told the FBI that they were boyhood friends and later studied together in Latzig (he must have meant Leipzig). He also told of Tisza's involvement with communist underground activities. Teller may or may have not known that Tisza had been debriefed by the FBI upon his arrival in the United States in 1941.[115]

Teller even made excuses for Tisza's involvement with the communists, saying that he had to choose between the Nazis and the communists. Teller then told about their continued interactions, including the time Tisza stayed with the Tellers in Washington upon his immigration. The above and similar information that Teller volunteered to the FBI sounds like a preemptive measure, because he must have suspected that his mail from

Hungary was read and his friendship with Tisza and other current or former communist sympathizers were known to the FBI. Subsequently, the FBI used Tisza to gather information about Teller's past.

When the FBI interviewed Frederick de Hoffmann on July 5, 1950, they asked him whether he knew about anyone at Los Alamos who might have relatives in communist-dominated Europe. De Hoffmann told the FBI about Teller, among others. It was obvious that he could not tell the FBI anything they would not know or anything that Teller would not have told them. Further narrative makes it clear that at the beginning of the 1950s, the FBI was still not convinced that the nuclear physicist Teller and the New York communist teacher were not the same person. Various investigations were under way or being planned in this connection, including the suggestion to show Teller's photograph to Whittaker Chambers, the Soviet spy turned American informer for possible identification.

In August 1950, the FBI received information about Mrs. Edward Teller's statement to a "reliable" informant that she and her husband were both members of the Communist Party. She talked about this during the time her husband as well as the informant's husband were attending a scientific meeting. Mrs. Teller was worried about the ongoing investigations of former communists in the United States. The FBI asked around about Teller and asked Eugene P. Wigner whether Teller ever demonstrated sympathy for communist philosophy or whether he was a member of the Communist Party. The FBI told Wigner that Teller "probably was forced to join a Communist Trade Union in Hungary in the year 1919." They may have failed to notice that Teller was eleven years old at the time of the short-lived communist dictatorship in Hungary in 1919. Wigner found either supposition impossible. Teller was also investigated about his relationship and interactions with Stephen Brunauer (see chapter 6).

Teller's possible past engagement as a teacher at the Communist Workers' School in New York City caused a lot of headaches for the FBI, and they never found any conclusive evidence to close down the investigation either way. The other persisting topic was Mrs. Teller's alleged 1949 statement about her and her husband's membership in the Communist Party. The interest in this issue also fizzled away gradually rather than reach a reassuring solution. There was one point when the FBI asked the State Department to withhold Teller's passport to prevent his travels abroad. Renewed investigations were conducted on the occasions of his various appointments to sensitive positions over the years, such as, for example, to the Scientific Advisory Board of the US Air Force or to be consultant to the Science and Technology Staff of the White House.

The FBI was interested not only in Teller's character, loyalty, and general standing but also in his scientific acumen to do good physics. The investigations covered his travels, phone calls, correspondence, partying habits, fondness for square dancing, his hotel rooms, luggage, his trash, and even what he carried in his pockets. When they uncovered some strange diagrams or scientific papers, they called on experts to interpret them.

There was another case of mistaken identity. Dr. Ludwig Teller, professor of law and a Democratic representative from New York, was mistaken for Edward Teller by reporters. When he was addressed as Dr. Teller he could not deny that he was Dr. Teller, and when he became tired of explaining the mistake, he began answering questions, although he refused to give autographs at the Geneva conference of atoms for peace, which he attended as a member of the US congressional group. Answers like "The matter requires further study" apparently caused no harm.[116] When Teller—our Teller—heard about the other, he promptly wrote him a gracious letter and expressed his hope that he would never be asked "questions of law or how to make the laws of the land."[117]

One of the most remarkable revelations in the above material is that Teller wanted to reserve his freedom to choose the people he associated with, even if they were communists. Some family members of one of Wendy Teller's best friends might have been card-carrying communists, and Wendy's parents knew about this but did not interfere.[118] This is fully consistent with Paul Teller's observation; Paul went to school at Berkeley, and many of his friends "were 'peaceniks,' people, who thought that we could and should make every effort to get along with the Soviets." Paul found that his father "really did practice what he preached about principles of freedom of expression" and had no problem with his friends.[119]

Wendy also remembers that FBI agents used to go through the Tellers' household trash. They found there an envelope from Hungary, but what really upset them was that the postage stamp displayed Stalin's portrait (which was a common stamp at the time in Hungary). The envelope carried an unknown name as addressee, and this was a great puzzle for the FBI. There was a considerable amount of discussion in the FBI reports about the addressee of the letter, but they never learned the identity. However, knowing about Teller's childhood, it is easy to determine that it was the Tellers' former governess back in Budapest (chapter 1), who had moved to Chicago and helped maintain contact between the American and Budapest wings of the Teller family. Both sides found it safer to correspond through her than directly. Incidentally, in a photograph much publicized by Livermore, Teller is seen with a book in his hand and, most probably unbeknownst to most, the book cover displays Stalin's portrait and a sufficient portion of Stalin's name is visible to see its Hungarian spelling.

Janos Kirz joined the Teller family in 1957 after he had escaped from Hungary in the wake of the suppressed revolution in 1956. It was an adventurous project because he did not even know the Tellers' address. When Janos was living with the Tellers, he did not correspond with his mother and grandmother directly either. They remained in Budapest and would get out only in 1959.

Although there was no specific instruction for Janos not to correspond with them directly, he understood that the Tellers were concerned not only for the safety of their Budapest relatives but also for themselves in California. In Janos's words, "the Tellers did not want to be seen as having contact with people behind the Iron Curtain lest they be considered a security risk."[120] Incidentally, upon his arrival in California in 1957, Janos was debriefed by the FBI, perhaps because they might have been concerned about his identity lest a secret agent had been planted into the Teller home. Janos was born in 1937, and the Tellers last visited Budapest in 1936, but Janos never for a moment felt any doubt on the Tellers' part about his identity.

Hoover Fellow John H. Bunzel described his many exchanges with Edward Teller in a personal remembrance a few months after Teller passed away. He also shared his experience with the FBI regarding Teller:

> A month or so before Edward died, I received a phone call from a retired FBI agent who had been contracted to do a security check on Dr. Edward Teller and wanted to know if I would be willing to answer a few questions. Incredulous, I asked if this was a joke. He explained that back at headquarters in Washington the clerk in charge had never heard of Edward Teller and that therefore the required "field check" had been assigned to him in San Francisco. He knew who Dr. Teller was, he assured me, and could understand my disbelief. He began by asking me, very apologetically, if I could testify to Dr. Teller's patriotism. "He should be arrested immediately," I said. "He's almost 96, he can't hear or see, and lives in a wheelchair, a perfect cover for his work as a spy." When I told Edward about this the next day, he smiled. "I prefer to think of myself as a dangerous mole," he said, relishing the thought.[121]

# Chapter 9
# FALLOUT AND TEST BAN

There are three kinds of physicists; theoretical, experimental, and political. Edward Teller is all three.

—LEWIS L. STRAUSS[1]

*During the mid-1950s, Teller's life settled at a plateau. Rather than feeling secure, however, he sensed that Livermore and his own position were challenged by various developments, both domestic and international. He found it especially threatening that the United States might find accommodation with the Soviet Union. He considered the Soviets and the communists more dangerous than Hitler and the Nazis. He wanted to avoid any situation in which an agreement would hinder the further development of American arms, since he believed that the Soviets would feel less obliged to honor the agreement. Teller fought against the test bans and debated their proponents with any means he found available, be it the promise of a clean bomb, assurance that the fallout was negligible and that birth defects were an acceptable price to pay for security, painting the Soviets as habitual cheaters, or telling the public that nuclear winter was survivable. Even those scientists who might have agreed with him refrained from supporting him lest they be ostracized.*

Once the United States (in November 1952) and the Soviet Union (in August 1953) started exploding thermonuclear devices and bombs, they embarked on a spiraling arms race. The March 1954 hydrogen bomb test at the Bikini Atoll led to tragic consequences, as we have seen. Eventually, the Soviet Union achieved the distinction of having blasted the largest-ever hydrogen bomb. At the same time, Western public awareness of the hazards of radioactive fallout necessitated the curbing of escalating weapons developments. Negotiations began in the framework of the United Nations Disarmament Commission in May 1955 for an international agreement to end nuclear tests. Five nations were involved initially: the United States, the Soviet Union, the United Kingdom, France, and Canada. Following are a few dates and major signposts along the way to the limited test ban. These are given to facilitate an understanding of Edward Teller's role in the development of events. In the opening quotation, Lewis Strauss refers to Teller as an experimentalist, but this was an exaggeration. Teller was a theoretician, but even this designation diminished as he became increasingly political.

On October 17, 1956, the Soviet Union suggested declaring a test ban without international control, but eventually it offered a test ban with international controls on June 14, 1957. There was no agreement in sight, but the Soviets announced that they were stopping the tests and appealed to the others to follow suit. President Eisenhower declined, citing the possibility of secret tests by the Soviets. The American (and British) tests thus continued. In the summer of 1958, experts discussed possible international controls, and the Americans demanded onsite inspection and proposed a suspension of testing for one year to be renewed on a year-by-year basis. Soviet leader Nikita S. Khrushchev renounced the proposal, and the Soviet Union resumed testing and continued it until November 3, 1958. It was then that a moratorium on testing began, and all sides kept to it through August 1961. The Soviet Union resumed testing on September 1, 1961, and the United States followed suit in mid-September.

The Soviet leadership initiated actions and employed propaganda to its advantage while it played around with moratoria and testing when they best served its political interests. On June 10, 1963, President Kennedy announced that the United States, the United Kingdom, and the Soviet Union would negotiate for a test ban in Moscow. Khrushchev declared three weeks later that the Soviet Union would be willing to outlaw testing in the atmosphere, in outer space, and underwater. The three-power meetings began on July 15, and ten days later a treaty was initialed by the three foreign ministers (the secretary of state on the US side). The US Senate approved the treaty on September 24, and President Kennedy signed it into law on October 7, 1963. This Limited Test Ban Treaty prohibited nuclear weapons tests or any other nuclear explosions, except those performed underground. This exception left open the possibility of considerable testing of nuclear devices and bombs, but the treaty removed the problem of inspection and verification.

## AGAINST THE TEST BAN

Teller opposed the ban of testing for two main reasons. First, he felt that tests produced new knowledge, and second, while he knew the Americans would observe the provisions of such a ban, he strongly believed that the Soviets might cheat and it would be difficult to catch them. He considered the increasing fear from radiation, as fallout from testing, to be exaggerated and baseless. He even claimed that low-level radiation might have beneficial effects.

Donald Glaser once hosted Teller and Szilard in his home, and the two got into a vigorous debate about the consequences of radiation. Teller said that certain flies lived longer if they were exposed to a low level of radiation. Szilard responded that the reason was that there were parasites on these flies and they were very sensitive to radiation. It was an intense discussion in which Szilard accused Teller of knowing the real reason why flies lived longer under the impact of radiation. But there was no yelling or screaming.[2] Teller had other formidable and less amiable opponents among scientists, above all, chemistry Nobel laureate and Caltech professor Linus Pauling.

In the mid-1950s, the Atomic Energy Commission conducted tests of atomic devices in the Nevada desert, and the radioactive fallout spread over the continental United States. Edward B. Lewis, a young Caltech biologist, became interested in the impact of fallout. He was a conscientious and meticulous scientist who would share the Nobel Prize in Physiology or Medicine in 1995 for his discoveries concerning the genetic control of early embryonic development. I talked with Lewis in the spring of 1998 about his fallout studies, which, although only a side project, were nevertheless referenced by the Nobel Prize citation.[3]

He collected information about the Japanese atomic bomb survivors and examined published data on adults treated with X-rays for *ankylosis spondylitis* (arthritis of the spine, a rheumatic disease) and on infants treated with X-rays for enlarged thymuses (lymphoid organs). In addition, he collected data on radiologists. Nobody before Lewis had put together such data, and his findings showed that every one of these groups experienced a significantly elevated risk of leukemia. Lewis spent many months on this topic and published his findings in the prestigious *Science* magazine.[4] He cited Hermann J. Muller's earlier discovery that X-rays can induce mutations in the germ cells of *Drosophila* (fruit flies),[5] for which Muller was awarded the Nobel Prize in Physiology or Medicine for 1946. Muller also anticipated that X-rays could produce cancers by inducing mutations in somatic (body) cells, not only in germ cells (responsible for development and reproduction).

Lewis's conclusion was that fallout could produce cancers if those cancers resulted from somatic mutations and if there were no threshold dose of radiation for their induction. As there was no evidence for a threshold dose—that is, no minimum amount—below which no germ mutations would be induced, there was no reason to suppose a threshold for somatic mutations either. This was not what Teller professed, but it was in agreement with Muller's argument, which had called attention to the genetic risks to the human race posed by indiscriminate use of ionizing radiations and, prophetically, argued that such uses would also increase the risk of cancers.[6]

Lewis thus came into contradiction with Teller's advocacy, although he did not directly participate in the debates. Nonetheless, his paper came to the attention of the Joint Congressional Committee on Atomic Energy, and Lewis was invited to testify in 1957. He told the committee about the expected number of leukemia cases based on the fallout levels in the United States. His numbers were then used extensively by Pauling, to whom Lewis had shown the draft of his paper. Lewis noted that "Pauling multiplied the risk estimates by rather a larger number of persons than would actually have been exposed to the estimated doses from fallout."[7] Apparently, Teller was not the only one prone to exaggeration.

Lawrence was Teller's ally in his resistance to the test ban, but, once again, he preferred to be less vocal than Teller. He considered testing safe and called fallout "a phony issue."[8] Teller tried to persuade Lawrence to make a public statement to balance the information about fallout that Adlai Stevenson espoused in his race for president in 1956. Lawrence first declined but eventually agreed to issue a joint statement with Teller in support of continuing testing, with some help from the university public relations people. There were two versions, a softer and a harder version, of the statement; the public relations people wrote the softer one and a dissatisfied Teller produced a stronger one. At Lawrence's insistence, the softer version was communicated to the press.

The idea of a so-called clean bomb was an important argument for the continuation of tests. Teller claimed that it would drastically reduce, almost eliminate, fallout. The fallout from hydrogen bombs came from the fission explosion, which was used to create the extremely high temperatures and initiate the thermonuclear reaction, which in itself did not add to the radioactive pollution. Teller, Lawrence, and two of their associates testified before the military applications subcommittee of the Joint Congressional Committee on Atomic Energy on June 20, 1957.[9] Teller and his colleagues stressed that testing was crucial to the development of the clean bomb: "A continuation of testing will make it possible to overcome the very thing that people are objecting to—radioactive fall-out." The chairman of the subcommittee interpreted the Soviets' proposal of a two- to three-year suspension of testing as a means to weaken the West's military preparedness.

Lewis Strauss, chairman of the AEC, spoke about a "humanized" thermonuclear weapon, and Teller referred to a "virtually clean bomb," but according to press reports, nobody knew yet how to make such a device.[10] In order to eliminate fallout, the fission part should have been removed from the hydrogen bomb, but it was the device that induced the thermonuclear reaction. The journalist compared the situation to "making an omelet without eggs, or an apple pie without apples." Following the scientists' testimony before the congressional subcommittee, Lawrence, Teller, and Mark M. Mills were invited to the White House to inform the president of their progress.[11]

A few weeks before, Eisenhower had stated during a press conference that the fallout had already been reduced by nine-tenths. Strauss topped this, stating that the reduction stood somewhere between nine-tenths and ten-tenths.[12] The president was greatly impressed by the scientists' report. Their small delegation was obviously headed by Lawrence, but the press reports were gradually giving more and more exposure to Teller. Eisenhower was so much taken by the idea of the clean bomb that he reversed his support for a possible agreement with the Soviet Union for a test ban. Rather, he encouraged work that aimed at producing a clean bomb. Although previous statements and reports indicated a near-complete achievement of the clean bomb, Teller's estimate for completing the task was five years and upward, which also meant continued testing for as many years.

Meanwhile, Andrei Sakharov, the Soviet nuclear scientist who had yet to achieve international fame, was working on an article showing that there was no such thing as a "clean" thermonuclear bomb. It was an intriguing task for him, because his superiors did not want him to implicate "conventional" thermonuclear weapons.[13] He went further than such an article of primarily political purposes would have required and carefully studied all the possible consequences of nuclear explosions. He would

## LINUS PAULING (1901–1993)

Linus Pauling was born and grew up in Oregon. His father was a pharmacist who died young, and his mother suffered from illness. Pauling was gifted, ambitious, and hardworking, but was not privileged to go to superior schools until he got to the California Institute of Technology (Caltech). He contributed to Caltech's rise to eminence, both as a student and as a professor. Having earned his PhD degree in 1925, Pauling spent some time in Europe, where he became acquainted with quantum mechanics and some new experimental techniques of structure determination.

Upon his return to the United States, he applied the new physical theories to chemistry and utilized the new experimental techniques to build up a body of hard data for a solid foundation of his theories. He published a large series of papers on the chemical bond, from which he developed his influential monograph *The Nature of the Chemical Bond*.

Early on Pauling recognized the importance of the structure of biological macromolecules, and he narrowed his interest to proteins, paying less attention to nucleic acids. He was triumphant in discovering the alpha-helix structure of proteins. His work on the understanding of the nature of the chemical bond and the alpha-helix discovery brought him the Nobel Prize in Chemistry in 1954.

By then he was increasingly interested in social and political issues. It was a long way from the 1930s when he was rather insensitive to the plight of would-be refugee scientists in Nazi Germany who had turned to him for help. Pauling developed leftist views and was accused of being a member of the Communist Party, but he was not. During the McCarthy era he was not allowed to travel abroad and so was hindered from being a more active participant in the flow of scientific information.

With proliferation of nuclear weapons, Pauling took up the challenge of fighting for a test ban. He was energetic and enigmatic if somewhat one-sided. After Pauling's Nobel Peace Prize in 1963, some of the Caltech trustees viewed his political activities disapprovingly, so he resigned from the school. During the remaining decades of his life, his scientific activities suffered from his other pursuits. His crusade for the advisability of large intake of vitamin C was met with skepticism. Still, his most important insight during this period was the recognition that illnesses have a molecular basis, a view that has had tremendous subsequent success.

Pauling was an excellent lecturer and engaging personality with a large following both at home and internationally.

become a conspicuous exception among his colleagues; apart from him, the Soviet Union had the advantage in that they had no domestic accountability or "outing" of their ways, whereas they fully played on the open debate in the free American society.

Important decisions about questions like a moratorium on nuclear testing in the Soviet Union were often taken purely on the basis of politics, without consulting the scientists. For example, when the Soviets declared their suspension of further testing and called on the United States to do the same, the Americans assumed that the Soviets had completed their forthcoming series of tests. However, Khrushchev's decision took the Soviet scientists as much by surprise as the rest of the world. The Soviet Union declared moral victory when the United States decided to carry out a set of tests that had been in planning and were deemed necessary. This was a time-honored approach by the Soviet leadership. The Soviets were also aware of the fact that President Eisenhower was eager to put a brake on the arms race. And so the negotiations were renewed in the summer of 1958. Teller was a formidable force, but so was Khrushchev, and often it seemed that the struggle was between the two of them rather than between Khrushchev and Eisenhower.

Wolfgang Panofsky played a vital role in the American negotiations with the Soviets. Once he chaired a committee in which both Hans Bethe and Edward Teller participated, and Panofsky considered it a distinction when the committee managed to produce a unanimous report in 1958, signed by both Bethe and Teller.[14] In connection with discussing the possibilities of cheating, Teller came up with various schemes that, ostensibly, the Russians might use to avoid detection. One of them was that it would be possible to send one rocket into outer space carrying a nuclear bomb and another carrying the diagnostic instruments to measure the consequences of the explosion, which would happen somewhere between ten thousand and a hundred thousand miles from Earth. Still, Panofsky did not find the possibility of such kind of cheating realistic, and his judgments were based on detailed technical analyses.

## TELLER VERSUS PAULING

Teller participated in direct debates about the test ban. Linus Pauling, the most vocal proponent of the test ban, welcomed the opportunity of a personal confrontation. Pauling had a petition signed by thousands of scientists from all over the world, including thirty-five Nobel laureates and

## ANDREI D. SAKHAROV (1921–1989)

Andrei Sakharov was born in Moscow; his father was a physics teacher and author. He studied at the Physics Department of Moscow State University between 1938 and 1942; the last portion of his studies took place in Ashkhabad, where the school had been evacuated because of the war. While a student, he worked at a munitions factory as an engineer and inventor and wrote research papers in physics.

He did his doctoral work under Igor E. Tamm at the Lebedev Physics Institute of the Soviet Academy of Sciences in Moscow, and in 1948 he became a member of Tamm's group working toward the creation of Soviet nuclear weapons. Sakharov and his family moved to Arzamas-16, the secret installation for Soviet nuclear weapons work. He has often been referred to as the father of the Soviet hydrogen bomb. For his achievements he was showered with awards and privileges by the Soviet state. He never expressed regret for his contribution to the Soviet nuclear might and never wavered in his conviction that nuclear power was preferable to traditional sources to generate energy.

He always considered his work on nuclear weapons as his patriotic duty. His first incident with the authorities was in 1955, when, toasting their successful test explosion, he expressed hope that their bombs would always explode over test sites and never over cities. Nikita Khrushchev let him know immediately that the scientist's duty was to create weapons, not to have a say about their deployment.

Sakharov became interested in the possible harmful effects of fallout from nuclear explosions and determined that for each (TNT equivalent) megaton of explosives tested, the number of fatalities might amount to ten thousand people. He determined that there was no threshold to possible biological effects and published his findings in 1957. He criticized the notion of a "clean" bomb and called it a fiction because even its idealized version produced plenty of the harmful radioactive isotope carbon-14, whose half-life was five thousand years. He got into direct conflict with Khrushchev when the Soviet leader announced the resumption of nuclear tests in 1961, which Sakharov did not find necessary.

Sakharov gradually became more and more involved with human rights issues and other topics that were considered subversive in the Soviet Union. Repercussions followed, especially when his writings began to appear in the Western media. From 1968, he was no longer an associate of Arzamas-16, but he could still continue doing physics at the Academy of Sciences in Moscow.

He became increasingly alienated from the Soviet establishment; a sad signal of this was when a large group of academicians condemned his activities. This and similar acts, many years later, might be used to compare his predicament to Teller's alienation from his fellow physicists. But the two cases were hardly comparable. The Soviet academicians followed governmental guidelines, and at least some of them acted under duress, whereas the American scientists' attitude toward Teller had nothing to do with American officialdom.

When the Soviet leadership found Sakharov's presence in Moscow intolerable and his interactions with the most diverse groups of dissidents harmful, they exiled him to the city of Gorky (today it is called by its old name, Nizhny Novgorod), where he and his wife had restricted contacts with the external world between 1980 and 1987. At the end of 1987, under Gorbachev, they were permitted back in Moscow, and for the remaining two years of his life, Sakharov became a significant player in Soviet political life.

Sakharov's influence and example extended far beyond nuclear weapons and physics; he became a major source and symbol of the painful changes the Soviet Union underwent during the last years of the 1980s. Even he could not have foreseen the collapse of the Soviet Union and the events in the 1990s, but he contributed to them more than most. It was Sakharov's scientific work more than anything else that made the Soviet Union a superpower, because the superpower status of this backward country was due more than anything else to its thermonuclear weapons.

many members of learned societies. The two-hundred-and-fifty-word petition stated, among other points:

Each nuclear bomb test spreads an added burden of radioactive elements over every part of the world. Each added amount of radiation causes damage to the health of human beings all over the world, and causes damage to the pool of human germ plasma such as to lead to an increase in the number of seriously defective children that will be born in future generations.[15]

The petition called for "an international agreement to stop the testing of all nuclear weapons." In response, Teller published an article in *Life* magazine.[16] A one-hour debate on the topic "Is Fallout Overrated?" was arranged at the studio of the public station KQED-TV San Francisco on February 21, 1958.[17] It is interesting to examine the debate for its contents and also for the attitudes and approaches of the two scientists.[18] One of Pauling's biographers pointed out the similarities between the two scientists: "Both held tight to a simplified political vision, both were dogged in their quests, both marshaled impressive scientific facts to support their political positions. They went to war with each other in the spring of 1958."[19]

On camera, the two men appeared very different. Pauling looked more elegant, but he lost his advantage as soon as he started talking. His high-pitched voice sounded shrill, whereas Teller was calm and witty with his resonating basso. Pauling engaged in detailed discussion of minor differences between the original text of his petition and what he considered to be its distorted presentation by Teller in the *Life* article, although the magazine piece conveyed the essence of the petition. The minute differences Pauling discussed were hard to follow in an oral presentation. Pauling started his performance by fiercely attacking Teller, whereas Teller began his by expressing respect for Pauling as a scientist and his pleasure of appearing with him. Whereas Pauling talked about differences, Teller stressed the points about which they agreed. Teller's approach was light and made his message accessible for his audience.

Each gave an opening and closing statement and had the chance to speak twice in between. Here are a few excerpts from Teller's opening statement:

> I would like to emphasize at the outset that there are many, many facts about which Dr. Pauling and I agree. . . . and the more you look into details the more you will find that the basic facts and also the basic assumption about what we all want *is* the same. . . . [emphasis in the original]
>
> The first points with which I would like to agree very strongly with Dr. Pauling are his quest for peace and his great appreciation for human life. . . . We want peace, and we have to work for it. It is absolutely clear that each human life has a high value and should not be sacrificed, except for very, very good reasons. . . .
>
> Peace cannot be obtained by wishing for it. We live in the same world with Russia, whose leader has said that he "Wants to bury us" . . .
>
> . . . if we stop developing our nuclear weapons, I do not believe that we can check whether the Russians have done the same. . . .
>
> . . .
>
> We must avoid war under all possible circumstances, except, in my opinion, one: when the freedom of human beings is at stake.
>
> . . .

If such [war] should happen, then it would be of great importance that these weapons should do as little damage [to] human life as possible. If a war of this kind has to be fought, then the danger from radioactivity will be very great indeed. I shall talk about the damage due to radioactivity from testing—it is something about which I continue to say one does not need to worry about, certainly not to the same extent as one needs to worry about a war. . . .

. . .

. . . there is no doubt that some radioactivity is spread throughout the world by nuclear explosions. Dr. Pauling says this causes damage. . . .

. . . this alleged damage which the small radioactivity is causing—supposedly cancer and leukemia—has not been proved, . . . It is possible that there *is* damage. It is even possible, to my mind, that there is *no damage*; and there is the possibility, furthermore, that very small amounts of radioactivity are helpful. . . . [emphases in the original]

There is . . . the question of genetics. We know enough about the mechanism of heredity to be sure that changes will be made in the germ plasm, just as Dr. Pauling has said, and many, very many, probably the great majority of these changes will be damaging. Yet without some changes, evolution would be impossible.

. . . If we proceeded in everything with as great a caution as we are proceeding in the case of nuclear testing, there would be very little progress in the world.

In his response, Pauling missed the opportunity to question Teller's assertion about the possibility of no damage from radioactivity let alone the possibility of its being helpful (and by "helpful," Teller was not referring to medical uses of radioactivity with well-defined purposes and within well-defined limitations). Further, he twice attempted to explain Khrushchev's infamous statement "We will bury you,"[20] which was a hopeless task. Pauling did not need to answer for the Soviet leader's rude carelessness, but somehow he felt he did. Pauling was also given to exaggerate what Teller had just said, which detracted from what could have been a reserved statesman-like approach.

Teller in his next turn played on that by saying that he would like Pauling to exercise the kind of magnanimity toward him that he demonstrated toward Khrushchev. Pauling did not do well in his next turn by returning one more time to the *Life* magazine article—not a winning issue in his first statement. This time he questioned the validity of one of Teller's arguments in which Teller had pointed out that no genetic damage had been observed in the population of Tibet, which is exposed more than any other to cosmic radiation. This was smart, because there were no reliable data to support Teller's claim. Pauling scored another point by warning against the dangers of nuclear proliferation.

Teller tried to counter Pauling's collection of scientists who had signed this petition by invoking Willard Libby's studies; in retrospect it was even a better argument than Teller might have known at that time, because in two years Libby won the Nobel Prize in Chemistry. The question of the peaceful use of nuclear explosions was an interesting one, and Pauling admitted its importance, but he bumped into the problem of how to develop it without testing in case of a test ban. Pauling's performance improved toward the end of the debate as he began applying Teller's approach in being more conciliatory.

Pauling started the debate, so Teller had the last word, and he stressed two points. One was that a ban of testing would be dangerous because while the Americans would not violate it, the Russians might. The other was his statement "I am passionately opposed to killing, but I'm even more passionately fond of freedom." He reminded his audience that the way they expressed their opinions would not be possible in Russia. To drive his point home, he mentioned that one of Pauling's great discoveries "was suppressed in Russia because it contradicted some kind of official philosophy in Russia." He was referring to the theory of resonance (see chapter 3).

The debate contained the elements that made Teller such an effective debater. Pauling declared that he never would debate him again. Years later he told an interviewer that although he saw Teller rarely, from what he read about Teller, he formed the impression that they were not far apart in their views on nuclear power plants but were still "diametrically opposed" about nuclear weapons.[21] Pauling drew a conclusion from his encounters with Teller and also with Wigner. He said, "In each case I felt that the person, Hungarian, with that sort of experience involving the Soviet Union was governed to such an extent by his emotional feelings that he was no longer rational when it came to discussing problems of that sort."[22] This was yet another attempt to reduce Teller's (and Wigner's) political stand to personal experience with communism and the Soviet Union, which hardly existed. Besides, in the Pauling–Teller debate, if anybody was emotional, it was Pauling, while Teller kept his calm throughout.

Although Teller proved to be Pauling's formidable opponent, nobody could have accused him of trying to destroy Pauling by his actions. It was very different from his controversy with Oppenheimer. The comparison comes to mind for a reason. There were attempts in Washington to glue various labels to Pauling's name, accusing him of being a communist sympathizer or a communist; of manipulating the signatures of his petition; of using funds of unknown (meaning illegal) origin for collecting the signatures; of acting at Moscow's order; and the like. When in the spring of 1952 Pauling was not allowed to travel abroad, Teller was among the signatories of a letter from leading atomic scientists at the University of Chicago that was sent to Secretary of State Dean Acheson in protest. The scientists asked, among other things, "What harm, what information, what tales could Professor Pauling take with him to England, even if he were so inclined, that can compare in damage to the incredible advertisement that this country forbids one of its most illustrious citizens to travel?"[23]

When Pauling thanked Teller for his support, he stressed that he was not a left-winger but rather a "middle-of-the-road-moderate."[24] It would have been difficult, though, to characterize Pauling as being moderate on the question of the test ban. There is a famous photograph of Pauling on the picket line in front of the White House carrying a big poster saying, "Mr. Kennedy, Mr. Macmillan, WE HAVE NO *RIGHT* TO TEST."[25] Mr. Khrushchev is conspicuously missing, and Pauling has been criticized for his one-sided approach. The next day Pauling joined the dinner reception at the White House, which President Kennedy gave for American Nobel laureates. A similar scenario in Moscow would have been unthinkable.

Teller knew that his debate with Pauling did not make him popular among Pauling's many followers. He told us a lighter story on this topic. "A few years later, a very pretty girl ran after me in New York, at the airport: 'May I have, please, your autograph, Professor Pauling?' I made a very ugly face and I said, no. That was the only time I was impolite to a pretty girl."[26] The debate with Pauling probably did not leave Teller with many pleasant memories either. A few months after the debate, the Federation of American Scientists wanted to initiate another Pauling–Teller debate, but Teller politely declined the invitation citing his heavy obligations at Livermore.[27] A few years later there was a choice between Pauling speaking alone at a meeting of the American Chemical Society or engaging in a second debate with Teller. In this case Teller would have opted for a debate—however reluctantly—because he did not want Pauling to have "a free chance to make propaganda in the Chemical Society."[28]

## CONTINUED OPPOSITION

In the spring of 1958, it appeared that a test ban agreement would be concluded with the Soviet Union within months. Teller received the news while vacationing with his family in Yosemite Park and he hurried back to Livermore to hasten the ongoing tests. The tests were needed for the Nobska Project, which would develop one-megaton fusion bomb warheads, light enough to be carried on missiles in submarines. To increase their pace, Livermore enforced a six-day work week. In the meantime, Teller continued his fierce advocacy to prevent ending the nuclear tests. For example, he testified before the Senate Disarmament Subcommittee on April 16, 1958, that by suspending testing, "we may be sacrificing millions of lives in a 'dirty' nuclear war later."[29] He meant that the amount of radioactivity from testing was still much less than it would be from nuclear bombs in war. Concerning the dangers of radioactive fallout from testing, he admitted that it might cause genetic mutations, but he argued that mutations were not necessarily bad.

When the moratorium on testing went into effect without any formal agreement, Teller continued his warnings that the Soviet Union could carry out blasts the size of the Hiroshima bomb without detection. The tests in underground cavities would be indistinguishable from the many hundreds of earthquakes that occur in Russia annually.[30] Bethe and Teller were both asked to testify before the Joint Congressional Committee on Atomic Energy on April 23, 1960. Bethe supported the ban; Teller opposed it. At the end of November 1961, Kennedy asked the same two scientists for advice and got the same diverging responses.[31] According to the famous columnist James Reston, it became "almost impossible to go to one of those interminable dinners in Georgetown, where the men disappear for forty-five minutes to solve the problems of the world, without running into the conflict between the Teller pessimists and the Bethe optimists."[32]

When the Soviets lifted the ban on tests and began vigorous testing in September 1961, Teller warned that the Soviets had carefully prepared them. In a letter to the *New York Times* on November 12, 1961, he noted that the danger from fallout was "quite small compared to the obvious and imminent danger of Soviet power."[33] Linus Pauling did not wait long with an answer, which appeared exactly one week after Teller's letter. Not surprisingly, Pauling disagreed with Teller on every point. He noted that both the Soviets and the Americans have amassed so many millions of tons of TNT-equivalent thermonuclear bombs that they were both capable of annihilating the other several times over. He called for negotiations, a moratorium on testing, and cooperation in disarmament.[34]

Henry DeWolf Smyth wondered about the public's dilemma of learning the truth when such eminent scientists as Pauling and Teller held so vastly different views. He observed that when discussing damages from fallout, "the same prediction . . . may sound trivially small expressed in terms of percentage of the world's population, but appallingly large translated into actual numbers of people. Not surprisingly, the advocates of continued bomb testing use percentages; their opponents cite actual numbers."[35]

A distressing example of juggling information is the comparison of the total radiation people receive annually from various sources as compared to worldwide fallout. In the Teller and Albert L. Latter book, *Our Nuclear Future*, the main conclusion was that the worldwide fallout carried less danger than many other radiation effects that have not worried people. One of the problems with this comparison is the absence of information about how the amount of worldwide fallout radiation was estimated to be below the statistically observable limit. It can be supposed that the estimation referred to the fallout evenly dispersing over the entire earth. In reality, the fallout caused much higher concentrations of radioactivity at much smaller, specific regions. This is why it is misleading when the Teller–Latter book states that drinking a glass of water or wearing a wristwatch with a luminous dial gives us orders of magnitude larger amounts of radiation than the nuclear explosions.[36]

Teller and Latter claimed that remaining unmarried reduces life expectancy by five years; being of the male sex, by three years; living in the city instead of the country, by five years; whereas the worldwide fallout reduces life expectancy by one to two days (this is also extrapolated to a lifetime, as, one should suppose, remaining unmarried or staying male or being a city dweller). This is very difficult to comment on with dignity. Other examples in the comparison, such as being overweight and smoking, are no joking matter, but translating the danger of worldwide fallout into their terms, Teller and Latter estimated that it would be the equivalent of being an ounce overweight or smoking one cigarette every two months.[37] This approach in assessing the hazards of worldwide fallout undermined their own argument.

By curious coincidence, it was at this time that Andrei Sakharov in the Soviet Union published his reports on the consequences of fallout from nuclear explosions. His findings were different from Teller's. Sakharov "showed convincingly the powerful damaging effect on hereditary structures of both the radiation at the moment of the explosion and the residual radiation."[38]

A few years later Teller would be less adamant about the absence of any consequence of worldwide fallout; rather, he would admit that it could influence heredity. Once it is acknowledged that there may be changes, the question arises as to whether they might be necessarily harmful or might even be beneficial. Here comes perhaps the most shocking of all of Teller's statements concerning fallout:

Cesium-137 [a radioactive isotope of cesium] in the fallout, by affecting reproductive cells, will produce some mutations and abnormalities in future generations. This raises a question: Are abnormalities harmful? Because abnormalities deviate from the norm, they may be offensive at first sight. But without such abnormal births and such mutations, the human race would not have evolved and we would not be here. Deploring the mutations that may be caused by fallout is something like adopting the policies of the Daughters of the American Revolution, who approve of a past revolution but condemn future reforms.[39]

In contrast, Andrei Sakharov regarded uncontrollable mutations to be an alarming prospect in his discussions about the consequences of nuclear explosions. Not only did Sakharov struggle against further testing of nuclear weapons; he wanted to investigate the effect of radiation upon heredity. Thus he found himself in conflict with the unscientific direction called *Lysenkoism* in the Soviet Union, which denied the validity of the science of genetics. When Sakharov first became interested in assessing the possibility of a "clean" bomb, his investigation determined that the possible harmful biological effects of fallout from tests were nonthreshold effects, that is, there was no minimum dosage below which damage would never occur.[40]

Sakharov felt confident in his estimates that about ten thousand casualties—new cancer cases or genetic disorders—would result for every megaton (TNT equivalent) of explosives used in testing. About two-thirds of them would result from carbon-14, which is formed in the explosions of "clean" as well as conventional hydrogen bombs, and about one-third would result from strontium-90 and cesium-137. The isotopes strontium-90 and cesium-137 had immediate harmful effects, whereas carbon-14, with a half-life of five thousand years, caused long-range damage. Further, Sakharov estimated that by 1957 already some fifty-megaton TNT-equivalent explosions had been tested, which meant about five hundred thousand casualties over the years.[41] For the time being, the Soviet leaders welcomed Sakharov's studies, and when he gave intervals of casualties, they preferred using his upper limits in their eagerness to condemn the American tests. Soon they would learn, to their chagrin, that for Sakharov, the number of casualties did not depend on whether it was a consequence of capitalist or communist explosions.

Teller issued a warning against the test ban on January 31, 1963, to the Republican Conference of the US House of Representatives. He claimed that the planned treaty would endanger American security and "help the Soviet Union in its plan to conquer the world."[42] The three major issues were (1) The ban would prevent the United States to further develop its nuclear arsenal; (2) It would not interfere with Soviet progress partly because their underground explosions and those in outer space would remain undetected; (3) The ban would induce the United States to pressure France—who would not be an original signatory to the treaty—to comply, thus causing tension within the NATO, whereas the Chinese would ignore any such agreement.

Teller also cautioned that during the previous moratorium, when the Soviets had felt they were ready for massive testing, they simply abrogated the moratorium and conducted testing on an unprecedented scale. Teller's warning about the possibility of the Soviets cheating appeared realistic in view of the recent missile crisis in Cuba. Initially, the Soviet representative denied point-blank to President Kennedy the presence of Soviet missiles on Cuban soil. The possibility of undetected underground explosions and Soviet reluctance to permit sufficient onsite inspection eventually led to the compromise of leaving the underground explosions out of the treaty.

From the time the test ban treaty was initialed in Moscow on July 25, 1963, the last possibility for its opponents was to prevent its approval in the Senate. During these weeks Teller was very active lobbying against the ratification of the agreement, and his efforts received ample press coverage. For example, on August 15, a long article quoted excerpts from his testimony before the Preparedness Investigating Subcommittee of the Senate Armed Services Committee. Teller listed a whole set of technical and military considerations to show why the treaty would hurt American interests and favor Soviet aspirations.[43]

The Kennedy administration took Teller seriously, and Robert McNamara argued against Teller's points. Glenn T. Seaborg, chairman of the Atomic Energy Commission, found Teller's points of opposition "not important enough to forego the treaty."[44] Just as Teller's testimony was released, an approval of the treaty was published with the signature of thirty-five American Nobel laureates, constituting almost two-thirds of all living American Nobel laureates. The statement of the laureates declared the treaty to be "a concise expression of our country's desire for peace," which was a vague consideration but could count on having a broad appeal.

Of course, there is limited value in such an appeal by Nobel laureates; such an appeal should be considered more of a public relations instrument than a document of scientific merits, but it was impossible to ignore it. We have seen above a similar mobilization of Nobel laureates. Teller might have taken it as a compliment to be opposed by such a collection of distinguished colleagues. Biologist James D. Watson, whose immediate interest was in genetics, considered the danger from fallout a minor reason for his signing the document; his main concern was to avoid nuclear war.[45]

On August 20, 1963, President Kennedy held a news conference and assured the senators that American national interest and preparedness would not be hurt by the treaty. He paid tribute to Teller by singling him out. Referring directly to his opposition, the president said, "It would be very difficult, I think, to satisfy Dr. Teller in this field."[46] He also noted that Teller appeared inflexible in the matter of the treaty. Kennedy did not stop at generalities but gave a list of various measures to maintain existing nuclear laboratories at Livermore and Los Alamos; to advance underground tests; to improve detection systems for monitoring treaty violations; and the like.

Teller did not give up his resistance. He reiterated his opposition to the treaty in a day-long testimony before the Senate Foreign Relations, Armed Services, and Atomic Energy Committees, and urged the senators to reject the treaty. He blamed the president for having limited the tests for political considerations. Kennedy rejected the accusation and stressed that they should never carry out more tests than the minimum necessary to avoid fallout. He cited the ninety-seven underground and thirty-six atmospheric tests performed recently. On the occasion of this Senate testimony, Teller lost his cool and spoke in "fervid, almost perfervid tones."[47] In the same article in which Kennedy and Teller figured, there was mention of two generals who were critical of the treaty. Once again, Teller and the US Air Force were on the same side of the battle.

One of Teller's main objections to the treaty was that it might prevent the United States from using nuclear weapons in case one of its allies was attacked by an aggressor. Former president Eisenhower also expressed his reservation in this connection while giving support to the treaty. He called for ensuring the right of the United States to resort to nuclear weapons to repel aggression. Others argued that such a right was implicit in the treaty, whereas changing its text would meet major obstacles. Former president Truman gave the treaty his unqualified endorsement.[48]

As we have seen, the US Senate ratified the test ban treaty with a great majority in spite of Teller's fierce opposition and the criticism by air force generals. Before Teller answered specific questions from the senators on August 20, 1963, he made a broad-ranging statement. He paid special attention to the harm the test ban would cause to the development of missile defense. He made bombastic statements about it, such as "missile defense may make the difference between the end of national existence and survival as a nation" and that it "may make the difference between peace and war." He also stressed the difficulties for a democracy in intelligence gathering as compared with a police state, such as the Soviet Union. Hence the information about Soviet nuclear preparations was very limited: "What we know firmly is only the great extent of our ignorance," he quipped.[49]

The question arises: How could Teller become such a major force in a debate of national and international importance? He was a recognized scientist, but there were others just as distinguished. He did not occupy any official position that would have justified the attention of the US Senate or even the president of the United States. He had, of course, personal traits that made him quite formidable, such as his unlimited dedication, his consistency, his perseverance, and his ability to attract media attention with his statements and arguments, to mention but a few. We have seen how hopeless his fight for the development for the American hydrogen bomb was initially, and how he carried it to completion. He must have, on some level, relished waging what he felt was the righteous fight against a large and mighty adversary or group of adversaries. Moreover, one of Teller's strengths was his ability to connect defense considerations with support for basic research while emphasizing the importance of a science education. He also had grown to like a good fight. It further helped him that he had no obvious personal gain from the issues he represented and that he could always present his points succinctly, without notes, and in terms that his audience could understand and appreciate.

Soon after the ratification of the Limited Test Ban Treaty, an article in *Science* magazine compiled the main points that Teller used in the debate.[50] Some of them were seemingly not related to the issue of the test ban. It is worthwhile to take a look at them because they appear as relevant almost half a century later as they were at the time. Concerning research, Teller stressed that it is a difficult field to direct because: "[t]o plan it in detail means to emasculate it. . . . Research is a game, led by curiosity, by taste, style, judgment, intangibles. It seems unreasonable to spend a great effort of the best people on play, yet . . . the decisive things are coming from that game, and always have."[51] It was a recurring complaint that he was not satisfied with the status of applied research at the American universities. He advocated the need for more inspiring and better-paid teachers. He lamented the fact that the students were twenty years old by the time they met exciting problems of science. He argued for greater investment in research and development from the federal government and called for a broader geographical distribution of research funds.

Teller received criticism for his misleading statements about fallout and about the necessity of atmospheric tests. The respected columnist Walter

Lippmann dismissed some of his arguments as "a romantic form of self-deception."[52] It was not beneath Teller to warn, cajole, intimidate, and do whatever it took to get his point across. At one of his Senate testimonies, he told the senators that if they ratified the test ban treaty, "you will have given away the future safety of our country and increased the dangers of war."[53] No wonder some officials in the Kennedy administration wished for Teller to disappear for a few weeks during the treaty debate.

During these years there was broad interest in antimatter due to the discovery of new antimatter particles. There was a piece of poetry, "Perils of Modern Living," in the *New Yorker* in 1956 about Teller meeting his antimatter counterpart and the two dissolving in a handshake. It was composed by a fellow physicist and one-time Livermore colleague, Harold P. Furth.[54] About two weeks after the appearance of the magazine issue Teller responded. He presented the difficulties of getting to visit an anti-galaxy where Anti-Teller might live, and described what would occur if Teller and Anti-Teller met. There was a lot of easy-to-digest science applied in his description. He concluded with a witty statement: "I was pleased that the *New Yorker* mentioned me. Come to think of it, only Anti-Teller was mentioned; but I am confident that somewhere in an anti-galaxy, the *Anti-New Yorker* devoted some pleasant lines to Yours sincerely, Edward Teller."[55]

Civil defense and the building of underground shelters were considered to be among the responses to the looming Soviet danger. At some point these activities became a national movement. Teller was very much interested in civil defense—it was, in fact, "one of his manias,"[56] which was also supported by Eugene P. Wigner. In 1979, Teller and Wigner published a joint article in which they expressed their envy of the Russian civil defense preparations and preparedness, and suggested enhanced expenditures for building shelters and perfecting evacuation plans.[57] Teller had already joined up with Willard Libby around 1960, proposing a big shelter program, whereas many of their peers felt that such a program would weaken rather than strengthen the United States. Wigner studied the civil defense of other countries during other wars. He knew about the efficient use of the extensive subway systems in London and Moscow during World War II. He supposed that the tunnels of the Budapest subway were dug so deep underground for civil defense purposes, although at least for some portions this had to be done because the tunnels crossed the Danube under the river bed.[58]

Even as late as 1984 Teller still complained about the absence of efficient civil defense in the United States in case of a nuclear attack. He noted, for example, that the big American cities evacuated to a considerable extent during every holiday weekend, which usually took a mere few hours, without the benefit of careful planning. Thus, Teller argued, it would take very little extra effort to prepare plans for orderly evacuation. He estimated that by employing all measures of civil defense, the number of human lives saved in case of a large-scale nuclear attack could be increased from about forty million to about one hundred and fifty million.[59]

The same day the test ban treaty went into effect, on October 10, 1963, the Nobel Prize Committee of the Norwegian Parliament announced its choice for the 1962 Nobel Peace Prize: Linus Pauling. The presentation speech stated unambiguously that Pauling was awarded for his struggle for the test ban treaty.

Otto Bastiansen, a chemistry professor and well-known public figure in Norway, had studied with Pauling as a postdoctoral associate.[60] He stressed that "Pauling's campaign against testing was the reason for the award." According to Bastiansen, "Pauling's greatest effort was that he calculated the harmful effects of radioactive fallout."[61] Pauling's lecture included a curious quote from Alfred Nobel, who wanted to invent "a substance or a machine with such terrible power of mass destruction that war would thereby be made impossible forever."[62]

## TELLER–SZILARD DEBATES

It was not only Nobel. Others also thought that the more terrible the weapons available to humankind, the less probable it might be for nations to resort to war to settle their disputes. Surprisingly, it happened during a public debate that Szilard brought up the possibility of enveloping the hydrogen bomb with a cobalt layer, thereby enhancing enormously the radioactive fallout from the bomb. It was called residual radiation and it came from the naturally occurring cobalt turning into its deadly isotope cobalt-60 when exposed to the radiation from the hydrogen bomb. Those who knew Szilard were appalled by his even mentioning this terrible possibility.

Teller and Szilard loved debating; they hardly agreed on anything substantive, but they had a unique style, which their audience loved. During a television debate the moderator asked Teller what he would do if he were president of the United States. Teller expressed hope that such a tragedy would never befall him, but should such a disaster happen, the first thing he would do would be to consult Szilard.[63] Teller and Szilard did not give up debating even during Szilard's bout with bladder cancer. In the spring of 1960, a team of television people, led by noted broadcasters Howard K. Smith and Edward

R. Murrow, descended on Szilard's hospital room at the Sloan-Kettering Memorial Cancer Center and taped several sessions of the interview.

Teller was a participant, and the two reminisced and at times hotly argued about issues of disarmament.[64] Teller upset Szilard when he compared his crusade for the hydrogen bomb with Szilard's for the atomic bomb. In his turn, Szilard suggested that they make a joint statue for Teller and Klaus Fuchs for their shared responsibility in both the United States and the Soviet Union having acquired the hydrogen bomb. Szilard, however, was mistaken in this statement, because the Soviet Union would have developed its thermonuclear weapon regardless of Fuchs.

On November 12, 1960, Teller and Szilard met again in front of the television cameras, this time on NBC's television program *The Nation's Future*. The one-hour program had a rigid structure; the opening statements by the two participants were followed by a discussion between themselves. Finally, members of studio audiences in New York, Chicago, and Los Angeles had the opportunity to ask questions. The passages quoted below illustrate the unique style of their exchange as much as its content.[65]

> Teller: . . . Some people want disarmament even if it's unilateral. I admire their courage. I disagree with them. There are others who hope for something and I'm afraid their hopes might be more dangerous because we might accept a package with a false label. The label says controlled arms reduction. The package contains unilateral arms reduction. . . . I'm talking about the real situation of nuclear test cessation where we know today that we cannot check by any technical means whether the Russians have stopped testing or not. . . . If we want disarmament we need two very difficult things. We need an open world. We need a world authority which has moral force and which has physical force. . . .

> Szilard: . . . We were in agreement that the danger was great, but Teller meant this danger is great if the US government should listen to me and I meant the danger was great if the US government should listen to him. . . .

At one point the moderator stopped Szilard's discourse, but Teller praised Szilard's foresight and yielded him some of his own time. This was a component of his style with which he won even his opponents' appreciation. The debate continued:

> Szilard: Let me ask you point blank, Teller, and I would like a yes or no answer. . . . If tomorrow morning the Russians give in on all these points, would you then be satisfied or would you demand that we then fail to ratify the agreement, and start nuclear testing?

Teller: I'm continuing to beat my wife . . . [laughter] . . . I can be quite explicit. . . . If we do that [stop testing], then I will tell you what I think the consequences will be. We will have recognized that for two years we do not know whether we can check nuclear tests. We will start on developing better methods for checking—and this is desirable. At the same time, from what I know, I am bound to tell you that even in two years it is exceedingly unlikely to find reasonable methods of checking underground and interspace tests. And if we, therefore, ratify an agreement of the kind that Szilard is describing—in that case we are embarked, for an indefinite period, on action which is essentially unilateral. . . .

Szilard: Mr. Teller is evading the question.

Teller: Sure, I'm evading the question. I think that you are right. I think that you should sum up and say it is obvious on past performance that we should trust the Russians and it is entirely unnecessary to police anything.

Szilard: He is not only evading the question, he imputes certain thoughts which I don't hold . . . [laughter]. I think, Teller, we should shake hands because maybe later on we don't . . . [laughter]

Teller: Szilard . . . Szilard, I pledge myself to this. That, to me, it will be always a pleasure to shake your hand. And I will make a prediction that our feelings will remain as before.

   . . .

Teller: You see there is just barely the possibility that when I say that I'm ignorant I'm telling the truth . . . [laughter] . . . I know this: if the Russians wanted to test in the last two years they could have done so easily and without our getting the slightest indication of it.

   . . .

Teller: There is a very important area of agreement between us. You are advocating a special way toward openness and every way toward openness, to my mind, is a way toward progress. I would like to make one very short final remark on my side. I do like to say this. There has been a proposal of a very serious reconsideration of disarmament and of the possibilities of peace. I think that this is an urgent and important thing. I think that what we are doing here, on television, should be done on the closed councils immediately throughout the next few months, without losing a day. I believe that these internal discussions should be put before the people as completely as possible. I think completely. And I think such discussions are important. All sides should be heard and we should clarify our desires and the price which we have to pay for peace.

The next day the columnist of the *New York Times* gave high marks to Szilard and Teller, saying that they "stand this morning as fairly formidable heroes of the television age." He gave a telling title to his piece, "TV: Victims of Production Nonsense."[66] He noted that "the nuclear physicists staged a rare revolt against the ludicrous aspects of the program . . . moreover with a combination of sternness and humor, they made it stick." When the moderator stopped Szilard, citing his time limitation, Teller magnanimously offered his own time to let Szilard finish. This gesture threw the moderator off his course, and his attempts to regulate the show according to his rules not only failed but made him appear ridiculous. Whereas the moderator was no match for the occasion, Szilard and Teller made the most of it almost at the expense of the contents, but they showed that even serious issues can be discussed without jumping at each other's throat. The report about the debate ended with this lesson: "able minds do not need the embellishment of the theatrical gimmicks."[67]

Two further debates, on June 3 and 10, 1962, took place on a CBS television program.[68] Each debate concerned the dangers of wars and how to avoid them; how the existence of nuclear bombs changed the nature and aims of wars; how to seek agreement between the United States and the Soviet Union; and how to seek agreement between Teller and Szilard. Again, the few excerpts quoted below are provided as much for the flavor of the debaters as for their contents.

Excerpts from the June 3 debate:

> Szilard: . . . the Russians want disarmament very much and that they want it mainly because of the economic saving in which it would result and which they need very badly. . . . Disarmament would not offer America any guarantee of peace and some mechanism must exist for preserving the peace in a disarmed world. . . .

> Teller: . . . Russia would like to see that we disarm. I did not see any signs that Russia, on her part, wants to disarm, or is willing to pay the necessary price for disarmament, which is openness, . . . Americans do not at this point know whether they do want disarmament and the reason they don't know that is because I don't know how the peace could be secured in a disarmed world. . . .

At a later point in the discussion, Teller suggested that rather than always expect the Americans to come up with proposals that might be acceptable to the Russians, the burden should be shifted to the Russians. Szilard, in his turn, explained that the burden should rest with the American government because the American citizens can influence what their government does, but the American citizens have no influence over the Russian government. (Szilard did not waste time explaining that the Russian people could have not influenced the Russian government.)

In the June 10 debate, the two men discussed, among other things, the question of how to limit a possible war involving atomic bombs. Szilard suggested that dropping atomic bombs should be limited by each side only within the lines that existed before the hostilities started, and within each belligerent's own limits. He expected from this limitation to favor defense rather than offense. Teller countered this by pointing out that this limitation would encourage the communist aggressors to try conquest, which may or may not succeed without risking too much for the aggressor.

> Teller: . . . [T]he Russians, while they want to conquer the world, are cautious. They don't want to take chances, and if they know that force will be met by force on our part in such a way that the consequences might be a real loss to themselves, even a loss in some of the territory and people which they control at the present time, I think this in itself will be a stabilizing factor, . . .

Szilard did not think that the Russians were primarily interested in territorial gains, rather, they were thinking in political terms.

Then, they discussed the desirability and possibility of forming a closer union among the nations of the West, such as the United States, Europe, Australia, and Japan, but Szilard observed—apparently with great foresight—that, for example, the Europeans were aiming at a tighter integration among themselves. Szilard thought that "[d]isarmament is not impeded by those that oppose it, it is impeded by the fact that so few people in America in responsible positions are wholeheartedly in favor of it." At some point, Teller hinted at Szilard's petition and he said something that came out as if he was rewriting his own history.

> Teller: You see, Szilard, I am in a very peculiar position, in an unusual one. There have been many cases where I listened to you with some doubt, because I thought that what you were proposing is not quite feasible. I listened to you, frequently with complete agreement. For instance, at the time when you proposed that we should not drop the bomb on Hiroshima, on Japan, I agreed with you completely.

Unfortunately, this was not true at the time; this became Teller's interpretation of the events with the benefit of hindsight. We have seen the fate of Szilard's July 1945 petition in Teller's hands at Los Alamos and his note to Szilard explaining why he would not sign it (chapter 4).

The Teller–Szilard debates were witty and substantive, and their light tone was well received by their audience. Teller's attitude toward Szilard in the debates and in general was always friendly and appreciative. He did not make derogatory statements about him as, for example, he would about his other partner in debate, Linus Pauling, with whom he was also polite and deferential during the debate itself. At the heart of the disagreement between Teller and Szilard was the irreconcilable difference in how they looked at the Soviet Union.

Szilard gave the benefit of doubt to the Soviets whereas Teller distrusted them deeply. It is curious that Szilard believed that if Teller only visited the Soviet Union he would change his mind and would trust the Soviets more. Szilard's visits and interactions with Soviet leaders should not have made him so trustful, but apparently he did not lose faith in them in spite of the fact that he knew he would not survive living under a Soviet regime. In August 1963, a mere few months before he died, Szilard drafted a statement about Teller that focused on Teller's unwillingness to visit the Soviet Union and Teller's refusal to give the Soviets the benefit of doubt. When Szilard died in 1964, it was a blow to Teller. He knew he lost his truest friend.

## "RELUCTANT REVOLUTIONARY"

Teller carried out a heavy lecturing and book writing program as part of his multifaceted life, dedicated to his self-imposed goals as defender of freedom. We single out here a lecture series and the little book that came out of it.[69] His approach is succinctly illustrated by this booklet. The timing of the lectures and the appearance of the booklet sandwiched the ratification of the test ban treaty between them.

He gave a set of three lectures at the University of Missouri in St. Louis, on April 8 and 9, 1963, and at the beginning of 1964, Teller revised the text for publication. The organizer of the lectures mentioned in the foreword "two important changes in the world situation during this intervening period—the weakening of the cold war and the adoption of the test ban treaty."[70] In spite of these changes, he noted, much of what Teller said in his lectures remained applicable, and the texts of the lectures did not need much changing. This might also be interpreted to mean that Teller's views were not altered by the events.

The lectures were closely related to recent events, and if Teller did not change their text that might be an indication of his clinging to his views before the test ban treaty and before "the weakening of the cold war." There were other events in between the lectures and the book. One was the assassination of President Kennedy, and Lyndon B. Johnson's accession to the presidency on November 22, 1963. The other was J. Robert Oppenheimer's Fermi Prize awarded on December 2 of the same year.

The book is not great literature but a political masterpiece for its accessibility, for its economy, and for the way it unobtrusively communicates the culture of its author. At the same time Teller does not seem to be able to tear himself away from his recent political motivation. The atmosphere of the first dozen pages of the first part, "The Noble Lie," is exalted, and the discussion appears to have universal validity. Teller examines the responsibility of the scientist, one of his favorite topics over the years. Once again, he comes to the conclusion that the scientist is expected to enhance knowledge, make discoveries, and educate the public about them. However, he contends that the scientist must not engage in recommending applications of the new knowledge and discoveries; rather he should let the public make its decisions about them.

Here again we see a sharp discrepancy between his assessing the scientist's tasks and his own activities in the test ban debate. In his discussion Teller even mentioned that "scientists have all too frequently given wrong advice in the past."[71] The impression is that he did not count himself among those to whom he referred. Teller cited Plato's *Republic*, and soon enough we see the purpose of bringing in Plato, according to whom "To the rulers of the state . . . it belongs of right to use falsehood, to deceive either enemies or their own citizens, for the good of the state."[72] When the lie is so effective that you believe it yourself, it becomes "the noble lie."[73]

In the second part, "Doomsday," Teller noted that the possibility of isolation for the United States and the safety associated with it disappeared after 1945. What Teller wrote in 1964 about the situation after 1945 is chilling in light of the terror attacks of 2001:

> . . . [O]ur frontiers are open to aggression. Past generations were sheltered; we find ourselves exposed in a hazardous world, and there is among us a great feeling of anxiety. . . . It is disturbing . . . and may turn into unconstructive channels. The fear of a catastrophe is particularly disturbing because what we are afraid of is something without precedent, is something that we cannot imagine and something against which for these very reasons it is difficult to make realistic preparations.[74]

At one point Teller raises the question of whether humankind must choose between being "Red or dead." This question resonates with another question asked two decades before when the dilemma arose as to whether to go ahead with the atomic bomb, even if it might ignite the atmosphere and the oceans and thereby draw the curtain on the existence of humans on Earth. Teller argued that this was not a realistic question, and he concentrated on the dangers from fallout, which his opponents, many of the proponents of the test ban, liked to cite, and which he, once again, found negligible.

He then went to the extreme to prove his case. Suppose that the fallout would be ten thousand times as much as before, which would mean nuclear explosions of the equivalent of several millions megatons of TNT. To get an idea of this amount, imagine a Hiroshima-type explosion on every square mile of all lands and seas, everywhere. Even 1 percent of this amount would suffice to destroy the United States, but even this would not mean the end of all humanity. Compared with this, fallout from testing is so negligible that it is even less than the natural radioactive background with which the human race has lived from times immemorial.

Teller also discussed missile defense and how nuclear explosives could be used to shoot down incoming missiles. This issue would return in the 1980s in much greater detail. A considerable chunk of this part of the booklet was devoted to civil defense.

Again he emphasized that he considered the Soviet leaders and communists more dangerous for the democracies than Hitler and his Nazis. They were madmen, whereas the communists were cautious and made consistent and realistic plans to conquer the world. According to Lenin, when facing difficulties, it was preferable to retreat one step before moving forward two. Mao Tse-tung (Mao Zedong) declared that only a fool would fight against odds, not a revolutionary. Hence, Teller's conclusion: Only a strong country could avert the communists from attack.

Teller considered the future in the third part of his booklet. He predicted that by the end of the twentieth century it would be possible to control the weather. He also called for better utilization of the oceans for food supply. Then he turned to the hydrogen bomb, which he called the best instrument for Project Plowshare—the peaceful utilization of nuclear explosions (see chapter 10). He concluded with a prediction that before the end of the twentieth century, law and order on an international scale would be established.

However, Teller was afraid that the strict world order might be established on communist terms. Compared with past empires, he thought that current technologies called for a yet-larger, all-encompassing empire and he considered the Russian communists as standing ready to act. In contrast, he found the United States vastly lacking preparedness. He then shared his experience with Russian scientists as the people who might facilitate the creation of freedom in Russia. Teller knew that even though the freedom of discussing scientific issues had been recently enhanced in the Soviet Union, still there was no political freedom there. His Russian colleagues, he noted, could not even fathom that Teller and others had such a freedom.

Teller's tour de force is the point at which he finds parallels between Socrates and Khrushchev in their views on artistic expressions. Socrates wanted to shun a new kind of music more than two thousand years ago "since styles of music are never disturbed without affecting the most important political institutions." Khrushchev condemned the avant garde paintings and sculptures in his impromptu and rude criticism at an exhibition in Moscow in 1962. He had described the exhibits as "dog shit" because they did not conform to the style of "socialist realism" as prescribed by the Communist Party.

Teller painted a gloomy picture of the world should the communists accomplish their goal—by force, of course—of universal domination. In contrast, democracy could be established all over the world by negotiations and peaceful means, but in this Teller saw little hope. It could not be done by a safe and slow process of evolution, only by revolution, and this is why he called himself a revolutionary, albeit a reluctant one.

## AGAINST OTHER TREATIES

There were two main disadvantages of the test ban from Teller's point of view. One was the loss of information and data that the American tests provided for their further work. The other was the loss of information and data that the analyses of air samples might have provided about the Soviet explosions. The test ban, however, was not aimed at reducing the quantity and strength of weapons. Negotiations on limiting strategic arms eventually led to the first and later to the second SALT—Strategic Arms Limitation Treaty, I and II, respectively. They limited the growth of missile arsenals in both superpowers. Teller opposed these treaties and many others. He kept reminding the Western democracies that Hitler had gradually made many conquests through treaties before he had to resort to armed aggression.

The first SALT, which was not only signed but was also ratified by the United States, limited the systems capable of delivering nuclear weapons. There could not be a limitation on the nuclear weapons themselves because there was no way to determine their numbers in the Soviet Union. SALT II made a first step in the direction of reducing arms; it was signed by President Carter and Soviet leader Leonid Brezhnev in Vienna on June 18, 1979. Brezhnev looked ill and disoriented, and it was hard to imagine that he understood what he was signing. It was an unreal scene as Brezhnev hugged Carter after having signed the documents. The United States never ratified SALT II, although it observed its provisions.[75]

Teller continued to clash with his peers about these treaties, as is illustrated in his debate with Wolfgang Panofsky, who was not only an important physicist but also director of the Stanford Linear Accelerator Center. He also served at various times as member of influential committees advising American presidents and administrations. The Panofsky–Teller debate was conducted in the pages of *Physics Today*.[76] Panofsky stressed in his introduction the broad involvement of many American institutions in drafting the SALT II agreement and that it had to be a compromise between the two superpowers. He then listed and discussed the achievements of the treaty. For Teller, beyond various specific points in SALT II, he was generally opposed to entering into any agreement with the Soviet Union concerning weapons and their developments. He was apprehensive of the weakening American positions, knowing that in order to maintain peace the United States had to be strong.

One of Teller's recurring topics was to warn of the dangers of the treaty—that it would mislead the American people about the necessity of remaining vigilant in matters of defense. He denied that the consequences of an arms race would have to be disastrous. He used one of his favorite arguments about the roots of World War I and World War II. According to him, an arms race among hostile countries resulted in war in the first case, whereas the tendency for the Western democracies to disarm was a decisive ingredient for the second. Teller enumerated a set of developments that he advocated to follow in place of a SALT-type agreement. It included increased attention to civil defense; strengthened alliance with the other democracies; enhanced use of computerization in which the West had gained an advantage; and modernization of the three basic components of American defense: bombers, nuclear submarines, and land-based missiles.

Another component of Teller's argument was his pointing out the fundamental difference between the social orders in the United States and the Soviet Union; the former being an open society and the latter a closed one.

He stressed that the Soviet Union had a clearly stated program to dominate the world, which in 1979 might have sounded a little dated and oversimplified. Panofsky pointed out that Teller's comments expressed his general opinion and were not specifically related to SALT II. He understood that Teller would consider any arms limitation agreement dangerous for American defense because it would lead to neglecting it. Panofsky shared Teller's observation that the Soviet Union had become stronger during the previous two decades, but he did not ascribe this change to arms limitation treaties. On the contrary, he expected enhanced security for the United States from SALT II and similar treaties. The two physicists went into a detailed discussion of the various points; the bulk of their debate was factual, but for Teller, his real territory was live debate.

After World War II, Albert Einstein called atomic energy a menace, but he did see its positive side, noting: "It may intimidate the human race into bringing order into its international affairs."[77] The policy of Mutually Assured Destruction, appropriately abbreviated as MAD, began taking shape many years later. Its premise was that no nuclear power would attack another because the retaliation would be just as devastating as the initial attack. The policy of MAD thus assumed that there would be no protection against a nuclear attack. Nonetheless, a lot of research and effort went into developing missile systems to destroy incoming intercontinental ballistic missiles (ICBM). An antiballistic missile (ABM) system was created in the United States, and similar developments took place in the Soviet Union. The antiballistic missiles were supposed to destroy the incoming missiles, even those carrying warheads with megaton-size nuclear explosives, at several miles height. They were supposed to be effective merely by using kiloton-size warheads. The explosion at sufficient height ensured the minimization of harmful effects on the ground.

The SALT I agreement was not only about limiting strategic offensive arms but also restricting strategic defensive missile systems. Both the United States and the Soviet Union were allowed to develop defensive systems for two sites, one for the capital and one for the silos used for launching intercontinental ballistic missiles. As we saw in the Panofsky–Teller debate, Teller opposed the ratification of SALT I. However, according to Teller, Henry Kissinger—then President Nixon's national security adviser—convincingly argued in 1972 that ratifying the treaty would be reasonable because regardless of whether there was or was not a treaty, there would be no money to expand the antiballistic missile system.[78]

The antiballistic missile systems, however, lost much of their efficacy with the appearance of further "improved" ICBMs that carried several warheads simultaneously. This was the Multiple Independently Targetable Reentry Vehicle (MIRV) system. This deficiency of the ABM system against the ICBMs made its further development yet more uneconomical. So there was little opposition to the protocol in 1974, which reduced the two sites defendable by ABMs in each country to one site. This one site was the Grand Forks Air Force Base in North Dakota in the United States, and Moscow in the Soviet Union.

Teller was unhappy about these developments, and he was further discouraged by the decommissioning of the North Dakota site as a consequence of budget constraints. Originally, Teller would have preferred to develop ABM protection for cities, whereas other scientists, Panofsky among them, opposed it.[79] It was conspicuous, and so noted by senators during ABM deliberations, that Teller was the only well-known scientist supporting the development of the ABM system.[80] Failing to develop defense means, and the elimination of existing ones, however ineffectual they might have been, gave further importance to the policy of MAD. However abhorrent it appeared, it helped maintain peace between the superpowers for decades.

Teller remained unhappy about the American–Soviet treaties and complained about massive violations primarily amounting to the Soviets improving their armaments. It takes time to read the actual texts of the treaties—written in legalese and specialized technical military language—to appreciate how difficult it was to understand what was and what was not permissible under their provisions. Teller saw the various treaties with the Soviet Union as barriers to existing weapons developments and to developments not yet envisioned for the future.[81]

## NUCLEAR WINTER

Teller was fascinated by the history of arms and war and at one time contemplated writing a monograph about it.[82] This interest may have brought him to the story about the invasion of Persia by Genghis Khan's armies in 1219, when their goal was to destroy all Persians. First they killed everybody they could find when they took a city, then they returned in a few days' time and killed those who had hidden during the first massacre. In spite of this careful design of annihilation, about 10 percent of the Persians survived, according to Teller.[83]

He was interested not only in man-made calamities but natural ones as well. He found it remarkable that humankind survived the black plague, which at times attacked people in subsequent waves. Teller's fascination with survival in general was a useful exercise for him for the debates that revolved around the question of whether humankind could survive a nuclear war. He was criticized for considering mere survival and ignoring the loss of the quality of life.[84] It is a bizarre consideration to discuss whether people might still be left alive in the Southern Hemisphere following a devastating nuclear war in the Northern Hemisphere.

Albert Einstein was repeatedly asked about the consequences of a possible nuclear war. In 1945, he did not think that it would be the end of civilization: "Perhaps two-thirds of the people on earth would be killed, but enough men capable of thinking, and enough books, would be left to start out again, and civilization would be restored."[85] In contrast, in 1949, he said, "I do not know how the Third World War will be fought, but I can tell you what they will use in the Fourth—rocks!"[86] Right after President Truman's decision about the hydrogen bomb, Einstein said, "If it is successful, radioactive poisoning of the atmosphere and hence annihilation of any life on earth will have been brought within the range of what is technically possible."[87]

The year 1983 was one of the lowest moments in late cold war history. President Ronald Reagan gave a speech to the National Association of Evangelicals in Orlando, Florida, on March 8, 1983. He urged his audience "to speak out against those who would place the United States in a position of military and mortal inferiority." Further, he urged his audience "to beware the temptation of pride—the temptation . . . [to] label both sides equally at fault, to ignore the facts of history and the aggressive impulses of an evil empire, to simply call the arms race a giant misunderstanding." He went on to ask his audience not to remove itself "from the struggle between right and wrong and good and evil."[88] This speech has become known as the "Evil Empire Speech."

Later in the same year, on November 20, ABC Television broadcast the movie *The Day After*, about the consequences of a nearby nuclear attack in Lawrence, Kansas. The movie was filmed to seem like a documentary and it shook the American audience despite its familiarity with television violence. My background made me feel more optimistic than my friends because there *was* a day after, in the first place, and the backbone of society did not appear broken. The film showed physical suffering and material loss to be sure, but missing was the psychic suffering. There was no occupying foreign army; no humiliation of the people; no uncertain fates; no torture; no terror of the mind; people were not induced to report on each other; there was no invading army raping women and girls; no hunting for young people to take them away, maybe forever, for forced labor; there were no medical experimentations on humans; all that was shown fell short of the total potential of such a disaster to inflict human suffering.[89] The consequences of the Soviet nuclear attack might have been those of a natural catastrophe and, in this, the movie belittled what a real war could cause. In view of considerable criticism of ABC for having shown such a terrible scenario, I felt silly because I did not find the story sufficiently frightening.

The concept of "nuclear winter" was advanced at about the same time. The term encompassed the climatic catastrophe that would result as a consequence of the soot and dust cooling and darkening the earth in the event of a nuclear war. A group of scientists published a report stating that the consequences of a nuclear war could be interpreted in its apocalyptic effects as being similar to the consequences of the asteroid collision sixty-five million years before.[90] Carl Sagan, well-known astronomer and author and, at that time, a Cornell University astrophysicist, brought together a coalition of scientists, associations, and foundations, and, using commercial advertising techniques, launched an attack on the nuclear arms race for disarmament. He published a long, persuasive article in *Foreign Affairs* about the climatic catastrophe of a nuclear war.[91] The article began with a Teller quotation: "It is not even impossible to imagine that the effects of an atomic war fought with greatly perfected weapons and pushed by the utmost determination will endanger the survival of man."[92]

Teller was inevitably drawn into the discussions, which he waged with obvious zeal. In this case, he could claim that his opponents exaggerated. In a long and detailed report solely under his name in the prestigious British magazine *Nature*, he presented his doubts about Sagan's "nuclear winter."[93] The paper was factual and well researched. His arguments focused on the possible climatic changes due to smoke and dust. He did not say much about fallout and ozone depletion, as they had been discussed already. Furthermore, he stressed the importance of informing the public about what was known and what was not known, and asserted: "Highly speculative theories of worldwide destruction—even the end of life on Earth—used as a call for a particular kind of political action serve neither the good reputation of science nor dispassionate political thought."

Sagan and Teller participated in a live debate on May 16, 1984; officially it was called a dialogue. Two congressmen acted as organizers. One was Newt Gingrich, a well-known Republican conservative and later Speaker of the House of Representatives. The other was Democrat Timothy E. Wirth from Colorado, later senator from Colorado, who worked with Al Gore on environmental issues. The dialogue took place on Capitol Hill, with no press present. It was an additional challenge for the seventy-six-year-old Teller that he had just undergone coronary bypass surgery. His partner in the dialogue was the fifty-year-old astrophysicist Sagan, who successfully popularized science and was known to hundreds of millions worldwide for his informative and accessible television programs. Each reiterated their points that had already been published. In typical Teller style, he said, "I am not telling you that Dr. Sagan is wrong in asserting that the possibility of nuclear winter should be taken seriously. He is wrong in saying that his assertions have been proven."[94] As the year was 1984, the dialogue soon turned toward various issues of the Strategic Defense Initiative (see chapter 11).

Teller received a lot of support at Livermore in his struggle, yet he was a lonely fighter, as no well-known scientists had joined him. There was sympathetic press coverage of his actions, noting possible deficiencies of some of Sagan's claims, but that did not substitute for support from Teller's peers. Russell Seitz's *Wall Street Journal* piece was sadly characteristic of the attitude surrounding the debate: "Many scientists were reluctant to speak out" in Teller's favor, "perhaps for fear of being denounced as reactionaries or closet Strangeloves." He mentions Freeman Dyson by name; Dyson called the publication of Sagan's group in *Science* "an absolutely atrocious piece of science" but also said "I quite despair of setting the public record straight. . . . Who wants to be accused of being in favor of nuclear war?"[95] When in 2008 I asked Dyson about the debate on nuclear winter, and why Teller was left virtually alone in that debate, he said, "It was a conspiracy of silence."[96]

It did not help that Teller had made belittling predictions as far as fallout from nuclear explosions was concerned. In the present debate, Teller sometimes referred to cleaner future weapons rather than to the currently existing ones. On the other hand, Teller realistically emphasized that the main dangers were "the direct and the intended effects of nuclear war, including the local effects of prompt radiation, while Sagan [concentrated] on side effects."[97] The direct interaction with a scientist, even if it was a collision, brought out a more scientific Teller than in his interactions with generals and politicians. Sagan and his associates continued their research and published yet another long and detailed paper in *Science* in 1990.[98] However, whereas their first *Science* paper appeared at the height of the late cold war period, the second one coincided with the great political changes and the demise of the Soviet Union. The interest in the vision of nuclear holocaust was diminishing.

It is worth noting, however, that regardless of the outcome of the Teller–Sagan debate, the horrors of a possible nuclear calamity should not be dismissed. Apparently, there is no limit to the degree of survival that people might still consider to be better than nothing. Swiss biologist and Nobel laureate Werner Arber formulated his non-anthropocentric view about the survival of life on the earth:

> I'm convinced that if living conditions would change so drastically that human life and the lives of some other highly developed organisms would no longer be feasible, life would still continue for a very long time because other organisms could exist under very different conditions. That makes me an optimist. I should add that I am not so much anthropocentric to say that for me only the human life form is what counts. For me, life as such is more important than the specific existence of human life. Therefore I have a big hope for the continuation of life on our planet as long as some living conditions continue to exist. We are just one life form among many and if some other organisms would survive, even if they would be very simple organisms, they could have a chance to develop into higher complexity again by the given means of biological evolution.[99]

Arber's considerations are from the early 2000s. Interestingly, as early as 1949 Teller made a suggestion to the Ford Foundation that discussed support for sending a rocket to Mars. The rocket would be loaded with algae, bacteria, and protozoa "in case an atomic war terminates life on earth."[100] But in 1961, Teller belittled the possible consequences of a nuclear war and appeared more worried about the possibility of a test ban treaty.[101]

# Chapter 10

# "A MONOMANIAC
WITH MANY MANIAS"

If you want peace—be strong
—TITLE OF BOOK PUBLISHED AT THE
RUSSIAN NUCLEAR WEAPONS INSTALLATION[1]

*In addition to his opposition to the test ban treaty, Teller was occupied with many other activities and fights from 1954 to 1983, earning him the nickname that titles this chapter—a nickname bestowed upon him by Enrico Fermi. He gave the impression that he was the self-appointed guardian of freedom and seemed to thrive on controversies. During this time, Livermore grew into a powerhouse for the national defense of the United States. The Plowshare program was meant to utilize nuclear explosions for constructive rather than destructive purposes, but it was not acceptable to the public. Teller had clashes with the environmentalists. He became engulfed in partisan politics and became a member of the political-military establishment. But he also initiated successful educational programs in science applications.*

Teller's third exile, from the community of theoretical physicists, began in 1954, and it was vastly different from the previous two. Teller wrote of his second exile, from Germany to the United States: "In my new land, everything had been unfamiliar except for the community of theoretical physicists." By the time the third exile came, "I was passably acquainted with my adopted country's customs, language, and attitudes. But the community of my fellow scientists was the only place that afforded me complete comfort . . ."[2] Contrast this with what he wrote a few years later to Maria Goeppert Mayer, saying that he had "43 minutes per week for physics,"[3] and "I am no longer a physicist."[4]

By the mid-1950s, important changes overtook the world. In the spring of 1953 alone, Stalin died; Elizabeth II of the United Kingdom was crowned; and two heretofore unknown scientists, James D. Watson and Francis Crick, discovered the double-helix structure of the substance of heredity. Fifty years later, in Teller's penultimate letter to me in August 2003, he mused about the greatest successes in science. He wrote that for him "it is important that the same four letters describe the DNA of all living creatures. This may bring us closer to the understanding of what life is" (see preface). He did not mention in this letter or in his *Memoirs* that, however superficially, in 1954 he was involved in the quest for the genetic code that followed the discovery of the DNA structure. It must have been a pleasant diversion from his otherwise troubled life.

## CODING EXCURSION

Watson and Crick published a one-page letter in *Nature* on April 25, 1953, in which they proposed a double-helix structure for the DNA molecule.[5] Their model proposed a copying mechanism for the genetic material, and thus the mechanism of heredity could be understood, at least qualitatively. Soon, Watson and Crick further elaborated the genetic implications of the double helix; their second paper was published on May 30.[6] A few weeks later they received a letter from George Gamow raising the question of information transfer from nucleic acids to proteins, the essence of the genetic code (although this expression was not used at that time). The question was, How do proteins come into existence on the basis of the genetic information stored in DNA molecules? Gamow did not solve the genetic code, but he raised crucial questions, pushing the researchers in the direction of looking for answers.

The interaction between Gamow and Crick and Watson led to a playful organization called the RNA Tie Club, "RNA" being the abbreviation for ribonucleic acid. There were to be twenty regular members, each assigned to one of the twenty naturally occurring amino acids (all of which had been identified), and four honorary members, each assigned to one of the four bases of DNA. Each member of the club was to have a customized tie and tie pin. Gamow chose all the members of the club.[7] In a nice gesture he selected Teller for the club, and this is how Teller's name appeared among the charter members of the club. Gamow did not consider Teller's exclusion from the community of scientists to be justified. He felt "Teller had the right to say what he believed about Oppenheimer." When Watson questioned Gamow about this, Gamow explained to him that "politics was dirty and nothing could be done differently."[8]

The RNA Tie Club was much more than a childish diversion. The members exchanged ideas that might have been premature for research papers, but they stimulated further work. The activities of the club peaked in the second half of 1954 and during 1955 then fizzled out. In Gamow's November 25, 1954, chain letter to the members, Teller was listed as number one, and leucine (LEU) was to be his amino acid.[9] Great names such as Richard Feynman, Melvin Calvin, Erwin Chargaff, Nicholas Metropolis, Max Delbrück, Sidney Brenner, Fritz Lipmann, and Albert Szent-Györgyi were among the members in addition to Gamow, Crick, and Watson. The world of DNA–RNA scientists was small at the time. It was not difficult to squeeze in almost anybody who was interested in the topic and still include a few "nonbiologist" friends, such as Richard Feynman, among them.

In the spring of 1954, when the Oppenheimer hearing was under way, Teller suggested a code in which each amino acid was defined by two bases and the previous amino acid. This code had the advantage of possessing a direction. Gamow nicknamed Teller's model the Russian-bath code and made a drawing to illustrate it.[10] It was also called a stepwise code. In Gamow's drawing, the code was represented by a stack of blocks; each block corresponded to a base in the nucleic acid; a man—an amino acid— was seated on each base, his back against a base behind him and his legs resting over the shoulders of the next man beneath.[11] Teller's code used the idea of the DNA molecule serving as a physical template, and others, including Gamow, made similar suggestions.

By all indications Teller forgot his brief excursion into biology. Not only did he not mention it in his *Memoirs*, but when an Australian scientist sent a query about it in the 1990s, he responded in the negative.[12] But his suggestion for a genetic code did not completely disappear into oblivion. Soviet nuclear physicist Igor Tamm in a lecture on the molecular mechanism of heredity at the University of Leningrad in 1957 mentioned Teller's unsuccessful attempt at creating a code.[13] Incidentally, Teller gave a talk on the relationship of quantum mechanics and biology at the Department of Biophysics of the Medical Center, University of Colorado in 1949.[14] His brief comment about DNA in his penultimate message to me also showed a deep-rooted interest in understanding the science of life (see preface).

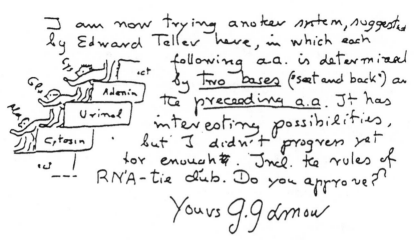

George Gamow's illustration of Teller's sequential model for the genetic code, in a letter dated May 27, 1954, from Gamow to Yčas. Courtesy of Igor Gamow, Boulder, CO.

## TASTE OF BIDDING

Teller's isolation following the Oppenheimer hearing was compounded by the death of three cherished friends. Enrico Fermi died in 1954. Teller had been fond of him and looked up to him in all matters related to physics. Fermi, in his turn, appreciated Teller and ignored what made Teller unpopular in others' eyes. John von Neumann was the first of the Martians to pass away, in 1957. Though he and Teller were friends, they were not very close and did not share jokes like von Neumann did with Ulam. Yet Teller could be sure of von Neumann's good will toward him and the soundness of his professional advice, which von Neumann was always ready to offer.

Ernest Lawrence, who died in 1958, had one of the most complicated relationships with Teller. Theirs was not a friendship, rather, it was a comradeship; they fought for common goals. Their relationship began with the development of the hydrogen bomb and culminated in the creation of Livermore. There was at times a one-sided nature to their interactions. Teller needed Lawrence more than Lawrence needed him. Lawrence was more secure and he was a better politician because he managed to dampen the negative sentiments of his political opponents, whereas Teller would often sharpen them.

Teller was fully devoted to building up the Livermore Laboratory among whose youthful population he considered himself an anomaly. He had dreamed about bringing together some famous physicists at Livermore in addition to the fresh, talented PhDs who came mainly from California schools. This did not happen, and he remained the sole truly famous physicist of Livermore for a long time, developing into a father figure and spokesman for the laboratory. Livermore soon matured into one of the most advanced research places in America for its computerization, thanks mainly to von Neumann, who repeated for Livermore what he had done for Los Alamos. His work was influential more through his visits than his becoming a member of the laboratory, but he operated the same way with Los Alamos. He was much concerned lest the Soviets, who were advanced in rocketry, combine nuclear warheads with missiles and thus overtake the United States in this most strategic area of military matters.

Teller's concern about Soviet expansionism and his responsibility toward keeping Livermore busy reinforced each other. He considered it his job to help the laboratory secure sufficient numbers of contracts from the defense segment. He developed strong bidding abilities and tactics. The bidding battle for the nuclear warheads small enough to be carried on missiles and fired from submarines was an early example of his projects. By the end of the 1950s, there was no question about nuclear-powered submarines—a far cry from the time when Fermi first contacted naval officers in 1939 and was unable to convince them of the feasibility and utility of nuclear power.

The US Navy started working on nuclear-powered submarines in 1946 and launched the first one, the USS *Nautilus*, in 1955. Admiral Hyman Rickover, an engineer by training, was their champion. The nuclear-powered submarines were the least vulnerable to a first strike from the Soviet Union. In 1960, Teller accompanied Rickover for an overnight cruise on a nuclear submarine, the *Patrick Henry*, which was the US Navy's second missile launcher.[15] A few years later, in 1964, Rickover was to be the first nonscientist recipient of the Fermi Prize. There were similarities between Teller and Rickover not only in their dedication to defense matters but in the fact that they both went outside of their original circle of colleagues to win over congressional support for their schemes.[16]

The next question to arise was about the kind of arms the nuclear-powered submarines should be equipped with. In 1956, the navy sponsored a study by the National Academy of Sciences on anti-submarine warfare at Nobska Point in Woods Hole, Massachusetts, hence the name Project Nobska. The navy's intention was to have a new missile developed that would be lighter than the existing missiles and that would cover a range up to fifteen hundred miles. The problem was that such a missile would not be able to carry the desired one-megaton thermonuclear warhead. Teller attended the Nobska meeting from Livermore along with Carson Mark, who came from Los Alamos. Teller was already known as a nuclear salesman, but this was the first big bidding battle in which he outbid his Los Alamos counterpart. They knew each other well: Mark was named head of the theoretical division of Los Alamos in 1947, a job Bradbury had originally offered to Teller. Mark was a cautious physicist and no match for Teller in a bidding war.[17]

Teller offered to develop a lightweight warhead of one-megaton strength within the prescribed five years. Mark thought that a warhead of half a megaton would be more realistic and he quoted a higher price and a longer deadline. Paradoxically, the navy considered Mark's cautious approach as evidence for the feasibility of Teller's offer. Livermore got the project, which was named the Polaris program. Almost four decades later Teller said, referring to Mark's performance, that it was "an occasion when I was happy about the other person being bashful."[18]

When Teller got back to Livermore and told the people who were supposed to make good on his offer, they were astonished by the boldness of Teller's promise. However, they proved Teller almost right when the Polaris submarine USS *Ethan Allen* fired a successful test of the Polaris A-1 missile system in the spring of 1962. The Livermore-designed megaton-class warhead exploded as it was meant to. Teller deliberated about the success of his bidding and used this experience in furthering his techniques.[19] He thought that Mark committed a grave error by offering an estimate that was more conservative than Teller's and that ended up emphasizing the general feasibility of the project. Teller thought that Mark would have been more successful if he had instead declared the project impossible or suggested further study—or declared Teller's offer exaggerated tenfold.

Teller's interactions with the US Navy were not limited to submarines. He spent some time, along with Lawrence, on the flagship of the American Mediterranean Fleet. His invitation was a response to his having expressed his doubts about the efficacy of aircraft carriers. He even got involved in some sea games that demonstrated the efficiency of the carrier. On this occasion, Teller noticed that most servicemen did not speak foreign languages and suggested to his host that learning foreign languages might enhance their effectiveness. His suggestion was brushed off, and the admiral even teased him by calling him a Don Quixote, but Teller took this with good grace.

## TRIGA

After Teller's subcommittee on reactor safety had been dissolved, he continued to serve on the replacement Advisory Committee for Reactor Safety (ACRS). In 1955, he resigned from this committee, citing lack of time to do a proper job. Once he was free from those obligations, he became involved in consulting for private companies. Even before this, he noted that a good consequence of the publicity he received following the Oppenheimer hearings was being offered an increasing number of contracts for consulting.[20] Once he told a colleague who had been lamenting the low salaries of university professors that "your salary is irrelevant, it's your consulting that will give you your income."[21] His participation in one of these ventures was especially noteworthy.

Frederick de Hoffmann, who was his closest associate at Los Alamos during the quest for the hydrogen bomb, had also left Livermore. He went on to spend some time at the General Dynamics Corporation in San Diego, which was concerned with peaceful uses of atomic energy. This company had a very strong advisory board, and at least on one occasion, in 1953, the attendees of its meeting included Theodore von Kármán, John A. Wheeler, Edward Teller, Eugene P. Wigner, George Gamow, and Richard L. Garwin, among others. After a while, de Hoffmann wanted to strike out on his own. He had been characterized as "an able and shrewd scientific aide [to Teller] of high managerial and political ability."[22] He founded a new company called General Atomic to build nuclear reactors for the open market. The first and only successful product of the new company was based on Teller's suggestion in 1956 for a small and "foolproof" reactor to be used primarily by hospitals and universities. The first reactor was put into operation in 1959.

The reactor was named TRIGA, which stands for Training, Research and Isotopes, General Atomic. Its principal goal was to provide short-lived radioactive isotopes for research and diagnosis. A short lifetime was essential so that the radioactive isotopes would not linger in the organism after they had helped find the pathway of various materials. Due to the short lifetime, the reactor had to be near the place where its radioactive isotopes would be used.

Teller made the most of his prior experience with reactor safety for the company, especially since any accident with the reactor would have meant the end of its acceptability. He defined the task of creating a safe reactor as one that "could be given to a bunch of high school children to play with, without any fear that they would get hurt."[23] There were two kinds of safety involved. One was "engineered safety," which made it impossible to pull the control rods out of the fission fuel suddenly, because such an action would cause a drastic accident. Then there was "inherent safety," which would be ensured by the laws of nature, meaning that any arbitrary action by someone ignorant of its workings would not lead to accident. Thus, if all the control rods were to be pulled out, rather than causing an explosion, the reactor would settle down to a steady operation.

Freeman Dyson had fierce debates with Teller almost daily in the company of their associates. Teller arrived almost every day "with some harebrained new idea." According to Dyson, "some of his ideas were brilliant, some were practical, and a few were brilliant and practical. I used his ideas as starting points for a more systematic analysis of the problem."[24] The two of them helped create the safety they were aiming for. Thus, Teller's lamentation that he ceased to be a physicist was an exaggeration; he was working with a different kind of physics for which he apparently made good use both of his fundamental knowledge and his imagination.

Niels Bohr was invited to put the first TRIGA reactor into operation, which was made into a spectacular event in June 1959—the time of Teller's directorship of Livermore. Teller's opponents attacked him for consulting for a private company while being in government employ, which was a fair criticism. Teller did not believe this was against the law, but nonetheless he resigned as an adviser to General Atomic.[25] The TRIGA program became a big success, but in the long run nuclear energy has not fulfilled its initial promise, given the public's fear of everything nuclear.

## POLITICS UNLIMITED

Teller warned scientists repeatedly to abstain from politics, but his heavy involvement in political decisions made him sound like a hypocrite or implied that he no longer considered himself a scientist. Teller was director of the Livermore Laboratory for a short period, but he never felt fit for such an administrative position. He had a strong enough influence at Livermore, so his title was superfluous. In fact, it limited his freedom of movement and action and curbed his usual outspokenness. Also, he had to give up most of his teaching duties as professor of physics at Berkeley, even to the point of having to vacate his office at the Physics Department. He retained only one course, introductory physics, which was especially close to his heart. He resigned his membership with the General Advisory Committee; by then the GAC did not play such an important role as it had in the hydrogen bomb debate in 1949, and Teller did not greatly regret his

Participants of the 1947 Shelter Island meeting of theoretical physicists. From left to right: Isidor Rabi, Linus Pauling, John Van Vleck, Willis Lamb, Gregory Breit, Duncan McInnes, K. Darrow, George Uhlenbeck, Julian Schwinger, Edward Teller, Bruno Rossi, Arnold Nordsieck, John von Neumann, John Wheeler, Hans Bethe, Robert Serber, Robert Marshak, Abraham Pais, Robert Oppenheimer, David Bohm, Richard Feynman, Victor Weisskopf, and Herman Fershbach. Courtesy of Marina von Neumann Whitman, Ann Arbor, MI.

Left: The "Mike" device of the thermonuclear test. Right: The mushroom cloud from the "Mike" explosion. Both photos courtesy of Los Alamos National Laboratory.

Left: Edward Teller in the early 1950s. Courtesy of Los Alamos National Laboratory. Right: Edward Teller in the late 1950s. Courtesy of Lawrence Livermore National Laboratory Archives and Research Center.

The Teller family in the mid-1950s. Photo by Jon Brenneis, courtesy of Lawrence Livermore National Laboratory Archives and Research Center.

Edward Teller being "knighted" aboard a military vessel in 1957. His signature reads: "Edward Don Quixote Teller." Ernest Lawrence is sitting at the right, partly obscured. Courtesy of Lawrence Livermore National Laboratory Archives and Research Center.

Glenn T. Seaborg, Edwin McMillan, Ernest O. Lawrence, Donald Cooksey, Edward Teller, Herbert York, and Luis Alvarez at the Berkeley Radiation Laboratory, 1957. Photo by Jon Brenneis, courtesy of Lawrence Berkeley National Laboratory.

Edward Teller in the spotlight. Courtesy of Lawrence Livermore National Laboratory Archives and Research Center.

Award ceremony of Teller's Fermi Prize at the White House, 1962. From left to right: Glenn T. Seaborg, Edward Teller, John F. Kennedy, and Mici Teller. Courtesy of John F. Kennedy Library, Boston.

Edward Teller arguing at a meeting in 1979. Courtesy of Lawrence Livermore National Laboratory Archives and Research Center.

Left: Linus Pauling at Moscow State University. Photo by and courtesy of Larissa Zasourskaya, Moscow. Right: Brunauer, Emmett, and Teller at a reunion. Courtesy of Burtron Davis, KY.

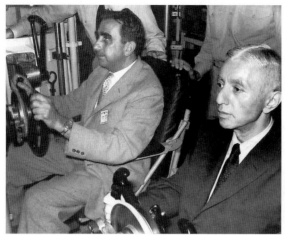

Left: Edward Teller and Admiral Hyman Rickover onboard a nuclear submarine. Courtesy of Oak Ridge National Laboratory. Right: Edward Teller with Lyndon B. Johnson in the Oval Office. Courtesy of Lyndon B. Johnson Library, Austin, TX.

Eugene P. Wigner and Edward Teller at Livermore. Courtesy of the Lawrence Livermore National Laboratory Archives and Research Center.

Soviet president Mikhail Gorbachev, President Ronald Reagan, and Edward Teller at the White House reception, 1987. Courtesy of Ronald Reagan Library, Simi Valley, CA.

Lowell Wood and Edward Teller. Courtesy of Lawrence Livermore National Laboratory Archives and Research Center.

Left: Edward Teller at the Paks Nuclear Power Station, Hungary. Photo by and courtesy of Károly Gottvald, Szekszárd, Hungary. Right: Edward Teller at the bust of John von Neumann on the campus of Budapest University of Technology and Economics. Photo by and courtesy of János Philip, Budapest.

Edward Teller with the author in the Tellers' home, 1996. Photo by and courtesy of Magdolna Hargittai.

departure from the committee. He also had to withdraw from his financially beneficial consulting position for the Rockefeller family, although his interactions with them continued.

Nelson Rockefeller built up a thorough advising network as part of his attempts to garner the Republican nomination for the presidency of the United States. Teller started his political life as a Democrat, but he changed his party affiliation to Republican in 1959 as part of his support for Rockefeller.[26] On September 8, 1960, the Rockefeller Brothers Fund issued a report recommending that the United States "define the national purpose and wrest the initiative in the 'cold war' from the Communists."[27] The report was prepared by a panel of thirty distinguished individuals, including Teller. He found himself in the company of such prominent members of the American Establishment as James R. Killian, chairman of the M.I.T. Corporation and President Eisenhower's science adviser; Henry R. Luce, the editor-in-chief of *Time, Life,* and *Fortune* magazines; Dean Rusk, president of the Rockefeller Foundation and soon to be President Kennedy's secretary of state; and others.

The fund was headed by Nelson A. Rockefeller until his election as governor of the state of New York and was supported by over a hundred million dollars of donations from the Rockefeller family. The fund aimed at strengthening the democratic institutions domestically and advocated strong military preparedness internationally. One of its objectives was to prevent "the world balance of power from shifting to the Soviet bloc."[28] Teller could fully identify with these goals and contributed to the project with great enthusiasm.

Through Rockefeller, Teller became involved more directly with politics. In 1964, he sided with Rockefeller against Barry Goldwater in Rockefeller's quest for the Republican nomination for president. Rockefeller lost to Goldwater, who was the most hawkish among the major presidential candidates of the era. Observers noticed the similarities between Teller's and Goldwater's views concerning nuclear testing and their distrust of the Soviets.[29] In a letter to Kissinger in October 1963, Teller admitted that he liked Goldwater and he was worried that many considered Goldwater more straightforward than Teller's and Kissinger's preferred Republican presidential hopeful, Nelson Rockefeller.[30]

There were speculations that in case of a Goldwater victory, Teller might become his science adviser. This sounded feasible not only because of the similarity of their views, but also because there were not many well-known scientists in sympathy with Goldwater's positions with respect to nuclear matters. Teller protested the rumors, saying that he was not interested in helping Goldwater.[31] Yet an interview in 1972 gave the impression that he viewed Goldwater with sympathy at least in comparison with Lyndon B. Johnson.[32]

In its turn, the Johnson campaign played on the general fear of a nuclear war. Nothing showed this more vividly than what has become known as "the most famous political ad on television."[33] It was broadcast on the evening of September 7, 1964, during an NBC movie. A little girl is shown counting aloud the petals of a daisy. Soon a man's voice takes over, counting down. As the camera zooms in on the little girl's eye, the countdown ends in an atomic explosion. Although Goldwater was not mentioned in the ad, his campaign protested, and the spot was pulled immediately. It was shown only that one time, though it subsequently appeared repeatedly on news programs. The powerful and dramatic ad—Tony Schwartz's creation—was credited to having considerably contributed to Johnson's landslide victory in 1964.

Richard Nixon was elected president in 1968 while Rockefeller was governor of New York. The governor put together a commission to work out recommendations for the future of his state. When Richard Nixon asked Rockefeller to extend the study to the whole nation, Rockefeller resigned the governorship and brought together the Commission on the Critical Choices for Americans. Forty-two distinguished people were asked to serve on the commission, and six panels worked on issues ranging from energy to national security, to the quality of life, and others. Teller became a member of three of the six panels.[34]

He prepared the report concerned with energy, which was made available publicly in 1975.[35] Predictably, Teller recommended a considerable increase in nuclear energy production by the year 2000. He admitted, however, that public acceptance was a grave problem with the spread of nuclear reactors and he recommended a thorough educational campaign in order to overcome it. He opined that the possibility of violent accidents, such as an explosion of a nuclear reactor, was unrealistic, and considered the fear of radioactivity to be exaggerated among the public. A book was published from Teller's panel activities with the title *Power & Security*.[36]

Rockefeller finally reached his highest office when Gerald Ford became president following Nixon's resignation, and Ford appointed Rockefeller to be his vice president. By the time of the Ford–Rockefeller administration, science advisers had lost much of their access and influence in the White House. Ford was a dark horse to the scientific community, but Rockefeller was welcomed as "a man who is well known for seeking out science advice, albeit mostly from Edward Teller."[37]

Teller's involvement in Rockefeller's projects opened a new world to him. It was no substitute for the world of physics, but it was a welcome diversion from politicians dedicated to immediate vote getting and from generals fighting their interservice battles. Teller spent the summer of 1974 at the Rockefeller estate, along with Mici, who was there part of the time. His two associates, Hans Mark, then deputy director of NASA, and John Foster, former director of defense research at the Pentagon, joined him.

Rockefeller again added some notable people to his orbit. They included Norman Borlaug, recipient of the Nobel Peace Prize in 1970 and developer of a semidwarf, high-yield, disease-resistant wheat variety that increased food production in third world countries and saved millions of people from famine. Another was Marina von Neumann Whitman, von Neumann's daughter, a renowned economist. She joined the team at Teller's recommendation, even though they were adversaries in the areas of politics and economics. Yet another was Dixy Lee Ray, a marine biologist who was to be the last chair of the Atomic Energy Commission before its portfolio was reorganized in 1974. She later became governor of the state of Washington. The task of making recommendations for the future felt like a comfortable enterprise for Teller.

His main avocation during these years was the activities and development of the Livermore Laboratory. After Lawrence's death, the lab became even more Teller's territory, and he felt full personal responsibility for its operations and performance. He wanted to be sure that Livermore had appropriate projects to work on. The international situation as well as domestic politics favored his intentions, except for the Soviets trying—as he saw it—to force a test ban on the United States.

When he reviewed Henry Kissinger's new book *Nuclear Weapons and Foreign Policy*, Teller welcomed Kissinger's strategy, no doubt having in mind the challenge such a strategy would present for Livermore.[38] Kissinger advocated developing simultaneous preparedness for an all-out war and limited wars, and for both kinds of war, the utilization of nuclear weapons. Kissinger's comprehensive approach prescribed the necessity of a whole range of nuclear arms from the smallest to the largest. He advised engagement in local wars answering Soviet provocations. It was different from the Soviet position according to which any application of nuclear weapons would inevitably lead to an all-out war. For developing a whole arsenal of different-size nuclear weapons, more testing was needed.

Teller liked Kissinger's book, but he did not like the way Kissinger presented his ideas. Teller was very good at popularization, so he used his review of Kissinger's book to explain Kissinger's ideas to a broader public.

He did not mention this in his published review, but he wrote a letter to Kissinger in which he bluntly stated that Kissinger put his "ideas in a language which will prevent many people from reading it and understanding it. As a consequence my review has become less of a review and more of a popularization of Dr. Kissinger."[39]

Teller was at the top of his career as a scientist–politician in 1957 when *Time* magazine put him on the cover of its November 18 issue and included accolades for Teller in its cover story.[40] Paradoxically, technological progress in the Soviet Union also contributed to his enhanced visibility. The Soviets sent up the first ever man-made satellite, Sputnik, into orbit around the earth on October 4, 1957. The Russian word *sputnik* means "travel companion" and has since become part of the international vocabulary. Teller took the situation symbolized by Sputnik seriously and he also saw in it an opportunity to further his aims. He wrote to Senator Henry M. Jackson: "The situation highlighted by Sputnik really requires much more cooperation and much more financial support. I hope that our Allies . . . will share our opinion: Not to belittle the recent achievement of the Russians but to take it as a powerful stimulus to increase our joint effort."[41]

The satellite had an electrifying effect on the United States. According to the *Time* story, "Of all US scientists on campus, in government, in industry, Teller worked hardest and most belligerently to send the warning that the Russians were coming." He was quoted as saying that he was less afraid of a military attack by the Soviet Union than their advancement in science. Teller observed signs of alienation among young people from science and complained about the fact that his son felt he had to conceal his interest in science from his teachers and his classmates,[42] lest his peers "would tease him and call him a 'square.'"[43]

The general atmosphere in American schools changed markedly in the following years. Teller realized that the bottleneck in American science education was the shortage of adequate science teachers. He appreciated the values of a broad education and wished to see more history and psychology taught and fewer routine mathematics and science assignments. He did not want to add "new irrelevancies to the old irrelevancies." He called for a careful selection of teaching material lest pupils were fed knowledge that only future bookkeepers would find useful.[44] He grasped every opportunity to advocate the importance of quality science education.

He spoke about "an emergency" in American education at a luncheon at the University Club at Columbia University on October 11, 1960.[45] He criticized low-quality science teaching, which led to the loss of many gifted students to other fields. He wanted to have higher mathematics and science and mathematics clubs, and recognition for teachers who produced superior students. He suggested making it possible for anyone to teach mathematics or science who had a degree in the field without the prerequisite of a teaching certificate.

Teller did not have a teaching certificate either, yet he acted as a very efficient pedagogue whenever he had the opportunity. When he was eighty-four and his daughter Wendy worried about his daunting travels and other obligations, she knew she could not persuade him just to stay idle. So, she devised a scheme for him to do some high school physics teaching. She found a willing partner in the Illinois Math Science Academy (IMSA), a public boarding high school for grades ten through twelve for the best and brightest high school students in Illinois. Starting in October 1992, Teller lectured to the students on physics every day for three weeks. He stayed with Wendy and her family, and the physics teachers of IMSA drove Teller to the school and back; they were most enthusiastic about the project. Both the students and Teller enjoyed the experience.[46]

The January 2, 1961, issue of *Time* magazine put US scientists on its cover as "Men of the Year" for 1960.[47] The cover featured fifteen portraits. Of them, only three were not Nobel laureates or soon to become one. Considering that there were well over fifty Nobel laureates alive in the United States at the time and that not all science fields were covered by Nobel Prizes, the three non-Nobel scientists, including Teller, were given exceptional honor. The *Time* choice reflected the heightened interest in and expectations of science and scientists in the post-Sputnik era in the United States. In any case, Teller was the best known in this very distinguished company, which was not due to his high visibility in science but rather to his involvement in politics.

At the time of this *Time* story it was not known that yet another blow for the Americans was in the offing: The Soviets put the first human, Yuri Gagarin, into orbit around the earth on April 12, 1961. It was a technological tour de force, and Khrushchev made the most of it for communist propaganda. Again, it served also as a stimulus for American aspirations. President Kennedy announced before a joint session of Congress on May 25, 1961, "that this nation should commit itself to achieving the goal, before this decade is out, of landing a man on the moon and returning him safely to the earth."[48] On July 20, 1969, the commander of Apollo 11, Neil Armstrong, stepped out of the Lunar Module onto the surface of the moon. The space efforts were, of course, much intertwined with defense developments.

Teller found satisfaction in Kennedy's announcement and in the success of the American space program. He warned, however, of the toughness of Soviet competition. There was a tendency in America to belittle their abilities and ascribe Soviet success to imported German scientists. In a speech at the 76th anniversary convocation at Rollins College in Winter Park, Florida, on November 4, 1961, Teller stated: "The great majority of Soviet scientific achievements have been made in Russia by Russians and not by captured German scientists as we would like to believe."[49] In spite of the Kennedy administration's interest in and support for research and development, for Teller and his projects, the ride became rougher. Yet his influence did not completely vanish; he remained a forceful voice in resisting the test ban under negotiations, as we have seen.

The young president and the new generation of politicians that arrived with him were expected to develop a more conciliatory relationship with the Soviet leadership than the previous administration. However, Khrushchev wanted to test the resolve of the inexperienced American president, and in his turn, Kennedy strove to demonstrate that he was sufficiently tough. Paradoxically, the Kennedy–Khrushchev relationship would become more confrontational than the relationship between Eisenhower and Khrushchev, which also suffered as a result of the U-2 spy plane incident.[50] The two superpowers used Cuba as a proving ground for testing each other, and with the Soviets placing missiles so close to American shores, the world was rapidly drifting toward the brink of war. The international situation should have strengthened Teller's position in his relationship with the new administration, but the president's advisers were as wary of Teller as the president himself was.

Teller continued to be much in demand for lectures and other appearances outside the Kennedy administration, though, and he published articles and books. He presented two sets of lectures in 1975, one at the State University of New York at Buffalo and the other at the Technion–Israel Institute of Technology, on the occasion of his prestigious Harvey Prize. From his lectures, he developed the book *Energy from Heaven and Earth*, which was published in 1979. It gave a panoramic picture of the origins, sources, utilization, and future of energy. Predictably, he advocated the use of nuclear energy, but he paid attention to other energy sources as well.

Teller remained a political figure throughout much of his career. His highest point was under President Reagan and his lowest was under President Carter. The irony of this low point was that Carter was supposed to be versed in nuclear matters, having earned a degree in nuclear engineering. Teller's frustration with politics reached such a level under the Carter presidency that he was considering running for office. Had he run, he would have had to challenge the Democratic senator Alan Cranston from California, who championed the cause of nuclear arms control and disarmament. By this time Teller was an established Republican, but he was prepared to change his party affiliation had Dixy Lee Ray, a Democrat, been willing to run for the presidency, challenging Carter, the incumbent Democratic president.[51]

In the 1984 presidential campaign for Reagan's reelection, Teller was active on many fronts, and in his zest he accused Democratic representative from California Tom Lantos of taking a stand "not in line with the true interest of the Nation."[52] This was a rather extreme accusation because politicians in the United States on either side of the aisle rarely questioned the patriotism of their opponents. Teller challenged Lantos to a public debate; he named five possible nights within the next three weeks and issued a threat of holding a press conference in case Lantos declined his offer. To make his letter look even more like an ultimatum, Teller gave Lantos a four-day deadline by which to respond. Lantos was a Hungarian–American, a Holocaust survivor, and the longtime chairman of the House of Representatives Committee on Foreign Affairs. When he died in 2008, President George W. Bush awarded him the highest civilian recognition, the Presidential Medal of Freedom. It is not known whether and how Lantos responded to Teller's challenge.

Teller was always on the alert as he watched incumbent and upcoming politicians alike. He was a keen observer of election campaigns and approached presidential, senatorial, and other candidates he admired, expressing his allegiance to them and giving them strategic advice. If the politician in question was elected, Teller's advice became more concrete and more detailed.

In the summer of the 1980 election year, he wrote a letter to Ronald Reagan in which he complained about the deteriorating US military under President Carter and listed what needed to be done, elaborating each of his suggestions with concise and easy-to-read explanations. He advocated conducting a technology race rather than an arms race with respect to the Soviet Union and stated in conclusion, "[T]echnology can be used or misused. In the hands of a thoughtful American administration, it can be used to defend freedom without risking any massive conflict."[53]

Soon after Vice President George Bush was elected president in November 1988, Teller sent him a letter that included a short sentence of congratulations then moved immediately to the appointments the new president should be making. He had suggestions—in fact, three alternatives—for the position of secretary of energy; one for NASA administrator; and two alternatives for the president's science adviser.[54] None of his advice was taken.

In the summer of 1999, a good fifteen months before the 2000 presidential elections, Teller wrote to Texas governor and presidential hopeful George W. Bush and expressed his wish to see him in the White House. This sentiment was expressed in the first, brief paragraph. The rest of the letter was about missile defense, including promotion of the Lawrence Livermore National Laboratory. There is a sentence in Teller's letter that sounds ominous in view of the 9/11 terror attack, although in context it does not seem to refer to terrorism: "It seems to me that the possibility of sudden damage is perhaps even more important than the possibility of big damage."[55] One might admire Teller's political acumen in his early prediction of the outcome of the 2000 presidential race. However, he sent similar letters to a number of candidates, thereby increasing the probability of making early contact with the eventual winner.

## "AND THEY SHALL BEAT THEIR PLOWSHARE INTO CHARIOT"

Teller and his associates at the Livermore Laboratory started early in their campaign for peaceful applications of nuclear energy. In these efforts, Teller found it strategically advantageous to involve private business. Nuclear matters had almost been the monopoly of the government; this is why it could be called an "island of socialism." In his review of the book *Atomic Energy for Your Business* by nuclear scientist and intelligence expert Arnold Kramish and former AEC commissioner Eugene Zuckert, Teller called for enhanced utilization of atomic energy in industry and elsewhere.[56]

In a rare show of unity, Bethe and Teller joined forces in advocating the use of nuclear power plants for easing the problem of energy shortage in the United States, as it was becoming increasingly at the mercy of imported fuel. Teller kept stressing the growing need of energy for the developing countries. His views on energy have not lost their importance after the political changes of 1989–1990 in Central and Eastern Europe. The dependence of Western Europe on Moscow for energy has hardly changed since the Soviet Union gave way to Russia. Teller's statement in 1982 retains its validity today: "If the Soviets choose to cut off the oil supply to the West, they could create the most severe depression that the industrialized world has ever faced."[57]

In addition to harnessing nuclear energy, Teller became increasingly involved in looking for peaceful applications even for atomic bombs and hydrogen bombs. The AEC program established in 1957 and named Project Plowshare, as noted earlier, referred to the biblical passage "And they shall beat their swords into plowshares, and their spears into pruning hooks; nation shall not lift up sword against nation, neither shall they learn war any more."[58] The name "Plowshare" for the AEC program came from a quip from Isidor Rabi, who was the chairman of the General Advisory Committee of the AEC at the time.[59] Over the years, the peaceful applications of nuclear science manifested themselves as power plants around the world, but not to the extent that had been initially hoped for. Project Plowshare was distinguished by the fact that in this case explosions rather than controlled processes would have been utilized for civilian applications.

As early as 1949, the Soviet Union had declared the intention of utilizing nuclear detonations for grandiose projects. The most spectacular undertaking would have been reversing the direction of the great Siberian rivers: making them flow from the north to the south and using their water for irrigating wasteful land in the Central Asian region. The advocates of these plans had so much confidence and government backing that they presented them at international forums. It was obvious that nuclear detonations would be involved in a major way in the necessary earth-moving construction work. Eventually, however, the projects were quietly shelved in the Soviet Union due to a confluence of various concerns and considerations.

Following the nationalization of the Suez Canal by Egypt in 1956, the idea came up at Livermore to dig a canal through the Sinai Peninsula in Israel from the Gulf of Aquaba to the Mediterranean Sea. No practical steps were taken, however, and it remained an exercise within the confines of the laboratory. However, it became the impetus of further plans, including one for a sea-level canal as an alternative to the Panama Canal. The latter is marred by several steps of changing altitudes that make sailing through it time consuming and costly. Teller was greatly interested in these possibilities for several reasons. One was the general preference for hydrogen bombs in such constructions, ultimately of "clean" bombs. Another was that such plans would provide suitable projects for the Livermore Laboratory and they would have softened the image of nuclear weaponry by blurring the line between military and civilian applications.

Project Plowshare led Teller to yet another unpleasant encounter with President Kennedy. When the president inquired about the proposed new canal on the Panama isthmus, he asked Teller about how long it would take to build. Teller's response was needlessly rude: "It will take less time to complete the canal than for you to make up your mind to build it."[60]

Kennedy was apprehensive of Teller, but not to the extent that he would not personally hand him the Fermi Award in December 1962. It was the highest recognition by the Atomic Energy Commission. The recommendation for it came from the General Advisory Committee and had to be approved by the AEC and by the president. The AEC gave its first award to Enrico Fermi in 1954, just weeks before he died. The next year the commission decided to name the award after Fermi, and it was first given to John von Neumann in 1956. In 1957, 1958, and 1959, Lawrence, Wigner, and Seaborg received it, respectively. Then, after having skipped a year, Bethe became the awardee in 1961. It was a prestigious list.

Alvin Weinberg, director of Oak Ridge National Laboratory, was one of those who nominated Teller for the award, and he stressed in his nomination Teller's achievements in reactor safety rather than his weapons work.[61] Teller's award citation read, "For contributions to chemical and nuclear physics, for his leadership in thermonuclear research and for efforts to strengthen national security."[62] The president gave a short speech at the award ceremony in which he remembered the contribution of the European refugee scientists to the development of nuclear science in the United States. He also praised Teller for his research and pedagogical activities.

Soon after the Suez crisis, Teller, Harold Brown, and Gerald Johnson proposed to the Atomic Energy Commission "to use nuclear explosions to reshape the earth."[63] Brown, Jimmy Carter's future secretary of defense, was head of Livermore's division for thermonuclear weapons design at that time. There was a secret meeting on the industrial uses of nuclear explosives at Livermore, February 6–8, 1957, bringing together the organizations and institutions that might be participants of the relevant projects. Teller mentioned the possibility of exploding a thermonuclear bomb near or right at the moon in order to learn more about the earth's satellite. Another idea concerned extracting water from the moon for future space travelers using nuclear explosions. Current explorations appear as a continuation of such efforts. (Water is prohibitively expensive to bring from Earth.)

Teller thought about a new discipline of geographical engineering to better the planet and make it more suitable for human use.[64] What Teller meant by this new discipline could be summarized by one of his succinct exclamations, "If your mountain is not in the right place, drop us a card."[65] At the time, this slogan symbolized the new possibilities; eventually, it became an illustration of Teller's recklessness.

The first ideas of geographical engineering foresaw projects in foreign countries. Curiously, these countries were not even informed about these plans. It was soon realized that a domestic example should precede any further extension of the project beyond the borders of the United States. Studies were carried out at Livermore, where they found an appropriate task to dig a deep-water harbor in northwest Alaska. It might have been a welcome undertaking with potential benefit for the economy. Alaska was just preparing to become the forty-ninth member of the Union. President Eisenhower would sign the official declaration on January 3, 1959.

Teller and his associates descended on Alaska on July 14, 1958, where he held a press conference in the capital of Juneau to announce Project Chariot, the name for the Alaska plan. He told his audience about the deep-water harbor to be created near Cape Thompson. When he saw that the local people had doubts about the location and had alternative suggestions, he displayed flexibility in such details. What he lacked in information about Alaska, he compensated with his engaging style, and the local people looked forward to what the project might do for their fledgling economy. Teller assured his audience that the Chariot Project would be carried out for the economic benefit of Alaska. This claim was eventually dropped. The Alaskans called the people of Livermore dealing with nuclear explosions the firecracker boys.

Initially, many Alaskans were quite impressed by the attention they received from Livermore and the AEC, and from Teller in particular. In 1959 and 1960, the University of Alaska offered honorary doctorates to some famous individuals, among them J. Edgar Hoover (who declined it); former Nazi rocket designer Wernher von Braun; Admiral Hyman Rickover; and Edward Teller. Teller gladly accepted the distinction and delivered a speech to the class of 1959 on May 18.

Years later, the author of *Firecracker Boys* listened to the tape recording of the speech, in which "Teller's words march in step, shouldering themselves in deep, r-rolling accents" reminded him of the great Hungarian warriors, although he gave no hint where he might have learned about what they were like.[66] Teller spoke about fallout and kept repeating that its harmful nature had never been proven. In June 1959, he returned to Alaska; he had grown so close to the state that this time he took Mici along. He was at the peak of his involvement with Chariot, so he gave speeches and held press conferences.

Soon, however, doubts started forming, especially in some scientists' minds. Teller assured them that the fallout from the explosion to be used for digging the harbor would be negligible; it would not amount to more than what people were already being exposed to as existing background radiation. This sounded very good until the argument was turned around and looked at from the point of view of the total radiation people would be exposed to after the explosion. It meant that the amount of radiation could double or nearly double. This was but one example that at the time did not stand out as being suspicious but that in hindsight might have raised a red flag for those involved.

The whole affair had broader implications than just Livermore and Alaska. It was part of a series of events of international dimensions. The summer of 1958 was a critical period for the fate of future tests of nuclear explosions and weaponry, as we have seen with the test ban negotiations in the previous chapter. In the negotiations continuing in Geneva, the first major issue was how to detect violations in case a test ban were to take effect. When there was a preliminary agreement about verification, it was declared that the negotiations about the test ban itself would continue. Eisenhower wanted to announce that the United States would suspend testing for the duration of the negotiations up to one year. He made an exception, though, at the insistence of the AEC that testing of explosions for peaceful purposes would be exempt from the suspension. Teller's foresight of the importance of the Alaska project seemed to be paying off.

During the long weeks before the Geneva talks resumed on October 31, 1958, both the United States and the Soviet Union frantically carried out nuclear explosions. After the closing date, there would be no tests of military importance, though it is difficult to imagine that any of the experts might have believed that the tests for peaceful purposes could have had no military application. In any case the Chariot project was judged to be of crucial importance and its public relations potentials were also to be utilized. Teller attended a big meeting on the peaceful uses of nuclear energy in Geneva between September 1 and 3, 1958, and gave a presentation on the "Peaceful Uses of Fusion."

The Soviet delegates attacked Project Plowshare as a military program in disguise, and many Western scientists were also skeptical about it. When the leader of the Soviet delegation, Vasily Emelyanov, denounced Plowshare as a "dirty capitalistic trick introduced only to justify continued work on weapons," Teller reminded him of similar Soviet plans.[67] Emelyanov should have known better because he was the director of the board for the peaceful uses of atomic energy in the Soviet Union. Andrei Sakharov characterized him as an Old Bolshevik, who was known for his dogmatic politics.[68]

In some aspects, the Plowshare program was like the SDI controversy in miniature, which would come about a quarter century later. Plowshare was pushed by the American leadership but hardly supported by mainstream American scientists. It was opposed by Soviet leadership while the Soviet Union was entertaining similar plans. Soon the need for thermonuclear explosions appeared unnecessary, and it was found prudent to replace them with fission explosions. Just before the October 31, 1958, deadline, an underground explosion test, called Neptune, was carried out in Nevada, and it resulted in unexpected observations. It was a fission explosion, yet it created a much larger hole than had been predicted, and the fallout was much smaller than a fission bomb was expected to produce. The experiment raised the question of whether fusion bombs were necessary to do the job of building the deep-water harbor in Alaska. Thus the original scheme became technically questionable in addition to its other uncertainties.

It was inevitable that sooner or later questions of the impact on the environment would be raised, especially in view of the precariousness of the environment under the conditions of Alaskan climate. There was then no delay in Teller and Livermore offering support for environmental research. Both the University of Alaska biologists and Livermore claimed to have been the first to come up with environmental considerations.[69] This did not mean that Teller was ready to deal with environmental issues. When, in June 1959, he was asked about them, he instead talked about his dedication to the cold war against the Soviet Union.[70]

By 1962, Teller and the AEC understood that Chariot would not reach fruition; they announced that instead of digging a harbor in northwest Alaska, they were looking for "more practical" projects.[71] They carried out a series of blasts in the Nevada desert. The test under the code name Sedan created a huge crater in seconds using a hundred-kiloton-TNT-equivalent hydrogen bomb. Teller made the announcement about it to a meeting of the American Nuclear Society. He enumerated several potential programs, such as digging harbors in Hawaii; creating a harbor in the Alaskan Panhandle for accessing coal deposits; connecting the Tombigbee River in Mississippi and Alabama with the Tennessee River; and others. He said that only "a lack of imagination, a lack of enterprise and some political timidity" could delay these projects.[72]

Teller made another big proposal for using hydrogen bombs to extract natural gas from gas-bearing rocks from which other approaches could not produce the fuel in an economical way.[73] Teller and his associates planned nuclear detonations in northern New Mexico, but their most ambitious plan was to put nuclear power stations deep underground that would be safe enough to possibly be put underground in New York City. Teller talked about these plans at a conference on air pollution.

In May 1973, there were underground hydrogen bomb blasts in the western slopes of the Colorado Rockies. These were attempts to find the appropriate technologies to extract natural gas trapped deep underneath the mountains.[74] The idea was to free a huge amount of gas that would meet America's needs for decades. Teller argued that oil could be extracted from shale, whereas others warned that the valuable shale reserves could be lost if the nuclear detonations fractured them or contaminated them with radioactivity. According to some estimates, the shale in the Rocky Mountains contained the world's largest oil reserves. Teller suggested a technology whereby the shale would be heated in its place and the oil would thus be freed. The heating would be accomplished by exploding numerous hydrogen bombs.[75]

Teller found a great ally for Plowshare in the person of Yuval Ne'eman.[76] Both Teller and the Israeli physicist-soldier-politician were political conservatives; many would call them extreme right-wingers, which goes to show how different the meaning of such terms can be in different corners of the world. The Plowshare projects might have fit Israel's needs for building a new canal and other ambitious constructions. However, the dangers from fallout for its concentrated population and the environmental hazards proved to be prohibitive, so none came to realization. Teller's other international attempts at Plowshare-like projects were sooner or later similarly aborted in Australia,[77] Thailand,[78] and elsewhere. One of the would-be projects planned to cut a canal across Greece using thermonuclear explosions, but the outspoken Queen Frederica said: "Thank you, Dr. Teller, but Greece has enough quaint ruins already."[79]

According to Teller's own account, he at some point had become a skeptic regarding environmentalist arguments.[80] It was after he had experienced another setback due to their resistance toward one of his projects. Following the first major blackout in New York City in 1965, plans were proposed for a hydroelectric power station near West Point, New York, which would have been based on a reservoir at the Storm King Mountain. The local activists protested and formed the Scenic Hudson Preservation Coalition.

Ironically, Teller once again faced one of the members of Oppenheimer's legal team, Herbert Marks. Teller was helped by two of his future closest associates, Lowell Wood and Harry Sahlin.[81] The case was eventually labeled a landmark case, and the project for the reservoir was defeated. A remarkable episode of this story occurred during a public hearing, when the environmentalists challenged Teller to disclose how much he was being paid by the utilities company that wanted to build the hydroelectric plant. Teller told them that he was not paid anything. Noticeably fewer consultants for the other side appeared at the next session, no doubt for fear of having to face a similar question.[82] Clearly, Teller never made decisions based on real or potential financial rewards for himself.

It would be wrong to remember only Teller's clashes with environmentalists. Especially in the framework of his association with Rockefeller's projects, Teller himself was involved in issues of protecting and improving the environment. Thus, for example, when Governor Rockefeller announced in 1965 his plans of an intensified attack on air pollution, he asked Teller to "mobilize the best scientific talent in the nation . . . to seek new breakthroughs in dealing with air pollution."[83] Teller also agreed to find the means to combat water pollution and to prevent electrical failures. At the time he was a distinguished visiting professor at the State University of New York at Buffalo.

Teller's suggestions to curb air pollution included reduction of sulfur content in fuels used for power generation. Teller's suggested regulations were stricter than New York City's already stringent antipollution rules and were anticipated to increase prices on power but were also to yield long-term benefits.[84] This kind of activity was fully consistent with Teller's advocacy of projects for applied research to be determined by the needs of the community. This was different from basic research, where outside interference was undesirable and where everything was left exclusively to the scientists to decide and direct. Teller spoke about these issues to congressmen, expressing his concern over whether they were prepared enough to make informed decisions about appropriations of many billions of dollars for scientific projects.[85]

One area of peaceful application of nuclear explosions would be destroying asteroids. In connection with this idea, Teller liked to mention the catastrophic collision of an asteroid, probably of ten or fifteen miles in diameter, with the earth sixty-five million years ago, an event that led to the extinction of many species, among them the dinosaurs. That catastrophe also contributed to the eventual development of humans on our planet. A much smaller collision happened about a hundred years ago in Siberia. Humankind may find it advisable to prepare for meeting more asteroids in the future, he warned. An asteroid large enough to disrupt life on the earth could be sighted one year before it hit the earth, but it is doubtful whether one year would suffice to prepare to destroy it.

It might yet be that Teller's weapon-building activities could find their most important and least controversial applications in protecting the earth from the catastrophe of colliding with a large asteroid. In January 1992, Teller was among the hundred participants at a meeting at Los Alamos to discuss new nuclear armaments. The event coincided with his eighty-fourth birthday. He called for producing the largest-ever bomb, ten thousand times larger than any that had hitherto been built; it would be so large that it could never be exploded on the earth but would be used to destroy asteroids. According to the report about the meeting, as Teller was describing this super super-bomb, his protégé at the Livermore Laboratory, Lowell Wood, could not contain his excitement and shouted from the audience, "Nukes forever!"[86]

David Morrison and Teller contributed a chapter to a multi-author volume, *Hazards Due to Comets and Asteroids*, published in 1994.[87] (David Morrison of NASA has prudently been warning of a potential asteroid collision with the earth for years.) The effects and the possible avoidance of meteorites and asteroids as well as comet impact on Earth was the main topic of a Russian–American meeting in Chelyabinsk, Russia, in the fall of 1994. In addition to Teller, there were six other Americans attending, including Lowell Wood. Teller argued for expanding the circle of participants of similar discussions in the future, making the potentially emerging projects multinational.[88] The considerations included earthquakes and hurricanes.

Edward Teller was compared to Voltaire's Dr. Pangloss, Candide's friend, whose optimism shone through all circumstances. He was labeled as an archetypal Pangloss for his "merging progress and science in one unwrappable package."[89] He equated Plowshare with progress, but he met with resistance from many of his fellow scientists. During the Vietnam War, because so much technology was being employed in that tragic conflict, it was easy to suggest connections between the war and his activities. Teller supported the American involvement in Vietnam; he was unhappy about the domestic perception of the war; and he considered practical steps to improve the situation.

He was concerned, for example, with the scarcity of Vietnamese-speaking American personnel and urged enhanced training to improve the military's performance and the pacification process in the Vietnamese countryside.[90] He mentioned the language problem in his letter to Peter Lax in 1968, but Teller's main message was—obviously in response to Lax's critical remarks about the Vietnam War—his lamentation about "the dreadful reaction on the home front, of which your short note is an example. I know that you do not jump to conclusions; yet in this case you were induced to jump. Please don't."[91]

At the December 1970 meeting of the American Association for the Advancement of Science, the platform on which Teller was to speak displayed "placards proclaiming him a war criminal."[92] The newly elected president of the association, Glenn T. Seaborg, was accorded a similar reception, and he fled the room. Teller elected to speak, and his opponents eventually alienated the audience with their aggressive behavior. Teller also caused himself some embarrassment, as he walked around during the meeting surrounded by bodyguards, whose presence proved unnecessary.[93]

The student unrest and the radical movements mushrooming on American campuses, and conspicuously at Berkeley, targeted Teller and also accused him of war crimes. The Vietnam War reflected a tragic misunderstanding of world communism in general and the situation in Vietnam in particular by the American leadership. It is not our call to discuss this most tragic chapter of American history, only to relate an episode at Berkeley in which Teller became a target of unrest. He undoubtedly had the ability to attract the wrath of his opponents in a focused way even when others might have been sharing it with him.

The students' anti-war movement at Berkeley was supported by some faculty members, including Charles Schwartz of the Physics Department. Schwartz and Teller had clashed, even if only indirectly in the circles of the American Physical Society, where Schwartz proposed a more active participation in discussing political issues within the framework of the society. Teller declared that pressure groups were a nuisance, ignoring the fact that he had often represented a one-man pressure group.[94] Schwartz conflated Vietnam with the Oppenheimer episode: "Oppenheimer was the angel and Teller was the devil."[95]

One of the expressions of campus unrest was the war crime trials following the examples of those conducted in Sweden by Bertrand Russell, the British Nobel laureate philosopher, and others. (Incidentally, in 1960, Teller and Russell participated in a televised debate.)[96] In early November 1970, a leaflet was distributed on the Berkeley campus, accusing Teller of working on the atomic bomb; of fathering the hydrogen bomb; of establishing the Livermore Laboratory; of advocating the arms race and nuclear arms; and of influencing Washington politics by his hawkish advice. His home address and a sketch of a map augmented the accusations. Then, a war crime tribunal was called for November 23, 1970. The list of the accused included several others, among them, again, Seaborg, who was also a former chancellor of Berkeley. As the speakers and audience worked themselves up, their anger was increasingly focused against Teller.

Teller did not respond to the protests, and in this he differed from his only surviving Martian friend, Eugene Wigner, in whom the protests induced a strong emotional reaction. The events reminded Wigner, just as they did Teller, of Germany in the 1930s, when there was a Nazi danger of world domination. Conversely, the protesters saw the American government with its use of napalm on civilian populations as the aggressor that mirrored Nazi Germany. Wigner was upset by the passivity of his colleagues, while he had the strong impression that democracy and the free world were once again being threatened.[97] When he faced anti-war demonstrators in a classroom, he unfolded a banner: "Thank you for the compliment; abuse by you is a compliment for me."[98]

The crowd at one rally became increasingly belligerent and at one point moved out of the auditorium and headed in the direction of the Tellers' home. Two colleagues of Teller, one of them Lowell Wood, attended the meeting out of curiosity and when they sensed that events might turn dangerous for the Tellers, they left to warn them. By then there was police protection around the house, and the demonstrators were stopped a couple of blocks from the Tellers'. They burned Teller in effigy, but somehow their energy had dissipated and they dispersed. The Tellers decided to acquire a dog and set up a fence around their home.

## SECRECY

Teller was often considered to be an enemy of democracy, but in fact he advocated the need to strengthen democracy through openness in certain areas of his activities. Such was the question of secrecy in scientific research. He used the folksy wisdom "Square pegs don't fit into round holes" to characterize the incompatibility of democracy and secrecy.[99] He remembered with nostalgia the pre-war times when secrecy was not an issue in anything in which he was involved. After the war, there was hardly any aspect of his activities that was not marred by secrecy. Throughout the decades he made many statements condemning secrecy, desiring its elimination, though he rigorously observed the secrecy regulations. Teller was first introduced to the concept of secrecy when Szilard brought it up in the wake of the discovery of nuclear fission. For Szilard, this was not unexpected, for he had practiced it years before. When he patented the nuclear chain reaction, he did not bring it out into the open; rather, he deposited it with the British Admiralty.

When the genie was out of the bottle in the form of the discovery of nuclear fission, and, even worse, when the discovery was made in Germany, Szilard found it imperative that further developments in the democracies should be kept secret. When Szilard and Teller discussed it with Niels Bohr, the Danish scientist first opposed it but eventually agreed to it reluctantly. The next to be convinced was Fermi, and Teller was dispatched to New York to talk with him. Fermi would agree only if other physicists would, too. Eventually, secrecy was observed, and the peculiarity of the situation was that the scientists initiated it before the government even learned about what should be classified (chapter 4).

During the war, the atomic secrets were well guarded against the enemy Germany but not at all successfully from allied Soviet Union. Espionage helped the soon-to-become enemy in embarking on its own nuclear program and in producing its first fission bombs. Teller called equating secrecy with security a "perilous illusion."[100] His confidence in losing little and gaining a lot from lifting secrecy stemmed partly from his recognition of the values of scientists in the Soviet Union, something that many politicians did not share with him. He stated in 1960: "The Russians are fully capable of unraveling the secrets of nature and putting them to effective use; there is probably no major scientific development of which the Russians are ignorant."[101]

Immediately after the war, Teller noticed that what was declassified and what remained a secret was difficult to distinguish: "Security policy at present is a little confusing. A number of things have been told, a number of things can be told, and a number of things must not be told, and as far as I can see there is little or no relation between the three groups."[102]

Eventually, Teller challenged the two major objections to lifting secrecy. One objection involved the danger of nuclear proliferation that would be facilitated without secrecy. However, in the process of peaceful application of nuclear reactors, fissionable fuel could be produced, and by possessing such material, it would be an easy step for anybody to construct a bomb. Teller felt that international law rather than secrecy would be a better guarantee to prevent proliferation. The other objection concerned the fact that lifting secrecy by the Americans while the Soviets kept theirs would provide an unbalanced advantage for the Soviets. This Teller found meritorious, but he turned the question around and suggested examining the costs of continuing secrecy. He found it alien to a free society, whereas it was an accepted way of life in a totalitarian society. Among the costs he mentioned was that many scientists declined working for secret projects, but in the Soviet Union there was no such deterrent, as secrecy was part of their lives. Secrecy, furthermore, limited cooperation among American scientists and more so between them and among scientists of allied countries, even within NATO. Teller found it disadvantageous to keep the general public in the dark about nuclear weapons and explosives.

In the dilemma over whether to abandon secrecy and thereby give the communists an advantage or retain secrecy and slow down the development of weapons, Teller was unambiguously for less secrecy. He made certain suggestions. In other fields people must justify their desire to keep certain findings secret, whereas in nuclear research the opposite was practiced. He suggested that in nuclear science, too, classification should be justified; otherwise all new results should be published. The general principles should not be kept secret except for a short while. For technical and engineering details, longer periods of secrecy should be observed. Some military and operational information, he felt, should obviously not be revealed. His advocacy was for a liberal policy, although his scientific openness was coupled with operational and technical secrecy. Teller did not expect the Soviet leadership to follow suit in lifting secrecy, but he counted on a gradual recognition of the advantages of openness by Soviet scientists. He hoped that they might eventually influence their government to ease the rigidity of their system. In this, he showed some departure from the rigidity of his views on disarmament, where he felt that an American example would not be followed by the Soviets.

By the 1960s, it was found that more than one million government employees were involved in making documents classified, and some agencies even made some newspaper clippings classified. The Pentagon files contained about twenty million classified documents.[103] Teller could be very biting when he wanted to get his point across. He published an article about three different examples of secrecy. Of the Russian secrets he listed the following: military, administrative, economic, scientific, and everything else. Of the American system, he noted that it did not work well and that it increased confusion among the citizens. When he addressed the secrecy of the press, he condemned the arbitrariness of the editors' decisions of what to publish and what to withhold. He noted that the press thrived on secrecy, and that the remedy for that would be less secrecy. He insisted on distinguishing secrets whose declassification would not harm national security from the real ones, such as the planned route of nuclear submarines or the codes used for communications with embassies.[104]

It is interesting to compare Teller's views in 1960 with the responses he gave on secrecy and how to reduce it in a 1972 interview. His interviewer was not satisfied with generalities, so he went after Teller's views on their practical application. They had a long conversation, considering that it revolved about one specific topic.[105] Teller described his frustration when, during the Manhattan Project, he was sent on a mission where he had to collect information from colleagues who were supposed to tell him everything and he could tell them nothing. He noted that the engineers tolerated this lopsided relationship better than the physicists. He suspected that General Groves was not above misusing the secrecy situation for possible political gains. When Teller left Los Alamos at the end of the war and joined the University of Chicago, he vowed never again to be part of secret projects; but, ironically, this time it was Fermi who convinced him otherwise. Apparently, not much changed between 1960 and 1972, and Teller still complained that the existing rules hurt science and did not serve the best interests of the defense of the United States.

He quoted explicitly that in spite of secrecy, the Soviets were ahead of the United States in nuclear weapons. In contrast, in the field of computers, where there was no secrecy, the Americans were ahead of the Russians. The interviewer pressed Teller about the practical realization of removing secrecy. Teller admitted that "abolishing secrecy completely is unworkable and unwise." However, he would restrict secrecy "to pieces of information that many people have to know in order to take action. In all other cases, let us use human confidence rather than formalized secrecy."[106] He added that if the Americans would be ahead in nuclear weaponry, secrecy would make sense, but under the circumstances, keeping these secrets did not. He stressed that bureaucracy tended to overclassify things; if there was an option between classifying something or not, a bureaucrat would find it safer to classify. Teller found it especially peculiar not to let the American public know what the Russians knew, although he admitted that there may be justification for that in some cases, for example, when the sources had to be protected. But he was for providing more information to the American people because that would facilitate making informed decisions.

Teller made an interesting comparison between the Russian and American societies in the same interview, and he grasped some of its fundamentals if only from a distance. He admitted that the Americans knew less about the Russians than the other way around, in spite of the virtually hermetic isolation of Soviet society from the outside world. By the 1970s, though, this isolation began to crack. Teller noted that secrecy was an anomaly in America and complained that even the limited secrecy rules remained unenforceable. In contrast, all existing rules were enforced in Russia—and some that did not even exist. This was good insight on Teller's part, as he realized that it was not only censorship that operated under communism, but self-censorship, too.[107] People followed the rules and, in addition, followed what they supposed might or might have been expected of them to follow. Teller repeated what he said twelve years before about openness facilitating cooperation among allies and that it would help the ideas of democracy to penetrate the iron curtain. In 1973 he urged Congress to pass a law that would require "all secret government scientific research to be declassified one year after it is completed."[108]

Teller's critics questioned his sincerity when he suggested the one-year limit for keeping secrets with the exception of certain kinds of information. They found his taking a stand against secrecy suspect after what he had done to Oppenheimer. But he probably genuinely felt the need for reducing secrecy, and this was not a unique situation when he demonstrated conflicting views and actions.

Teller was a member of the Task Force for Security set up by the Pentagon, which found that classified information was unlikely to remain secure after five years. It might become known to others even within one year, and not so much by spying or stealing, but by independent discovery. The task force criticized classification for its becoming a barrier to cooperation between nations and to the free flow of information within the United States. It further asserted that "more might be gained than lost if our nation were to adopt, unilaterally, if necessary, a policy of complete openness." But this was a declaration of general principle, which was immediately dampened by warning that such a policy was "not a practical proposal at the present time."[109] They recommended a 90 percent decrease of rendering scientific and technical information classified. Arthur Schlesinger found that the recommendations of the task force presupposed a different world. The gap between ideas and recommendations was wide and characterized not only Teller's approach to secrecy.

The Soviet scientists in one area of research showed remarkable openness—obviously not on their own initiative—and this area of research, energy production by controlled fusion, happened to be one of Teller's pet projects. The czar of the Soviet nuclear program, Igor Kurchatov, gave a lecture on controlled fusion at the Atomic Energy Research Establishment in Harwell, England, on April 25, 1956. Kurchatov was a member of the visiting Soviet delegation headed by Khrushchev and Bulganin. Whereas fission can be used for explosions as well as for energy production in a straightforward manner, fusion has been harnessed for detonation only. Keeping the thermonuclear reaction under control has still not yet been solved. The long-range research might have contributed to the Soviet decision to lift secrecy on their progress in this specific area. Teller was one of four scientists who were called to Washington to assess Kurchatov's lecture, and each of the attendees found that it indeed communicated genuine results. According to Teller, Kurchatov's findings were identical to those of the American scientists, which helped him convince Strauss to let them report their results at the Atoms for Peace Conference in 1958.[110]

For the United States, however, there was not a general lifting of secrecy on controlled fusion research, as Teller lamented as late as

1973.[111] Apparently, every time they had new findings, it had to be decided anew whether or not they could come out with them. In 1972, there was an international meeting on quantum electronics in Montreal, at which a new possibility related to controlled fusion was discussed by American scientists. They set up an experiment directing a laser onto a small droplet of a liquid deuterium–tritium mixture. The AEC's Division of Classification gave its blessing to the report, but the work had been done in secret. This gave Teller yet further impetus to call for abandoning secrecy in scientific and technical development.

Teller found the American situation in double jeopardy. He asserted that "the Russians know most of our secrets including quite a few that we haven't even discovered as yet."[112] A few years later, a furor erupted over some disclosures in the American press about information on how to make a hydrogen bomb. The source of the information, who labeled himself "a very conservative Republican," was puzzled because all his information came from the open press and from publicly accessible communications by American scientists, including Edward Teller.[113] Teller stated "that he had consistently advocated less secrecy but that it would be dangerous to take a 'capricious or haphazard approach' and 'I would not consider publishing anything to effect a reduction of secrecy in a piecemeal or unauthorized way.'"[114]

It was during President Eisenhower's administration that the United States proposed the "Open Skies" policy for enhancing openness in the airspace, including aerial surveillance flights, over the countries that would participate in appropriate treaties. When the discussion was initially about a comprehensive test ban, the open skies might have facilitated inspection. The policy would have been similar to the freedom of the seas, but it was different because the skies were above and not outside the territories of sovereign states. Teller supported the proposal of "Open Skies" and stressed its importance for weather forecasting, learning about crops, and discovering ore deposits.[115]

Teller reiterated the lack of progress in decreasing secrecy in an interview conducted in 1982 by the conservative William F. Buckley on his PBS television program *Firing Line*.[116] While maintaining the importance of short-term secrecy, Teller came out strongly for ending long-term secrecy, for ending the "secrecy of knowledge." On the other hand, he would not invite communist scientists to Livermore and would not want to teach them all the technical details in their constructions. He was visibly enjoying the occasion; for example, he called quantum mechanics his own best-kept secret, which he had tried to divulge to many students, yet it remained a secret. In another lighter moment, he mused that the ideal solution would be if they could tell their secrets to the American people while keeping them from the Soviets.

It was no consolation to Teller that the Soviets operated an even more security-conscious system than the Americans. Possession of state secrets was a pretext for not allowing Andrei Sakharov to travel to foreign countries well into Gorbachev's period of "glasnost" (which meant transparency). Sakharov adhered rigorously to observing the rules of secrecy despite the harshest conditions of attrition and exile he and his wife were suffering at the hand of the authorities, including constant eavesdropping. Nobody could be sure whether he was being serious or sarcastic when once, as he and a physicist colleague were about to discuss classified information, Sakharov stopped the conversation, saying that the KGB officers listening in might not possess the necessary security clearance.[117]

Although Teller advocated the need to reduce government secrecy, he appeared in his debates to be misusing the myth of inside information by saying that if only his opponents could know what he knew. . . . An example of this occurred during a debate organized by Stanford students between Teller and Sidney Drell, who was at the time deputy director of the Stanford Linear Accelerator Center (SLAC) and was much engaged in arms control issues. The debate was about the intermediate-range missiles, the Pershing II, and their deployment in Europe in the 1980s. Drell thought they were not necessary, but since the decision was made, he thought it would be better if these missiles were on submarines or ships—in any case, at sea—rather than on land.

That was when Teller said something like "You are wrong, but I cannot tell you why. If you knew what I know, you'd know that I'm right." The sense was that the missiles should not be put on submarines because the submarines might be vulnerable. Drell pointed out that Teller used that argument in the past and he was using it now, and that it was not true. He added that since Teller used that kind of argument, he—that is, Drell—would quit the debate, and he got up and left. A scandal followed, which was discussed at length in the student newspaper. Teller apologized in a letter, but his apology was limited to only being sorry that he may have hurt Drell's feelings, while he maintained that his statement was true. He insisted that he knew more about defense issues than Drell, whereas he conceded that Drell knew more about elementary particles than he did.[118] The elementary particles were not the issue of the debate, of course. Drell did not accept the claim that he was not informed sufficiently for conducting the debate and did not accept Teller's apology.[119] Looking back decades later, Drell added that by revealing the vulnerability of the American submarines—if Teller's statement was true—Teller committed a serious breach of security.[120]

In spite of his consistent criticism of secrecy, Teller was known to observe the law rigorously and could not have been expected to divulge classified information even when he found its classified designation meaningless. In this light, the following episode has remained puzzling. When Livermore physicist Stephen Libby started having long discussions with Teller about modern physics, their joint lunches did not provide enough time, so Teller invited Libby to visit him at home in Palo Alto on the weekend. On the first occasion Teller asked Libby about his work, which at that time focused on specific problems of the X-ray laser. Teller wanted to know the details, at which point Libby warned him that they were not supposed to discuss it outside the laboratory. However, Teller did not give up and assured his visitor that they were alone except for the housekeeper, who did not understand English. Yet Libby insisted that they should not discuss classified topics outside of Livermore. At this point Teller abruptly switched the topic to basic physics. Libby is not sure to this day whether Teller was testing him or would have let him breach the security rules.[121]

After a case of alleged spying by the Chinese of atomic secrets at Los Alamos in 1999, Teller reacted consistently with his previous stand regarding security questions, and compared the attitudes of the Truman and the Clinton administrations. When Klaus Fuchs was unmasked as a Soviet atomic spy, Truman did not impose additional measures of security; rather, he called for acceleration of work on all aspects of nuclear weapons. In the alleged Chinese spy case, Teller lamented that the response of the Clinton administration was only to call for more security and the prevention of the participation of people from abroad in the weapons programs. Teller quipped: "[S]ecurity is acquired by new knowledge rather than by conserving old knowledge."[122]

## THREE MILE ISLAND

Teller had been interested in the safety of nuclear reactors for a long time. The disaster at Three Mile Island—the worst nuclear accident in the United States—drew his attention, as it did the world's. The accident began on March 28, 1979, and lasted for a few days. Its immediate cause was a faulty valve, which was then compounded by human operators' errors. The power plant was located near Harrisburg in Pennsylvania, where about twenty-five thousand people lived within five miles from it. A significant amount of radioactivity was released into the environment, but there were no fatalities, not even injuries that could have been directly ascribed to the accident. Long-range consequences to human health could not be excluded but were expected to be so small as to make it impossible to identify them unambiguously as being due to the radioactive release.

At the time the accident took place, Teller's Reactor Safeguard Committee (RSC) had long been out of commission; even the Atomic Energy Commission no longer existed. According to the Energy Reorganization Act of 1974, two new institutions replaced it from 1975: the Energy Research and Development Administration assumed its function concerning research and development and the Nuclear Regulatory Commission was charged with regulatory and licensing functions. In yet another reorganization, the Department of Energy was established in 1977 to coordinate federal-level energy policies and programs of the United States. The stringent regulatory oversight that Teller and his RSC used to exercise had long before been relaxed.

The excellent safety record of the nuclear industry probably contributed to the relaxation. But there was a confluence of international interest that underplayed safety problems and even the potential for accidents by the advocates of nuclear power production. It is also worth mentioning that there was scarcity of information about the catastrophe in the Ural region in 1957, which was surrounded by silence not only in the secretive Soviet Union but mostly in the West, as well.[123]

President Carter appointed a commission to investigate the Three Mile Island accident with computer scientist and Dartmouth College president John G. Kemeny as its head. Kemeny was born and spent his childhood in Budapest. He started high school at the Berzsenyi Gimnázium, where former pupils included the Nobel laureate inventor of holography, Dennis Gabor, the longtime California congressman Tom Lantos, and the financier George Soros. The commission's report contained "a damning indictment of the official attitudes towards the safety of nuclear power."[124] Only a few years before did Teller claim during a public debate that he was like the famous consumer advocate Ralph Nader because of his early and dedicated interest in reactor safety. His claim was met with laughter from the audience, which only showed their ignorance.[125]

Yet during the debate that followed the Three Mile Island accident, Teller appeared to be belittling the problem. He had large-size advertisements published in the media, in which he declared, "I was the only victim of Three-Mile Island." Soon after the accident, he worked so hard to refute Jane Fonda and other opponents of nuclear power plants that he fell ill and was admitted to the Cedars-Sinai Hospital in Los Angeles. He had had a mild heart attack. Considering his age—he was seventy-one years old—his general condition was good.[126] He had to take it easy for a while, so he reduced his traveling and increased his writing assignments. He accepted George (Jay) Keyworth's invitation to spend a few months recuperating at Los Alamos. It was during this stay that he dictated to Keyworth his recollections of the implosion project for the plutonium bomb and the origin of the new concept for the hydrogen bomb.[127]

Teller's ad asserted the safety of nuclear power and the nuclear fuel cycle as excessive, which generated some very negative comments from different quarters. The head of the Nuclear Regulatory Commission, Peter A. Bradford, protested vigorously.[128] He summarized his objections to Teller's ad in three points. One was that the ad did not mention that Dresser Industries, the sponsor of the ad, was the one that produced the faulty valve. Another was that by proclaiming himself the only casualty, Teller was ridiculing the fears and hardships of those who lived near the power plant, not to mention the fact that Teller was nowhere near the place at the time of the accident. Finally, the third was the harmful consequences of considering the accident so casually. The report of the Kemeny Commission called attention to the dangerous mindset of complacency in regard to the safety of nuclear reactors. Teller would have been the first to fight such complacency when he was in charge of the Reactor Safety Committee.

Lately, yet another danger has appeared for nuclear power plants, and that is terrorism. Alas, when years after the September 11, 2001, terror attacks the Nuclear Regulatory Commission was asked about the possible consequences of a terrorist attack on a nuclear power plant, its spokesperson declined even to enter a discussion as to what scenarios the NRC would consider credible. She said: "In calculating the threat of accident, the commission takes into account the probability of the event, . . . but the commission has long argued that it is impossible to calculate the probability of a terrorist attack and thus it does not need to take that threat into account when approving installations."[129] We can be sure that Teller in his earlier days and his RSC would have been dissatisfied with such an approach.

The number of Teller's "manias" kept growing to the end of his life and included a diverse array of topics. Some of them were clearly outside of his area of expertise. For example, he wrote numerous letters and appeals discussing how to combat breast cancer; encouraging girls to pursue careers in engineering; encouraging students with poor family backgrounds to continue their education in engineering; and suggesting an increase of teachers' salaries; among other topics. One of his recurring subjects was a controversial one, hyperbaric medicine, that is, the medical use of oxygen at elevated pressure. He had received many treatments under an oxygen tent and served on the Advisory Board of the American College of Hyperbaric Medicine, for which he performed some lobbying.[130]

Cloning was yet another topic to which he committed himself; he took a stand explicitly disagreeing with President George W. Bush. Teller did so while stressing that he was a devoted Republican. He suggested "immediate and strong support for animal cloning and stem-cell research for medical purposes."[131] Furthermore, he did not shy away from the question of human cloning. He considered human cloning premature, but he insisted that the United States should get involved if it did not want to be left behind in worldwide efforts. He agreed that for the time being the next human generation should be produced through the institution of marriage, but he left open the possibility of cloning individuals through other means. He proposed that a small committee should be required to make a unanimous recommendation in such cases and made recommendations himself as to the composition of such a committee. Teller was past his ninety-fifth birthday when he wrote this letter, which demonstrated his alertness regardless of whether one agreed or disagreed with its contents.

His "Open Letter to the Citizens of the Year of 2100," which he penned in 1999, contained a strong positive statement about cloning: "It worries me that more people are talking about the dangers of cloning and practically no emphasis is given to a statement 'love your brother because he is like yourself.'"[132] He expressed hope that the problem of cloning would be solved by the year 2100 and in a way that would correspond to the meaning of the word "brotherhood." According to a report from 2002, cloning was about as much on Teller's mind as missile defense at the time. To the question of whether Teller would like to see his own clone, he responded in the affirmative, and he gave an explanation for his response: "I remember my early interest in mathematics. I would like to discuss mathematics with my clone and see if I could arouse his enthusiasm."[133] One could interpret from this response that Teller liked to contemplate his life taking a different turn than the one he had followed.

# Chapter 11

# WARRING THE STARS

... an anti-missile system ... its purpose is not to kill people, but to save human lives.
—ALEKSEI KOSYGIN, PRIME MINISTER OF THE SOVIET UNION, 1967[1]

Wouldn't it be better to save lives than to avenge them?
—RONALD REAGAN, 1983[2]

... public policy could ... become the captive of a scientific-technological elite.
—DWIGHT D. EISENHOWER[3]

*President Eisenhower issued two warnings in his farewell address. One was that "we must guard against the acquisition of unwarranted influence, whether sought or unsought, by the military-industrial complex." The other, quoted above, could be formulated in a different way, namely, that there should be some healthy skepticism toward replacing political solutions to political problems with technical solutions. This warning was not heeded in deciding on and executing the policy of the Strategic Defense Initiative (SDI). The costly and much-promoted technical solutions of missile defense—first the X-ray laser called Excalibur, then the rocket interceptors called Brilliant Pebbles—never came to fruition, yet SDI still may have contributed to or at least accelerated the demise of the Soviet Union. The septuagenarian-turning-octogenarian Teller continued to play an active role in the events, not so much as a physicist but as a promoter, and he threw himself into his role with zeal.*

We have referred to a Teller interview on June 15, 1982, by conservative journalist William F. Buckley. During the interview, the conversation turned quickly to the question of the international situation—the timing was a low point in the later period of the cold war—and Teller made a comparison between the adventurer Hitler and the conservative people in the Kremlin. Teller complained about lacking funds for research, and Buckley was puzzled because Ronald Reagan was now president, and both he and his secretary of defense, Caspar Weinberger, were sympathetic to Teller's ideas. Teller told Buckley: "But neither of them know [*sic*] enough about technology. Each of them tends to see in terms of an arms race or more and more of the same. . . . Policy is made by people who do not understand enough about technology."[4] When Buckley expressed surprise that Teller apparently could not get sufficient support from "his old friend," the president, Teller let him in on a non-secret: "May I tell you one little secret which is not classified? From the time that President Reagan has been nominated I had not a single occasion to talk to him."[5]

## GREAT ALLIANCE

Reagan and Teller met for the first time at the Hoover Institution in 1966. Reagan, a former movie star and labor leader, ran successfully for governor of California in 1966, and soon after he was elected Teller invited him to visit Livermore. Here is how Teller described their interaction to me in 1996:

> Reagan became Governor of California in 1967. Early that year I went to see him, and told him: here is a laboratory in your state, you should visit us and see what we are doing. He came, and at the end of 1967 he heard in great detail about the possibility of defense against missiles. It was quite clear from that discussion that it was a subject new to him but interesting to him. He is a man well known for his excellent propensity of shooting from the hip. My opinion is that he heard about this possibility in a serious way in 1967, and he did shoot from the hip in 1983. Throughout the years he talked with many other people, and was very slow to make up his mind. But when he made up his mind, he did so firmly, and, furthermore, he knew my position, and I was then in the position to be listened to by him.[6]

George (Jay) A. Keyworth, President Reagan's science adviser, considered the discussions between Reagan and Teller in 1967 important in shaping Reagan's views. By the time he became president, Reagan had developed his "deep opposition to the doctrine of Mutually Assured Destruction (MAD) and, in particular, to the SALT I ABM treaty."[7]

Teller campaigned enthusiastically for Reagan in the presidential race in 1980 and again in 1984 for Reagan's reelection. He immersed himself into actual campaigning when he accepted the largely but not entirely ceremonial leadership of the movement of Hungarian–Americans to elect and then reelect Ronald Reagan. Reagan was elected president in November 1980; assumed his office in January 1981; and declared his Strategic Defense Initiative (SDI) on March 23, 1983. Teller had opportunities to see the president in person between the Buckley interview on June 15, 1982, and the SDI speech, and he had other channels as well to exercise influence over Reagan.

One such channel allowed Teller to recommend that Keyworth be appointed to his post. It was at the time when Reagan's staff was being assembled and Teller was visiting the White House in early 1981. Teller did not meet with the president on this occasion, but he talked with Martin Anderson, whom he knew from the Hoover Institution and who was then the president's domestic policy adviser. Teller called Keyworth to prepare him to expect the invitation and told him, displaying his typical humor, "I want you to do me a favor. I want you to tell one tiny lie, just for me. Let's just call it a white lie. Tell them that you can, at least for a while, be a team player."[8]

The caveat to Keyworth about being a team player shows Teller's political acumen, despite his own foibles. In looking for a science adviser, the Reagan White House wanted a scientist who could advise the administration about matters related to science but not to be the usual liaison between the president and the scientific community. Being a team player meant someone who provided support rather than criticism. Keyworth was not the first choice, but better-known scientists had turned down the invitation for the appointment. After he took office, Keyworth set up a fifteen-member White House Science Council (WHSC) of which Teller became a member. The importance of the WHSC was limited in that its role was to advise Keyworth rather than directly the president.

When the Buckley interview was broadcast and Keyworth heard about Teller's lamentation, he arranged for a visit. He thought that it would be useful for Reagan to hear about Teller's experience with the Nuclear Winter movement. Keyworth did not want to leave anything to chance, so he and Teller went so far as to rehearse the meeting. However, at the actual meeting, Teller "dramatically altered the topic, and the tone of the meeting," and "delivered a passionate harangue on the topic of third generation weapons, warning that the Soviet Union was developing them and would soon be in a position to blackmail the U.S." When the president asked Teller what he should do, Teller suggested that he immediately transfer $250 million to the Livermore Laboratory.[9] Incidentally, during the Buckley interview, Teller said they needed $110 million.[10]

Strategic defense was so much on Teller's mind as he was entering the Oval Office that he was muttering, "third generation, third generation," according to someone at the meeting.[11] "Third generation" referred to the latest crop of nuclear weapons after the atomic bomb and hydrogen bomb that corresponded to the first and second generations. Teller did not find his visit with the president very successful. There were far too many people present in addition to the president to make him feel comfortable, and he felt he was interrupted too many times.[12]

## SDI

According to Keyworth, Teller "had no role in the President's decision to make his now-famous SDI speech on March 23, 1983," which may be literally so. His influence came through the members of Reagan's "Kitchen Cabinet," which consisted of the president's old friends. Apparently, there was substantial difference between Teller's interest in the third-generation nuclear weapons and Reagan's intention of reducing reliance on nuclear weapons. In Keyworth's words:

> Edward's thinking about missile defense in 1983 was largely, if not uniquely, focused on his bomb-pumped X-ray laser. Reagan, who had spent much of the previous two years focused on the many aspects, and tribulations, of selling to the public and the Congress his program to modernize strategic forces, from Trident-carrying submarines, to Stealth bombers and MX land-based missiles and, most important, a new command and control infrastructure, was simply on a different path than Edward. As the president grasped the intricacies of moving from simple MAD where the targets are cities, to the then-emerging counter force deterrence, where the missiles themselves are the prime targets, he came to realize that even the unacceptable but reasonably stable MAD was eroding into a less stable, meaning more risky balance, one where a preemptive strike could be more easily envisioned. So Ronald Reagan made his decision to write his SDI speech, calling for developing means to make nuclear weapons "impotent and obsolete" while Edward was trying to develop a new generation of directed-energy nuclear weapons. Edward

helped launch Ronald Reagan, beginning in the 1970s, on his opposition to MAD, but he had little influence on the SDI speech. In fact, Edward was unaware of it until I called him to invite him to the meeting of scientists and statesmen at the White House on the evening the President delivered the SDI speech. Even then, I couldn't tell him the subject of the speech—just that he would be pleased.[13]

In addition to his Livermore visit in 1967, Reagan visited the air force observation and communication facilities at Cheyenne Mountain in 1980, when he was already a presidential candidate. He was impressed by the capabilities of observing a missile approaching an American target but gravely distressed when he learned that apart from observing it, they could do nothing to destroy it.[14]

The question of defense became a central consideration for Reagan from the start of his presidency. His Kitchen Cabinet channeled various studies to him. One of them was from the conservative Heritage Foundation, which reflected Teller's views about defense with the most advanced means technology could provide. The X-ray laser was already being developed at Livermore and was called Excalibur, after King Arthur's sword, but it did not generate much enthusiasm in Keyworth. This is why he wanted Teller to speak about something else on the occasion of his meeting with Reagan in September 1982, but Teller was unstoppable.

The Excalibur belonged to the third-generation nuclear weapons. It was to be pumped by a nuclear bomb whose directed energy would reach its target with the speed of light. It was only one of the approaches promoted by various people dedicated to building up a new defense program. Teller was Excalibur's principal promoter along with his protégé, Lowell Wood. They did not stop at using one channel for exerting their influence but waged a program of informing congressional and military leaders about it too.

It is remarkable how small was the circle of people who were involved in the preparation of President Reagan's major address on March 23, 1983. Even Keyworth learned about the speech only a mere four days before it was delivered. The main thrust of the address was to be the continuing Soviet threat and a request to Congress for an increased budget for the Pentagon. Only a small group of the National Security Council headed by Deputy National Security Adviser Robert McFarlane helped Reagan in drafting the part about strategic defense against ballistic missiles. The Strategic Defense Initiative was almost an afterthought for the address; it was called the "insert."[15]

Opinions are divided as to who played what role in President Reagan's formulation of his SDI policy. That Teller's role was limited was indicated by Reagan's great emphasis on the nonnuclear nature of the SDI, which could hardly have come from him. Jerome B. Wiesner, science adviser to both Presidents Kennedy and Johnson, ascribed the primary influence over Reagan in formulating the SDI approach to retired air force general Daniel Graham.[16] The general had developed a science-fiction–like alternative to the policy of MAD, which he called "High Frontier," the core of which was a system of space-based interceptors against incoming missiles.

It is not expected of presidents to be versed in the intricacies of technology; this is why they have advisers and experts at their disposal, and also why they have unlimited means to gather the best technical information and alternatives to aid in their decision making. The question is whether President Reagan utilized the best minds in this respect before making the momentous decision about SDI. All signs indicate that he did not. Rather, he relied on a narrow group of people, mostly nonexperts, to formulate his policy. Of course, Teller was an exception because of his scientific brilliance; unfortunately there were hardly any other physicists of similar caliber close to President Reagan, who could have questioned Teller's often unchecked ideas.

After the Reagan presidency was over, McFarlane commented that Teller's stature was more important to Reagan than the device Teller was promoting.[17] Teller's participation in the conservative movement around Reagan was conspicuous because he was the only world-renowned scientist willing to speak up, travel, and use every opportunity to voice his anti-Soviet and pro–strong-defense politics. He contributed significantly to the political atmosphere in which Reagan's speech gestated and he could count on congressional support for the defense programs he advocated. Teller could argue by invoking science whether it was credible or not; he had scientific authority. He was not above using this authority in presenting such controversial issues as the clean bomb and the limited nuclear war.

In his speech, President Reagan reviewed the role of nuclear weapons in maintaining stability and preventing war. However, he called the reliance on mutual threat "a sad commentary on the human condition." And he asked: "Wouldn't it be better to save lives than to avenge them?" He proposed deterring a Soviet attack by intercepting and destroying the missiles before they could reach their targets. He turned to the scientists with the following appeal: "I call upon the scientific community in our country, those who gave us nuclear weapons, to turn their great talents now to the cause of mankind and world peace, to give us the means of rendering these nuclear weapons impotent and obsolete."[18]

Reagan referred to the American technology "that spawned our great industrial base and that has given us the quality of life we enjoy today." In this respect, a tremendous difference existed between the United States and the Soviet Union. There were many world-class scientists in the Soviet Union, but the means of transferring their discoveries into applications lagged far behind that of America. For example, in American life, computerization and miniaturization made great strides in the military and the civilian sectors alike, the two mutually reinforcing each other. This was not the case in the Soviet Union. It was also noteworthy that defense spending was a much smaller portion of the Gross National Product in the United States than in the Soviet Union. Had the technology of SDI been feasible, and had the United States adopted it, the Soviets could not have followed suit, not in any reasonable time frame.

There have been suggestions that a covert purpose of SDI was to bleed the Soviet Union to poverty by forcing it into overspending. Robert M. Gates, secretary of defense under Presidents George W. Bush and Barack Obama, wrote in 1996 that "SDI was a Soviet nightmare come to life . . . that would require an expensive Soviet response at a time of deep economic crisis."[19] By the time Gates wrote this he had served as director of the Central Intelligence Agency under the first President George Bush.

However, there appeared to be conflicting information about the Soviet opinion as to whether SDI might or might not work. While the Soviets declared that the SDI would not work, they were engaged in a similar development that was based on the belief that missile defense was possible. The Soviets had a ballistic missile defense system around Moscow, whereas nothing comparable existed in the United States. There were efforts in Soviet research facilities toward reaching similar goals to those of SDI, but the CIA also advised that the Soviets were hindered by their "relative backwardness in remote sensing and computer technologies."[20] Teller stressed that the Soviets had already engaged in missile defense long before Reagan's SDI, and repeatedly suggested that they call their own efforts Strategic Defense Response, SDR, rather than SDI, implying that in reality, the new policy was not an initiative.

The Soviet supreme leader at the time, Yuri Andropov, was quick to react to President Reagan's speech, condemning it as a means to fuel the continuation of the arms race. Andropov referred to Aleksei Kosygin's words from fifteen years before with which the Soviet premier introduced the Soviet deployment of antimissile defense. Kosygin in 1967 stressed that the purpose of their antimissile system was to save human lives rather than to kill people. At that time the United States interpreted the development of the Soviet antimissile system as another step toward accelerating the arms race. At the Glassboro Summit between President Lyndon B. Johnson and Kosygin in 1967, the American president argued against building a missile defense system either in the United States or in the Soviet Union.

Possessing a reliable strategic defense system is supposed to make a country invulnerable. However, there is a danger that it may tempt its possessor to launch a first strike without the fear of retaliation. The Soviet development of their antimissile system accelerated negotiations over the antiballistic missile treaty in 1972, which outlawed large-scale defensive systems. Teller quoted the Soviet premier's words (used as one of this chapter's opening quotations), which were very similar to one of Reagan's slogans, except that the Soviets had pursued a strategic defense system long before the Americans. Attention was called to similarities between the Soviet and American approaches by saying that "Moscow's reaction should be familiar: It used to be ours."[21]

The pursuit of SDI was fundamentally different from those of the Manhattan Project and the hydrogen bomb. When the Manhattan Project was set up, the technology of creating the atomic bomb had been worked out. When the theoretical physicists gathered at Berkeley in the summer of 1942, they talked more about the Super, which was not a goal for the current war, than about the atomic bomb, which had to be produced because the basic problems of the atomic bomb had been solved by then. When President Truman issued his directive for the continuation of the work on the hydrogen bomb on January 31, 1950, the technology was not yet in place, but there was an approach to it, which might or might have not worked, and it took one ingenious discovery to solve the problem.

As for SDI, not only did its scientific problems appear insurmountable, most of the very scientists to whom the president turned declined to participate in it. Again, this was different not only from the Manhattan Project but from the hydrogen bomb project, too, where even its staunch opponents joined the project however reluctantly—but once part of it, they gave their all to it. The scientists had not been consulted before Reagan's SDI address, but once it was delivered, they spoke up, and the eminent scientists who voiced their opinions opposed it. Once again, Teller found himself an outcast in the scientific community.

At the same time, there was scarcity of information from the Soviet Union, so it could be supposed that America's archenemy was already much advanced in its own strategic defense. It would also have been consistent with everything the Americans had heard about the Soviet Union—that its leaders would be ready and willing to make great sacrifices for carrying out their supposedly expansionist plans. Debates were not tolerated in the Soviet Union. Furthermore, superficially, the whole Eastern bloc appeared monolithic to the unsuspecting observer, although not to the careful scholars of the communist world.

President Reagan did not specify the actual technical approaches in his SDI speech except to say that it would render nuclear weapons "impotent and obsolete." Prior to the SDI speech, there were three competing concepts that had been discussed in the media.[22] Two of the three approaches involved lasers, and the third was an improved version of an earlier concept called BAMBI. The acronym stood for "Ballistic Missile Boost Intercept," which was supposed to unleash small rockets that would bump into enemy missiles during their initial flight and smash them. It was to be similar to an unmanned kamikaze action. One of the two laser solutions was large-size stations orbiting in space using chemical lasers to destroy enemy missiles. The other was the X-ray laser. Nothing better characterizes Teller's influence on Reagan's SDI dream than the fact that for years the president did not perceive—or perhaps he chose to ignore—the fact that the Excalibur was based on a nuclear bomb, all the while he advocated *nonnuclear* defense against nuclear weapons.

## VISION AND DEDICATION

Charles H. Townes, co-discoverer of the laser, was one of about a dozen prominent scientists invited to the White House to attend the president's speech and dine with him. Townes had received the Physics Nobel Prize in 1964, essentially for his work leading to the laser, and shared it with two Soviet physicists, Nikolai G. Basov and Aleksandr M. Prokhorov. The original discovery, which occurred at about the same time in the United States and the Soviet Union in the 1950s, was that of the maser, which amplified a microwave signal by stimulated emission. The name "maser" came from *m*icrowave *a*mplification of *s*timulated *e*mission of *r*adiation. The laser was one of its possible versions in which the microwave radiation was substituted by light.[23] Like "maser," "laser" is also an acronym and stands for *l*ight *a*mplification by *s*timulated *e*mission of *r*adiation. In spite of its being an acronym, a verb has evolved from it as well, "lasing."

The first laser was built by Theodore Maiman at the Hughes Research Laboratory in 1960.[24] It was announced at a press conference, and the initial reactions were not all enthusiastic. The *New York Times* announced its arrival as "Light Amplification Claimed by Scientist."[25] Another account suggested that "Hughes Labs invented a killer-ray gun with a future on battlefields to come."[26] Although the laser was not mentioned in Reagan's speech, laser physics and its applications had subsequently become prominent in the program. That very evening, Townes asked others around the White House whether they thought the SDI feasible, people like the national security adviser and the chairman of the Joint Chiefs of Staff. He received evasive responses, as if they had reservations about it, "but nobody wanted to disappoint the president."[27]

That evening Teller appeared very pleased to see Townes. In the course of the evening Townes asked Teller directly whether he had been talking to President Reagan about SDI; his response was, "I haven't met with the President in quite a while now."[28] I talked with Townes in 2002 about SDI, and he did not want to add anything to what he had written in his book *How the Laser Happened*. About Teller he said: "I have had substantial interactions with Teller. I believe he was sincere and well intentioned, but I disagreed strongly with him in a number of ways. He felt strongly against communism and I believe took somewhat extreme positions partly because of this."[29]

Teller had long advocated enhanced defense efforts. As early as 1945, when he wrote a report for the navy, he proposed defense against atomic weapons. That was at the time when the United States enjoyed a monopoly on atomic bombs. Thus, Reagan's policy was very much in accord with Teller's wishes, and perhaps Teller's relentless lobbying had contributed to it. Soon, the X-ray laser would become the centerpiece of SDI for some years, ensuring a leading role for the Livermore Laboratory. At this time Teller was no longer involved in technical details, but he had a firm grasp of them. The laser had never been a subject of his research, and in the 1950s he had not yet been interested in it. Of course, the laser did not yet exist then, but von Neumann wrote to him, on September 19, 1953, about his idea that was a progenitor of the laser, according to Townes. Alas, Teller never responded to von Neumann's query, and von Neumann did nothing further about it.

Teller's interest was kindled after the actual discovery of the laser and especially through his participation in the work of the US Air Force Scientific Advisory Board. One of the subcommittees of this board looked into the possible military applications of the laser, and Teller became interested in equipping large planes with laser weapons.[30] A great advantage of a laser in shooting down moving targets is that its beam travels at the speed of light, so the change of target position is negligible from the moment of firing to hitting. The Airborne Laser Laboratory (ALL) program was born in 1971 and was completed in 1983. Teller developed a broad knowledge of laser physics and its applications in chemistry. Similarly, the Livermore Laboratory became very active in laser research and in looking for possibilities of applying the laser in weapons development.

In 1972, the organizers of the prestigious chemistry conferences in Houston, Texas, under the auspices of the Robert A. Welch Foundation, invited Teller to give a talk on "Lasers in Chemistry."[31] The lively and informative talk was very well received, and the participants of the ensuing discussion were top scientists. At some point during his presentation, Teller brought up the topic of X-ray lasers, which did not yet exist. Still, he considered them potentially most important for applications. Nonetheless, he did not mention weaponry but rather a purely basic scientific use: molecular structure research in crystals. Teller recognized that the X-ray laser could become pivotal in solving some heretofore unsolved problems in crystal structure analysis. This observation by Teller about a technique that did not yet exist (and still does not) was a stroke of his genius.

The X-ray laser would produce high-intensity X-rays, which would be much more powerful than the conventional light laser because the much shorter wavelength of the X-rays carried much higher energy. Although there might be other ways to generate an X-ray laser, the SDI tests utilized radiation from a nuclear blast, which was directed onto rods whose length could be varied. The atoms in the lasing material lost some electrons, while the electrons that stayed got excited, meaning they jumped onto orbits of higher energy. When they returned to orbits of lower energy, they emitted electromagnetic radiation—X-rays, in this case—and the radiation left the rods in the direction in which the rods were pointing. All this was taking place during a very short time period, a fraction of a microsecond, before the whole device was annihilated by the hydrogen bomb explosion. The powerful beam—that had left the device before it disappeared—was being directed toward a target, and its immense heat destroyed it.

It was not the first time that using powerful beams would be considered for developing weapons. As early as 1950, Peter Kapitza wrote a letter to one of the Soviet party leaders, eventually Prime Minister, Georgii M. Malenkov, proposing to work on high-energy radiation that would be able to destroy incoming missiles.[32] The three physicists who compiled the book in 1990 in which the letter was reproduced wrote: "These ideas were somewhat similar to those of the present American Strategic Defense Initiative programme, although with the technology of 40 years ago, even more remote from practical realization than the SDI of today."[33]

Wolfgang Panofsky declined the invitation to attend the SDI speech and the White House dinner without knowing what it would be about.

After the speech he expressed doubts that a Manhattan Project–like effort could be organized for SDI. He was troubled by the request to the scientists to create "a workable defensive shield against ballistic missiles."[34] Victor Weisskopf, who attended the White House event, called SDI "extremely dangerous and destabilizing."[35] For Hans Bethe, who was also present, it was a "mystery" why they were invited to the event when so little had yet been understood. He called the briefing "very thin" and regretted that the scientists' questions could not be answered properly.[36] Other scientists expressed fear that the initiative would undermine the prospects for future arms control agreements with the Soviet Union. Yet others found the plans described in SDI unrealistic; one called it "Pie in the Sky." John Bardeen was distressed that the members of the White House Science Council were not even consulted about SDI in advance. He resigned his membership of the WHSC two weeks after the presidential announcement.[37] The reports in the media pointed to Teller as the only well-known scientist who supported the Reagan plan.

In exactly one week after President Reagan's speech, two op-ed pieces appeared side by side in the *New York Times* reflecting on SDI, one by Richard Garwin and the other by Teller. They were in an antisymmetric relationship with each other. The title of Garwin's article was "Reagan's Riskiness."[38] In establishing his credentials, Garwin introduced himself as the author of the virtually perfect design of the "Mike" shot in 1952, the first-ever thermonuclear explosion. Garwin stressed that defense can never be perfect, something Teller never claimed either, but the imperfection in this case would mean that a sufficient number of nuclear missiles could reach and destroy much of the United States. Garwin warned that "the Soviet Union will exploit every opportunity to blind, render dumb, decoy or attack proposed defenses that threaten to disarm it."

At this time, it was not yet clear whether or not the technologies considered in the SDI would work, so Garwin considered the possibility that they might. In that case, preference, he felt, should be given to destroying the missiles at the booster stage. Destroying incoming missiles would be hopeless, as it would be impossible to distinguish the real ones from the numerous decoys. Attacking the missiles at the booster stage, though, would also be hindered by various technological difficulties. Further, Garwin warned that war in space, which SDI would inevitably lead to, would only be a prelude to war on Earth. He favored the approach of mutual deterrence over the attractive but impossible dream of saving lives. He saw the tendency for increased military spending and the love of technology in SDI, and called the whole exercise a "dangerous misconception."

Teller's article was titled "Reagan's Courage."[39] He made a comparison between Franklin D. Roosevelt and the circumstances of the decision to develop the atomic bomb, on the one hand, and Reagan and SDI, on the other. According to Teller, the introduction of nuclear weapons during World War II gave predominance to offense instead of defense, while Reagan's proposal might lead to a "multiple shield . . . that *completely* protects America from even a sizeable nuclear attack." Here italics were added, because in many other instances Teller emphasized that there was no perfect defense, but even partial defense was better than no defense.

In a surprising twist, Teller contrasted the approaches by Roosevelt and Reagan. He said that Roosevelt did not ask questions and referred to his optimism in making his decision about developing the atomic bomb. The more skeptical Reagan asked questions of Keyworth and many other scientists (Teller included) before deciding. He added, "The result of Mr. Roosevelt's action was to furnish the strongest argument for weapons of aggression. With the help of high technology, the result of Mr. Reagan's action could be to prevent nuclear destruction by offering real defense. Mr. Reagan showed a true sense of historical perspective." There seems to be a thinly disguised criticism toward Roosevelt here.

It is odd that Teller contrasted the decision-making process leading to the Manhattan Project with that for SDI because the latter was still only a policy announcement. The subsequent choice of the X-ray laser could not have been made on the basis of detailed scientific considerations because they just did not exist. As for the atomic bomb, it is worth remembering Szilard's and Fermi's experiments at Columbia University, the protracted meeting of top theoreticians at Berkeley in the summer of 1942, and the construction of the first-ever atomic "pile" at Chicago before the construction of the atomic bomb commenced and the Los Alamos Laboratory was born. Teller may have meant President Roosevelt's decision to create the Uranium Committee, because he decided that on the spur of the moment upon having read Einstein's letter, but that was far from a commitment.

Teller then touched on a sensitive issue, the cost of defense, which must be lower than that of offense, otherwise defense would never catch up with offense. But if defense were more economical, Teller said, the other side would not be able to overcome it by building more offensive arms, and the whole approach of MAD would become obsolete. Then he brought up another delicate issue about whether nuclear weapons would or would not be part of the new defense initiative. Mr. Reagan's words could be interpreted as meaning that everything should be nonnuclear, whereas Teller did not interpret them as if nuclear means would be excluded. This was an important point because the X-ray laser whose development and eventual deployment he advocated contained a nuclear element: the X-ray lasers had to be initiated by a thermonuclear explosion. Toward the end of his essay, Teller mentioned the value of President Reagan's making sure "that the novel defense-oriented ideas proposed by an ingenious group of young scientists had a way of reaching him."[40] What he did not say was that the group was in Livermore and that he, Edward Teller, was the link between them and the president.

It is curious that Teller chose the Manhattan Project to compare with SDI and did not make a comparison rather between the decision of developing the hydrogen bomb and the pronouncement about SDI. In case of the hydrogen bomb the main argument was that the Soviet Union would develop it regardless of whether the United States would or would not. Stalin could then blackmail America and the Free World as the sole possessor of this most terrible weapon. Although he missed this opportunity in 1983, he made ample use of such a comparison in his book *Better a Shield Than a Sword*, which was published in 1987.[41] By then, he had faced fierce opposition toward SDI by many of his peers, which may have hurt him but did not change his approach. And to some degree, he relished his role by then, being the one righteous voice against the rest.

In his book, he attempts to make the division between supporters and opponents of SDI into a generational question. He says that the older scientists had been out of touch for a long time, whereas the new ideas were coming from young scientists. He ascribes their lack of recognition by the scientific community to the consequence of their work being classified. Further, he relates the division to political difference; the opponents are on the left and the supporters on the right. He finds yet another and even more ominous difference between the hydrogen bomb decision and SDI in the relative strengths of the United States and the Soviet Union. Whereas the military superiority of the United States in 1949 was unchallenged, in the mid-1980s, the Soviet Union had overtaken the United States in every aspect of military matters. Following Truman's decision about the hydrogen bomb, Teller stressed that scientists must create new knowledge and explain its possible applications, but the decision about deployment should belong to the entire community. Teller did not obey his own stipulation in the debate about the development of the hydrogen bomb, and in the SDI controversy he again did exactly the opposite of what he prescribed.

## EXCALIBUR

When announcing the policy of SDI, President Reagan did not yet make a commitment as to which technology the program would focus on. Teller lobbied among Reagan's official and unofficial advisers to make sure that the X-ray laser would be the choice. At that time, the only evidence pointing toward its feasibility was a single underground nuclear explosion in the Nevada desert, code-named Dauphin, on November 14, 1980. The X-ray laser, Excalibur, was expected to carry a tremendous amount of energy as a beam, a million times brighter than a thermonuclear bomb. One of the prominent participants in the X-ray laser program was a young physicist at Livermore, Peter L. Hagelstein. His original interest was in finding new biomedical applications for X-rays and for lasers, and he had no plans of having anything to do with nuclear physics and weapons. When he succeeded in combining the two—the X-rays and the laser—he found himself in the midst of nuclear physics, and it was applied to weaponry. Hagelstein quickly became fascinated with both. As it is popular to attach labels to inventors, journalists called him—mistakenly, as it happened—the "father of the nuclear-powered X-ray laser," an inaccurate title, because another Livermore physicist, George Chapline, was its principal inventor.

SDI became known by its critics as "Star Wars," a name that reflected the notion that it reminded many of the *Star War* movies; and it also pointed to the fact that the proposed program would weaponize space. For Teller, the military use of space was a long-considered concept. For example, in a long letter to Henry Kissinger as early as 1960 (!), he discussed such possibilities. He admitted that much of what he wrote must have sounded fantastic. Yet, he added, "I am more worried by the many things of which I am ignorant and which I have therefore omitted than by the improbability of those things which I have mentioned."[42] In a letter to then congressman Donald Rumsfeld in 1963 he called attention to the military applications of space and complained that they had been neglected. He considered it a mistake to separate the military and the peaceful applications. Rumsfeld had proposed a congressional select committee to investigate the issue, and Teller expressed support for Rumsfeld's proposal.[43]

Teller loved to talk about his SDI work and the inventions of his gifted associates. Lowell Wood was so close to him that some referred to Wood as Teller's second son, since he emulated his enthusiasm and dedication. Teller appeared rejuvenated by his Livermore environment. His advocacy of the X-ray laser for the centerpiece of SDI was going on at different levels, such as briefings of and letters written to administration officials and congressmen, interviews, newspaper articles, and more. The presentations whether in private or in the media were augmented with charts, models, and artists' renderings of the lasers and their actions. The public was being educated on the mechanism of the generation and action of the X-ray laser.

The impressed and willing journalists conveyed the boastful statements to the public, although many aired skeptical opinions as well. The media often succeeded in explaining rather complex physics and other technical features to an audience not versed in such matters. People learned that the X-ray laser was powered by a nuclear explosion and that it consisted of many narrow tubes whose strength was proportional to their length. The X-rays generated in these tubes reached their targets with the speed of light; thus once the laser was locked onto its target, it could not miss it, however fast the target was moving.

There were bombastic promises, like the one that such a nuclear explosion could power a sufficiently strong X-ray laser to destroy the entire force of strategic Soviet missiles. This in itself raised the question of whether such a weapon could be considered defensive only, and whether the Soviets would view it as such. Also, just as the X-ray laser would be capable of destroying an enemy missile, it could also destroy anything else. When confronted with the possibility that the X-ray lasers might be capable of destroying other targets, not just missiles, Teller did not consider it reasonable. "To use this expensive system to accomplish something as pedestrian as that, something that could be accomplished much more easily by methods already available, what kind of sense is that?" he asked.[44] However, his was not a convincing argument against the possibility that Excalibur might also attack, not only defend.

If considering other kinds of lasers as well, the potential for military applications broadens considerably. According to Garwin, chemical lasers could "shoot down airplanes or set millions of fires simultaneously all over the Soviet Union."[45] Henry W. Kendall of the Union of Concerned Scientists, an organization in opposition to SDI, noted that lasers might be used for "selective assassination," for example, killing "a whole row of top Kremlin officials watching the annual May Day military parade."[46]

Whereas there were doubts as to whether the X-ray laser if developed into a weapon would be capable of destroying enemy missiles, there were no doubts that it would be an excellent tool to destroy satellites and whatever else it might find in space. Penetrating the atmosphere could be a problem because of the dissipation of the energy of the X-rays on the molecules of the atmosphere. However, plans were being conceived to enhance the energy of the X-rays to make such a penetration possible. Alternatively, they could have been launched from shorter distances to the target, which would have saved some of their energy.[47]

The X-ray laser was a project of the Livermore Laboratory, and Teller promoted it with all his skills and experience. Perhaps due to insider knowledge, some of the early skeptics also came from Livermore. Teller did not appreciate dissenters and he had great authority within and outside the laboratory. For some, the thought of Teller's treating them as a nonperson sufficed to keep them silent. With Reagan's election Teller's autocratic style hardened.

American satirist Art Buchwald described the talent hunt for Excalibur in an article under the title "The Laser Brain Hunt." According to him, in this hunt loyalty to conservative politics counted more than professional credentials.[48] He may have found this characteristic of the Reagan administration, but it was not of Teller's environment. Livermore had excellent scientists, young and ambitious researchers, for whom weapons development was more than a job. It was their avocation, and they were willing to be engaged literally day and night. Teller did not need to go into the details of their work, but his insatiable curiosity served as catalyst. Thus there was a difference between the feasibility of SDI, including its grand-scale promotion and the actual research whose many elements were sound and were being carried out at the highest professional level.

The security of the carriers of Excalibur presented some problem from the start because their orbiting in space would make them vulnerable. One of the possible solutions was bringing them up high enough to where they could take a geosynchronous position above the equator and move in synchrony with the rotation of the earth. The tremendous distance from the earth would diminish their vulnerability to attack. Another solution was to make Excalibur "pop up" from a silo or a submarine into space with extraordinarily high speed at the moment a Soviet attack in the making was detected. This was the method Teller preferred and publicized. It was criticized, however, because such a pop-up X-ray laser could be made useless by fast-burn Soviet boosters achieved at minimal increase in cost.

In August 1984, a study by a group of scientists headed by Stanford physicist Sidney Drell declared SDI "technically unfeasible and strategically unsound."[49] The study recommended a freeze on the amount of existing funding, which was about $1.5 billion per year, and a limit on work to basic research. This was reasonable because without the proper understanding of the science of the X-ray laser, it would have been difficult to move it to applications. However, Teller and his associates were eager to declare the project sufficiently ripe for entering the engineering phase. The Drell report also warned that the planned space-based defense would violate the ABM Treaty, which, as we've seen, Teller had always found harmful to American aspirations.

By the spring of 1985, various problems marred the Excalibur program. One was the brightness, that is, the power of the X-ray laser. There were gradual improvements; first Excalibur Plus, then Super Excalibur came about with supposedly increased brightness by many orders of magnitude. One of the sources of the increased brightness was improved focusing of the X-rays. Even Hans Bethe was impressed, and he had been one of the principal critics of the program. Now Bethe stated that just because there was a nuclear component in the X-ray laser, it should not be disqualified. This was in contrast with President Reagan's intentions, who kept stressing the nonnuclear nature of SDI. When he met with the new Soviet leader Mikhail Gorbachev in Geneva in November 1985, he told him that "we are investigating non-nuclear defensive systems that would only threaten offensive missiles, not people."[50]

In spite of Bethe's endorsement of parts of the program, the Cornell physicist continued to consider SDI in its totality as nonsense. He was very critical of Teller in his general approach, pointing out that his friend— former friend would be more accurate—had more faith in technology than in diplomacy. Teller did not notice or did not want to notice that the Soviet Union had undergone significant changes since the days of Stalin, and had become more amenable to negotiations under its current leaders than ever before. On the other hand, there was the danger of the Soviet Union finding itself cornered by the technological advantage of the United States, deciding on a first strike. The situation was very complex indeed.

The most important blow to the Excalibur program came from Livermore's rival, the Los Alamos Laboratory, and it was purely technical. They uncovered some fundamental flaws with the information Livermore provided about the brightness of the X-ray laser. The charges were confirmed by independent studies at Livermore, and the revised data were duly reported in the summer of 1985 to the Department of Energy, which was the higher authority to both laboratories. The setback did not change Teller's optimism, and it did not seem to lessen his standing with the Reagan administration. It is more surprising that even Teller's congressional support did not dwindle. This meant that he had more than just Republicans backing him.

Still, some individual Democrats tried to curb his subsidies.[51] For

example, John Kerry, then a freshman senator from Massachusetts, proposed that there should be no increase in the SDI budget; rather, it should
be kept at $1.9 billion. Kerry noted that the program already had more
money than it could spend. He mentioned that the recruiters of the SDI program were "skulking the halls of MIT 'like vultures,' looking for blue
ribbon eastern scientists to give a little gloss to a program that was hatched
in the West in the febrile brain of Dr. Edward Teller, father of the H-bomb
and favorite of the president."[52] However, Kerry's proposal was defeated at
68 to 21. Future vice president Senator Al Gore, Democrat from Tennessee,
introduced a resolution in the Senate to limit research to areas that would
not violate the ABM Treaty; it was also defeated. Teller, like Reagan,
proved to be untouchable. At some point in 1986, it was suggested to
expose SDI to the scrutiny of an expert panel.[53] It was too late; Teller's
roaming imagination combined with Reagan's aims could not be stopped.

Teller still had his remarkable gift for debate. Nobel laureate physicist
Philip W. Anderson was Teller's opponent, and the two debated twice. One
of the debates was taped in New York City for a WNYC radio program
on NPR, but it was never aired as far as Anderson knows. Like his prede-
cessors in debating Teller, the Princeton physicist found it "a hopeless
task."[54] However, theirs was not an angry exchange. According to
Anderson "it could have sounded friendly, . . . I was just discouraged by
the charisma of the man and his ability to seem reasonable and friendly
when he's nothing of the kind."[55] Anderson was a strong opponent of SDI.
First he did not give too much importance to Reagan's SDI speech, but by
the summer of 1983 he and many of his colleagues realized that the president was serious about it, and they responded with a Pledge Campaign.
They made a pledge not to take money from the Star Wars program.[56]

Another Nobel laureate physicist, Arno Penzias, witnessed an
Anderson–Teller debate in Washington, DC, during an annual event. Several hundred high school seniors attended together with a mix of "successful adults from various walks of life including astronauts, billionaires,
CIA chiefs, directors, Emmy-winners, and film stars. Scientists were also
well-represented."[57] Anderson talked about some of the technical obstacles that would hinder achieving the goals of SDI and he also warned
about the dangers of upsetting the existing balance in strategic weapons.
But his well-reasoned account never had a chance against Teller's direct
appeal to the audience. Teller suggested to ignore the technical obstacles
and to doubt the "naysayers," who would be proven wrong one more
time. He stressed that "the present stability depended on nuclear arms,"
whereas SDI offered the only way out. Assuring the audience that science

would find a way out of the threat of nuclear destruction, Teller won a standing ovation from the assembled students.[58]

Even after various debacles, Teller's high standing in the Reagan administration continued. In spite of setbacks with the Excalibur program, he could personally appeal to the president and obtain a promise of $100 million of additional funding in 1985, but that money had to be collected from other programs. There was a meeting at the Pentagon on September 6, 1985, to decide how to distribute the extra funding. It was presided over by Lieutenant General James A. Abrahamson, who was head of the SDI Office. In addition to federal officials, representatives of the three laboratories involved—Livermore, Los Alamos, and Sandia National Laboratory of Albuquerque, New Mexico—were present. According to a preliminary agreement among the three laboratories, Livermore would receive $60 million, Los Alamos $30 million, and Sandia $10 million.

Just as Abrahamson was about to write down the sums next to the names of the laboratories on the display board, Teller arrived and took the floor. He said there were two reasons why the entire sum should go to Livermore. One was a somewhat detailed argument about the importance of the project at Livermore and the promise of the performance of its young associates. The other was, and he waited a little to increase the dramatic effect, "President Reagan told me I could have it."[59] If the verbatim quotation is accurate, his using first-person singular added further weight to his statement, which nobody questioned. According to an official who was present, "Do you really want to challenge someone who says he's talked to the President? Do you really want to risk your status by asking Reagan if that's what he really said?"[60] Teller walked away with the full amount.

Again, Teller was not a greedy man when it came to money. He had simple tastes and no desire for grand material possessions. But he felt a fierce loyalty to his brainchild—Livermore—and to the people who worked there, and he relished the power of presiding over the laboratory that was his creation. Still, the episode was painful for everybody involved, including Roy Woodruff of Livermore, who was responsible for all weapons research, development, and testing, as well as for weapons engineering. Woodruff was in a difficult position due to Teller's repeated promises, which Woodruff felt were irresponsible. He could not protest properly, partly because of Teller's standing at Livermore and with the administration and partly because of the secrecy of the program.

Eventually, the controversies between Woodruff and Teller made Woodruff's situation impossible, and he resigned. Teller did not need to do much personally to further the deterioration of Woodruff's position, work conditions, and even his salary, because the leadership of Livermore felt subordinate to Teller. Woodruff for some time became an internal exile at Livermore and was banished to a place of work onto whose door somebody pasted a sign "Gorky West"—an unmistakable reference to Andrei Sakharov's exile in Gorky.[61]

The impression is that under the circumstances of his unique situation both locally and nationally, Teller lost some self-control in dealing with his colleagues when handling the information he was transmitting about his project to Washington. A conspicuous example was when he claimed that the X-ray laser was entering the engineering phase, when in fact it was far from it. When a member of Congress initiated an investigation about whether the administration was misled about the X-ray laser, the General Accounting Office (GAO) obliged, and as a result of an investigation—which many considered biased favorably for Teller—it compiled a report.

Excuses were made for Teller, such as he was not an official member of Livermore. This may have been true literally, but for all practical purposes he was, and a very influential member at that. As for his exaggerations in claims and promises, it was argued that they stemmed from his optimistic nature, and it was further argued that all those who knew him knew how to take them without getting misled. Woodruff was glad that at least the documents quoted in the GAO report exonerated him. In other actions, he won his grievance and was reinstated as associate director of the Livermore Laboratory with an assignment corresponding to his qualifications.[62] In 1990, he finally left Livermore and moved to Los Alamos.

By the end of 1985, sufficient information leaked out about the highly classified project to warrant a *New York Times* article with the weighty title "A 'Star Wars' Cover-Up."[63] The journalist, Flora Lewis, quoted a Livermore physicist who said about the program, "The public is getting swindled by one side that has access to classified information and can say whatever it wants and not go to jail, whereas we [the skeptics] can't say whatever we want. We would go to jail, that's the difference." The time-honored approach was to denounce the doubters as acting against national interest.

Flora Lewis crammed a large amount of information into her article about problems with the X-ray laser. She scored with a clever locution when she wrote that President Reagan's dream "could be a devastating nightmare, sapping the authenticity of American science."[64] Her most devastating comment likened Teller to Trofim Lysenko, the unscientific czar of Soviet biology and agricultural science under Stalin and Khrushchev. The parallel was telling even if mainly inaccurate. Lysenko was uneducated whereas Teller was a world-renowned physicist with scores of important achievements in science from his earlier research work. But the comparison should have given Teller and his backers pause, because Lysenko exercised unprecedented power in a crucial area of science in the Soviet Union. He decided what was right and what was unacceptable—something unthinkable in science—and destroyed his opponents, which often led to their physical annihilation.

Teller responded to the article, lamenting that secrecy prevented him from refuting the accusations of technical nature. He accused the press, in this generalized way, of focusing on the issue of space rather than on the true purpose of the SDI, which was to deter aggression. He praised his young associates for having discovered the X-ray laser, an achievement worthy of the Nobel Prize, and the possibility of making breakthroughs in studying the living cell in addition to stopping weapons of mass destruction. He found it "probable that Russians have anticipated us in this remarkable achievement." In addition to his complaints about the press, he faulted Congress as well and found praise only for the administration.[65] There was no comment about Lysenko. On this occasion, Teller's letter did not sound very convincing, and even his usually snappy style was missing.

Teller did not defend the secrecy of the program even if it shielded him from disclosures that might have shown that some of his reports and statements had been premature at best. He realized—what he had seen in previous projects and lamented often—that recruiting suffered from the project's being classified. And at the same time, some of the scientists involved with research in SDI were leaving the program. Teller felt he needed to take action. Freeman Dyson told me that at some point Teller asked him to accompany him on a visit to the Pentagon. During a meeting with General Abrahamson, Teller requested lifting the secrecy of the project. According to Dyson, Teller told the general that if they wanted the project to actually work, they had to bring in all kinds of criticism from the outside and get everybody thinking about it rather than just a few people thinking about it in secret.[66] Dyson thinks that Teller chose him as a companion for the visit because he was publicly in favor of the SDI rather than one of the critics. Dyson's name is well respected among physicists, but he is known even to a broader audience due to his successful popular science books.

Abrahamson promised Teller and Dyson that everything would be declassified in a few weeks' time. This was a strange promise since everything could have not been declassified, but this was the promise the general made to them. Nothing happened in a few weeks' time or ever. The story is interesting in that it shows that Teller did not feel nearly as confident about the program as his public conduct might have led people believe. Dyson remarked in his Teller obituary, "Teller remained publicly supportive of SDI but privately furious at the general for deceiving us."[67]

I asked George Keyworth whether Teller's activities in any way misled the administration, and this is how he responded:

I must confess that Edward was not helpful to the President with his effort to sell X-ray laser. I had to ask the President specifically if his vision of SDI included nuclear weapons orbiting in space to shoot down incoming missiles. His unequivocal response was that he sought to reduce reliance upon nuclear weapons, not increase it. So I had to draw a line, in public, distinguishing between Edward's bomb-pumped laser and the President's program. Moreover, the sheer difficulty of designing and building the X-ray laser made it, to me, an unlikely prospect in any event.[68]

After some initial hesitation, Keyworth was a committed supporter of Ronald Reagan's vision of SDI and he became the principal spokesman for the program. Of course, SDI was not the same thing as an X-ray laser, even if the X-ray laser was its centerpiece for some time. At some point, congressional support for SDI started eroding.

There was a long-awaited underground test named Goldstone in the Nevada desert on December 28, 1985. Its results were in sharp contrast with the boastings by Teller about the brightness of the laser—that it would be a trillion times brighter than a hydrogen bomb. It proved to be rather inefficient, certainly incapable of doing what he had expected. At a certain point, Teller must have realized that he could not ignore the criticisms indefinitely, yet he was waging some rear-end battle. If brightness did not work, he emphasized now the importance of focusing. Initially there was promise on this front, but further tests in 1986 proved it illusionary as well, and the X-ray laser project reached a quiet end in the early 1990s.

The very end could be pinpointed to the July 1992 decision by Energy Secretary James D. Watkins.[69] There was some sad irony here because Watkins, who was secretary of energy under President George Bush, was the former Admiral Watkins who served as chief of naval operations under President Reagan. He was one of the Pentagon officials with whom Teller had shared his vision of the third generation of nuclear weapons. Watkins played a pivotal role in the Joint Chiefs of Staff becoming early devotees of SDI. In the framework of his lobbying efforts Teller visited Watkins on January 20, 1983, and made him an enthusiast for Excalibur. Watkins either heard from Teller the expression "instead of avenging people, protecting them," or it might have been his brainchild—and thus may have been the source for President Reagan to carry it over into his SDI speech.

Whether it was deception or self-deception, Teller remained optimistic, or displayed optimism, to the end and did not slow down in his public appearances to discuss the X-ray laser adventure. On the contrary, he came up with new ideas and suggestions, like launching mirrors over Western Europe shortly before a Soviet attack and destroying the Soviet short-range missiles with a ground-based visible or infrared laser beam reflected off these mirrors.[70] He thought that such a system could be put into operation within a few years' time. This was in 1986, and he hoped that he would be able to demonstrate the antimissile defense system before President Reagan left office, which, at that point, was less than three years away. The problem with the X-ray laser was not that it was not technically feasible as a weapon, because it might have been. The problem was the enormous cost, the time frame, and most of all that it would have enhanced the arms race to an unprecedented level. The main problem with it was whether it was needed at all in the first place.

In addition to his optimism, Teller also kept up his vigilance, defending everything related to SDI. In February 1986 he attended the Davos Symposium, where T. J. Watson, the head of IBM, gave a talk in which he was negative about SDI. Teller could not be present because he did not feel well, but he acquired the necessary information about the talk afterward and decided to take immediate action. He called his secretary at Livermore and asked her to organize a rebuttal by Lowell Wood to be ready for Teller's signature upon his arrival. He wanted to read the draft in the car as he was driven from the airport to Livermore. He felt that Watson was negative about SDI because he fell under the spell of IBM associate Richard Garwin, who had been very critical of SDI. Teller specifically instructed Wood not to mention Garwin's name in the rebuttal; instead, he wanted Wood to refer to associates at IBM in more general terms. Teller further asked Wood to get some famous computer scientist to co-sign the letter with Teller, whose name would enhance the weight of the letter in Watson's eyes.[71]

Teller was dismissive and vitriolic in his reaction to a letter urging Congress to slow down SDI-related spending. Sixteen hundred scientists signed the letter, including some who worked at Livermore and Los Alamos. According to a press report, Teller said, "I wonder if there is any statement for which you couldn't get 1,600 signers from that type of people. In science we would call that a fluctuation and a fluctuation is not the wave of the future."[72] This was in the summer of 1986.

Reading his *Better a Shield Than a Sword*, published in 1987, one could have hardly guessed that there might have been problems with SDI.

Nonetheless, for some time Teller must have realized that if he and Livermore wanted to stay in SDI action, it was imperative for them to change course. It was paradoxical that in the meantime, the significance of SDI in the international arena had been enhanced.

## WORLD POLITICS

By the mid-1980s there were decisive changes in the Soviet Union. Following the reign of three old and sick leaders, a younger and energetic man came to power, Mikhail Gorbachev. In spite of contrary expectations, Reagan and Gorbachev hit it off at their first encounter in Geneva in 1985. Although no concrete agreement was reached at that meeting, their personal relationship was a good omen for future success in decreasing tension and, possibly, coming to an agreement on some disarmament. The Geneva Summit was a preliminary meeting, providing an opportunity for the two leaders to size each other up and to get acquainted. However different the two might have appeared, they realized that they could do business with each other.

More than just being a new leader, Gorbachev brought substantial changes to the priorities of Soviet politics. Building up an expensive equivalent to the American SDI in the Soviet Union would have torpedoed Gorbachev's economic reforms. The next summit took place in Reykjavik in 1986, for which expectations were heightened. Perhaps because there was a general feeling that Reagan might not be up to Gorbachev, a group of influential political personalities sent him a letter expressing their worries lest the president let SDI fall victim to a broader agreement with the Soviet Union. Teller was not only one of the signatories; he may have put the final editorial touches on the letter.[73] However, the fear that Reagan would sacrifice SDI proved unfounded.

Regardless of its ultimate fate, SDI played an important if somewhat unexpected role at the Reykjavik Summit. Both leaders seemed to be enamored with SDI for different reasons. Robert Gates noted: "Amid countless skeptics that such a defensive umbrella could ever be built, there were two small groups of people who believed it probably could. The first was Ronald Reagan and a small group of his advisers. The second was the Soviet leadership."[74] The summit took place on October 11 and 12, 1986, and has been characterized in various ways indicating its uniqueness, for example, as "the most bizarre summit in the history of the Cold War."[75]

As a prelude to the summit, Gorbachev gave a long speech at the 27th

Congress of the Communist Party of the Soviet Union in the spring of 1986. He invoked something Reagan told him in Geneva: If extraterrestrials tried to conquer the earth, the United States and the Soviet Union would have to join forces to avert the attack. Even if it made common sense, it was an extraordinary suggestion and a clear departure from the status of the two superpowers being mortal enemies. Gorbachev went even further when he declared that the dangers of nuclear arms, the ecological hazards, and the need to help developing nations were similar challenges to extraterrestrial attack. He called for competition and cooperation to replace belligerence, and promulgated "durable peace" rather than merely "peaceful coexistence" as their common goal. He called for "the realization that it is no longer possible to win an arms race, or a nuclear war for that matter. . . . The task of ensuring security . . . can only be resolved by political means."[76]

Gorbachev's speech did not mean that the Soviets would give up their position in the arms race. They were no less vigilant than the Americans and they were aware of the doctrine that they had to be strong to maintain an adequate position in the negotiations. Yet the Soviet Union was changing, and many in the United States did not realize this or did not want to realize it. Teller did not change his rhetoric to any noticeable extent during these years.

The cold war and the Reykjavik Summit in particular have a growing literature, and the summit itself has generated further studies and publications.[77] Here we single out one feature of this summit, namely, the role of SDI in the tremendous proposals and dramatic counterproposals on the table. The proposals and counterproposals were of such extraordinary weight that their acceptance would have shaken the standing order of the world, might have contributed to world peace, or might have introduced irreparable instability. There is no consensus in judging what really took place in all the details of the discussions. However, there is consistency in that at the end of all demands and proposals from the Soviet side, there was a recurring stipulation that the United States should confine all work on antimissile systems, that is, SDI, "to the laboratories." Significantly, nobody ever defined what "the laboratories" exactly meant.

There were unbelievable exchanges that might have meant the elimination of all nuclear arms by the two superpowers, for example, which might have led to an imbalance in regard to their conventional forces. These examples only serve to show that SDI was elevated to unprecedented importance in these negotiations. It was Reagan's inflexibility about SDI and Gorbachev's insistence on his demand that—depending on one's viewpoint—saved the world from or deprived the world of an agreement in Reykjavik. In the words of former secretary of defense James R. Schlesinger, SDI "has probably now done more to protect the United States and its allies than it will ever do in the future."[78]

Reagan refused to even consider the slightest concession about SDI; not that his aides would have dared to suggest anything substantial in this respect. Secretary of State Shultz, however, made a slight attempt when he hinted, "So we should give them the sleeves from our vest on SDI and make them think they got our overcoat."[79] Although Reagan welcomed the idea, eventually he rejected it. In the end, the two leaders left Reykjavik with empty hands to the great disappointment, or the relief, of many.

SDI may have seemed to be a bargaining chip in Reykjavik, but it was not one in the final account. There have been suggestions that Teller saw it as a bargaining chip, especially when he told the following story: he picked up a big stick while on a hike and a friend asked him what it was for. Teller said it was to protect him from grizzly bears. When the friend protested that the stick would not suffice to protect him, Teller responded that he knew that, but he hoped that the grizzly bears did not. This sounded less innocent when on one occasion his talk about ballistic missile defense was introduced by this story.[80]

The momentum of negotiations between the two superpowers, though, could not be stopped. Soon the Soviet Union agreed to decouple the question of disarmament and their demand of confining SDI to the laboratory. In this, Sakharov's influence may have played a role, as he found such coupling counterproductive. When he spoke at a conference in Moscow in February 1987, he called SDI the "Maginot line in space," meaning that it could be easily evaded.[81]

There was another summit in 1987, this time in Washington, DC, where the stakes were more modest than in Reykjavik and the negotiations ended in an agreement signed by Reagan and Gorbachev. It was the Intermediate Nuclear Force (INF) Treaty. The agreement was celebrated by a state dinner for the Gorbachevs at the White House on December 8, 1987. The press reports described all the important and not so important features and happenings of the event. It was labeled "pure Americana,"[82] and the guest list included "126 stars of business, science, sports, politics, and the arts," among them Edward Teller.[83] At the dinner, he was seated at President Reagan's table.

The evening started slowly because there was more conversation in the receiving line than usual. Raisa Gorbacheva especially showed interest in the guests and she exchanged some pleasantries with Teller, too. Then

Teller moved forward to the head of the receiving line. Although not even the minutest detail of the event escaped the notice of reporters, a story that Teller eventually conveyed about the evening did not make the media.

According to the *Memoirs*, President Reagan introduced Teller to Gorbachev, saying, "This is Dr. Teller." The old scientist put his hand out for a handshake, but the Soviet leader did not move his hand and did not say a word. The president then repeated, "This is the famous Dr. Teller." Gorbachev, keeping his hands at his sides, said, "There are many Tellers." To which Teller responded, "I agree."[84]

If the story is true, Gorbachev's behavior—as Reagan's guest—was rude, but it was also a sure sign that he was aware of Teller's importance. I was intrigued by this episode and tried to get independent confirmation of the story. My letter and e-mail message to Mr. Gorbachev have remained unanswered. Former state secretary George Shultz responded that he "has no recollection of the incident described by Dr. Teller in his *Memoirs*."[85] Hugh Sidey's description of Reagan introducing Teller to Gorbachev merely noted that Gorbachev's response to Reagan's introduction was "minimal."[86] There was no sign of Teller's feeling hurt by any incident, as evidenced by a much-publicized photograph of this event that shows Gorbachev and Teller engaged in an amicable exchange with President Reagan standing by.

Teller sent a letter of acknowledgment to the president and Mrs. Reagan thanking them for allowing him to be "present on the historic occasion," which he found "more than equivalent of one year of college education."[87] When the INF Treaty came before the Senate, Teller was asked to testify and he publicly supported its ratification, despite whatever reservations he might have expressed privately.

It is possible to spin Gorbachev's declining to shake Teller's hand as being in contrast to Robert Christy's refusal to shake hands at Los Alamos in 1954. Still, two other handshakes come to mind—both with Robert Oppenheimer. When Teller finished his damaging testimony against Oppenheimer at the hearing on April 28, 1954, he shook Oppenheimer's hand. Then, on December 2, 1963, after President Johnson handed over the Fermi Award to Oppenheimer in the White House, Teller again stretched out his hand to Oppenheimer, who, once again, accepted it. These handshakes could have become symbolic of reconciliation, but they did not. The Oppenheimer handshakes were not embraced by the physics community, and Oppenheimer likely came off simply as the more classy of the two. Of all these four episodes, Christy's denied handshake was the only one that mattered.[88]

## BRILLIANT PEBBLES

As the Excalibur project was losing ground, Teller and Wood were getting ready to launch an entirely different project in its stead. Before introducing it, let us look at an excerpt from a letter by Defense Secretary Caspar Weinberger: "Although we appreciate your optimism that technicians will find the way and quickly, we are unwilling to commit this nation to a course which calls for growing into a capability that does not currently exist."[89] This sounds like something that could have been addressed to Teller about Excalibur, but it was not. The excerpt is from a letter to General Graham in 1982, declining support for his antimissile plan based on small rocket interceptors. Graham never worked out his ideas in detail, but it could have been considered to be the predecessor of the next scheme by Teller and Wood called Brilliant Pebbles—there has never been a shortage of fancy names.

The original suggestion came from a Los Alamos physicist, Gregory Canavan, based on a decades-old prior suggestion. The details were worked out by Lowell Wood and his team at Livermore. Teller and Canavan published an in-depth study about the technical questions and cost-effectiveness of strategic defense for the 1990s.[90] They noted that the United States had adopted Brilliant Pebbles for development and potential deployment, so the purpose of their paper was clarification rather than an attempt to convince anybody of the usefulness of the project.

Teller noted the change in the relationship between the Soviet Union and the Free World, but he called for continued vigilance because the rocket forces of the Soviet Union (as it would be known for a few more months) remained the strongest in the world. Furthermore, he pointed to the proliferation of rockets in the third world. One of the remarkable features of his argument—in light of his prior advocacy of the X-ray laser weapon and especially of his prior warnings about the vulnerability of space-based rockets for defense—is that Brilliant Pebbles could be protected from beam weapons (that is, the X-ray laser, in particular) whose development he did not anticipate anytime soon.

The concept was to populate space with intelligent small rockets operating independently. Their purpose would be to knock out Soviet missiles by the sheer kinetic energy of their collision before they reached American territory. Teller's and Wood's campaign to promote Brilliant Pebbles mirrored the Excalibur campaign, and President Reagan embraced it. The new project had a great advantage over Excalibur because it was truly nonnuclear. There were dramatic scenes such as Teller's and Wood's unveiling the model of a Brilliant Pebble in the White House in the presence of the president. To ensure that the photographers taking pictures of the meeting did not capture a shot of the classified model, it was covered with black cloth. The published photographs showed only the jubilant faces of the attendees.

It was at the beginning of the George Bush presidency when Brilliant Pebbles became the centerpiece of SDI. The elder Bush followed Reagan's approach of not waiting for expert opinions and scientific advice to make his decision. Critics noted that Brilliant Pebbles came from Teller, who had not long before promised to build something the size of an office desk that could knock down all Soviet ICBMs. By then, Teller was increasingly playing the role of a high-level lobbyist with excellent access to important officials for whom it no longer seemed to matter what he was lobbying for. Nobody in a decision-making position seemed to care that Brilliant Pebbles conspicuously displayed some features that Teller had fiercely criticized not long before. On the contrary, his switch was looked at in a complimentary way, as proof of his flexibility.

However, even as late as 1988 Teller formulated his stand on the X-ray laser in such a way that made it seem as if it were an existing weapon whose single module could produce one hundred thousand independently aimable beams toward enemy targets. This was the interpretation that came through in a CBS *60 Minutes* interview with Mike Wallace on August 10, 1988. Teller filed a strong protest against the way the interview was edited.[91] Upon re-reading the edited and unedited versions, there is no reason to believe that the average viewer might have had a different interpretation of what Teller had said. However, Teller, as noted earlier, had "a standing rule regarding television interviews. That is, they use my entire statement or they use nothing." He knew that there were rigorous time limitations in television programs and he was willing to tailor his statements to any length of time he had been given in advance.[92]

## SUMMITS

At the end of 1988, just as Reagan's presidency was coming to end, a different kind of summit took place in Washington. It was between Teller and Sakharov, the two respective fathers of the American and Soviet hydrogen bombs. The occasion was a banquet on November 16, 1988, honoring Teller for his contribution to supporting the United Nations human rights policy. The Ethics and Public Policy Center organized the event in a hotel with about 750 attendees. Sakharov came but could not stay long. First he and Teller had a private chat for about fifteen to twenty minutes.

When Sakharov addressed the gathering, he made some critical remarks about SDI, and left immediately after he finished. He called SDI a great error and cited its enormous economic cost. He warned that it could trigger a nuclear war and that it would hinder arms control. The conservative crowd welcomed him with enthusiasm but remained cold to his statement.[93] For this crowd, Teller was still a hero, and some referred to him as the father of SDI, which was remarkable after so many failures and the fact that it had not progressed much in five years. Later in the evening, Teller responded to Sakharov's criticism in a dismissive way, noting that his Soviet colleague could not have had proper information about the American program.

Unbeknownst to Teller, Sakharov had done some extra homework in preparation for discussing SDI. When he arrived in the United States, after his visit to his doctor, the next person he chose to see was Arno Penzias.[94] He asked the Nobel laureate physicist to visit him while he was staying with his stepdaughter in Newton, Massachusetts. Penzias dutifully came from California and spent a whole afternoon satisfying Sakharov's curiosity. The Russian physicist wanted "to hear the arguments being presented in favor of SDI from someone he [could] trust, in order to prepare for them."[95]

Penzias had not investigated the issues of SDI in detail; he only had instinctive doubts about its feasibility. However, he trusted the experts of the American Physical Society, who concluded that the X-ray laser would unlikely be able to do what it was expected to do. On the other hand, Penzias had met Reagan and Teller on separate occasions and understood that they did an excellent job of enhancing the public image of the program. Penzias was very helpful in presenting the pro-SDI position to Sakharov, a position he himself did not agree with but that Sakharov wanted to be informed of.

With some simplification, Penzias listed the following points for Sakharov: (1) SDI proponents offer a world without nuclear weapons; (2) SDI is about research rather than deploying a working system; (3) There are outstanding scientists among the supporters of SDI, not only among its opponents; (4) There are important technology spin-offs from the research for SDI.[96] In connection with the third point, Edward Teller was practically the only one who fell into this category, but Penzias warned that he represented a formidable force. Penzias witnessed the Anderson–Teller debate (see above), and the next morning he was having breakfast with Teller. On this occasion, Penzias told Teller the following parable to illustrate how he viewed SDI:

Imagine that you and I are a pair of cowboys with six shooters pointed at one another's chests. Neither of us would dare to shoot, because any shot would cause the death of both parties, since the first bullet's impact would cause a reflexive pulling of the victim's trigger. Not a very comfortable situation. If either one of us sneezed, we'd both be killed. Now, suppose I announce that I'd been working on a bullet proof vest, and that I promise to show you how to make one of your own as soon as I got mine on. How would you react to that? Would you give me the opportunity of ending our stalemate once and for all by shooting you with impunity as soon as the odds tilted in my favor? Wouldn't you at least consider other options, like trying to shoot the gun out of my hand?[97]

Teller did not accept the validity of Penzias's parable for characterizing SDI.

At their November 1988 meeting, Sakharov told Penzias that the comparison of relative costs of SDI for the American economy was much lower than its implementation would cost for the Soviet Union, considering its economy. Sakharov added that the Soviet leadership was apprehensive of the American SDI and did not find it possible to create a comparable system for their country.

Penzias raised the possibility that since the Soviets were unable to match the American SDI, "won't they just change the rules of the game by applying pressure somewhere else?" Sakharov thought that although there was always such danger, he found it more probable that the Soviet leaders "will announce a program to calm their people and then negotiate." Penzias had no doubt that Sakharov felt the weight of these heavy thoughts as he was considering the risks of SDI, but that he also felt—apparently with very good insight in view of the Soviet Union's collapse within a few years—that "Ronald Reagan's bet would most likely pay off."[98]

As for Teller, he was charging ahead. He understood the signs of the time and called attention to potential uses of and benefits from Brilliant Pebbles much beyond the military. He painted a panoramic view of their capabilities.[99] Brilliant Pebbles could be turned into Brilliant Eyes if equipped with radars, lasers, modern optical equipment, antennae, sensors, and whatever else high technology could offer. Besides the solution of defense against incoming missiles, the small rockets could monitor everything from troop movements to pollution. They could collect and digest information and transmit the data to satellites and to the ground. They could forecast the weather and earthquakes, hurricanes, floods, tidal waves, volcanic eruptions, even global warming, and provide damage assessments. They could also watch traffic and report congestions and accidents.

Once again, Teller excelled in making his point in an accessible way. He contrasted the potentials of computerization with his work on his thesis six decades before, in which each arithmetic operation took one second on his primitive calculator. The other major change he emphasized was miniaturization. In his typical approach, his description was about what could be expected to be achieved in a few years' time, and he put a price tag on it: $1 billion annually for about five years to put the system in place, and the same amount per year to maintain it.

By 1995, the eighty-seven-year-old Teller had become so enamored of the Brilliant Pebbles program that "he spoke almost derisively about the X-ray laser fantasy. . . . 'Lasers may be useful some day, but meanwhile we must destroy missiles as David slew Goliath.'"[100] But Brilliant Pebbles ended soon, just as Excalibur had, although it would be premature to consider the concepts having completely disappeared. Reagan's missile shield has been referred to as "An Idea Whose Time Will Not Go."[101] An air force colonel commented on the Airborne Laser (ABL) Program in 2002, "The 30-year-old vision of Edward Teller is on the threshold of becoming a reality."[102] This forecast sounded similar to Teller's promises.

Artist's representation of Edward Teller and Planet Earth.
Drawing by and courtesy of István Orosz, Budakeszi, Hungary.

The Strategic Defense Initiative has a huge literature, and recently interest has grown in looking at it from the Soviet point of view.[103] Our concern here is whether and to what extent Teller and his contribution to SDI played a role in the collapse of the Soviet Union. He certainly influenced President Reagan in formulating SDI regardless of whether or not he had a direct input in the president's March 23, 1983, speech. It seems very popular to deny any role of SDI in the demise of the Soviet empire, and through this, any achievement of Teller's.

According to Philip Anderson, "the Soviet Union was a paper tiger" at the time of President Reagan's SDI, and its military was characterized by "fake equipment and poor morale."[104] However, the collapse of the Soviet Union had been predicted ever since it had been formed. The Soviet Union would have been strained to find an adequate response to SDI, had it been feasible, but it would have tried. Arno Penzias allows some contribution of SDI to the collapse of the Soviet Union by saying that all the proponents "did with SDI was they saved 10 years at best. The Soviet Union was totally going downhill."[105] Of course, there is no difference between ten years and a few tens of years in historical measures. Yet Penzias was angry with Reagan's risk taking:

> If the Soviet Union was brought down by the threat of an antimissile system, that threat risked a nuclear war. Ronald Reagan bet that they would back down, but if he had lost his bet, our entire species would have been destroyed. He bet the entire planet, and no one has the moral right to bet the entire planet on that kind of confrontation. I would rather live with the Russians than risk a nuclear war. Nuclear stalemate between us and the USSR was based upon a set of treaties set up to insure that neither side could win by launching a pre-emptive strike on the other. If successful, Reagan's Strategic Defense Initiative would have allowed the side possessing an antimissile system to launch such a pre-emptive strike.[106]

Penzias's considering living under the Russians—that is, in this context, under communism—brings back the dilemma Arthur Compton raised about living in slavery under the Nazis rather than drawing the final curtain on mankind.[107] So, how would Penzias, originally a Jewish refugee from Nazi Germany, consider such a dilemma? He was unambiguous about it: "If the difference was that some part of the world would live under the most horrible regime, even if I were in that part of the world, I would opt for it rather than risk nuclear war."[108]

Whether Reagan was gambling with the planet or not, he won his bet: the Soviet Union collapsed; and it would be impossible to exclude at least some contribution to such a development by SDI and thus, however indirectly, to Edward Teller. The Strategic Defense Initiative might have been impossible to accomplish, and the Soviet leadership eventually recognized that SDI could be countered at a fraction of its cost. But initially they could have not afforded not to take it seriously. There was a dividend from SDI regardless whether or not it was feasible in that it contributed to or at least accelerated the historic changes of the political landscape around 1989–1990 in the Soviet Union and in Eastern-Central Europe in particular, and, consequently, through the whole world. Teller felt satisfaction regardless of the fate of the program, but fact and fantasy were blending in his *Weltanschauung*. At one point he wrote to President Reagan, "To establish SDI as then protector of the free world, with the help of our allies, would make your role more remarkable than that of Lincoln. He saved the Union. You have set out to save freedom."[109]

# Chapter 12

# FINAL THOUGHTS

. . . in the end, events might prove Teller to have been right—
for the wrong reasons.

—Leo Szilard[1]

We can forgive a man for making a useful thing as long as he
does not admire it.

—Oscar Wilde[2]

*The political shifts in 1989–1990 brought tremendous and welcome changes for Edward Teller. The cold war was over, and he felt he had played a considerable role in its victorious ending. He traveled to Hungary and Russia where he was accorded a hero's reception, and he became "a born-again Hungarian." The last decade of his life, however, hardly influenced the prevailing judgment of his oeuvre. As one of the architects of mutual deterrence between the two superpowers, he contributed to avoiding a third world war. But his personality and his controversial activities in subsequent events made it difficult to appreciate even what should have been acknowledged. His negative deeds and his aggressive image seem to have dominated how he has been viewed and likely will be viewed by many, at least for some time to come.*

The political changes in Eastern Europe in 1989–1990 were a unique gift for Teller beyond their political importance. They came at a time when he was increasingly marginalized in America not only because of old age but also because of the changing political climate. In President Clinton's congratulatory note of January 14, 1998, to Teller, the absence of the topic of thermonuclear weapons and strategic defense was more conspicuous than what the letter contained.[3]

Dear Dr. Teller:

I am pleased to join your family, friends, and colleagues in wishing you a very happy 90th birthday.

As you celebrate this milestone, you can reflect with pride on a life and career dedicated to science, public service, and education. From your pathbreaking work on the Manhattan Project to your pivotal role in the creation of the Lawrence Livermore National Laboratory, you have brought together many disciplines and some of the world's finest scientific minds to confront and overcome the great challenges facing our nation in this century.

In your emphasis on the importance of advanced computations, you helped usher in the era of supercomputing and the technology of the information age. As an educator and mentor of a new generation of scientists, you have shown how we can engage the young minds that will shape America in the next century.

On behalf of your fellow Americans, I thank you for your extraordinary contributions to the security and success of our nation. Best wishes for a wonderful birthday celebration and every happiness in the years to come.

Sincerely,

Bill Clinton

Nonetheless, it was a great recognition from a president whose administration could have hardly harbored very friendly feelings toward him and it singled out four significant areas of Teller's oeuvre.

In 2002, the George Washington University in Washington unveiled a plaque dedicated to Teller. It stressed "his outstanding contribution to molecular physics, to the understanding of the origin of stellar energy, to the theory and application of fusion reactions, to the field of nuclear safety, and for his continued leadership in science and technology." Then it added: "His colleagues remember the precise, profound, and prodigious character of Professor Teller's mind." Finally, in 2003, George W. Bush awarded him the Presidential Medal of Freedom shortly before Teller's death. The award had to be brought to him; Teller no longer traveled. He could not even walk on his own.

## HOMECOMING

Teller visited Hungary and also Russia in the early 1990s. His home-coming made him a "born-again Hungarian," to use his own expression. He was surrounded by attention and admiration and met heckling from environmentalists only on rare occasions. He enjoyed the interest in him, visited schools and nuclear plants, gave advice and lectures, and talked with politicians and scientists. Initially bodyguards were assigned to him by the Hungarian government as they would be to any very important person. The measures for his protection were relaxed when it turned out that nobody seemed intent on harming him.

Teller had strong feelings of being Hungarian all his life. It was a natural feeling for him and a necessity—a sense of belonging that his clashes with his peers in America could not take away from him. Besides, he never lost his old Hungarian friends—the other Martians—regardless of whether they were politically on the same side or the opposite. In particular, Szilard never abandoned him. For his part, Teller did not shy away from sending a powerful letter to Senator Tydings on behalf of Brunauer when the latter was being persecuted during Senator McCarthy's campaign. Fred Ausbach noted that, as a newcomer to Los Alamos, one day at lunch he noticed that at a neighboring table, five or six men started singing the Hungarian National Anthem. When he walked over he learned that one of the men was Edward Teller.[4]

Yet the Hungarian Revolution in 1956 did not induce Teller to make any statement, lobby for American or international assistance, or condemn the Soviet aggression. At the time the Hungarian Revolution was being crushed by the Soviet Union, Teller was actively interacting with Ernest Lawrence, who was particularly disturbed by the Hungarian events. Lawrence took a strong stand in the Hungarian question, independent of Teller.[5] Given Teller's well-known anti-Soviet sentiments and his strong Hungarian feelings, his silence at the time might have been explained only by his fear for his mother, sister, and nephew, who were still in Budapest. It is more puzzling that he devoted a mere one paragraph to the Hungarian Revolution and its suppression in his *Memoirs*, which were written at a time when no restraint existed for him anymore. He mentions that Eisenhower "often talked publicly about pushing back the Iron Curtain, but he laid no plans to accomplish that goal. Hungarians, taking his words at face value, felt encouraged to revolt; when they did, they were not supported."[6]

However, the revolution of October 23, 1956, did not happen under the influence of the United States. Misleading media information about possible American assistance came mostly during the following days and when the Soviets attacked on November 4. Hungarian resistance was being encouraged to hold out until after the American presidential election. When the election was over (Eisenhower was elected for a second term), nothing happened—as nothing could have under any realistic assessment of the situation. Still, many in Hungary felt betrayed. As for the armed resistance, Teller writes in a simplistic manner about freedom fighters trapped at the Budapest radio station. His assessment displayed appalling ignorance for a man who claimed deep concern for Hungarian affairs. The armed resistance was much more widespread than just being localized to one place. What was yet a stronger expression of resistance was the months-long nationwide strike afterward. The impression is that Teller was too much preoccupied with his struggles against the nuclear test ban at the time to follow the Hungarian events in 1956–1957, and he did not feel inclined to learn about them afterward. He may have realistically estimated the hopelessness of the Hungarian Revolution from the start and did not think it was worthy of his closer attention.

There was one aspect of the Hungarian Revolution of 1956, however, that Teller was keenly aware of. About two hundred thousand people left Hungary in its aftermath, among them his nephew. Many of the refugees were young people eager to either find an occupation adequate to their education or to acquire an education. The United States expedited the immigration of many of them and set up a refugee camp at Camp Kilmer in New Brunswick, New Jersey. The relatively large number of young professionals induced the National Academy of Sciences to operate an onsite office to facilitate proper counseling and distribution. Teller visited the camp and invited a number of interested students to Berkeley.[7]

His identification with more recent Hungarian events came across in force after the political changes in 1989–1990, when his attention was much less absorbed by other matters. At this time his image in the Hungarian media underwent a rapid and drastic overhaul. Under the communist regime, he was considered a reckless warmonger and potential war criminal. First, he was transformed into the well-known atomic physicist, then to the world-renowned scientist of Hungarian extraction, finally, to one of the greatest Hungarians of the twentieth century.

His initial, brief visit to Budapest was a side trip in 1990 from a conference in Switzerland. The Soviet-built Hungarian nuclear power plant in Paks in central Hungary was his first destination. His praise for the plant and especially for the professionalism of its associates set the tone for his future statements about everything he saw. The next day he appeared at the general assembly of the Hungarian Academy of Sciences to accept its honorary membership. On this occasion, he gave a moving speech in impeccable Hungarian, using short, punctuated sentences and pregnant pauses:

> It is not easy for me to begin, but I must speak. This is an unbelievable event for me. I am deeply grateful, especially grateful.
>
> I came to Hungary to do something that interests me the most. I heard that it is considered very important here.
>
> In our century, it meant a great change that we understood the atoms. We created a unified science from physics and chemistry, at least in principle. We ought to welcome this change with hope rather than fear as is widespread all over the world. It is our special task to find the road, a cautious one, which leads on without fear.
>
> Of course, I am thinking of atomic energy. Yesterday I visited Paks. There is no doubt in my mind that atomic energy is important for Hungary, but it is also important for *the whole world*. Without atomic energy, the enormous difference cannot be bridged between developed and emerging nations. Atomic energy provides perspective for developing a harmonic world. But we must be cautious. I have been working on this for decades because this is important and because I am interested in it. Energy is essential for Hungary. If we are not scared and are willing to work rationally, Hungary can take a leading role in this field in the world.
>
> I came home. This was an enormous shock for me. I have been asked about my feelings. I could not say. I am feeling so much that I am unable to put it into words. Maybe, I will be able to do so in two weeks. What I feel is part enthusiasm. It stems from my hope for a great future; the hope that the word Hungarian will mean a lot, and not only for Hungarians, but for everybody. Words are especially important in the modern world because they can change it overnight and change it in a most frightful way. But they must not change it in a frightful way! Let us see to it that they change it so that the people conceive, feel, and understand that our future may be without fear in the most beautiful meaning of the word. Thank you.[8]

Subsequently, Teller visited the Budapest University of Technology and its experimental nuclear reactor, and the university bestowed an honorary doctorate upon him on January 23, 1991. This event was long overdue. The Ministry of Education and Religion in 1946 forwarded a request from the Hungarian Embassy in Washington to the university "to confer honorary doctorates to the following among the Hungarian scientists living in the United States and highly respected there and who had been induced to leave to some extent by the oppressive atmosphere of the anti-Semitic regime." The list contained six names: von Kármán, Szilard, von Neumann, Teller, Wigner, and Brunauer. This request, however, was not followed up. Von Kármán was conferred the doctorate on the occasion of his Budapest visit in 1962, not long before he died. Almost half a century passed from the original request before Teller finally received his award.

Teller was treated at an exceptional level during his first visits, but even a head of state would not be approached with so much humility by the media in the United States as he was in Hungary. It was more reminiscent of how ordinary people approached Emperor-King Franz Joseph I when he held a rare audience with them in his palace. The people could turn to the sovereign with their complaints but by the time they had been prepared for the etiquette, they forgot their grievances and had only one wish, to get it over with. The journalist who first interviewed Teller gave a vivid description of his anxiety and anguish as he was preparing for the meeting with the famous expatriate.[9] The result, of course, could not be anything but admiration and adulation. Teller reciprocated with praises of Hungary.

Teller did not get involved in actual politicking, but that does not mean that there were no attempts to get him involved. His own rhetoric and the political aura of the time predestined the political right to consider him as an oracle of the new Hungary. It happened at an election rally in 2002 held by the leading right-wing party that a message from Teller was announced. He cleverly formulated his support for the right in the form of two questions rather than in direct statements, but his message was unambiguous.[10] This was a paradox, because much of the right in Hungary has considered itself to be the followers of the ruling right over Hungary between the two world wars from which Teller had escaped. It is impossible to judge whether Teller understood this paradox, for all his love of paradoxes. We can only guess that he might have. There was then outright fraud in September 2003. According to the socialist daily, Teller dictated a memorandum about his concerns embracing a whole set of problems in Hungarian internal politics that could have been written by a leftist journalist and, as it turned out, it was.[11]

One of the concerns expressed in the hoax was that a leading right-wing environmentalist advocated shutting down the Paks nuclear power plant. Teller, ostensibly, lamented that this politician did not understand nuclear energy. The politician, without waiting to see if the memorandum was authentic or not, responded with a statement: "Mr. Teller is right: I have no idea how to produce atomic and hydrogen bombs, but the deaths of millions liquidated in mass murders do not burden my conscience either."[12]

Teller addressed the crowd on the occasion of one of his first visits to Hungary in the early 1990s with the phrase "Véreim, Magyarok!" meaning, "My blood brother Hungarians!" In doing so, he must have lived out a fantasy. He could have never used such an expression while he was actually living in Hungary. Being a Jew, he was an outcast in Hungarian society in the 1920s. In the 1990s, such an address sounded dated, but not quite, since for many, the change in the political system meant a return to the time between the two world wars. Teller could have not welcomed such a return, but, inadvertently, he created the impression that he did. In addition to bringing his fantasy to life, his address served another purpose, to please his audience. It probably did, especially the extreme right, which at the beginning of the 1990s was only yet probing the limits of its latitude. In a few years' time it would exercise no restraint. In the 2000s, their anti-Semitism has become so overt and vicious that one is reminded of the Nazis with the difference that today this behavior exists with the knowledge of Auschwitz, whereas in the late 1930s and early 1940s, Auschwitz was yet to come.

Teller did not have a high opinion about the Hungary of the period between the two world wars, and he lamented about the foundering Hungary to Laura Fermi.[13] In contrast, he was full of praise for Hungary in the early 1990s, and not only for its science education, but much else, as well. The question arises: What made Teller change his mind from referring to Hungary's foundering to his high praise? There were forty years of communism between his lamenting and his praising. Was it merely a bias for a nation—forget about forty years of communism—that welcomed him without reservation?

## BEING JEWISH

Teller was not religious and the Teller children did not receive religious education. For him, being Jewish was not so much a matter of religion or ethnicity, rather, it was one of background. It was not something for him to change either, and he remained the only one among the five Martians who did not convert. He was conscious of his being Jewish, and had he forgotten it, the world would have reminded him. He complained about the anti-Semitism he experienced in Hungary during his youth but refrained from citing specific incidents. This was characteristic not only of him but of the other Martians as well—the topic was too painful to revisit. The anti-Semitism was harsh during Teller's life in Hungary; it would have become yet harsher for him had he remained in Germany after 1933. Even upon the arrival of the Tellers in the United States, these reminders did not disappear overnight, as there was anti-Jewish discrimination—though not persecution, to be sure.

One of the painful effects of anti-Semitism in America was the barriers preventing many from entering graduate school or from becoming professors. These barriers in academia came down following World War II. Yet anti-Jewish discrimination in the society at large was slower to disappear. When the two dozen theoretical physicists gathered at the inn called Ram's Head on Shelter Island in June 1947, they noticed a sign declaring that the inn catered to a "restricted clientele," meaning that Jews would not be welcome.[14] Half of the participants were Jewish, but the sign did not apply to them—they were esteemed scientists and heroes who helped end the war—but the reminder was there.

In the beginning period of Teller's life in America, most of his friends were refugees from Europe, like himself, and thus Jewish. As the Tellers were settling in Washington in 1935, an associate of the Rockefeller Foundation cautioned him not to have only Jewish friends. When writing his *Memoirs*, Teller took stock of his friends in the United States. He consciously sifted through who was and who was not Jewish among them.[15] He surprised his daughter when he once asked of her date "Is he Jewish?" Her surprise may have been the reason why he never asked her such a question again.

In 1957, Rabbi Herschel Levin of the Temple Beth Sholom in Flushing, New York, asked Teller for a few words describing his attitude toward Judaism and the Jewish people. Each year the rabbi "selected and preached about the 'Jewish Man of the Year,' the Jew who has done the most for humanity during the past 12 months." In 1957, obviously in keeping with the general atmosphere of appreciating science in America, the rabbi selected the Jewish scientist and he wanted to symbolize his choice through Teller's example. He asked Teller whether he considered himself a Jew in either religious, cultural, or any other sense.[16] Teller's response mirrored both his commitment and his hesitations:

> As you know, I am of Jewish origin. I have been brought up in the Jewish religion. However, religion in any formal sense of the word did not play an important part of my life. While I have the deepest respect for religion and, in particular, the Jewish religion I feel that I should not get involved by any act, statement or opinion in questions of religion. My wife is not of the Jewish religion and my children are not being brought up in the Jewish tradition.
>
> Very many of my friends are Jewish. This, together with my family background, amounts to at least a strong component of Jewish culture.
>
> If you want to use my name as that of a Jewish scientist, I shall be greatly honored, but in view of what I have just written to you, I feel that

it is necessary that you do not refer to my religious or cultural connections in any detail.

In a general sense I am proud to be identified with the Jewish community. In some ways, however, I could imagine that you could select a
better candidate. There are many outstanding scientists who are indeed
more closely tied to the Jewish religion and Jewish culture, and it would
seem to me that you might find it more proper to mention one of those
outstanding men.[17]

There is no reason to suppose that Rabbi Levin would have changed
his selection of Teller.

As for Israel, we have seen how Teller's interest in education, defense,
and the Plowshare program, and his friendship with Yuval Ne'eman in the
early 1960s brought him close to Israel. And he grew very devoted to the
pioneer country. The Strategic Defense Initiative injected another strong
motivation into Teller's close ties with Israel, whereby he appreciated the
importance given to science and its applications. He was greatly impressed
by what the Israelis achieved in a historically short period of time in
building up a sophisticated new nation on the background of centuries-
long persecution and the Holocaust that were both part of Teller's and his
family's lives.

## RUSSIAN CONNECTION

The impression is that Teller had a flexible value system based on selec-
tivity, which manifested itself most of all in his attitude toward various
individuals. He made two visits to Russia and met there with Yulii
Khariton, who might be described, if not quite accurately, as the Soviet
Oppenheimer. He could be considered even more to be Teller's Soviet
counterpart—in Sakharov's absence—in view of his long record in creating
the Soviet nuclear might. Khariton was the scientific director of the Soviet
nuclear project, the secret installation called Arzamas-16. Teller thought
that Khariton, being the topmost surviving Soviet nuclear scientist, should
be given the Fermi Award, the topmost American recognition for nuclear
work. Khariton was an outstanding physicist on a world scale and head of
the Soviet nuclear project for many years, yet recognizing him with the
Fermi Award would have been a curious move. Had the cold war and its
arms race been a sporting event, this might have been an expression of fair
sportsmanship, but the cold war could have hardly been considered just
another sporting event.

Teller nominated Khariton for the Fermi Award in a long letter of January 17, 1995, to President Clinton's secretary of energy, Hazel O'Leary.[18] He expressed satisfaction that the political obstacles to such recognition had been removed and so the Russian contributions could be discussed publicly. He stressed the international character of the Fermi Award, which was true in principle, although there had only been a trickle of international recipients, and in exceptional cases. He emphasized Khariton's contributions to the understanding of the nuclear chain reaction and his collaboration with Igor Kurchatov, the late Soviet atomic czar, which led to the first Soviet nuclear explosion.

Teller ascribed correctly the protection of Soviet physicists from Stalinist purges due to the success of the Soviet nuclear project. He praised the independent Soviet weapons work, noting that he had classified information about this, which he could not divulge—a time-honored Teller approach for enhancing the weight of his words. He added that he had information about Khariton's contribution to the Soviet laser–related fusion project. He referred to the highest Soviet honors Khariton had earned and the backing of his nomination by the Soviet nuclear laboratories. Teller also said that Khariton's integrity during the Soviet times was a precondition for his nomination and that it had been vouched for. He praised Khariton for his participation in war-related technical programs and hastened to clarify that he meant the programs against the Nazis (lest anybody think he meant against the United States). Teller thought that Khariton's American award would be of the same character for cooperation with Russian scientists as the Marshall Plan was after World War II for cooperation with Western Europe. He urged timely consideration of this nomination since the Fermi Award was given only to living awardees and Khariton was already ninety years old. His efforts for getting an American award—and a very high one at that—for Khariton appeared somewhat bizarre.

Less than two years before, the *Bulletin of the Atomic Scientists* communicated a long paper about the Soviet nuclear bombs program titled "The Khariton Version."[19] The article by Khariton and one of his associates described the story and achievements of the Soviet nuclear weapons with considerable pride. It put unmistakable emphasis on Soviet priorities in the quest for the hydrogen bomb: "Today, many people realize that it was Soviet physicists who first developed thermonuclear weapons." They noted that "American nuclear physicists' research in this area was mistaken and unsuccessful." There was a brief mention in the article of the slave laborers: "Of course there was little joy in watching the columns of prisoners who built the installations initially. But all that receded into the background, and people had little regard for the difficulties of everyday life—they were trying to achieve success in the best and quickest way." There was then a long section describing the experience of working with the infamous secret police chief, Beria, whom Khariton found efficient and productive.

While Khariton admitted the crucial role of espionage in creating the first Soviet atomic bomb, he denied any such thing in the development of the Soviet hydrogen bomb. In fact, Khariton went out of his way to be candid about Fuchs's service, which was in contrast with his insistence on claiming that the Soviet Union had defeated the United States in the race for a deliverable hydrogen bomb. Siegfried Hecker was a member of the American delegation visiting Arzamas-16 in 1992, and he was quite taken aback when Khariton told them, unsolicited, about Fuchs.[20] Before that, while there was compelling evidence about Fuchs's activities, there was never an admission from the Soviet side. Looking back, an explanation for Khariton's forwardness might be that he meant to make it more credible when he categorically denied any benefit from spying for the Soviet thermonuclear weapons program.

It is curious that the information that exists in the open literature with regard to either the existence or the absence of espionage for the Soviet hydrogen bomb comes exclusively from Soviet–Russian sources. The material of the April 1946 Los Alamos meeting on the Super is still classified in the United States, but if Fuchs conveyed it to the Soviets, as he most probably did, it is not classified for the Russian experts. If radiation implosion was mentioned at the 1946 Los Alamos meeting, which is very likely, it may have benefited the Russian program, whereas it was no help to the American program, as Teller had to reinvent it in early 1951.

Teller was a master—sometimes rather transparently—in turning sins into merits and vice versa in order to achieve his goals. An example of this is when he ascribed the lack of controversy concerning the development of nuclear weapons in the Soviet Union at least in part to Khariton. He commented on the difference between the American and the Soviet stories: "In our case, there was strong controversy whether or not to proceed on nuclear weapons. In the Soviet case, there was no controversy and this may have been in part due to reasonable administrative work done by people like Khariton."[21] If this was a merit, one might as well have suggested Stalin for a distinction. It is hard to imagine that Teller was not aware of the ruthless suppression of any kind of dissent in the Soviet Union at the time of the American debate about the hydrogen bomb.

The sponsor of Teller's trips to Russia in the early 1990s was the Strategic Defense Initiative Organization (SDIO), which was created in 1984 in the Pentagon and was made responsible directly to the secretary of defense. There were discussions between the Americans and the Russians, and representatives from the Ukraine, Kazakhstan, and Lithuania also attended. The main topic in 1992 was the possibilities and implementation of an agreement about the use of satellites for early warning of aggression, weather conditions, environmental contamination, and natural disasters. Teller completed an exhausting program that included visits to Hungary, Italy, Israel, and Germany.[22] The principal location of his 1994 visit was Chelyabinsk, from which a photograph of Teller posing with the world's largest-ever hydrogen bomb emerged as the most publicized result. The trip also included visits to Italy and Scotland.[23]

Teller became popular among the Russian scientists. Their interactions were based on mutual respect and did not suffer from the difficulties Teller had developed with much of the scientific community in the United States. The Russians might even have written off the negative information about Teller in their disbelief of anything earlier Soviet politics had tried to instill in them. Despite Teller's lobbying, Khariton was nevertheless not given the Fermi Award.

However friendly Teller's disposition toward the Soviet–Russian nuclear physicists may have been, he had a more critical view of Russia in spite of the euphoric atmosphere about the political change of its regime. Teller did not exclude—as it should not be excluded today—that "Russia may return, if not to Communism, at any rate to an imperialistic policy." This is why he urged President Clinton to foster the NATO membership of Eastern European countries, including Hungary.[24]

## ASSESSMENT

Teller made an invaluable contribution to world peace with his work on the hydrogen bomb, however contradictory this statement might sound. But when Einstein alluded to the nuclear weapons forcing mankind to reevaluate international relations, he anticipated the policy of Mutually Assured Destruction, even though he did not use that expression. When Linus Pauling quoted Alfred Nobel's dream of creating a terrifying weapon that would scare away mankind from war, he could have meant something like the hydrogen bomb. When Lev Landau was asked whether there would ever be war again, he responded with confidence. "No," he said,

"Physicists provided an enormous service to mankind: they invented so terrible a weapon that war became impossible."[25] Peter Kapitza said: "I think that from the time Russia possesses a thermonuclear weapon, atomic war is absolutely excluded."[26]

In his article about the history of the hydrogen bomb, Bethe stated, "[T]hat the Russians might obtain an H-bomb was of course the most compelling argument for proceeding with our thermonuclear program. It was, in my opinion, the *only* valid argument" (emphasis in the original).[27] Further, Bethe posed the question in 1954, which was published only in 1982, asking "whether the Russians were indeed already engaged in a thermonuclear program by 1949?"[28] Today we know that they were and also that it would have been unthinkable for Stalin to abandon his nuclear program for the sake of an international agreement.

Thus, looking back, Teller deserves credit for his contributions to developing the American thermonuclear weapons. Of course, it was "the work of many people," including Ulam, but there is no doubt that if anyone can be singled out, it is Teller, just as in the Soviet Union it is Andrei Sakharov who should be. If we look at the list of Nobel Peace Prize winners, where controversial figures appear alongside purists, a shared Nobel Peace Prize for Teller and Sakharov would have been justified. Sakharov's Nobel Peace Prize in 1975, which was given to him for different reasons, was fully deserved by him to receive alone. For the hydrogen bomb, no single individual would have deserved the award alone. Only the two superpowers *both* having this most terrible of weapons could serve as instrument for the policy of Mutually Assured Destruction.

As long as the two superpowers existed there was no war between them, and this could only be appreciated if one tried to imagine what an armed conflict between them would have been like. From this it follows that Teller, the father of the American hydrogen bomb, has to go down in history as having made a positive contribution to humankind. Bethe, while finding the hydrogen bomb unnecessary, understood the reverse point as well: "Truman had no choice in the political atmosphere of the time. Had Russia developed the H-bomb and the US not, he and the scientific community that opposed it would have been considered traitors."[29] This was an important point, particularly by Bethe, since Bethe's prior criticism hurt Teller more than anybody else's. In 2009, Steven Weinberg allowed that "[it] may or may not have been just as well that he pushed the hydrogen bomb project," although according to Weinberg "that doesn't make his behavior rational."[30]

In connection with assessing the role of the hydrogen bomb in maintaining the peace between the superpowers for decades during the twentieth century, a caveat is warranted. It is not known—though it cannot be ignored—that the Soviets might have benefited from espionage in creating their hydrogen bomb. In this, Klaus Fuchs, possibly conveying all the details of the discussion of the 1946 meeting on the Super in Los Alamos, may or may not have played a pivotal role. There may have been another act of espionage following Teller's invention in early 1951. There is no formal recognition of such information in Russian sources, only ample gossip to this effect. If this is what happened, Teller might have been the "father of *the* hydrogen bomb" and not just the father of the American hydrogen bomb. Suppose the Russians were familiar with the contents of the 1951 Teller–Ulam paper, would it then be superfluous to continue keeping it classified? Not at all, provided the Russians continued keeping it classified as well. Declassifying the report might make it possible for other nations or groups to develop their own hydrogen bombs.

In addition to the development of the hydrogen bomb, Teller's activities during the second half of his life can be grouped into three principal areas: the creation of the Livermore Laboratory, his lobbying against various test ban treaties, and his role in the Strategic Defense Initiative. Regardless of personal motivations by Lawrence and Teller for its creation, the second laboratory can be judged only by its products, which have become an integral part of American might. There is no way to judge what would have happened had Teller's (and Lawrence's) argument not been accepted and Los Alamos remained the only major weapons laboratory in the United States. Incidentally, the Soviet Union copied the United States in establishing a second laboratory, which became Chelyabinsk-70 from about 1955. The development of the Livermore Laboratory demonstrates how influential Teller was.

His wrongheaded activities with respect to test ban treaties, his performance in connection with various peaceful applications of nuclear explosions, and in particular his claims about the consequences of fallout or lack thereof showed how far he was willing to go to get what he wanted. A conspicuously low point in his activities was his advocacy that genetic mutations—deformed babies—might be an acceptable price to pay for testing. As noted earlier, Teller's friend, George Gamow, expressed his disapproval when he portrayed Teller as Dr. Tallerkin in his book *Mr Tompkins in Paperback*. Tallerkin, the busy professor, explained atomic physics, fission and fusion, to a lay audience. He admitted that his group had not been able to create a "pure" (meaning clean) hydrogen bomb yet. Then came a memorable exchange:

*Final Thoughts*   439

'Dr. Tallerkin,' asked somebody from the audience, 'what about those fission products from the bomb tests which produce harmful mutations in the population of the entire globe?'

'Not all mutations are harmful,' smiled Dr Tallerkin, 'a few of them are leading to the improvement of progeny. If there were no mutations in living organisms, you and I would still be amoebae. Don't you know that the evolution of life is entirely due to natural mutations and the survival of the fittest?'

'You mean,' shouted a woman in the audience hysterically, 'that we all have to produce children by dozens, and, keeping a few of the best, destroy the rest of them?!'

'Well, Madame,—' started Dr Tallerkin, 'but . . .'[31]

Tallerkin could not complete his response, because his busy schedule caused him to abruptly leave the gathering. By using this plot device Gamow took mercy on Tallerkin–Teller and saved him from having to explain something that would have been impossible to explain.

Then came the Strategic Defense Initiative (SDI), whose essence Teller had been advocating for quite some time. When it became part of the Reagan administration's agenda, he was more a promoter than a creative participant. This is not to say that his probing questions and relentless desire to understand were not factors in encouraging his young associates. If we look back, it is astonishing that the efforts for Excalibur and then Brilliant Pebbles received funding at the scale they did—billions of dollars—while they were far from an advanced level of preparedness for defense application. In this, Teller's abilities to paint the future as if it were around the corner must have played a role. Incredible as it is, there was "no time or possibility to give a technical review of the whole SDI effort."[32]

Throughout his career Teller touted his visions as being a few years away, not decades, to keep them sounding attractive. One of his favorite slogans, "It is impossible to lie about the future"—because anything may happen—served him well when he argued for his projects. As Steven Weinberg expressed it, Teller "was in love with technology."[33] He believed in the ability of humans to make things happen once there was an idea that they might. Some referred to this almost blind belief in technology as "the Edward Teller syndrome."[34] He often distracted his critics—who wanted him to account for unkept promises—by saying that his sin was being an incorrigible optimist. That there were tremendous amounts of money spent on his projects did not seem to bother him.

Once there was an idea, it was Teller's belief that the corresponding technology could be found. If it was a difficult problem, the efforts had to

be enhanced, but sooner or later, the solution would be found. Furthermore, Teller believed that *any* problem could be reduced to a problem of
technology and thus solved by technology. As early as 1947 he wrote, "We
need a world government which makes war between the participating
nations *technically* impossible" (emphasis added).[35] This belief in the
almighty technical answer could have been a reason why Teller never
seemed to seek political solutions and ignored the human sides of the problems. In view of his difficulties with human interactions, it is not surprising
that technical answers would seem to him more promising.

There have been many different statements about Teller, both complimentary and critical. Eugene Wigner said, "Teller's imagination was more
fertile than that of anyone else I have ever known,"[36] to which Wigner
liked to add that he also knew Einstein. Of the negative statements, none
has been as devastating as what Isidor I. Rabi told Teller's biographers in
the 1970s and repeated in another interview in 1981: "He is a danger to
all that is important. I do really feel it would have been a better world
without Teller . . . I think he is an enemy of humanity."[37] Unfortunately,
the interviewers on both occasions failed to ask Rabi to elaborate. Because
of its timing, Rabi could not have meant SDI. He must have meant the
hydrogen bomb and/or, possibly, Teller's struggles against the test bans,
and in any case he must have also considered Teller's testimony in the
Oppenheimer case.

Rabi could be harsh in his criticism and was not always just. In referring to the Hungarian scientists' approaching Albert Einstein to warn President Roosevelt about the danger of a possible German atomic bomb, Rabi
proclaimed, "The Germans owed a lot to Szilard."[38] Rabi thought the
atomic bomb project would have taken off much more quickly if the Hungarians had gone to Lawrence rather than to Einstein. Szilard admitted
that they were not well acquainted with the American system (see chapter
4), but blaming Szilard, Teller, and Wigner seems excessive, given they
were the only ones who took *any* action in the first place.

Rabi was not only a great physicist; he also understood the dangers of
Nazism. When Szilard asked him in 1939 to enlist Enrico Fermi in activities to alert US authorities about the possibility of an atomic bomb to be
developed in Germany, Rabi did so. However, when Fermi initially
declined because he did not consider the danger to have high enough probability, Rabi did nothing further. It took additional prodding by Szilard to
mobilize Rabi to convince Fermi. This episode points to an important difference in the approach to questions of defense by Szilard and by Rabi.
Rabi was not an activist, whereas Szilard was, and so was Teller.

Similarly, Teller's relentless fight for developing the American hydrogen bomb in 1949 must have seemed alien to Rabi. On top of this, Rabi considered the hydrogen bomb immoral because it was so devastating. He thought that to prevent the development of the hydrogen bomb, it would suffice to call for an international pledge not to develop it. This was expressed in the minority opinion by Rabi and Fermi when the GAC decided to give advice against the development of the hydrogen bomb (chapter 5).

Rabi's and Fermi's naïveté may have been excusable then, but today we know that the Soviet Union was already developing the hydrogen bomb while it was still being debated in the United States. As for Teller's testimony in the Oppenheimer hearing, Rabi, of course, must have greatly resented it. There was sharp difference in Rabi's approach to Oppenheimer and to Teller. Both Oppenheimer and Teller were controversial figures. However, whereas Rabi tended to overlook Oppenheimer's weaknesses, he overlooked none of Teller's; rather, he amplified them.

In addition to the criticism of Teller's pushing for the hydrogen bomb, another recent accusation has been that his optimism and his insisting on the classical Super delayed the development of the hydrogen bomb. This accusation is difficult to refute because it is rather vague and may have some validity. However, until the Teller–Ulam approach came along, there was no alternative. The calculations by Ulam and Fermi and others were not definitive about the failure of the classical Super. The June 1951 meeting of the GAC at Princeton did not demonstrate that the classical Super would have been discarded quickly in spite of the availability of the new solution that everybody found promising. On the contrary, for some time it seemed that tests were being scheduled for both the classical and the new solutions. Finally, another criticism suggests that the development of the hydrogen bomb was superfluous because it was never used. But this is exactly its merit that because *both* superpowers possessed it, the terrible but highly effective policy of "Mutually Assured Destruction" has ensured peace for decades.

Teller had a different role than before in promoting SDI and in particular Excalibur in its ever-improving versions and, later, in the project of Brilliant Pebbles. He no longer functioned as a creative scientist; rather, he became an operator, utilizing his intellectual capital and past successes. As stressed before, the *feasibility of SDI* and the *dividend from SDI* in helping to end the cold war were two different issues and thus should be distinguished. There were many components facilitating the collapse of the Soviet Union and the communist systems under its thumb throughout the rest of Europe. We ascribe a role to SDI in this collapse without ignoring the many other factors, and it would be impossible to assign percentages to them. We are not discussing here the underlying reasons for Soviet-type economies malfunctioning in the long run; the Soviet system should have collapsed long before, and it did not. Its demise had been predicted many times over since the formation of the Soviet Union, but when it happened, it took the experts by surprise. The events in Berlin in 1953, in Budapest in 1956, in Prague in 1968, and in Poland over the years contributed to this process. So did the policy of the so-called loosening that, according to the communist leadership, the West exercised by making Western culture and ideas of democracy penetrate the communist world. Whatever the combination, it was successful in the long run because it added to the slow but determined decay of the communist system.

It was debatable whether President Carter's withdrawing from the Moscow Olympics in 1980 was the right move to protest the Soviet intervention in Afghanistan. However, in Eastern Europe it was perceived as very effective in hurting Soviet pride. Or take President Kennedy's powerful statement "Ich bin ein Berliner." Consider the debate at Oxford University about the respective merits of the two political regimes in which Caspar Weinberger argued about the message a system sends by building a wall *not* to keep out its enemies but to prevent its own people from escaping. Another powerful moment came when President Reagan issued his call to the Soviet leadership in the shadow of the Berlin Wall: "Mr. Gorbachev, tear down this wall!" These were but a few nails among many in the coffin of Soviet communism—as was SDI, if a recklessly expensive one.

## LABELS

According to the simplified classification of hawks and doves, Teller was obviously a hawk. However, this is not what alienated many of his friends; rather, it was his attitude. Teller's colleagues considered Lawrence to be more extreme on the right politically than Teller, yet Lawrence managed to avoid alienating his peers. Teller did not seem to have this talent, and in his later life he did not seem to care about honing such skills. If anything, he may have thrived on feeling he was the one righteous voice in combating the wrongs of the world. He had been exiled from Hungary, Germany, and from the community of physicists—even if not in the direct sense of the term—and to some extent he may have seen himself as David against Goliath.

Still, Teller never made outright aggressive political statements in the sense that von Neumann did when the mathematician advocated preemptive strikes on Moscow. We know of no proposals by Teller for blowing up the enemy, like the one Sakharov made when he suggested exploding a huge thermonuclear device in an enemy port, presumably New York City.[39] Interestingly, Oppenheimer may have considered the possibility of deploying atomic bombs in the Korean War, whereas Teller would have none of it. Teller consistently advocated maintaining a strong United States to exclude the possibility of becoming victim of the enemy's blackmail. He was a hawk to be sure, but a hawk for defense rather than for offense. Yet Teller's image has been considered to be more hawkish than those of Lawrence, von Neumann, Sakharov, and Oppenheimer.

Teller's attitude about the label he was most often associated with—"Father of the Hydrogen Bomb"—varied. In most cases, he protested it, saying "I am the father of Paul and Wendy," and he would occasionally even stop an interview if the interviewer used that label. When my wife and I talked with him in 1996, he did not seem to mind it. On some occasions, the recognition of his preeminence in creating the hydrogen bomb appeared important to him, especially in relation to Ulam's contribution to it. It would seem to be a dubious title, if one considers the ghastliness of the bomb. However, the authorship of even the most horrible thing may be contested, and, as John von Neumann has said, "Some people profess guilt to claim credit for the sin."[40] As for the authorship of the hydrogen bomb, according to Hans Bethe, "[I]t is more precise to say that Ulam is the father, because he provided the seed, and Teller is the mother, because he remained with the child. As for me, I guess I am the midwife."[41] George Gamow also refers to Teller and Ulam as the mother and father of the hydrogen bomb, respectively.[42]

A devastating comparison, as noted earlier, was made between Teller and the agronomist Trofim D. Lysenko. Lysenko destroyed Soviet biology for decades, not just advocating but forcing his unscientific theories onto everybody and ruthlessly destroying his opponents. He denied the chemical basis of genetics and believed that acquired traits were inheritable. He maintained his domination under such different leaders of the Soviet Union as Stalin and Khrushchev. He promised innovative practices that would lift agricultural production within a few years' time and when they failed to materialize, he would come up with new promises and new deadlines. The transition from Excalibur to Brilliant Pebbles recalls what was noted about Lysenko, "He could deny today what he had said yesterday without anyone asking whether he had been wrong at least one time."[43] Still, Teller was a first-rate scientist, while Lysenko was a charlatan.

The comparison with Lysenko may not carry too much weight, even though he was a real person. The name of the imaginary Dr. Strangelove is, however, widely recognized and has gained a life of its own by now, more or less independent of Stanley Kubrick's 1964 movie, *Dr. Strangelove or: How I Learned to Stop Worrying and Love the Bomb*. The film is about the start of a nuclear holocaust and involves an insane ex-Nazi warmonger scientist, by the name of Dr. Strangelove, who works among the American leadership. Teller has often been called Dr. Strangelove, though other figures have also been given the title, including former secretary of state Henry Kissinger and former Nazi rocket engineer turned American missile designer Wernher von Braun.

When the movie *Dr. Strangelove* first came out, Kissinger was not yet well known, and von Braun never achieved the prominence that would have justified such identification. By 1964, Teller had already been through the debate about the hydrogen bomb, however little justified it would be to hold it against him, especially in retrospect. His role in the test ban debates is a different matter, but that might have not been widely known. Teller's role during the decade of SDI might have brought him closer to what Dr. Strangelove symbolized, but that came about long after the movie. It is more probable that Dr. Strangelove was a composite figure in which Kubrick combined the negative traits of several individuals, Teller's among them. At least two recent books connected Dr. Strangelove and Teller.[44] But another author wondered "whether Edward Teller, whatever else he was, might also have been our collective scapegoat."[45] Certainly, it appears unfair to identify Teller with Dr. Strangelove.

Acclaimed author Salman Rushdie found part of Andrei Sakharov in Robert Oppenheimer, but another part in Edward Teller.[46] And we may wonder how Teller would have fared under the conditions of Soviet life, in the country that was for him the ultimate enemy. It is, of course, impossible to know what one's attitude would be under a different set of conditions unless one existed under those conditions. We might, however, consider a comparison with persons who had existed under those different conditions. Teller might have had the life of Lev Landau in that he might have been arrested and rescued, and then forced to work on the Soviet nuclear project. Whereas Landau quit right after Stalin's death, Teller might have made the best of it and stayed on. He might also have been a Khariton, directing the Soviet efforts. It is difficult though to imagine Teller's exceptionally imaginative and creative personality in a rigorously hierarchic system.

Perhaps most difficult to imagine is Teller as Sakharov in the second half of the latter's life—when he threw away everything in his involvement in creating weapons and devoted himself to human rights issues. It is easier to imagine Teller as a powerful leader in the Soviet program, respected by many and feared by more. Freeman Dyson, who remained friendly toward Teller in spite of many crises and the animosity of others, agrees that Teller might have made "a good Soviet citizen."[47]

It seems pertinent here to issue a caveat. Being a Soviet citizen, even a "good" Soviet citizen, and even—God forbid—a member of the Communist Party of the Soviet Union (CPSU), did not necessarily mean being committed to communist ideology, let alone to its Stalinist kind of implementation. Nonetheless, Sakharov never joined the CPSU, yet he was as devoted a Soviet citizen as anybody in the mid-1950s. Teller could have been like that.

It is interesting to note that Teller remained immune to communist ideas from his very youth, although he was exposed to them. He had a communist math teacher and a fascist math teacher in high school, and although he did not like either, he would have preferred the former to the latter. He was well acquainted with the work of the communist playwright Bertolt Brecht while in Germany. He had communist, or fellow traveler friends like Laszlo Tisza, Lev Landau, and George Placzek—and he witnessed their disillusionment. In Germany of the late 1920s and early 1930s, a growing segment of society became polarized between the Nazis and the communists. Teller could have not considered Nazism, but he never drifted toward communism either. During this period, however, there is no evidence of his anticommunism either, that is, until the early 1940s.

Teller may be criticized for not changing his attitude toward the Soviets when the Soviet leadership changed from Stalin's to Khrushchev's and to Brezhnev's, and even when it changed to Gorbachev's (and he traveled to Russia only following the collapse of the Soviet Union). He recognized, though, that the Soviet leaders would not have initiated thoughtless attacks on the free world without carefully weighing their consequences. He recognized the difference between Hitler and the Soviet leaders, but he did not fully appreciate the difference between Stalin and the rest. In 1991, perhaps belatedly, he noted that after Stalin's death, "Soviet policy, both internal and international, became more moderate and never resumed fully its formerly vicious character."[48]

In the new Russia, there have been steps toward democratization. The process has been hesitant, and no responsible democratic leader could yet be sure that Russia would continue on the road of further enhancing democracy. This has to be kept in mind especially in view of the huge arsenal of nuclear weapons in Russia. In this sense, Teller's tenacity could be somewhat justified.

Dyson wrote to his parents about Teller in 1949 and reprinted his words half a century later in his biographical memoir of Teller. Dyson thought he had solved the Teller puzzle and painted him an idealist, but he added a strong caveat: "I feel that he is a good example of the saying that *no man is so dangerous as an idealist*" (italics added).[49] This comment resonates with a statement made by Valentine Telegdi about Teller fifty-three years later: "Edward Teller suffers from a disease . . . he thinks that he is the messiah. That he, Edward Teller, personally has to reveal how to save the world and who is deeply convinced that he knows what's good for the rest of mankind."[50]

Looking back, Dyson feels that the attacks against Teller were unjustified. He was Teller's friend and he did what he could to counter his friend's negative image. He wrote a chapter about him called "Prelude in E-Flat Minor," in his book *Disturbing the Universe*.[51] The Dyson story is charming if not truly convincing.

The story takes place 1955 when the Dysons rented a house in Berkeley for their summer visit, not far from the Tellers. One day the Dysons went for a walk and left the house open. Upon returning they heard someone playing Bach's Prelude No. 8 in E-Flat Minor on their piano and doing it with heart and soul. It turned out to be Teller, who had come by with an invitation for a party. Dyson had not seen him since 1949, and in the meantime the making of the hydrogen bomb and the Oppenheimer hearing had occurred. Teller's beautiful playing convinced Dyson to decide "that no matter what the judgment of history upon this man might be, I had no cause to consider him my enemy."[52] This vignette evinces a side of Teller that few knew—a man with a poetic and a soulful side—the man who wrote passionately of his yearnings and desires to Maria Goeppert Mayer.

Of course, admiring him for his beautiful piano playing could not have negated Teller's image problems, but Dyson told me in our March 2008 conversation that narrating this story was the best he could do for Teller.[53] Perhaps his friend's cautious approach demonstrates the degree to which Teller was alienated from his fellow physicists.

## ADVISING

Teller repeatedly advocated that scientists should do everything to make discoveries and explain them to the public so that the public could make informed decisions through its elected representatives about their applications. And he kept warning that the scientist should never try to influence the public's decisions. We have seen that Teller constantly violated his own maxim. According to Leo Szilard, "The main aim of the scientist is to clarify. The main aim of the politician is to persuade."[54] In this sense Teller moved from scientist to politician quite early in his career.

It is debatable, however, whether the scientist should be considered just another member of society when new discoveries of enormous complexity are concerned. Secrecy then further complicates this matter because it prevents the scientist from his second task, as prescribed by Teller, of preparing the public for making an informed decision. Secrecy forces even a democratic society into acting in a way that resembles totalitarian regimes. If this is then compounded with a failing mechanism of science advising, the situation becomes hazardous and dangerous.

The channels of science advising worked properly under some administrations but not others. The Eisenhower administration was an example of the former, in terms of PSAC; the Reagan administration was an example of the latter. At the time of SDI, the role of the science adviser was to shore up support for the president's policies rather than channel the scientists' independent analyses to the American leadership. Teller managed to avoid these channels even under Eisenhower, when he presented his "clean" bomb idea directly to the president without subjecting it to the scrutiny of his colleagues.

It was a loss for everybody that Teller did not operate through the proper channels of advising, but it was not entirely his fault. With the benefit of hindsight one might argue that the GAC recommendations in 1949 should have dealt more with scientific issues and less with political and moral ones. The GAC should have recommended thorough studies of the feasibility of the creation of the hydrogen bomb rather than considering whether the United States should or should not demonstrate an example in restraint. The politicians, such as Dean Acheson, were rightly appalled by the naïveté of the GAC scientists. The GAC created a vacuum by failing to recommend an intensified quest for finding out whether or not a hydrogen bomb would be possible. In contrast, Teller's stand skipped the question of feasibility and recommended a crash program, merely filling the vacuum that the GAC had left gaping. Teller's presence in the GAC might have led to a more balanced representation of what the scientists could have contributed to solving the dilemma of how to approach the Soviet menace. Instead, the controversy between the GAC and Teller developed into a rift, which further deepened as a result of Teller's later actions.

Teller became a pariah due to subsequent events. He no longer felt restraint by his peers from whose circles he had been excluded, first by the symbolism of a denied handshake, then by becoming persona non grata at Los Alamos. Later, not being a member of PSAC and finding himself in his third exile allowed him to take advantage of acting on his own and at the same time becoming the darling of the "military-industrial complex." When he resigned from the directorship of Livermore it was not only because he lacked the talent for administering but more because he felt freer in embarking on his crusades without being locked into a position in the hierarchy of the establishment. His responsibilities toward keeping Livermore busy, his anticommunism, and his involvement in conservative politics provided a cooperative effect: strong defense necessitated the development of new weapons of ever-increasing sophistication, which Livermore could generate. His alienation from his peers and his welcome by the industrial-military complex strengthened his beliefs in his own instincts, views, and actions. By the time of SDI there was nothing to restrain him. The scientific community might have not been able to restrain Teller even if it had tried keeping him in its midst, but by having excommunicated him, it did not even give itself the means of trying.

The impression is that lacking the control of his peers Teller lost his self-control, at least on occasion, and he behaved in a way that was distinctly contrary to his prior efforts to be one among the collegial crowd of fellow scientists. The quiet astrophysicist Chandrasekhar had an experience with Teller that illustrates such behavior. Teller became friendly with Chandrasekhar during his sojourn in England in the mid-1930s. Their interests in physics diverged, but even then Teller had an insatiable curiosity and enjoyed talking about astrophysics with his Indian colleague. When they both worked for the University of Chicago in the postwar period, they renewed their relationship. After the Oppenheimer hearing, however, their relationship soured, and the reason was Teller's attitude, for which Chandrasekhar gave the following specific example.

Chandrasekhar was a member of a committee convened to advise the AEC on some questions of fusion research. At a meeting, the directors of national laboratories made presentations, among them Teller, as he was then in charge of Livermore. After the presentations, there was to be an executive meeting of the committee for which the laboratory directors were not supposed to be present, but Teller refused to leave. When asked why he wanted to stay, Teller made disparaging remarks about the knowledge of the subject matter among the members of the committee. As a consequence resignations followed, among them Chandrasekhar's.[55]

In 1990, Teller was asked to single out a low point and a high point in his life.[56] For the low point he mentioned observing John von Neumann dying when he was suffering from losing his most important commodity—his extraordinary quick and profound manner of thinking. Von Neumann used Teller to probe the speed of his own thinking and realized that he could no longer keep up with Teller, whereas when he was healthy he used to be much faster than his friend. For Teller, it was a devastating experience.

The high point was Livermore, because when he lost his lifelong friends in the wake of the Oppenheimer hearing, the Livermore Laboratory embodied his new and good home. Teller stated that his work on nuclear weapons was unavoidable. Then he came to the conclusion that "if I claim credit for anything, I think I should not claim credit for knowledge, but for courage. It was not easy to contradict the great majority of the scientists, who were my only friends in a new country, having left almost everybody behind me in Hungary and in Europe."[57]

Viewing a broader picture and looking beyond SDI, the strong military buildup and in particular the deployment of modern technology—automation, computerization, miniaturization—in defense under President Reagan changed the relationship between the two superpowers. While creating an astrodome-like shield to protect the United States from massive missile attacks proved to be nonsense, a sophisticated defense that would have made it possible to unmask cheaters of international agreements made a great deal of sense. In the preparations for the Reykjavik Summit, Gorbachev warned his politburo colleagues of the danger of being "drawn into an arms race that is beyond our strength. We will lose, because now for us that race is already at the limit of our possibilities."[58] Teller may have advised Reagan to embrace SDI for the wrong reasons, but it certainly contributed to the demise of the Soviet Union.

## TWO TELLERS (AT LEAST)

When the soundness of Teller's advising is to be judged, it should be done without bias toward his personality, though that is difficult to do. And from numerous descriptions, and even from my limited personal experience with him, what emerges is far from an unambiguous image of his personality. There seems to have been a few Tellers. To the outside world, and especially during the second half of his life, he was an arrogant, self-assured, overbearing, intolerant individual who appeared arbitrary on occasion in his decisions, misused his real or perceived influence over military and political leaders, and won every debate he entered. There was then the private Teller, who was full of self-doubt and who craved the approval of his superiors and especially the friendship and appreciation of his peers. This is the Teller who was revealed most conspicuously in his letters to Goeppert Mayer. At different periods and under various circumstances, these two sides of his character dominated his behavior to different extents. Teller changed for the worse during and after the war, that is, during his stay in Los Alamos. A third Teller was charming and kind, exercising self-deprecating humor, courteous, and magnanimous. This side was also revealed in those letters. In my many encounters with people who knew him I have developed a notion that he must have led a compartmentalized existence in which there was little overlap between his different selves. However, his charming side was often used in a Machiavellian way to further his causes, as in his public debates.

Teller had a troubled childhood, and it took a lot of effort to make friends. Then, in Germany and later in the first decade of his American life, he was well liked. He was a creative physicist, but he was at his best when working with another individual. The breaking point came in the early 1950s when he antagonized most of his peers with his relentless lobbying for the hydrogen bomb; with his struggle for a second weapons laboratory; and most conspicuously with his testimony against Robert Oppenheimer. The change of his circumstance impacted him negatively at more than one level, including that of his scientific creativity and output.

Undoubtedly, Teller had an exceptionally strong willpower, determination, and stamina, for which we have seen ample examples—among them how he treated his physical handicap of attaching a wooden foot every day—for most of his long life. He carried some terrible baggage with him from his Hungarian youth, during which he was conditioned that he could survive only if he outdid everybody else around him. This background helped drive him to develop his boundless ambition and his efforts for overachievement. This type of ambition often fuels the achievements of recent immigrants and often subsides in the next generation. For a Jewish Hungarian, like Teller, his experience in Hungary and his experience as a recent immigrant reinforced each other in developing his response to the challenges he faced. His youth programmed him for a defensive attitude under the conditions of a belligerent society. Then he proved unable to reprogram himself and continued his no-holds-barred attitude under the conditions of a democratic society. This attitude netted him many temporary victories but damaged him beyond repair in the long run.

Teller's ambitions drove him to beat down adversaries who hindered his efforts in achieving his goals or whom he simply viewed as hindering them.

Oppenheimer and Woodruff (of Livermore) were conspicuous examples. The recklessness of Teller's behavior in such cases was in stark contrast with his behavior toward people whom he did not consider adversaries or threats. Consider Heisenberg and Teller's exaggerated closeness to him. Heisenberg did not give Teller the highest grades when he was his graduate student; he did not choose Teller to be his assistant (in spite of the *Memoirs* stating that he did); and Heisenberg was noticeably reserved in his contribution to the volume for Teller's sixtieth birthday. But in spite of Heisenberg's work for the Nazis, Teller never saw him as a threat in any way.

It occurred repeatedly that before taking some actions Teller sought the approval of others, preferably his superiors. When he had overcome the resistance of the Rockefeller Foundation to his marriage, he still wanted the approval of Niels Bohr (in whose institute he was working at the time) before marrying Mici in 1934. In July 1945 at Los Alamos, he asked Oppenheimer about the petition he received from Szilard, and then subsequently blamed Oppenheimer for advising him against signing it. This act of shifting the blame gradually developed into the legend that Teller had actually opposed the atomic bombing of Japan. When in 1952 he was invited to attend the "Mike" shot in the Pacific, a decisive test that for the first time utilized the Teller–Ulam design, he asked Duane C. Sewell at Livermore for his thoughts. Sewell advised him against going, and he did not go.[59] In the summer of 1954, after he had testified that he would not give Oppenheimer security clearance, he contemplated issuing a statement that he had made a grave mistake. He asked Lewis Strauss about it and let himself be talked out of doing it. He consulted the dying Fermi about his article "The Work of Many People," and published it only after Fermi told him to do so. And there were other, similar cases where Teller appears as someone seeking advice and amenable to such advice—an image that is not very compatible with that of a tough person who was adamant and unshakeable in choosing the course of his actions.

Looking for the approval of others, preferably those of higher authority, is related to Teller's fervent respect for the law. This quality was inculcated in him in his childhood by his revered maternal grandfather, who taught him that laws must be obeyed without exception. He never accepted that personal responsibility might override the law even if it worked against one's conscience. He never subscribed to the American tradition of preferring to break a law rather than doing something against one's conscience. This is why it is uncomfortable to imagine how Teller might have conducted himself in a Soviet or Japanese environment had he chosen to immigrate eastward rather than westward after the Nazi takeover in Germany.

Teller's ability to be an amicable and caring friend to some and a ruthless adversary to others seemed to coexist within him without any problem. Not so for the outside world. He paid a high price for his ruthlessness in human relations and his unscrupulous pursuit of his goals. He lost the friendship of the people he needed most at more than one level. The community of fellow scientists had eased the pain of his immigration to America, and those scientists were the means for accomplishing his creativity in science. His alienation from them could not have been compensated by the physicists who could be summoned to his office at his whim at Livermore. The handshakes denied to him by Christy and Rabi in the summer of 1954 at Los Alamos could be ascribed to the immediate reaction to his testimony at the Oppenheimer hearing. He profoundly felt that loss.

When Teller's father abruptly terminated their friendly chess games—which had ceased to be very friendly—Teller felt a rejection by someone who might have been an authority for him but whom he could no longer trust. His ambivalent feelings toward authority figures whose approval he sought and then repelled might have been a factor throughout his life. As much as he craved fellowship, this inner turmoil may have contributed to his sabotaging relationships—leaving him with only himself to trust and without partners with whom to do science.

It was symbolic of his loneliness how he and Bethe met without recognizing each other, on an occasion in Los Alamos in 1992. The episode was witnessed by physicist Ralph W. Moir of Livermore, who found the event important enough to jot down for posterity. For an outsider, what Moir observed at the Los Alamos Inn on the morning of July 22, 1992, would have been a nonevent; for Moir it was significant:

> Bethe comes to breakfast approaching the hostess at the restaurant. He and I talk (we talked the night before briefly) casually. He doesn't invite me to join him. I am seated at the next table closer to the entrance. After food arrives, Edward Teller comes in carrying a set of music with Beethoven in big letters. He is seated and walks within one meter of Bethe. Neither acknowledges even seeing the other. Teller is seated with one table in between his and Bethe's. Teller finishes his breakfast quickly and leaves first, trudging up to pay his bill and walks out into the lobby. As he passes Bethe neither acknowledges even seeing the other (obvious snubs). Immediately, wonderful piano music (Beethoven) emerges from the lobby. This goes on and on. Bethe gets up, pays his bill and heads through the lobby not turning his head even the slightest to look at the piano player as he heads straight for the door.[60]

Both men had to be present for this scene to be meaningful. However, the largely isolated Teller craved Bethe's recognition, whereas Bethe, a much-revered figure among his peers, did not crave his.

## LEGACY

Teller created first-rate science, and his discoveries in science were non-controversial. The interest in and applications of some of his results are still on the rise, which is not a small feat in rapidly progressing modern science. As for long-term legacy, research scientists in general are not in as good a position as creative artists. Beethoven's Ninth Symphony could not have been composed without Beethoven, and Picasso's *Guernica* could not have been painted without Picasso. However, most scientific discoveries would be made sooner or later by others if those who made the discovery had not done so.

Teller made important discoveries, although they were not at the pinnacle of twentieth-century science. He and others, like Bethe, Gamow, and Oppenheimer, were destined more to pick up the pieces left over by such giants as Planck, Einstein, and Heisenberg. There is higher probability that the names of the latter will be remembered in the annals of science, whereas the others will fade away into oblivion. This can be said without belittling their significant contributions. As for their recognition, some of Teller's contributions were worthy of winning the Nobel Prize in Physics or in Chemistry. But that was not to be. The Nobel Prize committees have made much more striking omissions, such as of Lev Tolstoy in Literature, Dmitrii Mendeleev in Chemistry, or Oswald Avery in Physiology or Medicine (for recognizing that DNA was the substance of heredity).

Teller's missing prize is not among the conspicuous ones, but had he been awarded one, nobody could have justly protested it on the basis of scientific merit. Nobel Prizes in Physics and Chemistry are hardly ever awarded undeservedly, whereas many scientists with discoveries of similar caliber to those recognized go without the prize. Teller himself thought that it was the BET equation that should have been singled out for such recognition, and he told us so in our recorded conversation in 1996, although he later requested to delete this utterance from the printed version of the interview.[61]

Teller's idea of compressing the fusion fuel by electromagnetic radiation (X-rays) was as surprising to Hans Bethe as the discovery of nuclear fission (see chapter 6). By analogy of Otto Hahn's Nobel Prize, this discovery could have been considered for a Nobel Prize. Apparently, it was not. (Nominations for Nobel Prizes can be studied at the archives of the respective Nobel Prize committees for prizes that had been awarded fifty years before or earlier.) A light comment about a possible Nobel Prize was in Teller's book *The Pursuit of Simplicity*. He described his imaginary space trip and lamented the loss of ten years from his life. As a consolation, he "can hope to get the Nobel Prize and a ticker-tape parade on Fifth Avenue, at least."[62]

Teller's greatest monument is perhaps his contribution to creating a bomb that helped maintain peaceful coexistence between the two superpowers for decades, and hence prevented a World War III. It is, however, a difficult concept to accept that living with the most terrible bombs imaginable has made humankind more secure than living without them. There can be no control experiment. Teller's Livermore Laboratory made important contributions to the success of this policy, but, again, lacking a control experiment, it is difficult to assess whether the United States could have built up its modern defense without it. It is another question whether or not the amassed weapons arsenal was necessary. The most controversial project of all is the Strategic Defense Initiative. Based on what has been accomplished from such grand plans, one would be tempted to think it was a waste of resources, both human and material. On the other hand, even if it is debatable to what extent SDI contributed to the collapse of the Soviet Union, that it accelerated the process is probably true.

There is no scientific method to determine how influential Teller really was. My impression from my reading of his writings and the writings of others about him, as well as from numerous conversations, is that his influence may have been overrated. This may be an unexpected assessment, though I believe his importance was still considerable. One might think his actions were decisive in steering the United States toward developing the hydrogen bomb, but President Truman made the decision, not Teller. Once Truman knew about the possibility of the hydrogen bomb—and that by the same token the Soviets could also develop it—it was inevitable for him to make the decision to go ahead with it, and he did. Likewise, President Reagan decided to pursue SDI, not Teller. Once Reagan got the notion of building a protective shield for the United States, however illusory that notion was, he put it into the center of his defense policy. It was very helpful for this technically uninformed politician to have a world-renowned scientist giving him unconditional support and doing so in the most conspicuous way.

Even the creation of Livermore might have not been possible without Lawrence's input. Once the laboratory was established, Teller had a key role in developing it into a powerful institution. Lawrence had important influence but operated more behind the scenes, whereas Teller aimed at maximum visibility. For someone who felt somewhat insecure, he liked to be in the headlines. He had the right personality for it. And due to the hydrogen bomb's finality, it was easy to identify Teller with something awesome in its power that was universally known. His protest of the label "father of the hydrogen bomb" often sounded more like undue modesty than denial. In crucial instances he was unambiguous about his decisive role in creating the hydrogen bomb.

His correspondence also shows that he felt more important or wanted to appear more important than he realistically could have been. A point in case was that whenever a Republican presidential nominee appeared, much less was elected, Teller fired off a letter to him. Typically in such a letter, a sentence or a paragraph of congratulation was followed by the most concrete advice on what the potential president or acting president should be doing, how he should shape his defense policy, and whom he should appoint to key posts in his cabinet and as his advisers. It is well known that President Reagan's science adviser was appointed according to Teller's suggestion, but it is doubtful that Teller's many other suggestions to Reagan and others had ever been followed. Nevertheless, he gave the appearance that he was giving advice and suggestions regarding crucial questions and appointments at the highest level.

In spite of his involvement in politics, Teller was not a good politician, considering his disregard of the consequences of his actions and statements. According to Bethe, "When the decision came down against Oppenheimer, not a surprise in the McCarthy era, the majority of the scientific community blamed the witnesses they knew best, particularly Edward Teller."[63] Teller was also victim of his own rhetorical skills. The way he formulated his doubts about and distrust of Oppenheimer sounded so shrewd and calculated that his statements have been quoted over and over again. In this case Teller fell into the trap of his own clever words; we have seen in his homecoming speech at the Hungarian Academy of Sciences that he ascribed frightening power to words in the modern world. There were other scientists testifying against Oppenheimer in 1954, but they escaped the wrath of their colleagues. Teller did not.

Teller was worried about the growing apprehension among the population toward technology, science, and progress—and cloning in particular. Given the popular mood away from science and technology, he ascribed particular merit to the Lawrence Livermore National Laboratory, which he considered his legacy. This is rather reasonable. When James D.

Watson, the co-discoverer of the double-helix structure of DNA, was asked about his legacy, he did not mention his discovery but pointed to his books and to Cold Spring Harbor Laboratory, which he developed into a world-class research facility.[64] Teller penned over a dozen books, but he could not have been sure of their lasting impact, and neither can we, but he had no doubt about the importance of Livermore.

When Teller was asked in 1999 to write an "Open Letter to the Citizens of the Year 2100," he dealt with the question of his legacy, if indirectly. He saw a great difference between the first and second halves of the twentieth century in that only a small percentage of the fifty million killed in the first half were killed in the second. "The explanation of the difference lies in the fact of weapons in possession of those who did not want to use them."[65] There were scientific and technical advances that made it possible for the United States to win the cold war and to win it without bloodshed. There are other indications that he was concerned about his legacy, and he often talked about it with some of his closest colleagues at Livermore.

Edward Teller, who survived three exiles, contributed to the way the twentieth century played out on a worldwide stage, even if the extent of his influence is debatable. He had extraordinary willpower and triumphed over many obstacles, including his physical handicap. He waged struggles that many of his peers would have refrained from. He thrust himself into battles that often engulfed him as they would have engulfed anybody. He seemed at times invincible, and he was a dedicated fighter for the containment of communism to an extent few others were. His dedication often appeared as obsession and his schemes as irrational. While he was attempting to save the Free World, he was also trying to impose his will upon it. John A. Wheeler, who was friendly with him throughout his life, noted of him that "he fought obstinately for what he believed in. I may have disagreed with his tactics but never with his goals."[66]

There he was, this extraordinarily gifted physicist cutting a tragic figure. Even his most important contribution both as a physicist and as a self-appointed guardian of freedom—the thermonuclear bomb—has generated more animosity than admiration. He aimed at forward-looking achievements of technology but operated with medieval approaches in human relations. He could have been the right man for the right purposes, but he often emerged as the wrong man for the wrong purposes. He would have cherished leaving behind a paradox. But instead it is controversy and scorn—deserved and undeserved—that would form his legacy.

# AFTERWORD

## *Richard Garwin*

Over the years from 1947 until his death, my path and that of Edward Teller crossed many times, joining on occasion, intersecting at large angles, or even running antiparallel. I first met Edward (but didn't call him that at the time) in 1947 when I went to the University of Chicago for graduate work in physics, where I soon became an informal assistant to Enrico Fermi in his laboratory and then a formal PhD student. I received that degree from the Department of Physics in December 1949 and joined the physics faculty where I participated fully in all of the activities until December 1952, when I left to join a new IBM Laboratory at Columbia University where I was to work on liquid helium, superconductors, and many other aspects of physics and technology.

My interactions with Edward were already strong in the context of the weekly faculty seminar of the Institute for Nuclear Studies, which was chaired by Willard F. Libby. As Hargittai tells it well, Fermi always had a major role in these seminars, several times coming prepared to discuss a brand-new theory of high-energy particle collisions or of the origin of cosmic rays that within a few days would be a fundamentally new paper in the *Physical Review*. Teller was full of less significant ideas.

During my first of many summers at Los Alamos in 1950, where I went to work on nuclear weapons, I initially read about Teller's activities at Los Alamos during the war in the weekly progress reports from various groups. I was in a good position to understand his current work at the time—which was a longtime fixation of his—with a few associates, to devise a means for configuring the "classical Super" hydrogen bomb. But calculations showed that the reaction would die out rather than propagate well. I helped Edward Teller by responding to his suggestions about diagnostic means for nuclear explosions, some of which were first implemented in the George shot of the Greenhouse series of nuclear tests in the Pacific, May 9, 1951. But Teller's obsession was the full-scale thermonuclear weapon.

As Hargittai tells it, when I returned to Los Alamos in June 1951, I learned from Edward Teller of the idea of "radiation implosion," revealed in a still-secret publication at Los Alamos of March 9, 1951. Edward wanted an experiment that would leave no doubt that this was a feasible path to a thermonuclear weapon, and I responded on July 25, 1951, with a four-page technical memorandum and a large design sketch of what was to be detonated November 1, 1952, as the MIKE shot at eleven megatons of nuclear yield.

I was unaware at this time that in 1949, still with the goal of the classical Super in mind, Teller had identified me as one of twelve people who should be recruited to work on the Super at Los Alamos, two of the others being Enrico Fermi and Hans Bethe.

*Judging Edward Teller* is an illuminating volume, from which I learned a great deal, not only concerning Teller but also in regard to philosophy. Hargittai discusses at great length the role of Teller and Ulam in the concept of radiation implosion. In 2003 I had access to a letter from Stanislaw Ulam to John von Neumann dated February 23, 1951, in which Ulam describes his idea of having an "auxiliary bomb" that prepares a main charge through compression by means of the shock wave of the "primary" explosive—a term that was used later. Those of us who knew Stan Ulam understand that he would not have done any of the calculations in the Teller–Ulam joint paper. Moreover, Teller had at times gone out of his way to claim that he had already told people about the concept of radiation implosion as early as December 1950. Some have asked that if Teller had not already worked it out, how could he respond immediately in the initial conversation with his colleague Ulam that essentially "It wouldn't work," and besides he had a better way—the use of the energy radiated from the nuclear explosion and confined by what came to be termed a "radiation case." Knowing both men and being intimately familiar with the technology of the time, I am persuaded that Teller's years-long immersion in the classical Super blinded him to a concept that he and anyone else could have worked out the consequences of in a few minutes—the utility of compression of the thermonuclear fuel. Lest the untutored in weapon technology say, "Of course," it is necessary to state in Teller's own words of 1979 that he had long had a "theorem" that compression would not help. So, according to Teller, if you couldn't burn thermonuclear fuel at normal liquid density, you couldn't burn it at a thousand times liquid density, because the rate of energy loss would be enhanced by just the same factor as the rate of energy gain.

Only when Teller was about to explain to Ulam why his suggestion was wrong, in order "not to waste a lot of time" with Ulam did he consider committing his theorem for the first time to paper. It dawned on him that he had ignored one crucial fact—that with compression the total energy in the particles per unit volume increased linearly with the compression factor, but the maximum energy in the electromagnetic radiation was dependent on temperature and not on compression.

Edward was stuck in a rut and in my opinion would not have thought of radiation implosion for years, although anyone new to the field, in learning the fundamentals of reaction rates, energy loss, inertial expansion, and the like, had a reasonable probability of coming up with the idea. It had not yet happened, and Hans Bethe proclaimed the invention a "miracle," while the head of the Theoretical Division, Carson Mark, judged that anyone trying to work out the Ulam approach of shock compression would have found that the radiation got there first and was a much more suitable approach.

Hargittai asks why my contribution to the thermonuclear program and to the design of the MIKE shot was unknown to the public. It was not even very well known at Los Alamos. The paper of July 25, 1951, had a small distribution list, and when I spoke at Los Alamos in a classified historical discussion in 1993 I recalled the paper very well; yet only the copy that had been sent to Enrico Fermi, also a consultant in the summer of 1951 at Los Alamos, could be located. The detailed proposal was presented probably in early August 1951 to the appropriate committee at Los Alamos chaired by Bethe at the time, and, as Teller states, was thoroughly criticized and then endorsed and built essentially as I had proposed. It was important *what* was being achieved, and not *who* had proposed it, and I had by then sensed what has been a guiding principle in my own life, "You can either get something done or get credit for it, but not both."

In his book, Hargittai questions why it was left for a consultant to integrate the ideas floating around at the time. I tried hard to get the very capable Los Alamos low-temperature physics group to do the preliminary cryogenic design of MIKE, but I was told explicitly that the group was "burned out" because of all of the efforts they had put into the Greenhouse George test, to be shot May 1951, and that they needed to return to doing physics. So I did it myself.

I want to add one fact to the Teller life narrative. Hargittai writes, "In the spring of 1963, President Kennedy decided to award Oppenheimer the Fermi Award, which was handed to him by President Johnson on December 2, 1963, the anniversary of the first nuclear reactor, only days after Kennedy's assassination. Teller had been given the Fermi Award the previous year." My role in this has been untold, because it was known only by the eighteen members of the President's Science Advisory Committee and its executive secretary, Dr. David Z. Beckler. Teller's nomination for the Fermi Award had been approved by the General Advisory Committee (GAC) of the AEC, and was forwarded to President Kennedy for his approval. Naturally, it was transmitted through the president's science adviser at that time, Dr. Jerome B. Wiesner, and was presented as an agenda item for discussion by PSAC.

While there was no dissent that Teller's accomplishments warranted the award, the preponderance of feeling was that the award should not be made because of Teller's role in the Oppenheimer trial. In one of the few political acts of my life, I commented that I abhorred what I judged to be Teller's attack on Oppenheimer, but that I thought that Oppenheimer certainly merited, himself, the Fermi Award and that it could only happen if Teller received it first. The argument apparently carried the day, and the awards were made in successive years to Edward Teller and to Robert Oppenheimer.

Hargittai has meticulously and successfully judged Edward Teller. I came away from reading the manuscript feeling enriched not only in knowledge but in wisdom.

# TIMELINE

## Selected Events in Edward Teller's Life

| | |
|---|---|
| 1908. | Born Ede Teller in Budapest, January 15 |
| 1917–1925. | Attends Minta Gimnázium |
| 1925. | Enters Budapest Technical University |
| 1926–1928. | Leaves Hungary for Germany; attends Karlsruhe Technical University |
| 1928. | Attends University of Munich, spring |
| 1928. | Trolley accident, July 14 |
| 1928. | Joins University of Leipzig, November |
| 1928–1930. | Leipzig, PhD student under Werner Heisenberg; acquires doctorate |
| 1930–1931. | Assistant at the University of Leipzig |
| 1931–1933. | Assistant at the University of Göttingen |
| 1934. | Rockefeller Fellow, Copenhagen; marries Auguszta Schütz-Harkányi (Mici) |

| | |
|---|---|
| 1935. | Visiting Scientist at University College London |
| 1935–1946. | Professor of Physics, George Washington University, Washington, DC (on leave from 1941) |
| 1941. | Naturalized citizen of the United States |
| 1941–1942. | Visiting Professor at Columbia University, New York City |
| 1942–1943. | Metallurgical Laboratory at the University of Chicago |
| 1943. | Son Paul born |
| 1943–1946. | Manhattan Project, Los Alamos Laboratory in New Mexico |
| 1946. | Daughter Susan Wendy born |
| 1946–1952. | Professor of Physics, University of Chicago |
| 1948. | Elected to membership of the National Academy of Sciences of the USA |
| 1949–1952. | Los Alamos, Consultant; Assistant Director, for varying durations |
| 1952–2003. | Associate of the University of California Radiation Laboratory at Livermore, later, Lawrence Livermore National Laboratory |
| 1953–1960. | Professor of Physics, University of California at Berkeley |
| 1954. | Testifies at the hearings of the Personal Security Board of the Atomic Energy Commission (AEC) on the matter of J. Robert Oppenheimer, April |
| 1954–1958. | Associate Director, Lawrence Livermore National Laboratory |
| 1956. | August 1956–August 1958, member of the AEC General Advisory Committee |

1958.　　　Albert Einstein Award (Lewis and Rosa Strauss Memorial Fund)

1958–1960.　Director, Lawrence Livermore National Laboratory

1960–1970.　Professor of Physics-at-Large, University of California

1961.　　　American Academy of Achievement Golden Plate

1962.　　　Enrico Fermi Award (AEC)

1963–1966.　Professor Emeritus and Chair, Department of Applied Science, University of California at Davis and Livermore

1970–1975.　University Professor, University of California

1975.　　　Harvey Prize (Technion, Israel)

1975–2003.　University Professor Emeritus, University of California; Director Emeritus and Consultant, Lawrence Livermore National Laboratory; Senior Research Fellow, Hoover Institution, Stanford University

1977.　　　Albert Einstein Medal (Technion, Israel)

1982.　　　US National Medal of Science (in 1983)

1989.　　　Presidential Citizens Medal

2002.　　　Department of Energy Gold Award

2003.　　　Presidential Medal of Freedom

2003.　　　Teller dies in Stanford, California, September 9

# ACKNOWLEDGMENTS

I owe a lot of thanks to the following:

To my wife, Magdi (Magdolna Hargittai), who is present in all my projects whether or not we do them jointly.

To Linda Greenspan Regan of Prometheus for her tireless editorial guidance from the inception of this project to its completion.

To Richard L. Garwin (Scarsdale, NY), Balazs Hargittai (Loretto, PA), and András Simonovits (Budapest) for reading the whole manuscript at various stages of its preparedness and for the many valuable suggestions they made.

For information, illustrative material, advice, and other kindnesses, my thanks to Joy L. Anderson, Oak Ridge National Laboratory, Philip Anderson (Princeton, NJ); Judit Balázs and Csaba Sárosi, Archives of the Ministry of Justice and Law Enforcement (Budapest); Krisztina Batalka, Archives of Budapest University of Technology and Economics, Beate Bauer-Renner (Dorum, Germany); Günther Beer, Museum der Göttingen Chemie (Göttingen, Germany); Jens Blecher, University Archives Leipzig (Leipzig, Germany); Alan B. Carr, Los Alamos National Laboratory, Geoffrey Chew (Berkeley, CA); Burtron H. Davis (Lexington, KY); Edgar and Charlotte Edelsack (Washington, DC); Tibor Frank (Budapest); Igor Gamow (Boulder, CO); Christine Colburn, Julia Gardner, and Barbara Gilbert, Special Collections Research Center, University of Chicago Library; János Gács (Budapest); Richard Garwin (Scarsdale, NY); Marvin

Goldberger (San Diego, CA); Boris S. Gorobets (Moscow); Edina Gyor-gyevics, Trefort Ágoston Gimnázium (formerly Minta Gimnázium, Budapest); Diana Hay, Archives of the Hungarian Academy of Sciences (Budapest); Roald Hoffmann (Ithaca, NY); Klára Katona, Hungarian State Archives (Budapest); Michael Jahn (London); Roy Kaltschmidt, Lawrence Berkeley National Laboratory; George (Jay) Keyworth (Piedmont, CA); Janos Kirz (Berkeley, CA); Andreas Kleinert (Halle, Germany); Mária Kolonits (Budapest); Éva Sz. Kovács, Historical Archives of the Hungarian State Security (Budapest); Robert Labedzke and Bernard Nunner, Deutsche Physikalische Gesellschaft (Bad Honnef, Germany); Peter D. Lax (New York City); Carol Leadenham, Hoover Institution Archives (Stanford, CA); Stephen Libby, Lawrence Livermore National Laboratory; Margaret May (London); Chris Peterson, Oregon State University (Corvallis, OR); Kathleen Richards, *East Bay Express* (Emerville, CA); Alexey Semenov (Moscow); Judith Shoolery (Half Moon Bay, CA); George P. Shultz (Stanford, CA); Yurii N. Smirnov (Moscow); Randy Sowell, Harry S. Truman Library (Independence, MO); Manfred Stern (Halle, Germany); Csaba Sükösd, Budapest University of Technology and Economics; Judit Szűcs (Budapest); Wendy Teller and Richard Weyand (Naperville, IL); Paul Teller (San Diego, CA); Maxine Trost, Lawrence Livermore National Laboratory Archives; Zoltán Varga (Budapest); Attila Vértes (Budapest); Zsolt Vizhányó and András Koltai, Gimnázium of the Piarist Fathers (Budapest); Doron Weber, Alfred P. Sloan Foundation (New York City); Steven Weinberg (Austin, TX); Jocelyn K. Wilk, Columbia University Archives (New York City); Richard Zare (Stanford, CA).

To our interviewees in the *Candid Science* series in whose conversations Edward Teller came up (those by Magdolna Hargittai are marked "by MH"), including Edward Teller (Stanford, 1996); Alexei A. Abrikosov (Lemont, IL, 2004); Harold Agnew (Budapest, 2003); Philip W. Anderson (Princeton, 1999); Seymour Benzer (Pasadena, 2004); Owen Chamberlain (Berkeley, 1999); James W. Cronin (Chicago, by MH, 2003); Freeman J. Dyson (Princeton, 2008; by MH, 2000); Val L. Fitch (Princeton, by MH, 2002); Jerome I. Friedman (Cambridge, MA, 2002); Richard L. Garwin (Scarsdale, NY, 2004; 2008); Vitaly L. Ginzburg (Moscow, 2004); Donald A. Glaser (Berkeley, 2004); Maurice Goldhaber (Brookhaven, 2001; 2002); Nicholas Kurti (London, 1994); Yuan T. Lee (Lindau, Germany, 2005); Benoit B. Mandelbrot (Stockholm, 2000); Matthew Meselson (Woods Hole, MA, 2004); Yuval Ne'eman (Stockholm, 2000); Markus L. E. Oliphant (Canberra, 1999); Wolfgang K. H. Panofsky (Palo Alto, by MH, 2004); William H. Pickering (Pasadena, 2004); Norman F. Ramsey (Cambridge,

MA, by MH, 2002); Burton Richter (Palo Alto, by MH, 2004); Jack Steinberger (Geneva, Switzerland, by MH, 2004); Valentine L. Telegdi (Budapest, by MH, 2002); Laszlo Tisza (Budapest, 1997); Charles H. Townes (Berkeley, 2004); John A. Wheeler (Princeton, by MH, 2000; 2002).

To Arnold Kramish (Washington, DC) for having brought us together with the late Clarence and Jane Larson, who had collected a large number of video interviews with outstanding technologists and scientists and who had donated their original cassettes to us. Their relevant interviews in the 1980s that helped me in this project were recorded with Luis Alvarez, Martin Kamen, Edwin McMillan, Rudolf Peierls, Glenn T. Seaborg, Emilio G. Segrè, Edward Teller, Eugene P. Wigner, and Herbert York.

To all interviewees during our visit to California at Stanford University, Lawrence Livermore National Laboratory, and elsewhere in April 2009: George Chapline, David Dearborn, Sidney D. Drell, Siegfried S. Hecker, Malvin H. Kalos, Stephen Libby, Ralph Moir, Richard More, John Nuckolls, Balazs Rozsnyai, Neal Snyderman, Abraham Szoke, and Mort Weiss.

And to the following institutions for generously supporting this project: Budapest University of Technology and Economics, Hungarian Academy of Sciences, and Alfred P. Sloan Foundation (New York City).

# BIOGRAPHICAL NAMES

The asterisks following some of the names indicate that there are sidebars in the text with more information. Abbreviations: AEC: Atomic Energy Commission; CIA: Central Intelligence Agency; GAC: General Advisory Committee; LLNL: Lawrence Livermore National Laboratory; LANL: Los Alamos National Laboratory; SDI: Strategic Defense Initiative.

Abelson, Philip H. (1913–2004); American physicist and science writer

Abrahamson, James A. (1933– ); US Air Force general; former director of the SDI Organization

Abrikosov, Alexei A. (1928– ); Soviet–Russian–American physicist; Nobel laureate (2003)

Acheson, Dean G. (1893–1971); President Truman's secretary of state (1949–1953)

Adamson, Keith F.; US army lieutenant colonel; the army representative in the Uranium Committee

Agnew, Harold (1921– ); American physicist; third director of the LANL

Alpher, Ralph A. (1921–2007); American physicist; George Gamow's student; with Robert Herman, first to predict the background temperature in outer space

Alvarez, Luis W. (1911–1988); American physicist, Nobel laureate (1968)

Anderson, Carl D. (1905–1991); American physicist, Nobel laureate (1936)

Anderson, Herbert L. (1914–1988); American physicist

Anderson, Jean; American physicist, Herbert Anderson's wife

Anderson, Martin (1936– ); American politician

Anderson, Philip W. (1923– ); American physicist, Nobel laureate (1977)

Andropov, Yuri V. (1914–1984); Soviet politician; leader of the Soviet Union (1982–1984)

Arber, Werner (1929– ); Swiss biologist; Nobel laureate (1978)

Armstrong, Henry E. (1848–1937); Fellow of the Royal Society (London)

Armstrong, Neil A. (1930– ); American aviator and astronaut

Ausbach, Fred; Los Alamos scientist

Avery, Oswald (1877–1955); Canadian–American biomedical scientist; principal discoverer that DNA is the substance of heredity

Bacher, Robert F. (1905–2004); American physicist; commissioner of the AEC (1946–1949)

Bacon, Francis (1561–1626); English philosopher and scientist

Balogh, Thomas (1905–1985); former Minta pupil; British economist

Bard, Joseph; Miksa Teller's law clerk

Bardeen, John (1908–1991); American physicist; twice Nobel laureate (1956; 1972)

Barton, Derek H. R. (1918–1998); British chemist; Nobel laureate (1969)

Baruch, Bernard M. (1870–1965); American statesman

Basov, Nikolai G. (1922–2001); Soviet physicist; Nobel laureate (1964)

Bastiansen, Otto (1918–1995); Norwegian chemist

Bednorz, J. Georg (1950– ); Swiss physicist; Nobel laureate (1987)

Benford, Gregory (1941– ); American physicist; science fiction writer

Benzer, Seymour (1921–2007); American physicist and molecular biologist

Bergström, Ingmar (1921– ); Swedish physicist

Beria, Lavrentii P. (1899–1953); Soviet politician; chief of secret police; supervisor of the Soviet nuclear program

Bernal, J. Desmond (1901–1971); British crystallographer; professed communist; mentor to Nobel laureate scientists

Bernstein, Barton J.; American historian of science at Stanford University

Bersuker, Isaac B. (1928– ); Soviet–Moldavian–American chemist; scholar of the Jahn–Teller effect

Bethe,* Hans A. (1906–2005); German–American physicist, Nobel laureate (1967)

Beveridge, William Henry (1879–1963); Lord Beveridge; British economist

Biden, Joseph (1942– ); American politician; vice president (2009– )

Bischitz, György; see Hevesy, Georg(e) von (de)

Bloch, Felix (1905–1983); Swiss–American physicist; Nobel laureate (1952)

Bohr, Niels H. D. (1885–1962); Danish physicist; Nobel laureate (1922)

Borden, William L.; Former aide to Senator McMahon; author of letter to J. Edgar Hoover on J. Robert Oppenheimer in 1953

Borlaug, Norman E. (1914–2009); American agronomist; Nobel laureate (1970)

Born, Max (1882–1954); German physicist; Nobel laureate (1954)

Bradbury, Norris E. (1909–1997); American physicist; second director of the LANL (1945–1970)

Bradford, Peter A.; American scientist; former head of the Nuclear Regulatory Commission

Bradley, Omar N. (1893–1981); five-star general of the US Army; chairman of the Joint Chiefs of Staff at the time of the hydrogen bomb debate

Braun, Wernher von (1912–1977); German–American rocket scientist

Brecht, Bertolt (1898–1956); German playwright and theater director

Breit, Gregory (1899–1981); Russian–American physicist

Brenner, Sidney (1927– ); South African–British biologist; Nobel laureate (2002)

Brezhnev, Leonid I. (1906–1982); Soviet politician; leader of the Soviet Union (1964–1982)

Briggs, Lyman J. (1874–1963); American scholar with degrees in physics and agriculture; director of the National Bureau of Standards (1933–1945); director of the Uranium Committee (1939–1941)

Brown, Harold (1927– ); Livermore physicist; US secretary of defense (1977–1981)

Browne, John C. (1943– ); American physicist; former director of the LANL (1997–2003)

Brunauer, Esther; see Caukin, Esther

Brunauer,* Stephen (István) (1903–1986); Hungarian–American chemist; naval officer

Buchwald, Art (1925–2007); American journalist and author

Buckley, Oliver E. (1887–1959); American electrical engineer

Buckley, William F., Jr. (1925–2008); American political writer and journalist

Bukharin, Nikolai I. (1888–1938); Soviet–Russian politician and ideologue; victim of Stalin's purges

Bulganin, Nikolai (1895–1975); Soviet politician; prime minister (1955–1958)

Bunzel, John H.; American political scientist at the Hoover Institution, Stanford University

Bush, George H. W. (1924– ); American vice president (1981–1989); American president (1989–1993)

Bush, George W. (1946– ); American president (2001–2009)

Bush, Vannevar (1890–1974); American engineer, science administrator

Caldwell, Miriam; computing assistant at Los Alamos during the development of the hydrogen bomb

Calvin, Melvin (1911–1997); American chemist; Nobel laureate (1961)

Camon, Ferdinando (1935– ); Italian author

Canavan, Gregory H. (1943– ); American physicist

Carr, Alan B. (1978– ); American historian at the Los Alamos National Laboratory

Carter, Jimmy (1924– ); American president (1977–1981)

Casey, William J. (1913–1987); director of the CIA (1981–1987)

Casimir, Hendrik B. G. (1909–2000); Dutch physicist

Caukin, Esther (1901–1959); former US State Department official; Stephen Brunauer's wife

Chadwick, James (1891–1974); British physicist; Nobel laureate (1935)

Chamberlain, Owen (1920–2006); American physicist; Nobel laureate (1959)

Chambers, Whittaker (1901–1961); American writer; Soviet spy turned American informer

Chandrasekhar, S. (1910–1995); Indian–American astrophysicist; Nobel laureate (1983)

Chargaff, Erwin (1905–2002); Austrian–American biochemist

Charpak, Georges (1924– ); French physicist; Nobel laureate (1992)

Cherwell, Lord; see Lindemann, Frederick A.

Chevalier, Barbara (1907–2003), and Haakon (1901–1985); J. Robert Oppenheimer's friends

Chew, Jeffrey (1925– ); American physicist

Christy, Robert F. (1916– ); American physicist

Churchill, Winston L. S. (1874–1965); British politician; prime minister; Nobel laureate author (1953)

Clinton, Bill (William) J. (1946– ); American president (1993–2001)

Cockcroft, John D. (1897–1967); British physicist; Nobel laureate (1951)

Compton, Arthur H. (1892–1962); American physicist; Nobel laureate (1927)

Conant, James B. (1893–1978); American chemist and science administrator

Corbino, Orso Mario (1876–1937); Italian physicist and politician

Cranston, Alan (1914–2000); US senator from California (1969–1993)

Crick, Francis H. C. (1916–2004); British physicist turned biologist; Nobel laureate (1962)

Critchfield, Charles (1910–1994); American mathematical physicist

Czeizel, Endre (1935– ); Hungarian geneticist and author

Dayan, Moshe (1915–1981); Israeli politician and military leader

Dean, Gordon (1905–1958); American lawyer; chairman of the AEC (1950–1953)

Debye, Peter (1884–1966); Dutch–German–American chemist; Nobel laureate (1936)

de Gaulle, Charles (1890–1970); French president (1959–1969)

Delbrück, Max (1906–1981); German–American physicist and biologist; Nobel laureate (1969)

Deutsch, Ilona (1891–1985); Edward Teller's mother

Dirac, Paul A. M. (1902–1984); British physicist; Nobel laureate (1933)

Diven, Benjamin (Ben) (1919– ); American physicist

Donnan, Frederick G. (1870–1956); British physical chemist; head of the Chemistry Department of University College London (1928–1937)

Doolittle, James (Jimmy) H. (1896–1993); US Air Force general

Draper, Charles S. (1901–1987); American engineer and physicist

Drell, Sidney D. (1926– ); American physicist

Dreyfus, Alfred (c.1859–1935); French officer, unjustly accused of espionage

Droste, Gottfried von (1908–1992); German physical chemist

DuBridge, Lee A. (1901–1994); American physicist

Dulles, Allen W. (1893–1969); director of the CIA (1953–1961)

Dyson, Freeman J. (1923– ); British–American physicist and author

Ecker, Allan B.; American lawyer

Ehrenfest, Paul (1880–1933); Dutch physicist

Ehrenhaft, Felix (1879–1952); Austrian physicist

Einstein, Albert (1879–1955); German–American physicist; Nobel laureate (1922, for 1921)

Eisenhower, Dwight D. (1890–1969); American military leader; American president (1953–1961)

Eliot, Charles W. (1834–1926); American academic

Eltenton, George; British chemical engineer and member of Soviet atomic spy ring

Emelyanov, Vasily S. (1901–1988); Soviet engineer and nuclear administrator; director of the Board for the Peaceful Uses of Atomic Energy in Moscow

Emmett, Paul H. (1900–1985); American chemist

Eötvös, Loránd (1848–1919); Hungarian physicist

Ernst, Richard (1933– ); Swiss chemist; Nobel laureate (1991)

Ertl, Gerhard (1936– ); German physicist; Nobel laureate (2007)

Eucken, Arnold (1884–1950); German physical chemist

Evans, Ward V.; American chemist; member of the Personal Security Board of the AEC for the Oppenheimer hearing

Everett, Cornelius J. (1914–1988); American mathematician

Ewald, Paul P. (1888–1985); German–American physicist

Ewald, Rose; P. P. Ewald's daughter, Hans Bethe's wife

Farkas, Vladimir (1925–2002); officer of the Hungarian secret police in the early 1950s

Fermi,* Enrico (1901–1954); Italian–American physicist; Nobel laureate (1938)

Fermi, Laura, née Capon (1907–1977); Italian–American author; Enrico Fermi's wife

Feynman, Richard (1918–1988); American physicist, Nobel laureate (1965)

Finletter, Thomas K. (1893–1980); secretary of the US Air Force (1950–1953)

Firor, John (1927–2007); American physicist and environmental scholar; former director of the National Center for Atmospheric Research in Boulder, Colorado

Fitch, Val L. (1923– ); American physicist; Nobel laureate (1980)

Ford, Gerald R. (1913–2006); American president (1974–1976)

Ford, Kenneth W. (1926– ); American physicist and author

Foster, John S. (1922– ); American physicist

Fowler, William A. (1911–1995); American physicist; Nobel laureate (1983)

Franck, James (1882–1964); German–American physicist; Nobel laureate (1926)

Frank, Hans (1900–1946); German Nazi politician, the "Butcher of Poland"

Franz Joseph I (1830–1916); emperor of Austria (1848–1916) and king of Hungary (1867–1916)

Frayn, Michael (1933– ); English playwright

Friedell, Hymer L. (1911–2002); American radiologist; medical officer at the Metallurgical Laboratory in Chicago

Friedman, Jerome I. (1930– ); American physicist; Nobel laureate (1990)

Friedmann, Alexander A. (1888–1925); Russian–Soviet mathematician and cosmologist

Frisch, Otto Robert (1904–1979); German–British physicist

Fuchs, Klaus (1912–1988); German–British physicist and Soviet spy

Fuchs, Rudolf; co-winner of high school competitions in Hungary in 1925; murdered in the Hungarian Holocaust in 1944

Fujioka, Yushio; Japanese physicist

Furth, Harold P. (1930–2002); Vienna-born American plasma and fusion physicist at Berkeley, Livermore, and Princeton, author of the Teller–antimatter poem "Perils of Modern Living"

Gabor, Dennis (1900–1979); Hungarian–British physicist; Nobel laureate (1971)

Gagarin, Yurii A. (1934–1968); Soviet cosmonaut (1961); first man to travel in space

Galileo Galilei (1564–1642); Italian scientist

Gamow,* George (Georgii A.) (1904–1968); Russian–American physicist and author of popular science books

Garrison, Lloyd K. (1897–1991); J. R. Oppenheimer's leading counsel at his security hearing

Garwin, Richard L. (1928– ); American physicist

Gates, Robert M. (1943– ); CIA operative; secretary of defense under Presidents George W. Bush and Barack Obama

Gell–Mann, Murray (1929– ); American physicist; Nobel laureate (1969)

Gingrich, Newt (1943– ); American politician

Ginzburg, Vitaly L. (1916– ); Soviet–Russian physicist; Nobel laureate (2003)

Glaser, Donald (1926– ); American physicist; Nobel laureate (1960)

Goeppert (Göppert) Mayer,* Maria (1906–1972); German–American physicist; Nobel laureate (1963)

Goethe, Johann Wolfgang von (1749–1832); German writer

Goldberger, Marvin L. (1922– ); American physicist

Goldberger, Mildred (1923–2006); American mathematician and economist; Marvin Goldberger's wife

Goldhaber, Maurice (1911– ); American physicist

Goldman, Eric F. (1915–1989); American historian

Goldwater, Barry M. (1909–1998); American politician

Gömbös, Gyula (1886–1936); racist Hungarian politician; prime minister (1932–1936)

Goodchild, Peter; British journalist, author

Gorbachev, Mikhail S. (1931– ); the last supreme leader of the Soviet Union (1985–1991); Nobel laureate (1990)

Gorbacheva, Raisa M. (1932–1999); Mikhail Gorbachev's wife

Gore, Albert (Al) A. (1948– ); American politician; vice president (1993–2001); presidential candidate (2000); Nobel laureate (2007)

Göring, Hermann (1893–1946); German Nazi leader

Gorobets, Boris S. (1942– ); Russian scientist and author

Goudsmit, Samuel A. (1902–1978); Dutch–American physicist; scientific head of the Alsos Mission

Graham, Daniel O. (1925–1995); retired US Air Force general; leading architect of Ronald Reagan's SDI

Gray, Gordon (1909–1982); American lawyer and politician; chaired the "Gray Committee" (Personal Security Board) in the Oppenheimer hearing

Griggs, David T. (1911–1974); American geophysicist, chief scientist of the US Air Force

Grigory, Margit; Edward Teller's bilingual (English and Hungarian) assistant at the end of his life

Gromyko, Andrei A. (1909–1989); Soviet politician

Groves, Leslie R. (1896–1970); US Army general

Haber, Fritz (1868–1934); German chemist; Nobel laureate (1919, for 1918)

Hagelstein, Peter L. (1955– ); American engineer and educator

Hahn, Otto (1879–1968); German chemist; Nobel laureate (1945, for 1944)

Hall, Theodore Alvin (1925–1999); American physicist

Hansen, Charles R. (Chuck) (1947–2003); American historian of nuclear weapons and author

Harkányi, Ede; see Schütz-Harkányi, Ede

Hassel, Odd (1897–1981); Norwegian chemist; Nobel laureate (1969)

Hauptman, Herbert A. (1917– ); American mathematician; Nobel laureate (1985)

Hecker, Siegfried S. (1943– ); American metallurgist; former director of the Los Alamos National Laboratory

Heisenberg,* Werner (1901–1976); German physicist; Nobel laureate (1933, for 1932)

Heitler, Walter (1904–1981); German physicist

Henry, Patrick (1736–1799); American statesman

Herring, Conyers (1914–2009); American physicist

Herrington, John S. (1939– ); American politician

Hertz, John D. (1879–1961); American businessman and philanthropist; he was born as Sándor Hertz in the Kingdom of Hungary of Austria–Hungary

Herzberg, Gerhard (1904–1999); German–Canadian physicist; Nobel laureate (1971)

Heslep, Charter (1904–1963); American broadcaster and AEC staff member

Hesz, Magda; later Mrs. Jacob Schutz; governess for the Teller family in Budapest; later, she lived in Chicago and transmitted letters between the Budapest and American wings of the Teller family

Hevesy, Georg(e) von (de) (1885–1966); earlier, Bischitz, György; Hungarian–Swedish chemist, Nobel laureate (1944, for 1943)

Himmler, Heinrich (1900–1945); German Nazi leader

Hirohito (1901–1989); emperor of Japan

Hoffmann, Frederick de (1924–1989); Austrian–American physicist

Hoffmann, Roald (1937– ); American chemist; Nobel laureate (1981)

Holloway, Marshall G. (1912–1991); American physicist at Los Alamos

Hoover, Gilbert C.; navy representative in the Uranium Committee

Hoover, Herbert C. (1874–1964); American president (1929–1933)

Hoover, J. Edgar (1895–1972); director of the Federal Bureau of Investigation (1924–1972)

Horthy, Miklós (Nicholas) (1868–1957); authoritarian head of state of Hungary (1919–1945)

Hückel, Erich (1896–1980); German physicist

Hund, Friedrich (1896–1997); German physicist

Hunyadi, Dalma, also Dalma Hunyadi Brunauer (1924– ); Stephen Brunauer's wife from 1961

Hutchins, Robert M. (1899–1977); president of the University of Chicago

Illés, Zoltán (1961– ); Hungarian politician

Ioffe, Abram (1880–1960); Russian–Soviet physicist; mentor of many renowned Soviet physicists

Ivanenko, Dmitrii D. (1904–1994); Soviet physicist

Jackson, Henry M. (1912–1983); American politician

Jahn,* Hermann A. (1907–1979); British physicist

Jánossy, Lajos (1912–1978); Hungarian physicist

Javan, Ali (1926– ); Iranian–American physicist

Jensen, J. Hans D. (1907–1973); German physicist; Nobel laureate (1963)

Johnson, Edwin C. (1884–1970); US senator (1937–1955)

Johnson, Gerald W. (1919–2002); US Air Force general

Johnson, Louis (1891–1966); US secretary of defense under President Truman

Johnson, Lyndon B. (1908–1973); American president (1963–1969)

Joliot (Joliot-Curie), Frédéric (1900–1958); French chemist; Nobel laureate (1935)

Jordan, Pascual (1902–1980); German physicist

Joseph (József), archduke of the Hapsburgs (1776–1847); palatine of Hungary

Kádár, János (1912–1989); Communist leader of Hungary (1956–1988)

Kalckar, Fritz (1910–1938); Danish physicist

Kaldor, Nicholas (1908–1986); former Minta pupil; British economist

Kapitza, Peter L. (1894–1984), Soviet–Russian physicist; Nobel laureate (1978)

Karle, Jerome (1918– ); American scientist; Nobel laureate (1985)

Kármán [Kleinmann], Mór or Maurice von (1843–1915); Theodore von Kármán's father

Kármán,* Theodore (Tódor) von (1881–1963); one of the five "Martians"; aerodynamicist; principal scientist for the American air forces

Károlyi, Mihály (1875–1955); president of Hungary (January–March 1919)

Katona, Peter (1937– ); Janos Kirz's classmate and friend

Kekulé, Friedrich August (1829–1896); German chemist

Kemeny, John G. (1926–1992); Hungarian–American computer scientist

Kendall, Henry W. (1926–1999); American physicist; Nobel laureate (1990)

Kennan, George F. (1904–2005); American diplomat and historian

Kennedy, John F. (1917–1963); American president (1961–1963)

Kennedy, Joseph (1916–1957); American chemist

Kerry, John F. (1943– ); American politician; presidential candidate of the Democratic Party in 2004

Keyworth, George A. (Jay) (1939– ); American physicist; President Reagan's science adviser (1981–1985)

Khariton,* Yulii B. (1904–1996); Soviet physicist; head of the first Soviet nuclear weapons laboratory, Arzamas-16

Khrushchev, Nikita S. (1894–1971); supreme leader of the Soviet Union after Stalin until 1964

Killian, James R. (1904–1988); President Eisenhower's science adviser (1957–1959)

Kipling, Rudyard (1865–1936); British author

Kirz, András (1900–1945); Teller's brother-in-law; held a law degree; was murdered in the Hungarian Holocaust

Kirz, Emma; see Teller, Emma

Kirz, Janos (János) (1937– ); Edward Teller's nephew; American physicist

Kissinger, Henry A. (1923– ); President Nixon's secretary of state; Nobel laureate (1973)

Kistiakowsky, George B. (1900–1982); Ukrainian–American chemist; President Eisenhower's science adviser

Klein, Georg (1925– ); Hungarian–Swedish tumor biologist

Klug, Leopold (1854–1944); professor of geometry in Budapest

Koestler,* Arthur (1905–1983); Hungarian–British reporter, author

Konopinski, Emil (1912–1990); American physicist

Kornberg, Arthur (1918–2007); American biomedical scientist; Nobel laureate (1959)

Kőrösy, Ferenc (1906–1997); Hungarian–Israeli scientist

Kosygin, Aleksei N. (1904–1980); Soviet politician; prime minister (1964–1980)

Kotz, Sándor (1974– ); winner of Eötvös competitions (1993)

Kovács, László (1942– ); Hungarian physicist–historian

Kramish, Arnold (1923–2010); American physicist and author

Kun, Béla (1886–1938); leader of the Hungarian "Soviet" Republic (1919)

Kurchatov, Igor V. (1903–1960); Soviet physicist; head of the Soviet nuclear programs

Kurti, Nicholas (Miklós Kürti) (1908–1998); former Minta pupil; Hungarian–British physicist

Lamb, Willis (1913–2008); American physicist; Nobel laureate (1955)

Landau,* Lev D. (1908–1968), Soviet physicist; Nobel laureate (1962)

Landsberg, Peter T.; British scientist

Landshoff, Rolf (1912–1999); German–American physicist

Langmuir, Irving (1881–1957); American chemist; Nobel laureate (1932)

Lansdale, John, Jr. (1912–2003); American intelligence officer

Lantos, Tom P. (1928–2009); Hungarian–American politician

Larson, Clarence E. (1909–1999); American chemist; commissioner of AEC (1969–1974); initiated large interviewing program of scientists and technologists

Larson, Jane (1922–2008); American artist; co-producer and wife of Clarence Larson

Latimer, Wendell M. (1893–1955); American chemist

Latter, Albert L. (1921–1997); American physicist

Laue, Max von (1879–1960); German physicist; Nobel laureate (1914)

Laughlin, Robert B. (1950– ); American physicist; Nobel laureate (1998)

Lauritsen, Charles C. (1892–1968); American physicist

Lawrence,* Ernest O. (1901–1958); American physicist; Nobel laureate (1939)

Lax, Peter (Péter) (1926– ); Hungarian–American mathematician; Abel laureate (2005)

LeBaron, Robert; chaired the military liaison committee of the AEC at the time of the hydrogen bomb debate

Lee, Tsung-Dao (1926– ); Chinese–American physicist, Nobel laureate (1957)

Lee, Yuan T. (1936– ); Chinese–American chemist; Nobel laureate (1986)

LeMay, Curtis E. (1906–1990); US Air Force general

Lengyel, Béla (1844–1913); Hungarian chemist

Lenin, Vladimir I. (1870–1924); Russian revolutionary; first head of the Soviet Union

Levi, Primo (1919–1987); Italian author
Levin, Herschel; American rabbi
Lewis, Edward B. (1918–2004); American biologist; Nobel laureate (1995)
Lewis, Flora (1923–2002); American journalist
Lewis, Gilbert N. (1875–1946); American physical chemist
Libby, Leona Marshall (Leona Woods) (1919–1986); American physicist
Libby, Stephen B. (1951– ); American physicist
Libby, Willard F. (1908–1980); American chemist, Nobel laureate (1960)
Lifshits, Evgenii M. (1915–1985); Soviet–Russian physicist
Lilienthal, David E. (1899–1981); first chairman of the AEC
Lincoln, Abraham (1809–1865); American president (1861–1865)
Lindemann, Frederick A. (later Lord Cherwell) (1886–1957); Winston Churchill's science adviser
Lipmann, Fritz A. (1899–1986); American biomedical scientist; Nobel laureate (1953)
Lippmann, Walter (1889–1974); American journalist
London, Fritz W. (1900–1954); German–American physicist
Longmire, Conrad L.; American physicist at Los Alamos in the hydrogen bomb project
Lossow, Otto von (1868–1938); German general
Lossow, von (first name unknown); German surgeon who operated on Edward Teller after his trolley accident in 1928; Otto von Lossow's brother
Lovett, Robert A. (1895–1986); US secretary of defense (1951–1953)
Luce, Henry R. (1898–1967); American publisher
Lukacs, John (1924– ); Hungarian–American historian and author
Lwoff, André (1902–1994); French biomedical scientist; Nobel laureate (1965)
Lysenko, Trofim D. (1898–1976); Soviet agronomist
Macmillan, Harold (1894–1986); British politician; prime minister (1957–1963)
Magos, Gábor (previously Gábor Schütz); Mici Teller's half-brother
Maiman, Theodore H. (1927–2007); American physicist
Malenkov, Georgii M. (1901–1979); Soviet politician; prime minister (1953–1955)
Manley, John (1907–1990); American physicist
Mao, Zedong (Mao Tse-tung) (1893–1976); Chinese communist leader
Mark, Hans (1929– ); American physicist and science administrator
Mark, Herman F. (1895–1992); Austrian–American chemist
Mark, J. Carson (1913–1997); Canadian–American physicist; director of

the theoretical division of the Los Alamos Laboratory at the time of the development of the hydrogen bomb

Marks, Herbert S.; American lawyer; member of Oppenheimer's defense team

Marshak, Robert (1916–1992); American physicist

Marshall, John (1917–1997); American physicist

Marshall Libby, Leona; see Libby, Leona Marshall

Marvin, Cloyd H. (1889–1969); president of George Washington University (1929–1959)

Marx, George (György) (1927–2002); Hungarian physicist

Mather, John C. (1946– ); American physicist; Nobel laureate (2006)

May, Andrew J. (1875–1959); American politician

May, Michael (1925– ); French–American physicist

Mayer, Joseph E. (1904–1983); American chemist; Maria Goeppert Mayer's husband

McCarthy, Joseph R. (1909–1957); US senator (1947–1957)

McCloy, John J. (1895–1989); American lawyer and presidential adviser

McFarlane, Robert (1937– ); National Security Adviser to President Reagan (1983–1985)

McMahon, Brien (1903–1952); US senator (1945–1952)

McMillan, Edwin M. (1907–1911); American chemist; Nobel laureate (1951)

McMillan, W. G.; American physical chemist

McNamara, Robert S. (1916–2009); US secretary of defense (1961–1968)

Medvedev, Zhores A. (1925– ); Russian biologist and historian

Meir, Golda (1898–1978); Israeli politician; prime minister (1969–1974)

Meitner, Lise (1878–1968); Austrian–German physicist

Mendeleev, Dmitrii I. (1834–1907); Russian chemist

Merrifield, Bruce (1921–2006); American biochemist; Nobel laureate

Meselson, Matthew (1930– ); American biochemist

Metropolis, Nicholas C. (1915–1999); American atomic scientist and computer pioneer

Mills, Mark M. (1917–1958); American physicist

Moir, Ralph W. (1940– ); American physicist

More, Richard (1942– ); American physicist

Morrison, David (1940– ); American astronomer

Morrison, Philip (1915–2005); American physicist

Muller, Hermann J. (1890–1967); American biologist; Nobel laureate (1946)

Müller, K. Alex (1927– ); Swiss physicist; Nobel laureate (1987)

Mulliken, Robert S. (1896–1986); American chemist; Nobel laureate (1966)

Murphy, Robert D. (1894–1978); American diplomat

Murray, Thomas E.; American inventor; commissioner of AEC

Murrow, Edward R. (1908–1965); American journalist

Mussolini, Benito (1883–1945); Italian fascist dictator

Nader, Ralph (1934– ); American consumer advocate

Nagy, Imre (1896–1958); Hungarian politician; prime minister during the Hungarian Revolution in 1956; tried and executed

Nambu, Yoichiro (1921– ); Japanese–American physicist; Nobel laureate (2008)

Neddermeyer, Seth (1907–1988); American physicist

Ne'eman, Yuval (1925–2006); Israeli military leader, physicist, and politician

Neumann,* John (János) von (1903–1957); one of the five "Martians"; father of the modern computer

Neumann Whitman, Marina von (1935– ); economist; John von Neumann's daughter

Nielsen, J. Rud (1894–1979); Danish–American physicist

Nirenberg, Marshall W. (1927– ); American biomedical scientist; Nobel laureate (1968)

Nisbet, Lois; Edward Teller's one-time assistant

Nishina, Yoshio (1890–1951); Japanese physicist

Nixon, Richard M. (1913–1994); American president (1968–1974)

Nobel, Alfred (1833–1896); Swedish chemist, engineer, and inventor

Noddack, Ida, née Tacke (1896–1978); German chemist

Nordheim, Lothar (1899–1985); German–American physicist

Nuckolls, John (1930– ); American physicist; former director of LLNL

Obama, Barack (1961– ); American president (2009– )

Olah, George A. (1927– ); Hungarian–American chemist; Nobel laureate (1994)

O'Leary, Hazel (1937– ); US secretary of energy (1993–1997)

Oliphant, Mark (1901–2000); Australian–British physicist

Oppenheimer, Frank (1912–1985); American physicist, Robert Oppenheimer's brother

Oppenheimer, J. Robert (1904–1967); American physicist; first director of the Los Alamos Laboratory

Ortvay, Rudolf (1885–1945); Hungarian physicist

Ossietzky, Carl von (1889–1938); German journalist; Nobel laureate; perished in German Nazi concentration camp

Ostwald, Wilhelm (1853–1932); Latvian–German chemist; Nobel laureate (1909)

Packard, David (1912–1996); American engineer and politician

Pais, Abraham (1918–2000); Dutch–American physicist turned science historian

Panofsky, Erwin (1892–1968); German–American art historian

Panofsky, Wofgang K. H. (1919–2007); American physicist

Pauli, Wolfgang (1900–1958); Swiss physicist; Nobel laureate (1945)

Pauling,* Linus (1901–1994); American chemist; twice Nobel laureate (1954; 1963, for 1962)

Pegram, George B. (1876–1958); American physicist

Peierls, Rudolf (1907–1995); German–British physicist

Penzias, Arno (1933– ); American astrophysicist; Nobel laureate (1978)

Pétain, Henri-Philippe (1856–1951); French military leader; head of state of Vichy France (1940–1944)

Peters, Bernard (1910–1993); German–American physicist

Petrovich, Elek; Hungarian museum director

Pike, Sumner T. (1891–1976); commissioner (1946–1951) and chair (1950) of the AEC

Pitzer, Kenneth S. (1914–1997); American chemist

Placzek, George (1905–1955); Czech–American physicist

Planck, Max (1858–1947); German physicist; Nobel laureate (1919, for 1918)

Plato (c.429–c.347 BCE); Greek philosopher

Pogány, Béla (1887–1943); Hungarian physicist

Polanyi, Michael (1891–1976); Hungarian–British physical chemist turned philosopher

Power, Thomas S. (1905–1970); US Air Force general

Powers, Francis Gary (1929–1977); American pilot of the U-2 spy plane

Prokhorov, Aleksandr M. (1916–2002); Soviet physicist; Nobel laureate (1964)

Rabi, Isidor I. (1898–1988); American physicist; Nobel laureate (1945, for 1944)

Ramsey, Norman F. (1915– ); American physicist; Nobel laureate (1989)

Ray, Dixy Lee (1914–1994); American politician

Reagan, Ronald W. (1911–2004); American president (1981–1989)

Reines, Frederick J. (1918–1998); American physicist; Nobel laureate (1995)

Renner, Rudolf (1909–1991); German physicist

Reston, James (1909–1995); American journalist

Rhodes, Richard (1937– ); American author

Richtmyer, R. D. (1910–2003); American physicist

Rickover, Hyman G. (1900–1986); American engineer; admiral of the US Navy; "father of nuclear submarines"

Robb, Roger (1907–1985); government representative ("prosecutor") in the Oppenheimer hearing; later, he was a US federal judge

Rockefeller, Nelson A. (1908–1979); American politician

Rolander, Carl A. (1920– ); AEC official at the time of the Oppenheimer hearing

Röntgen, Wilhelm Conrad (1845–1923); German physicist; Nobel laureate (1901)

Roosevelt, Eleanor (1884–1962); President F. D. Roosevelt's wife and advocate for civil rights

Roosevelt, Franklin D. (1882–1945); American president (1933–1945)

Rosenbluth, Arianna W.; Marshall N. Rosenbluth's wife and co-worker

Rosenbluth, Marshall N. (1927–2003); American physicist

Rosenfeld, Leon (1904–1974); Belgian physicist

Rotblat, Joseph (1908–2005); Polish–British physicist; left Los Alamos when it became certain that Germany would not have an atomic bomb; Nobel laureate (1995)

Rowe, Hartley; member of the GAC at the time of the hydrogen bomb debate

Rozsnyai, Balazs; Hungarian–American physicist

Rumsfeld, Donald (1932– ); American politician

Rushdie, A. Salman (1947– ); British–Indian author

Rusk, Dean (1909–1994); US secretary of state (1961–1969)

Russell, Bertrand (1872–1970); British philosopher

Rutherford, Ernest (1871–1937); British physicist; Nobel laureate (1908)

Sachs, Alexander (1893–1973); American economist; delivered Albert Einstein's letter to President Franklin D. Roosevelt

Sagan, Carl (1934–1996); American astronomer and author; popularized science

Sahlin, Harry L.; American physicist

Sakharov,* Andrei D. (1921–1989); Soviet physicist; Nobel laureate (1975)

Schlesinger, Arthur M., Jr. (1917–2007); American historian

Schlesinger, James R. (1929– ); US secretary of defense under Presidents Nixon and Ford; secretary of energy under President Carter

Schmidt, Helmut (1918– ); German politician; West German chancellor (1974–1982)

Schriever, Bernard A. (1910–2005); US Air Force general

Schrödinger, Erwin (1887–1961); Austrian physicist; Nobel laureate (1933)

Schutz, Jacob, Mrs.; see Hesz, Magda

Schütz, Aladár; Mici Teller's stepfather

Schütz, Gábor; see Magos, Gábor

Schütz, István; Mici Teller's half-brother

Schütz-Harkányi, Auguszta (1909–2000); Edward Teller's wife (Mici)

Schütz-Harkányi, Ede (1908–1944); Edward Teller's brother-in-law; murdered in the Hungarian Holocaust

Schwartz, Anthony (Tony) (1923–2008); American advertising expert

Schwartz, Charles (1931– ); American physicist

Schwinger, Julian (1918–1994); American physicist; Nobel laureate (1965)

Seaborg, Glenn T. (1912–1999); American chemist; Nobel laureate (1951)

Segrè, Emilio G. (1905–1989), Italian–American physicist; Nobel laureate (1959)

Seitz, Frederick (1911–2008); American physicist and science administrator

Seitz, Russell (1947– ); American physicist

Semenov, Nikolai N. (1896–1986); Soviet chemical physicist; Nobel laureate (1956)

Seneca, Lucius Annaeus (c.4 BCE–c.65 CE); Roman philosopher

Serber, Robert (1909–1997); American physicist

Sewell, Duane C. (1918– ); American physicist

Shoolery, Judith (1935– ); American science teacher and editor; Edward Teller's longtime co-worker and the co-author of his *Memoirs*

Shubnikov, Lev V. (1901–1937); Soviet–Russian physicist; victim of Stalin's purges

Shultz, George P. (1920– ); US secretary of state (1982–1989)

Sidey, Hugh (1927–2005); American journalist

Silverman, Samuel J. (1908–2001); member of Oppenheimer's defense team

Simon, Francis (1893–1956); German–British physicist

Smith, Cyril S. (1904–1992); American metallurgist

Smith, Howard K. (1914–2002); American journalist

Smyth, Henry DeWolf (1898–1986); American physicist and administrator; commissioner of the AEC (1949–1954)

Snyderman, Neal (1949– ); American physicist

Socrates (469–399 BCE); Greek philosopher

Solzhenitsyn, Aleksandr I. (1918–2008); Russian writer; Nobel laureate (1970)

Sommerfeld, Arnold (1868–1951); German physicist

Soros, George (1930– ); Hungarian–American businessman

Speer, Albert (1905–1981); German architect; Nazi official

Sponer, Hertha (1895–1968); German–American physicist; James Franck's second wife

Sproul, Robert G. (1891–1975); American educator

Staudinger, Hermann (1881–1965); German chemist; Nobel laureate (1953)

Steinberger, Jack (1921– ); German–American–Swiss physicist; Nobel laureate (1988)

Stern, Philip M. (1926– ); American author

Stevenson, Adlai E. (1900–1965); American politician

Stimson, Henry L. (1867–1960); American statesman; secretary of war (1940–1945)

Strassmann, Fritz (1902–1980); German chemist; co-discoverer of nuclear fission

Strauss,* Lewis L. (1896–1974); American financier; admiral; chairman of the AEC (1953–1958)

Stur, Judit; Hungarian PhD, MD; Mici Teller's temporary helper after her brain surgery

Symington, W. Stuart (1901–1988); first secretary of the US Air Force (1947–1950); US senator from Missouri (1953–1976)

Szent-Györgyi, Albert (1893–1986); Hungarian–American biochemist; Nobel laureate (1937)

Szilard,* Leo (1898–1964); one of the five "Martians"; Hungarian–American scientist

Szily, Kálmán (1875–1958); Hungarian politician of education between the two world wars

Tamm, Igor Y. (1885–1971); Soviet physicist; Nobel laureate (1958)

Taylor, Theodore (Ted) B. (1925–2004); American physicist

Telegdi, Valentine (1922–2006); Hungarian–American physicist

Teleki, Pál (1879–1941); Hungarian politician; scholar; prime minister (1920–1921; 1939–1941)

Teller, Edward (1908–2003); one of the five "Martians"; Hungarian–American physicist

Teller, Emma (1905–2001); Edward Teller's sister

Teller, Ludwig (1911–1965); member of US Congress (1957–1961)

Teller, Mici (1909–2000); Edward Teller's wife

Teller, Miksa (Max) (1871–1950); Edward Teller's father

Teller, Paul (1943– ); Edward and Mici Teller's son

Teller, Susan Wendy or Wendy (1946– ); Edward and Mici Teller's daughter

Thirring, Hans (1888–1976); Austrian physicist

Tisza, Laszlo (1907–2009); Hungarian–American physicist

Tolman, Richard C. (1881–1948); American physicist

Tolstoy, Leo (Lev N. Tolstoy) (1828–1910); Russian author

Tomonaga, Sinitiro (1906–1979); Japanese physicist; Nobel laureate (1965)

Topley, Bryan; Edward Teller's associate at University College London; reader in the Chemistry Department (1933–1936)

Townes, Charles H. (1915– ); American physicist; Nobel laureate (1964)

Trefort, Ágoston (1817–1888); Hungarian politician

Trotsky, Leon (1879–1940); Russian communist revolutionary; executed by Stalin's agents while in exile

Truman, Harry S. (1884–1972); American president (1944–1952)

Turner, Louis A.; American physicist

Tuve, Merle (1901–1982); American scientist

Tydings, Millard E. (1890–1961); US senator (1927–1951); member of the Armed Services Committee (1949–1951)

Ulam, Françoise; Stanislaw Ulam's wife

Ulam,* Stanislaw M. (1909–1984); Polish–American mathematician; developed the Monte Carlo method

Urey, Harold C. (1893–1981); American chemist; Nobel laureate (1934)

Van Allen, James A. (1914–2006); American space scientist

Van der Waerden, Bartel L. (1902–1988); Dutch–German–Swiss mathematician

Van 't Hoff, Jacobus (1852–1911); Dutch chemist; Nobel laureate (1901)

Van Vleck, John H. (1899–1980); American physicist, Nobel laureate (1977)

Vavilov, Nikolai I. (1883–1943); leading Soviet geneticist; victim of repression

Vernadskii, Vladimir I. (1863–1945); Soviet–Russian geologist; initiator of the Soviet Uranium Commission

Virágh, Pál; Teller's friend; pupil of the Gimnázium of the Piarist Fathers

Voltaire (François-Marie Arouet) (1694–1778); French philosopher

Wallace, Mike (1918– ); American journalist

Wallenberg, Raoul (1912–?); Swedish diplomat; rescued Jews in Budapest in 1944; perished in Soviet captivity

Walton, Ernest T. S. (1903–1995); Irish–British physicist; Nobel laureate (1951)

Watkins, James D. (1927– ); American admiral at the time of the SDI; later secretary of energy

Watson, James D. (1928– ); American biologist; Nobel laureate (1962)

Watson, Thomas J., Jr. (1914–1993); American businessman

Weaver, Warren (1894–1978); American science administrator

Weinberg, Alvin M. (1915–2006); American physicist

Weinberg, Steven (1933– ); American physicist; Nobel laureate (1979)

Weinberger, Caspar (1917–2006); American politician

Weiser, Ella; Mici Teller's mother

Weisskopf, Victor F. (1908–2002); Austrian–American physicist

Weizsäcker, Carl Friedrich von (1912–2007); German physicist and philosopher

Weizsäcker, Ernst von (1882–1951); Nazi diplomat and convicted war criminal; Carl Friedrich's and Richard's father

Weizsäcker, Richard von (1920– ); Carl Friedrich's brother, president of the Federal Republic of Germany (1984–1994)

Weyand, Richard F. (1953– ); Wendy Teller's second husband

Wheeler, John A. (1911–2008); American physicist

Whitman, Marina, née von Neumann (1935– ); American economist; John von Neumann's daughter

Wieman, Carl E. (1951– ); American physicist; Nobel laureate (2001)

Wierl, Raimond; German physicist, Herman F. Mark's co-worker

Wiesner, Jerome B. (1915–1994); American engineer and scientist

Wigner,* Eugene P. (Jenő Pál) (1902–1995); one of the five "Martians"; Hungarian–American physicist; Nobel laureate (1963)

Wilde, Oscar (1854–1900); Irish playwright

Willstätter, Richard (1872–1942); German chemist; Nobel laureate (1915)

Wilson, Robert R. (1914–2000); American physicist

Wirth, Timothy E. (1939– ); US senator (1987–1993)

Wood, Lowell (1941– ); American physicist

Woodruff, Roy; American physicist

Woods, Leona; see Libby, Leona Marshall

Yang, Chen Ning (Frank) (1922– ); Chinese–American physicist; Nobel laureate (1957)

York, Herbert (1921–2009); American physicist; first director of LLNL

Yukawa, Hideki (1907–1981); Japanese physicist; Nobel laureate (1949)

Zacharias, Jerrold R. (1905–1986); American physicist

Zel'dovich, Yakov B. (1914–1987); Soviet physicist

Zuckert, Eugene (1911–2000); US Air Force official

# NOTES

## PREFACE

1. Charles W. Eliot, ed., "Introductory Note," in *Prefaces and Prologues to Famous Books* (New York: P. F. Collier & Co., 1938).

2. For historical accuracy: The Tellers' official address was 3 Szalay Street and their apartment was on the third floor of the building. The building extends between Honvéd Street and Vajkay Street through Szalay Street. The memorial plaque is on the wall of the building on Honvéd Street because the secondhand bookstore happened to be there. The Tellers' home was on the other side of the building.

3. István Hargittai, *The Martians of Science: Five Physicists Who Changed the Twentieth Century* (New York: Oxford University Press, 2006; 2008).

4. Francis Crick, *Life Itself: Its Origin and Nature* (New York: Simon & Schuster, 1981), pp. 13–14. Some authors eventually broadened the designation to include everybody who at different times emigrated from Hungary to the West and became famous in any field. George Marx, *The Voice of the Martians* (Budapest: Akadémiai Kiadó, 1997).

5. Edward Teller with Judith L. Shoolery, *Memoirs: A Twentieth-Century Journey in Science and Politics* (Cambridge, MA: Perseus Publishing, 2001), p. v.

6. István Hargittai, Magdolna Hargittai, and Balazs Hargittai, *Candid Science I–VI* (London: Imperial College Press, 2000–2006).

7. István Hargittai, *Our Lives: Encounters of a Scientist* (Budapest: Akadémiai Kiadó, 2004), pp. 17–19.

8. Ibid., pp. 75–82.

9. Ibid., pp. 27–32.

10. The latest edition of this book: Magdolna Hargittai and István Hargittai, *Symmetry through the Eyes of a Chemist*, 3rd ed. (Springer-Verlag, 2009).

11. The other Martians changed their names upon becoming émigrés as well. Leo Szilard only dropped the "accents" (which are not really accents but indicate modifications in how the vowels sound) in his name; so Leó Szilárd became Leo Szilard, which was an easy enough change, because Leo is an international name. Incidentally, in Hungarian, the surname comes first, followed by the first name. Theodore von Kármán's first name changed from Tódor to the German Theodor, then to Theodore. Curiously, his original first name was not even Tódor; it was Tivadar. John von Neumann was born Neumann János, then he became Johann before becoming John. Both in von Kármán's and von Neumann's name, the "von" served to indicate nobility according to the German usage (it would be "de" in French). In Eugene Paul Wigner's name, the first and middle names changed; first they were Jenő Pál, then Eugen, finally, Eugene Paul.

## CHAPTER I

1. In writing this chapter I relied, in part, on the first chapter, "Arrival and Departure," of my book *The Martians of Science*.

2. Edward Teller with Judith Shoolery, *Memoirs: A Twentieth-Century Journey in Science and Politics* (Cambridge, MA: Perseus Publishing, 2001), p. 5.

3. Conversation with Wendy Teller in Naperville, IL, May 10, 2008.

4. Conversation with Janos Kirz in Budapest, January 16, 2008.

5. See, e.g., Norman Macrae, *John von Neumann: The Scientific Genius Who Pioneered the Modern Computer, Game Theory, Nuclear Deterrence, and Much More* (Providence, RI: American Mathematical Society, 1999), pp. 39–41.

6. From a speech by Béla Lengyel as quoted in Mihály Beck, *Than Károly élete és munkássága* (Károly Than's Life and Oeuvre) (Piliscsaba, Hungary: Magyar Tudománytörténeti Intézet, 2008), pp. 10–11.

7. See, e.g, John Lukacs, *Budapest 1900: A Historical Portrait of a City and Its Culture* (New York: Grove Press, 1988); István Hargittai, *The Martians of Science: Five Physicists Who Changed the Twentieth Century* (New York: Oxford University Press, 2006; 2008); Kati Marton, *The Great Escape: Nine Jews Who Fled Hitler and Changed the World* (New York: Simon & Schuster, 2006).

8. *Nobel Lectures Including Presentation Speeches and Laureates' Biographies: Physics 1942–1962* (Singapore: World Scientific, 1998), pp. 20–21.

9. Abraham Pais, *The Genius of Science: A Portrait Gallery of Twentieth-Century Physicists* (Oxford: Oxford University Press, 2000), pp. 264–79; J. S. Rigden, *Rabi: Scientist and Citizen* (New York: Basic Books, 1987), p. 17.

10. Tibor Frank, *Double Exile: Migration of Jewish–Hungarian Professionals through Germany to the United States, 1919–1945* (Bern: Peter Lang, European Academic Publishers, 2009).

11. Lukacs, *Budapest 1900*, pp. 188–89.

12. Ibid., p. 14.

13. Today, with a decreased population in what is a considerably smaller country, Budapest's population of about 1.8 million appears out of proportion, a fact that is further emphasized by the hundreds of thousands commuters on weekdays.

14. Lukacs, *Budapest 1900*, p. 91.

15. Gyula Kornis, *Kultúrpolitikánk irányelvei* (Guiding Principles of Our Culture Policy) (Budapest: Athenaeum, 1921), pp. 21–22, 28.

16. Ignác Romsics, *Hungary in the Twentieth Century* (Budapest: Corvina/Osiris, 1999), p. 149.

17. See, e.g., Zoltán Varannai, "Numerus clausus," *Rubicon* (in Hungarian), no. 1 (1997). This legislation was introduced under the first premiership of Count Pál Teleki who, during his second premiership two decades later, introduced hugely devastating anti-Jewish legislation. Teleki was a much-revered scholar in geography. While prime minister, he committed suicide in 1941 when Hungary joined Germany in an aggression against Yugoslavia, with which Teleki had signed a treaty of eternal friendship just a few months before.

18. István Hargittai, "Beszélgetés Marx Györggyel" ("Conversation with George Marx"), *Magyar Tudomány* 48 (2003): 883–889, 884.

19. Stanislaw M. Ulam, *Adventures of a Mathematician* (Berkeley: University of California Press, 1991), p. 111.

20. István Hargittai, *Candid Science II: Conversations with Famous Biomedical Scientists*, "George Klein" (London: Imperial College Press, 2002), pp. 416–41, 425.

21. Donald Michie, "The Genius Phenomenon," *Nature* 292 (1981): 91–92, 92.

22. Ibid.

23. Eugene P. Wigner with Andrew Szanton, *The Recollections of E. P. Wigner as Told to Andrew Szanton* (New York: Plenum Press, 1992), pp. 221–22.

24. Walter Laqueur, *Generation Exodus: The Fate of Young Jewish Refugees from Nazi Germany* (Hanover, NH: Brandeis University Press, 2001), p. xiii.

25. See, e.g., István Hargittai, *Our Lives: Encounters of a Scientist* (Budapest: Akadémiai Kiadó, 2004), pp. 51–53, 115, 116.

26. E. Czeizel, *Tudósok, gének, tanulságok: A magyar természettudós géniuszok családfaelemzése* (Scientists, Genes, Lessons: Genealogical Analysis of Hungarian Geniuses in Science) (Budapest: Galenus, 2006).

27. Clarence and Jane Larson's video interview with Edward Teller, 1984.

28. Letter of November 21, 1957, from Joseph Bard of London to Edward Teller, Hoover Archives. Mowgli is a famous fictional character in Rudyard Kipling's writings.

29. Letter in April 1948 from Edward Teller to Maria Goeppert Mayer, UCSD Library/Hoover Archives (No. 3 in 1948).

30. Teller, *Memoirs*, p. 33.

31. Conversations with Judith Shoolery in Half Moon Bay, CA, February 2004.

32. The Minta is now named after Ágoston Trefort, a progressive minister of education who lived some time during the last third of the nineteenth century.

33. At the time, there was male domination in the Hungarian high school; the first gimnázium for girls opened in 1896.

34. His original Hungarian name was György Bischitz, which he changed a few years after high school to György Hevesy. He converted during the last year of his high school studies.

35. Theodore von Kármán with Lee Edson, *The Wind and Beyond: Theodore von Kármán, Pioneer in Aviation and Pathfinder in Space* (Boston: Little, Brown, 1967), p. 20.

36. In 1941, when Peter Lax was fifteen years old, he left Hungary with his family for the United States. The Abel Prize of Norway is arguably the highest international prize in mathematics.

37. Lukacs, *Budapest 1900*, pp. 145–46.

38. István Hargittai, "Last Boat from Lisbon: Conversations with Peter D. Lax," *Mathematical Intelligencer* (2010), in press.

39. Edward Teller Interview for the Academy of Achievement, Palo Alto, CA, September 30, 1990.

40. Larson's video interview with Edward Teller.

41. Teller, *Memoirs*, pp. 31–38.

42. Ibid., p. 57.

43. Ede Teller and László Zeley, *A biztonság bizonytalansága* (The Uncertainty of Security) (Budapest: Relaxa kft, 1991), p. 33.

44. The rest were Lutherans, 4 (9 percent); Calvinists, 3 (7 percent); and Eastern Greek Catholics, 2 (4 percent).

45. Teller, *Memoirs*, p. 18.

46. Ibid., pp. 32–33.

47. Letter in late September 1949 from Edward Teller to Maria Goeppert Mayer, UCSD Library/Hoover Archives (No. 13 in 1949).

48. Hargittai, *Our Lives*, pp. 67–70.

49. Larson's video interview with Edward Teller; see also Teller, *Memoirs*, pp. 22–23.

50. Teller, *Memoirs*, p. 38.

51. Ibid., p. 37.

52. Today, high school students of any grade also participate in these competitions.

53. Gyula Radnai, "Teller Ede az Eötvös-versenyen" ("Edward Teller at the Eötvös-Competition"), *Debreceni Szemle* 12 (2004): 548–60. In 1993, Teller gave a talk at Eötvös University in Budapest. Following his talk, he was introduced to Sándor Kotz, a nineteen-year-old student who had just won the most recent competitions in both mathematics and physics.

54. S. R. Weart and G. Weiss Szilard, eds., *Leo Szilard: His Version of the Facts. Selected Recollections and Correspondence* (Cambridge, MA: MIT Press, 1978), p. 5.

55. Balazs Hargittai and István Hargittai, *Candid Science V: Conversations with Famous Scientists*, "Charles H. Townes" (London: Imperial College Press, 2005), pp. 94–137.

56. Magdolna Hargittai and István Hargittai, *Candid Science IV: Conversations with Famous Physicists*, "Laszlo Tisza" (London: Imperial College Press, 2004), pp. 390–403, 392.

57. The school was named after Archduke Joseph (1776–1846) of the Habsburgs, the palatine of Hungary at the time, acting as surrogate for the king.

58. Radnai, "Edward Teller at the Eötvös-Competition," pp. 553–54.

59. Document in Folder K-1547/1, pp. 41–43; Historical Archives of the Hungarian State Security Services (HAHS).

60. See, e.g., Peter Goodchild, *Edward Teller: The Real Dr. Strangelove* (London: Weidenfeld and Nicolson, 2004), p. 197.

61. We lived in a small town from which people were not deported; rather, it was among the destinations for deportees, but not the worst destination. One day a well-dressed family came to see us; they talked with my parents who gave them some money and household items. When they left, I asked why they needed help because they looked better off than we were. My parents explained to me that they had to leave their home with a small suitcase, and the only clothing they had was what they were wearing.

62. Conversation with Janos Kirz.

63. This friend, Peter Katona, now of Washington, DC, was the president of a successful foundation until recently. Conversation with Janos Kirz.

64. Documents in Folder K-1547/1, pp. 63, 111–12, HAHS. This all sounds painfully real to me. When I was—for a while—denied the chance to attend high school at the age of fourteen in 1955, the authorities explained to my mother that my excellent grades made me especially dangerous for society to let me study (see Hargittai, *Our Lives*, pp. 75–78).

65. Document, November 3, 1951, in Folder K-1547/1, p. 65, HAHS.

66. Document in Folder K-1547/1, p. 100, HAHS.

67. Document, December 18, 1951, in Folder K-1547/1, p. 66, HAHS.

68. Examining their reports and making a mental comparison with FBI reports (see chapter 8), one has the feeling of déjà vu, as if the agents in the two faraway organizations had a lot in common. Their reports contained repetitions of many significant and hardly significant findings, and they goofed in ways that showed ignorance of the highest degree. They interpreted "nuclear" as something to do with cells and referred to John von Neumann as a New York senator. In time, though, the Hungarian reports showed marked improvement.

69. Document, January 3, 1952, in Folder K-1547/1, pp. 69–70, HAHS.

70. Document, April 21, 1954, in Folder K-1547/2, p. 12, HAHS.

71. Document, March 4, 1954, in Folder K-1547/1, pp. 168–70, HAHS.

72. Document, October 4, 1954, in Folder K-1547/2, pp. 75–76 (three pages), HAHS.

73. Document, July 20, 1954, in Folder K-1547/2, p. 123, HAHS.

74. See, for example, Document, October 4, 1954, in Folder K-1547/2, pp. 75–76 (three pages), HAHS.

75. Letter of about September 25, 1946, from Edward Teller to Maria Goeppert Mayer, UCSD Library/Hoover Archives (No. 14 in 1946).

76. Letter of April 11, 1956, from Lewis Strauss to Edward Teller, Hoover Archives.

77. Letter of October 18, 1957, from Edward Teller to Georg Hevesy, Hoover Archives.

78. Leona Marshall Libby, *The Uranium People* (New York: Crane Russak and Charles Scribner's Sons, 1979), p. 236.

79. Lajos Jánossy's letter of December 24, 1958, to Leo Szilard, Archives of the Hungarian Academy of Sciences, Document 526/958/Sz. This was no small feat because following Miksa Teller's death in 1950—and especially from the time of their deportation to Tállya—Teller's mother, sister, and nephew tried to get out of the country. They submitted an application to emigrate from Tállya to Israel. Subsequent applications indicated Israel and then Sweden, and the target country always provided the appropriate certificates that once the Tellers would possess permission to leave Hungary they would be issued the necessary visas. After Janos's escape at the end of 1956, the passport applications were limited to Teller's mother and sister. The decisions of the passport office of the Interior Ministry were invariably "refusal." There were two interesting deviations from this straightforward response. In 1957, there was a suggestion from a lower-ranking official to grant permission, but this was overruled by a higher official. In between the two notes, there is an explanatory note, saying that Emma was in contact with her atomic researcher brother living in the United States and also that Emma used to work at the American Embassy. These two facts were never mentioned in Emma's applications, in which she was supposed to give a detailed account of her story. At least as late as May 1958, their applications were rejected. Then there was a sudden change, and there were two identical notes, one each on their acts, dated December 20, 1958, saying that she must be summoned by December 22, and that her passport must be given to her. It seems that there was a deadline for them so that Jánossy could send out the information to Szilard about their release. The archives of the former Ministry of Interior, currently, the Ministry of Justice and Law Enforcement of Hungary, stored the respective documents in folders labeled E2225 for Mrs. Miksa Teller (Teller Miksáné) and E2337-6 for Mrs. András Kirz—Emma Teller (Kirz Andrásné). No other document could be found pertaining directly to the permission in the archival material of the Communist Party and the Ministry of the Interior, according to the Hungarian State Archive, which investigated the case according to my request. It was often the practice of party leaders to give instructions to subordinated organs, like the passport office, by a phone call rather than by written documents.

## CHAPTER 2

1. W. F. Bynum and Roy Porter, eds., *Oxford Dictionary of Scientific Quotations* (Oxford: Oxford University Press, 2005), p. 405.

2. William Lanouette with Bela Silard, *Genius in the Shadows: A Biography of Leo Szilard, the Man Behind the Bomb* (Chicago: University of Chicago Press, 1993), p. 96.

3. Victor Weisskopf, *The Joy of Insight: Passions of a Physicist* (New York: Basic Books, 1991), p. 30. It was the same Professor Thirring who, in 1946, would describe the hydrogen bomb in a German-language book (see chapter 5).

4. See, e.g., John F. Kennedy, *A Nation of Immigrants* (New York and Evanston: Harper & Row, 1964), p. 74.

5. István Hargittai, *Road to Stockholm: Nobel Prizes, Science, and Scientists* (Oxford: Oxford University Press, 2002; 2003), p. 17.

6. Ingmar Bergström, "Lise Meitner och atomkärnans klyvning (Lise Meitner and Nuclear Fission)," *Årsberättelse 1999* (Stockholm: Kungliga Vetenskapakademien, 2000), pp. 17–25.

7. John Cornwell, *Hitler's Scientists: Science, War and the Devil's Pact* (New York: Viking, 2003), p. 7.

8. István Hargittai, *Candid Science: Conversations with Famous Chemists*, "Erwin Chargaff" (London: Imperial College Press, 2000), pp. 14–37, 24.

9. Richard Willstätter, *From My Life: Memoirs of Richard Willstätter* (W. A. Benjamin, 1965); German original, *Aus meinem Leben* (Weinheim: Verlag Chemie, 1949; reprint edition of the 2nd edition, Wiley–VCH, 1973).

10. Herman F. Mark, *From Small Organic Molecules to Large: A Century of Progress* (Washington, DC: American Chemical Society, 1993), pp. 61–63.

11. Joseph Borkin, *The Crime and Punishment of I. G. Farben* (New York: Pocket Books, 1979), back cover.

12. William Taussig Scott and Martin X. Moleski, *Michael Polanyi: Scientist and Philosopher* (New York: Oxford University Press, 2005), p. 25.

13. Enclosure to a letter of September 4, 1997, from Edward Teller to Yossi Korazim-Kőrösy, Kőrösy's son, Hoover Archives.

14. Herman F. Mark, "Polymeric Materials for Extreme Conditions," in Hans Mark and Sidney Fernbach, eds., *Properties of Matter under Unusual Conditions (In Honor of Edward Teller's 60th Birthday)* (New York: Interscience Publishers, 1969), pp. 105–117. Mark times his encounter with Teller at the Technical University of Karlsruhe as taking place in 1929, when Teller was already in Leipzig. Their interaction in Karlsruhe probably took place sometime in 1927 (see Mark, *From Small Organic Molecules to Large*, pp. 33–35). Teller moved to Karlsruhe in January 1926 and stayed there through April 1928; Mark moved to Karlsruhe from Berlin in the beginning of 1927.

15. Edward Teller with Judith L. Shoolery, *Memoirs: A Twentieth-Century Journey in Science and Politics* (Cambridge, MA: Perseus Publishing, 2001), p. 45.

16. Ibid.

17. Clarence and Jane Larson's video interview with Edward Teller, 1984.

18. Odenwald School was founded in 1910 and had to close down during the Nazi regime. See Dennis Shirley, *Politics of Progressive Education: The Odenwaldschule in Nazi Germany* (Cambridge, MA: Harvard University Press, 1992). Today the school flourishes again.

19. When he notes that Landau and Mici liked to play tennis together, he mentions, as a rare reference to his handicap, that due to his accident he could not compete with them. Teller, *Memoirs*, p. 106. He did not play tennis, but he continued playing table tennis.

20. István Hargittai and Magdolna Hargittai, *Candid Science VI: More Conversations with Famous Scientists*, "Donald A. Glaser" (London: Imperial College Press, 2006), pp. 518–53, 548.

21. Otto R. Frisch, *What Little I Remember* (Cambridge: Cambridge University Press, 1979), p. 175.

22. One day Dr. von Lossow disappeared, and Teller connected this event to the fact that the doctor's relative, Otto von Lossow, was a military leader in Bavaria at the time of Hitler's Beer Hall Putsch in Munich on November 8 and 9, 1923. General von Lossow was captured by the Nazis, but he misled Hitler by offering allegiance, and when he was freed, he led the legitimate forces in putting down the Nazi coup d'état. Teller thought that his doctor feared for his life because of his family connection to Otto von Lossow (Teller, *Memoirs*, p. 50). However, Otto von Lossow stayed in Germany, where he died of natural causes in 1938. See Anthony Read, *The Devil's Disciples: The Life and Times of Hitler's Inner Circle* (London: Jonathan Cape, 2003). I could not find any information about the fate of the surgeon von Lossow.

23. The degree Diplom-Physiker was introduced in Germany in 1942 according to the official publication of the Imperial Ministry of Science, Education, and Culture, *Deutsche Wissenschaft, Erziehung, Volksbildung* (Ministerblatt des REM), August 7, 1942, Vol. 8, 319. The new regulations were justified by the increasing demands from the military and the economy; the new degree was lower than the doctorate. It may be that the ministry did not want young people to spend additional years in academic learning, but it also may have been a ploy by the bureaucrats to keep themselves busy.

24. Scott and Moleski, *Polanyi*, p. 55.

25. Lanouette with Szilard, *Genius in the Shadows*, p. 100.

26. Werner Heisenberg, "The Concept of 'Understanding' in Theoretical Physics," in Mark and Fernbach, *Properties of Matter under Unusual Conditions*, pp. 7–10, 7.

27. Jens Blecher and Gerald Wiemers, "Edward Teller in Leipzig 1928 bis 1930," in *Festschrift Werner Heisenberg 1901–1976. Beiträge, Berichte, Briefe. Festschrift zu seinem 100. Geburtstag*, edited by Christian Kleint, Helmut Rechenberg, and Gerald Wiemers (Leipzig/Stuttgart: Sächsische Akademie der Wissenschaften zu Leipzig, 2005), pp. 114–121.

28. Hans A. Bethe, "Wie die Kernspaltung die Physiker entzweite," *Süd-*

*deutsche Zeitung* (December 11, 2001) (Bethe used the expression "intellektuelles Zuhause").

29. Larson's video interview with Edward Teller.

30. Eugen Wigner, *Gruppentheorie und ihre Anwendung auf die Quanten-mechanik der Atomspektren* (Braunschweig, Germany: Vieweg, 1931), English translation: Eugene P. Wigner, *Group Theory and Its Application to Quantum Mechanics of Atomic Spectra* (New York: Academic Press, 1959).

31. Blecher and Wiemers, "Edward Teller in Leipzig 1928 bis 1930."

32. Teller writes about this appointment in his Hungarian-language curriculum vitae in December 1932 (Hungarian State Archives, Folder 1932–1936–68-6, Rockefeller Fellowships). In his *Memoirs*, he states that he was Heisenberg's assistant, p. 64.

33. Max Born, *My Life: Recollections of a Nobel Laureate* (New York: Charles Scribner's Sons, 1978), p. 250.

34. Records of registration at Göttingen City Archives (Einwohner Meldekarte Stadtarchiv Göttingen). Personal communication from Günther Beer, June 2009.

35. Edward Teller's notes in 1965, Special Collections Research Center, University of Chicago Library, James Franck Collection, Box 24, Folder 23.

36. According to the Records of Registration at Göttingen City Archives (see above), Teller was away in Rome between February 29 and April 25, 1932.

37. Born, *My Life*, p. 237.

38. Teller, *Memoirs*, p. 93.

39. Heisenberg, "The Concept of 'Understanding' in Theoretical Physics," pp. 7, 10.

40. Conversation with Wendy Teller, Naperville, IL, May 10, 2008.

41. Edward Teller, "Seven Hours of Reminiscence," *Los Alamos Science* (Winter/Spring 1983): 190–195.

42. Edward Teller, "Hiroshima—The Psychology of a Decision," *Fizikai Szemle*, no. 5 (1995): 176–81, 179. (This issue of the Hungarian magazine appeared in English.)

43. Teller, *Memoirs*, p. 61.

44. E. Y. Hartshorne, *The German Universities and National Socialism* (Cambridge, MA: Harvard University Press, 1937), p. 112. A variation of this is on the back cover of the book *Hitler's Gift*: "If the dismissal of Jewish scientists means the annihilation of German science, then we shall do without science for a few years!" See Jean Medawar and David Pyke, *Hitler's Gift: Scientists Who Fled Nazi Germany* (London: Piatkus, 2000).

45. Cornwell, *Hitler's Scientists*, p. 27.

46. Teller, *Memoirs*, p. 230.

47. Ibid.

48. Ibid., p. 231.

49. Letter in June–July 1946 from Edward Teller to Maria Goeppert Mayer, UCSD Library/Hoover Archives (No. 3 in 1946).

50. Teller, *Memoirs*, p. 244.

51. Letter of October 10, 1948, from Edward Teller to Maria Goeppert Mayer, UCSD Library/Hoover Archives (No. 18 in 1948).

52. Letter in mid-February 1949 from Edward Teller to Maria Goeppert Mayer, UCSD Library/Hoover Archives (No. 4 in 1949).

53. Samuel A. Goudsmit, *Alsos* (New York: Henry Schuman, 1947).

54. Ian Buruma, *The Wages of Guilt: Memories of War in Germany and Japan* (London: Vintage, 1995), p. 143.

55. Paul L. Rose, *Heisenberg and the Nazi Atomic Bomb: A Study in German Culture* (Berkeley: University of California Press, 1998), p. 306.

56. The Dutch newspaper *Haagse Post* published Goudsmit's comments on July 7, 1973. The English translation is among Teller's papers in the Hoover Archives.

57. Letter of March 23, 1981, from Edward Teller to Carl Friedrich von Weizsäcker, Hoover Archives.

58. Letter of July 23, 1999, from Edward Teller to Thomas Powers, author of the book *Heisenberg's War*, Hoover Archives.

59. Letter of January 17, 1996, from Edward Teller to Dieter Michel, dean of the Faculty of Physics and Geosciences, University of Leipzig, Hoover Archives.

60. Patricia Rife, *Lise Meitner and the Dawn of the Nuclear Age* (Boston: Birkhäuser, 2007), p. 250.

61. Rife, *Lise Meitner*, p. 262.

62. Ruth Lewin Sime, *Lise Meitner: A Life in Physics* (Berkeley: University of California Press, 1996), p. 350.

63. Ibid.

64. Nancy Thorndike Greenspan, *The End of the Certain World: The Life and Science of Max Born, the Nobel Physicist Who Ignited the Quantum Revolution* (New York: Basic Books, 2005), p. 197.

65. Ibid., p. 178.

66. Born, *My Life*, p. 269.

67. Rose, *Heisenberg and the Nazi Atomic Bomb*, pp. 236–37.

68. Ibid., p. 237.

69. Medawar and Pyke, *Hitler's Gift*, p. 175.

70. Rose, *Heisenberg and the Nazi Atomic Bomb*, p. 270.

71. Cornwell, *Hitler's Scientists*, p. 326.

72. G. Kuiper to a Major Fischer, June 30, 1945, University of Arizona Library, Kuiper Papers, Box 28; quoted in David C. Cassidy, *Beyond Uncertainty: Heisenberg, Quantum Physics, and the Bomb* (New York: Bellevue Literary Press, 2009), pp. 345–46.

73. Cornwell, *Hitler's Scientists*, p. 407.

74. Rose, *Heisenberg and the Nazi Atomic Bomb*, p. 306.

75. Ibid., p. 304.

76. Dong-Won Kim, *Yoshio Nishina: Father of Modern Physics in Japan* (New York: Taylor & Francis, 2007), pp. 151–55.

77. Hiroki Sugita, "Hiroshima, Nagasaki Sobered Japan's A-bomb Project Head," *Japan Times* online, August 14, 2007.

78. Sybe Rispens, *Einstein in Nederland* (Amsterdam: Anbo, 2006); Martin Eickhoff, *In namaan der Wetenschap? P. J. W. Debye en zuijn carrière in nazi-Duitsland* (Nederlands Instituut voor Oorlogsdocumentatie).

79. The letter by the chair of the Department of Chemistry and Chemical Biology of Cornell University and the dissenting statement by Roald Hoffmann have appeared in *Chemical & Engineering News* 84, no. 30 (July 24, 2006): 4–6, 30.

80. Edward Teller, "Die Verantwortlichkeit des Wissenschaftlers in der Welt," in *Quanten und Felder*, edited by H. P. Dürr (Braunschweig, 1971), p. 93.

81. Michael Lennick, "Edward Teller," *American Heritage* (June/July 2005): 54–63, 58.

82. Letter of December 11, 1980, from Edward Teller to Finn Aaserud, Hoover Institution.

## CHAPTER 3

1. J. Walsh, "A Conversation with Eugene Wigner," *Science* 181 (1973): 527–533, 532.

2. See, e.g., *The Complete Illustrated Works of Lewis Carroll* (London: Chancellor Press, 1982).

3. Laura Fermi, *Atoms in the Family: My Life with Enrico Fermi* (Albuquerque: University of New Mexico Press, 1982), p. 221.

4. When the future Nobel laureate George A. Olah was a refugee from Hungary in 1957, the AAC helped him find a job; see George A. Olah, *A Life of Magic Chemistry: Autobiographical Reflections of a Nobel Prize Winner* (New York: Wiley-Interscience, 2001), p. 65. The seventy-fifth anniversary of the institution was celebrated with a two-day conference in London in December 2008.

5. H. E. Armstrong's letter "Foreign Scientists in Britain. Professor Kapitza's Recall to Russia," *Times*, May 7, 1935. See also L. Badash, *Kapitza, Rutherford, and the Kremlin* (New Haven, CT: Yale University Press, 1985), pp. 121–23.

6. H. E. Armstrong, "Physical Chemistry in the University of Manchester," *Nature* (July 8, 1933): 67.

7. Stanislaw M. Ulam, *Adventures of a Mathematician* (Berkeley: University of California Press, 1991), p. 121.

8. Letter of June 2, 1956, from Edward Teller to R. J. Blin-Stoyle, Hoover Archives.

9. Spencer R. Weart and Gertrud Weiss Szilard, eds., *Leo Szilard: His Version of the Facts. Selected Recollections and Correspondence* (Cambridge, MA: MIT Press, 1978), pp. 13–14.

10. Letter of September 24, 1948, from Edward Teller to Maria Goeppert Mayer, UCSD Library/Hoover Archives (No. 16 in 1948).

11. Teller sent a detailed letter to his father; we know of the letter's arrival

500 NOTES TO PAGES 94–99

date: December 12, 1932. Miksa Teller prepared a one-page summary of the letter. Hungarian State Archives, Folder 1932–1936–68-6, "Rockefeller Fellowships."

12. Hungarian State Archives, Folder 1932–1936–68-6, "Rockefeller Fellowships."

13. Ibid.

14. Edward Teller, "James Franck," in *Remembering the University of Chicago: Teachers, Scientists, and Scholars*, edited by Edward Shils (Chicago and London: University of Chicago Press, 1991), pp. 130–37.

15. The information about Mici's ancestry and her family came from the following sources: the collection of physics historian László Kovács (Szombathely, Hungary) and his students; E. Czeizel, *Tudósok, gének, tanulságok: A magyar természettudós géniuszok családfaelemzése* (Scientists, Genes, Lessons: Genealogical Analysis of Hungarian Geniuses in Science) (Budapest: Galenus, 2006); and Wendy Teller, private communication, 2008.

16. Alíz Halda, "Magos Gábor (1914–2000)," *Élet és Irodalom* (February 4, 2000).

17. Document of March 26, 1955, in Folder K-1547/2, pp. 125–27 (five pages), Historical Archives of the Hungarian State Security Services (HAHS).

18. Letter of June 30, 1960, from Edward Teller to John R. Pellman, Hoover Archives.

19. Alden Whitman, "A Brilliant Scientist," *New York Times*, April 3, 1968.

20. Edward Teller with Judith L. Shoolery, *Memoirs: A Twentieth-Century Journey in Science and Politics* (Cambridge, MA: Perseus Publishing, 2001), p. 99.

21. Edward Teller, "A Few Memories," in *Niels Bohr: A Centenary Volume*, edited by A. P. French and P. J. Kennedy (Cambridge, MA: Harvard University Press, 1985), pp. 181–82, 181.

22. Edward Teller, "Niels Bohr and the Idea of Complementarity," in *Great Men of Physics: The Humanistic Element in Scientific Work*, edited by Emilio G. Segrè, Joseph Kaplan, Leonard I. Schiff, and Edward Teller (Los Angeles: Tinnon-Brown, 1969), pp. 76–97, 77.

23. Michael Polanyi, *Personal Knowledge: Towards a Post-Critical Philosophy* (Chicago: University of Chicago Press, 1962), p. 49.

24. Teller, "Niels Bohr and the Idea of Complementarity," p. 78 (note).

25. Teller, *Memoirs*, p. 108.

26. Lev Landau and Edward Teller, "Zur Theorie der Schalldispersion (Theory of Sound Dispersion)," *Physikalische Zeitschrift der Sowjetunion* 10 (1936), no. 1:34–43.

27. F. Kalckar and E. Teller, "Theory of the Catalysis of the Ortho-para Transformation by Paramagnetic Gases," *Proceedings of the Royal Society* 150A (July 1, 1935): 520–533.

28. Roger H. Stuewer, "Niels Bohr and Nuclear Physics," in French and Kennedy, *Niels Bohr*, pp. 197–220, 208–10.

29. Letter of December 11, 1980, from Edward Teller to Finn Aaserud, Hoover Institution.

30. George Gamow, *My World Line: An Informal Autobiography* (New York: Viking Press, 1970).

31. Letter of October 2, 1969, from P. A. M. Dirac to Edward Teller, Hoover Archives.

32. Blumberg and Owens, *Energy and Conflict*, p. 53.

33. Hans Mark and Lowell Wood, Foreword to *Energy in Physics, War and Peace: A Festschrift Celebrating Edward Teller's 80th Birthday*, edited by Hans Mark and Lowell Wood (Dordrecht, Holland: Kluwer Academic, 1988), pp. 1–2.

34. Ibid.

35. Gerhard Herzberg, "Molecular Spectroscopy: A Personal History," *Annual Review of Physical Chemistry* 36 (1985): 1655–1684, 1664. The interactions between Herzberg and Teller continued well after World War II and have been recorded in a Herzberg biography: Boris Stoicheff, *Gerhard Herzberg: An Illustrious Life in Science* (Ottawa: NRC Press, 2002).

36. Eduard Teller and Bryan Topley, "On the Equilibrium and the Heat of the Reaction $C_2H_4 + H_2 = C_2H_6$," *Journal of the Chemical Society* (July 1935): 876–85. Note that Teller listed his first name as Eduard, the German equivalent of Edward.

37. For Derek Barton and Odd Hassel in 1969.

38. Stephen Brunauer, P. H. Emmett, and Edward Teller, "Adsorption of Gases in Multimolecular Layers," *Journal of the American Chemical Society* 60 (1938): 309–19.

39. Edward Teller, "The History of the BET Paper," in *Heterogeneous Catalysis: Selected American Histories*, edited by B. H. Davis and W. P. Hettinger Jr., ACS Symposium Series Vol. 222 (Washington, DC: American Chemical Society, 1983), pp. 227–231, 227.

40. Burtron H. Davis, "Brunauer, Emmett and Teller—The Personalities Behind the BET Method," *Energeia* 6, no. 1 (1995): 1, 3–4.

41. Burtron H. Davis, "B, E, & T: The Scientists Behind Surface Science," *CHEMTECH* (January 1991): 19–25, 25.

42. H. A. Jahn and E. Teller, "Stability of Polyatomic Molecules in Degenerate Electronic States I—Orbital Degeneracy," *Proceedings of the Royal Society A* 161 (1937): 220–35.

43. The title of Renner's dissertation was "Zur Theorie der Wechselwirkung zwischen Elektronen- und Kernbewegung bei stabförmigen, dreiatomigen Molekülen."

44. Rudolf Renner, "Lebenslauf des cand. phys., " Göttingen, 1. XI. 1933, in the collection of Universitätsarchiv Göttingen.

45. Letter of November 11, 1980, from Rudolf Renner to Edward Teller and Teller's response (undated), Hoover Archives.

46. Correspondence with Rudolf Renner's daughter-in-law, Beate Bauer-Renner, Dorum, Germany, April–May 2009. See more on Renner and Jahn: Magdolna Hargittai and István Hargittai, "Hermann Jahn and Rudolf Renner of the Jahn–Teller and Renner–Teller Effects," *Structural Chemistry* 20 (2009): 537–40.

47. P. T. Landsberg, *Bulletin of the London Mathematical Society* 12 (1980): 383–86.

48. Magdolna Hargittai and István Hargittai, *Candid Science IV: Conversations with Famous Physicists*, "Edward Teller" (London: Imperial College Press, 2004), pp. 404–23, 416.

49. Robert Englman, *The Jahn–Teller Effect in Molecules and Crystals* (London: John Wiley-Interscience, 1972).

50. Boris S. Gorobets, *Krug Landau: Fizika voini i mira* (Landau's Circle: Physics of War and Peace) (Moscow: URSS, 2008); *Krug Landau: Zhizn' geniya* (Landau's Circle: The Life of a Genius) (Moscow: URSS, 2009); *Krug Landau i Lifshitsa* (Landau's and Lifshits's Circle) (Moscow: URSS, 2009).

51. J. Georg Bednorz and K. Alex Müller, "Perovskite-Type Oxides—The New Approach to High-$T_c$ Superconductivity," in *Nobel Lectures Physics 1981–1990* (Singapore: World Scientific, 1993), pp. 424–57.

52. Edward Teller, "Questions for the New York State Institute on Superconductivity Newsletter: Answers Provided by Dr. Edward Teller," typescript of October 2, 1990, Hoover Archives.

53. Letter of June 19, 1985, from Edward Teller to Ralph G. Pearson, Hoover Archives.

54. Edward Teller, "Deterrence? Defense? Disarmament? The Many Roads Toward Stability," in *Thinking about America: The United States in the 1990s*, edited by Annelise Anderson and Dennis L. Bark (Stanford, CA: Hoover Institution, 1988), pp. 21–32, 29.

55. Edward Teller, "Foreword," *Structural Chemistry* 2 (1991): vii.

56. Isaac Bersuker, personal communication by e-mail, September 2008. The book finally appeared as I. B. Bersuker and V. Z. Polinger, *Vibronnie vzaimodeistviya v molekulakh i kristallakh* (Vibronic Interactions in Molecules and Crystals).

57. G. Nordheim, H. Sponer, and E. Teller, "Note on the Ultraviolet Absorption Systems of Benzene Vapor," *Journal of Chemical Physics* 8 (1940): 455–58.

58. Philip Anderson, "Who or What Is RVB?" *Physics Today* (April 2008): 8–9. (RVB = Resonance Valence Bond.) Anderson was not aware of Teller's prior work; I learned this from our e-mail exchange in the summer of 2008.

59. *Sostoyanie teorii khimicheskogo stroeniya v organicheskoi khimii* (State of Affairs of the Theory of Chemical Structure in Organic Chemistry) (Moscow: Izdatel'stvo Akademii nauk SSSR, 1952). For more on the great Soviet controversy related to Pauling's theory of resonance, see, e.g., István Hargittai and Magdolna Hargittai, *In Our Own Image: Personal Symmetry in Discovery* (New York: Kluwer Academic/Plenum Publishers, 2000), pp. 84–88.

60. See, e.g., Thomas Hager, *Force of Nature: The Life of Linus Pauling* (New York: Simon & Schuster, 1995).

61. Teller, *Memoirs*, p. 132.

62. Nuel Pharr Davis, *Lawrence and Oppenheimer* (New York: Simon & Schuster, 1968), p. 129.

63. Stephen B. Libby and Morton S. Weiss, "Edward Teller's Scientific Life," *Physics Today* (August 2004): 45–50, 48.

64. George Gamow and Edward Teller, "Selection Rules for Beta-Disintegration," *Physical Review* 49 (1936): 895–99.

65. Stewart D. Bloom, "On the Prevalence of the Gamow–Teller Transition," in *Energy in Physics, War and Peace*, edited by Mark and Wood, pp. 15–37, 35.

66. George Gamow and Edward Teller, "Rate of Selective Thermonuclear Reactions," *Physical Review* 53 (1938): 608–609. This paper built on the prior work of R. E. Atkinson and F. G. Houtermans, "Zur Frage der Aufbaumöglichkeit der Elemente in Sternen," *Zeitschrift für Physik* 54 (1929): 656–65.

67. Arthur I. Miller, *Empire of the Stars: Obsession, Friendship, and Betrayal in the Quest for Black Holes* (Boston and New York: Houghton Mifflin, 2005), p. 161.

68. Peter Garrity, *The Galloping Gamows* (Peter Garrity, 2007), p. 63.

69. Ralph A. Alpher, Hans A. Bethe, and George Gamow, "The Origin of Chemical Elements," *Physical Review* 73 (1948): 803–804.

70. Ralph A. Alpher and Robert Herman, *Genesis of the Big Bang* (New York: Oxford University Press, 2001), p. 76.

71. Alpher and Herman, *Genesis of the Big Bang*, p. 76; Samuel L. Marateck, "Alpher, Bethe, Gamow," *Physics Today* (September 2008): 11–12.

72. Martin Harwit, "Ralph Asher Alpher," *Physics Today* (December 2007): 67–68.

73. G. Gamow and E. Teller, "The Expanding Universe and the Origin of the Great Nebulæ," *Nature* 143 (1939): 116–17.

74. G. Gamow, "On the Origin of Galaxies," in Mark and Fernbach, *Properties of Matter under Unusual Conditions*, pp. 11–22.

75. Eugene P. Wigner, "An Appreciation of the 60th Birthday of Edward Teller," ibid., pp. 7–12.

76. Stephen Libby, "Compilation of Edward Teller's Publications," to be published; private communication, 2008.

77. Clarence and Jane Larson's video interview with Edward Teller, 1984. This paragraph, even if somewhat paraphrased, appears also in Teller's *Memoirs*, p. 137.

## CHAPTER 4

1. Albert Einstein, a co-signed statement quoted in the *New York Times*, June 12, 1953; Alice Calaprice, comp. and ed., *The Expanded Quotable Einstein* (Princeton, NJ: Princeton University Press, 2000), p. 183.

2. W. F. Bynum and Roy Porter, eds., *Oxford Dictionary of Scientific Quotations* (Oxford: Oxford University Press, 2005), p. 566 (from a speech by Democratic presidential candidate Adlai Stevenson in 1952).

3. Otto Hahn and Fritz Strassmann, *Naturwissenschaften* 27 (1939): 11–15; Lise Meitner and Otto R. Frisch, "Disintegration of Uranium by Neutrons: A New Type of Nuclear Reaction," *Nature* 143 (1939): 239–40.

4. Spencer R. Weart and Gertrud Weiss Szilard, eds., *Leo Szilard: His Version of the Facts. Selected Recollections and Correspondence* (Cambridge, MA: MIT Press, 1978), p. 56.

5. Letter of January 25, 1939, from Leo Szilard to Lewis Strauss, in Weart and Weiss Szilard, *Leo Szilard: His Version*, p. 62.

6. Weart and Weiss Szilard, *Leo Szilard: His Version*, pp. 63–64.

7. Edward Teller with Judith L. Shoolery, *Memoirs: A Twentieth-Century Journey in Science and Politics* (Cambridge, MA: Perseus Publishing, 2001), p. 143.

8. Letter of February 17, 1939, from Edward Teller to Leo Szilard, in Weart and Weiss Szilard, *Leo Szilard: His Version*, p. 66.

9. Teller, *Memoirs*, p. 142. Szilard said this in Hungarian, which he used for conspiratorial reasons; otherwise he was known to insist on speaking in English even with his Hungarian friends.

10. István Hargittai, *The Martians of Science: Five Physicists Who Changed the Twentieth Century* (New York: Oxford University Press, 2006), pp. 188–95.

11. See, e. g., Winston S. Churchill, *The Second World War: Abriged Edition with an Epilogue on the Years 1945 to 1957* (London: Pimlico, 2002), p. 192.

12. Weart and Weiss Szilard, *Leo Szilard: His Version*, pp. 82–84, 90–94.

13. Patchogue is on Route 27 in the direction of the southern prong of the fork in eastern Long Island, whereas Cutchogue is on Route 25 leading to the northern prong. Peconic is adjacent to Cutchogue.

14. Einstein was staying at Dr. Moore's cabin on Old Cove Road, a short, little blind street close to Horseshoe Cove.

15. Teller, *Memoirs*, p. 146.

16. Eugene P. Wigner with Andrew Szanton, *The Recollections of Eugene P. Wigner as Told to Andrew Szanton* (New York: Plenum Press, 1992), p. 199.

17. Ralph E. Lapp, "The Einstein Letter That Started It All," *New York Times* August 2, 1964.

18. Letter of July 19, 1939, from Leo Szilard to Albert Einstein, in Weart and Weiss Szilard, *Leo Szilard: His Version*, pp. 90–91. The correspondence was in German, and the last sentence sounded very warm in German: "Er ist besonders nett."

19. Document 55 in Weart and Weiss Szilard, *Leo Szilard: His Version*, pp. 94–96.

20. Weart and Weiss Szilard, *Leo Szilard: His Version*, p. 84.

21. Ibid., pp. 84–85.

22. Richard G. Hewlett and Oscar E. Anderson Jr., *A History of the United States Atomic Energy Commission, Volume I 1939/1946: The New World* (Washington, DC: US Atomic Energy Commission, 1972), p. 20.

23. Ibid.

24. Teller, *Memoirs*, p. 148.

25. Hewlett and Anderson, *New World*, p. 32.

26. Weart and Weiss Szilard, *Leo Szilard: His Version*, pp. 115–16.

27. Ibid., p. 117.

28. William Lanouette with Bela Szilard, *Genius in the Shadows: A Biography of Leo Szilard, the Man Behind the Bomb* (Chicago: University of Chicago Press, 1992), p. 248.

29. Letter of June 1, 1940, from Louis A. Turner to Leo Szilard, in Weart and Weiss Szilard, *Leo Szilard: His Version*, p. 132.

30. Teller, *Memoirs*, p. 149.

31. Laura Fermi, *Atoms in the Family: My Life with Enrico Fermi* (Albuquerque: University of New Mexico Press, 1982), p. 220.

32. *The Public Papers and Addresses of Franklin D. Roosevelt with a Special Introduction and Explanatory Notes by President Roosevelt, 1940, Volume: War—And Aid to Democracies* (New York: Macmillan, 1941), pp. 184–87.

33. Hans A. Bethe and Edward Teller, "Deviations from Thermal Equilibrium in Shock Waves," report in 1941, reprinted in *Selected Works of Hans A. Bethe, with Commentary* (Singapore: World Scientific, 1997), pp. 295–345.

34. Freeman Dyson, "Hans A. Bethe (1906–2005)," *Science* 308 (2005): 219.

35. A. H. Compton, *Atomic Quest: A Personal Narrative* (New York: Oxford University Press, 1956), p. 120.

36. E. Teller, Introduction to L. R. Groves, *Now It Can Be Told: The Story of the Manhattan Project* (New York: Harper, 1962), pp. iii–ix.

37. Compton, *Atomic Quest*, p. 113.

38. S. Goldberg, "Groves and the Scientists: Compartmentalization and the Building of the Bomb," *Physics Today* (August 1995): 38–43.

39. Teller, Introduction to Groves, *Now It Can Be Told*, p. v.

40. See, e.g., A. C. Brown and C. B. MacDonald, eds., *The Secret History of the Atomic Bomb* (New York: Dial Press, 1977).

41. Otto Frisch, *What Little I Remember* (Cambridge: Cambridge University Press, 1979); Rudolf Peierls, *Bird of Passage: Recollections of a Physicist* (Princeton, NJ: Princeton University Press, 1985).

42. David C. Cassidy, *J. Robert Oppenheimer and the American Century* (New York: Pi Press, 2005), pp. 218–19.

43. C. P. Snow, *Variety of Man* (New York: Charles Scribner's Sons, 1966), pp. 118–19.

44. Teller, *Memoirs*, p. 151.

45. Robert Serber with Robert P. Crease, *Peace & War: Reminiscences of a Life on the Frontiers of Science* (New York: Columbia University Press, 1998), p. 54.

46. Teller told about the first Citizenship Examination in a letter of November 12, 1940, to James Franck, Special Collections Research Center, University of Chicago Library, James Franck Collection, Box 9, Folder 6.

47. Fermi, *Atoms in the Family*, pp. 167–69.

48. Letter of August 6, 1941, from Dean George B. Pegram to Columbia University president Nicholas M. Butler, Columbia University Archives, Butler Library, Central Collection, Box 335, Folder 1.

49. Letter of May 11, 1942, from Dean George B. Pegram to Columbia University president Nicholas M. Butler, Columbia University Archives, Butler Library, Central Collection, Box 335, Folder 1. Teller's salary of $6,000 was $2,000 dollars less than Fermi's at the time of his initial appointment at Columbia. For the period from 1942 to 1943, Teller's salary stayed the same as before, but Fermi's rose to $9,000. By this time, though, Fermi had left Columbia and moved to Chicago.

50. Letter of July 23, 1942, from Dean George B. Pegram to Columbia University president Nicholas M. Butler, Columbia University Archives, Butler Library, Central Collection, Box 335, Folder 2.

51. Ibid.

52. Eugene P. Wigner, "An Appreciation of the 60th Birthday of Edward Teller," in *Energy in Physics, War and Peace: A Festschrift Celebrating Edward Teller's 80th Birthday*, edited by Hans Mark and Lowell Wood (Dordrecht, Holland: Kluwer Academic, 1988), pp. 7–12, 9.

53. Leona Marshall Libby, *The Uranium People* (New York: Crane Russak and Charles Scribner's Sons, 1979), p. 234.

54. Magdolna Hargittai and István Hargittai, *Candid Science IV: Conversations with Famous Physicists*, "Maurice Goldhaber" (London: Imperial College Press, 2004), pp. 214–31, 221.

55. Eileen Welsome, *The Plutonium Files: America's Secret Medical Experiments in the Cold War* (New York: Dial Press, 1999), p. 60.

56. Fermi, *Atoms in the Family*, p. 230. Leona Marshall Libby remembered that in Chicago, Teller used to greet everybody with a "Good Morning," even in the late afternoon and evening, because being a night owl it was when the day was starting for him (Marshall Libby, *Uranium People*, p. 235). Teller was not the only inconsiderate member of this community; Richard Feynman played his bongo drum for hours in the night.

57. Nuel Pharr Davis, *Lawrence & Oppenheimer* (New York: Simon & Schuster, 1968), p. 179.

58. Teller, *Memoirs*, pp. 176–77.

59. Victor Weisskopf, *The Joy of Insight: Passions of a Physicist* (New York: Basic Books, 1991), pp. 134–35.

60. Rudolf Peierls, "The Case for the Defence," *Nature* 328 (1987): 583.

61. Teller, *Memoirs*, p. 178.

62. I am grateful to Richard L. Garwin for a copy of this memorandum.

63. Letter of June 18, 1982, from Edward Teller to Alvin W. Trivelpiece, director, Office of Energy Research, US Department of Energy, Hoover Archives.

64. Edward Teller, appendix I in Blumberg and Owens, *Energy and Conflict*, pp. 455–58.

65. Arthur Koestler, *Darkness at Noon* (New York: Macmillan, 1941).

66. Teller, *Memoirs*, pp. 181–83.

67. Letter of February 9, 1970, from Edward Teller to Alice Smith, Hoover Archives.

68. Hargittai and Hargittai, *Candid Science IV*, "Edward Teller," pp. 404–23, 409.

69. Teller, *Memoirs*, p. 183.

70. Koestler, *Darkness at Noon*, p. 47.

71. Ibid., p. 134.

72. Ibid., p. 191.

73. Ferdinando Camon, *Conversations with Primo Levi*, translated by J. Shepley (Marlboro, VT: Marlboro Press, 1989), pp. 45–46. Incidentally, Levi's father worked for a long time in Hungary and had a very different experience there than in Italy concerning anti-Semitism. While "no one in Italy spoke of ostracizing the Jews," this was not the case in Hungary (or in France, where he had similar experiences). Levi's father witnessed the short-lived but brutal communist dictatorship in Hungary and was afraid both of communism, which was headed by a communist Jew in 1919, and of the reaction to communism (ibid., pp. 5–6).

74. Letter of February 2, 1976, from Edward Teller to Golda Meir, Hoover Archives.

75. John S. Rigden, *Rabi: Scientist and Citizen* (New York: Basic Books, 1987), p. 298.

76. Hargittai and Hargittai, *Candid Science IV*, "Edward Teller," p. 407.

77. Ibid., "Laszlo Tisza," pp. 390–403, 394.

78. Koestler, *Darkness at Noon*, p. 123.

79. Ibid., p. 132.

80. Edward Teller, "Science and Morality," *Science* 280 (1998): 1200–1.

81. Letter of February 1, 2002, from Edward Teller to Sara Stewart, ed., *The Week* (New York), Hoover Archives. Shakespeare's *Anthony and Cleopatra* is third; John Toland's *The Rising Sun* is fourth; and two books share fifth place: Jules Verne's *Around the World in Eighty Days* and H. G. Wells's *The Man Who Could Work Miracles*.

82. Joseph Rotblat left. He was awarded the Nobel Peace Prize in 1995 for his later efforts for nuclear disarmament, jointly with the Pugwash movement.

83. Letter of July 4, 1945, from Leo Szilard to Edward Teller, Hoover Archives.

84. Teller, *Memoirs*, p. 205.

85. Document 107 in Weart and Weiss Szilard, *Leo Szilard: His Version*, pp. 211–12.

86. Teller, *Memoirs*, p. 206.

87. Letter of July 2, 1945, from Edward Teller to Leo Szilard, Hoover Archives. It appears that Teller's letter was dated two days earlier than Szilard's letter to which it was responding. Szilard had sent his letter to others as well and apparently indicated a later date than when the letter was actually sent.

88. Mary Palevsky, *Atomic Fragments: A Daughter's Questions* (Berkeley: University of California Press, 2000), p. 31.

89. Clarence and Jane Larson's video interview with Edward Teller, 1984.

90. The report of the Science Panel of the Interim Committee is reproduced in Cynthia C. Kelly, ed., *The Manhattan Project: The Birth of the Atomic Bomb in the Words of Its Creators, Eyewitnesses, and Historians* (New York: Black Dog & Leventhal, 2007), pp. 290–91.

91. Hewlett and Anderson, *New World*, p. 367.

92. Teller, *Memoirs*, p. 216.

93. Ibid., p. 217.

94. Alexei B. Kojevnikov, *Stalin's Great Science: The Times and Adventures of Soviet Physicists* (London: Imperial College Press, 2004), p. 297.

95. Margit Szöllösi-Janze, *Fritz Haber 1868–1934: Eine Biographie* (Munich, 1998), p. 450.

96. William Van der Kloot, "April 1915: Five Future Nobel Prize-Winners Inaugurate Weapons of Mass Destruction and the Academic–Industrial–Military Complex," *Notes and Records of the Royal Society of London* 58 (2004): 149–60.

97. Van der Kloot, "April 1915," p. 155.

98. Teller, *Memoirs*, p. 219.

99. See, e. g., Edward Teller with Allen Brown, *The Legacy of Hiroshima* (Garden City, NY: Doubleday & Co., 1962), p. 20.

100. Document 97 in Weart and Weiss Szilard, *Leo Szilard: His Version*, p. 163.

101. Luis W. Alvarez, *Alvarez: Adventures of a Physicist* (New York: Basic Books, 1987), p. 150.

102. Wigner, *Recollections*, p. 249.

103. Hargittai and Hargittai, *Candid Science IV*, "Eugene P. Wigner," pp. 2–19, 16.

104. William L. Laurence, "Would You Make the Bomb Again?" *New York Times*, August 1, 1965.

105. Walter Pincus, "Debates, Doubts among the Creators," *Washington Post*, July 21, 1985.

106. Richard Halloran, "Teller Says Unrest on Campus Perils Defense Research," *New York Times*, July 25, 1970.

107. Hargittai and Hargittai, *Candid Science IV*, "John A. Wheeler," pp. 424–39, 426–27.

108. Ibid., "Philip Anderson," pp. 586–601, 594–95.

109. Tsuyoshi Hasegawa, *Racing the Enemy: Stalin, Truman, and the Surrender of Japan* (Cambridge, MA: Harvard University Press), pp. 298–303, quoted in Kelly, *Manhattan Project*, pp. 401–402.

110. Hiroki Sugita of *Kyodo News* in *Japan Times*, August 14, 2007.

111. Theodore von Kármán with Lee Edson, *The Wind and Beyond: Theodore von Kármán, Pioneer in Aviation and Pathfinder in Space* (Boston: Little, Brown, 1967), p. 133.

112. István Hargittai, "Edward Teller—Guardian of Freedom or Dr. Strangelove?" *Hungarian Quarterly* (2008): 108–22.

## CHAPTER 5

1. Andrei D. Sakharov quoted in F. Shcholkin, *Apostoli atomnogo veka. Vospominaniya, razmishleniya* (Apostles of the Atomic Era. Reminiscences, Ideas) (Moscow: DeLi Print, 2004), p. 132.

2. Philip Shenon and Irvin Molotsky, "Blowup in the Family," *New York Times*, October 4, 1988.

3. There are conflicting references to the date of this conversation over whether it took place in the fall of 1941 or in the spring of 1942. What is certain is that by the time Oppenheimer convened the meeting of theoretical physicists at Berkeley in the summer of 1942, the idea had already been around.

4. The lightest element is hydrogen. The nucleus of hydrogen consists of one proton. The nucleus of heavier hydrogen, deuterium (D), consists of one proton and one neutron and that of tritium (T), one proton and two neutrons. Hydrogen is unique in that its isotopes have different names.

5. George Gamow, *My World Line: An Informal Autobiography* (New York: Viking Press, 1970), p. 121. Bukharin's name was mentioned in the discussion of Koestler's *Darkness at Noon* in the preceding chapter.

6. Arthur Holly Compton, *Atomic Quest: A Personal Narrative* (New York: Oxford University Press, 1956), p. 128.

7. Sara Lippincott, "A Conversation with Valentine L. Telegdi—Part II," *Physics in Perspective* 10 (2008): 77–109, 82.

8. Ibid., p. 81.

9. István Hargittai and Magdolna Hargittai, *Candid Science VI: More Conversations with Famous Scientists*, "Jack Steinberger" (London: Imperial College Press, 2006), pp. 732–51, 737.

10. Magdolna Hargittai and István Hargittai, *Candid Science IV: Conversations with Famous Physicists*, "Jerome I. Friedman" (London: Imperial College Press, 2004), pp. 64–79, 70.

11. Freeman Dyson, "Edward Teller 1908–2003," in *Biographical Memoirs of Members of the National Academy of Sciences of the U. S. A.* (Washington, DC: National Academy of Sciences, 2007), pp. 2–20, 12.

12. Conversation with Freeman Dyson at the Institute for Advanced Study, Princeton, NJ, March 31, 2008.

13. Valentine L. Telegdi, "Enrico Fermi in America," *Physics Today* (June 2002): 38–43, 42.

14. *Nobel Lectures: Physics 1942–1962* (Singapore: World Scientific, 1998), p. 404.

15. *Nobel Lectures: Physics 1963–1970* (Singapore: World Scientific, 1998), p. 39.

16. Emilio Segrè, *Enrico Fermi: Physicist* (Chicago: University of Chicago Press, 1970), p. 169.

17. Balazs Hargittai and István Hargittai, *Candid Science V: Conversations*

*with Famous Scientists*, "Harold Agnew" (London: Imperial College Press, 2005), pp. 300–15.

18. Letter in late August of 1947 from Edward Teller to Maria Goeppert Mayer, UCSD Library/Hoover Archives (No. 4 in 1947).

19. Letter in February–March of 1948 from Edward Teller to Maria Goeppert Mayer, UCSD Library/Hoover Archives (No. 1 in 1948).

20. Letter in August 1948 from Edward Teller to Maria Goeppert Mayer, UCSD Library/Hoover Archives (No. 11 in 1948).

21. Letter in early August 1946 from Edward Teller to Maria Goeppert Mayer, UCSD Library/Hoover Archives (No. 6 in 1946).

22. Private communication from Marvin Goldberger by e-mail, June 2009; private communication from Geoffrey C___ by e-mail, June 2009.

23. See, in C. W. Chu, "A Theoretical Alchemist," *Proceedings of the International Symposium on Frontiers of Science, 2002, Beijing, in Celebration of the 80th Birthday of C. N. Yang*, edited by Hwa-Tung Nieh (Singapore: World Scientific, 2003), pp. 514–17.

24. John Firor, "Former Student Remembers Teller and Fermi with Gratitude," *Physics Today* (February 2005): 10.

25. Blumberg and Panos, *Edward Teller*, p. 92.

26. Maurice Goldhaber and Edward Teller, "On Nuclear Dipole Vibrations," *Physical Review* 74 (1948): 1046–49.

27. B. L. Berman, "Nuclear Giant Resonances—a Historical Review," in *Energy in Physics, War and Peace: A Festschrift Celebrating Edward Teller's 80th Birthday*, edited by Hans Mark and Lowell Wood (Dordrecht, Holland: Kluwer Academic, 1988), pp. 49–85.

28. Stanley A. Blumberg and Gwinn Owens, *Energy and Conflict: The Life and Times of Edward Teller* (New York: G. P. Putnam's Sons, 1976), p. 193.

29. Hargittai and Hargittai, *Candid Science IV*, "Val L. Fitch," pp. 192–213, 198.

30. Private communication from Geoffrey Chew by e-mail, June 2009.

31. Conversation with Janos Kirz in Budapest, January 16, 2008.

32. Conversation with Ralph Moir at Lawrence Livermore National Laboratory, April 23, 2009.

33. Maria G. Mayer and Edward Teller, "On the Origin of Elements," *Physical Review* 76 (1949): 1226–31.

34. Eugene P. Wigner, "An Appreciation of the 60th Birthday of Edward Teller," in Mark and Wood, *Energy in Physics, War and Peace*, pp. 7–12, 11.

35. Letter in October 1949 from Edward Teller to Maria Goeppert Mayer, UCSD Library/Hoover Archives (No. 15 in 1949).

36. Letter in mid-February 1949 from Edward Teller to Maria Goeppert Mayer, UCSD Library/Hoover Archives (No. 4 in 1949).

37. Letter in the summer of 1949 from Edward Teller to Maria Goeppert Mayer, UCSD Library/Hoover Archives (No. 9 in 1949).

38. Letter in October 1949 from Edward Teller to Maria Goeppert Mayer, UCSD Library/Hoover Archives (No. 15 in 1949).

39. Arthur Koestler, *The Act of Creation* (New York: Macmillan, 1964).

40. Letter in October 1950 from Edward Teller to Maria Goeppert Mayer, UCSD Library/Hoover Archives (No. 19 in 1950).

41. Letter of September 8, 1946, from Edward Teller to Maria Goeppert Mayer, UCSD Library/Hoover Archives (No. 9 in 1946).

42. Letter in mid-September 1946 from Edward Teller to Maria Goeppert Mayer, UCSD Library/Hoover Archives (No. 11 in 1946).

43. Letter in August 1948 from Edward Teller to Maria Goeppert Mayer, UCSD Library/Hoover Archives (No. 9 in 1948).

44. Letter in August 1948 from Edward Teller to Maria Goeppert Mayer, UCSD Library/Hoover Archives (No. 10 in 1948).

45. Letter in August 1948 from Edward Teller to Maria Goeppert Mayer, UCSD Library/Hoover Archives (No. 11 in 1948).

46. Letter in late January of 1950 from Edward Teller to Maria Goeppert Mayer, UCSD Library/Hoover Archives (No. 2 in 1950).

47. Letter in January of 1955 from Edward Teller to Maria Goeppert Mayer, UCSD Library/Hoover Archives.

48. Nancy Thorndike Greenspan, *The End of the Certain World: The Life and Science of Max Born, the Nobel Physicist Who Ignited the Quantum Revolution* (New York: Basic Books, 2005), p. 158.

49. Letter of February 28, 1963, from Edward Teller to Lewis L. Strauss, Hoover Archives. The Albert Einstein Award in theoretical physics was endowed by the Lewis and Rosa Strauss Memorial Fund and was administered by the Institute for Advanced Study. It was first awarded in 1951. Teller received it in 1958. There are a number of other awards named after Einstein, but this is the oldest one.

50. Maria Goeppert Mayer, "On Closed Shells in Nuclei," *Physical Review* 74 (1948): 235–39.

51. Letter of February 28, 1963, from Edward Teller to Lewis L. Strauss, Hoover Archives.

52. Leona Marshall Libby, *The Uranium People* (New York: Crane, Russak, 1979), p. 236.

53. Personal communication from Wendy Teller by e-mail, March 24, 2009.

54. Marshall Libby, *The Uranium People*, p. 236.

55. Conversation with Judit Stur in Budapest, March 23, 2009.

56. Personal communication from Wendy Teller by e-mail, March 24, 2009.

57. Conversation with Judit Stur in Budapest, March 23, 2009.

58. Edward Teller with Judith Shoolery, *Memoirs: A Twentieth-Century Journey in Science and Politics* (Cambridge, MA: Perseus Publishing, 2001), pp. 224–25.

59. Letter in December 1948 from Edward Teller to Maria Goeppert Mayer, UCSD Library/Hoover Archives (No. 20 in 1948).

60. Teller, *Memoirs*, p. 224.

61. He gives emphasis to the importance of international distribution of iso-

tope tracers in a letter in July–August 1946 to Maria Goeppert Mayer, UCSD Library/Hoover Archives (No. 4 in 1946).

62. Leslie R. Groves, *Now It Can Be Told: The Story of the Manhattan Project* (New York: Harper, 1942; republished, New York: Da Capo Press, 1983 with an introduction by Edward Teller), p. 389.

63. Edward Teller, "State Department Report—A Ray of Hope," *Bulletin of Atomic Scientists of Chicago* (April 1946): 10, 13.

64. Alice L. Buck, *A History of the Atomic Energy Commission* (Washington, DC: US Department of Energy, 1983).

65. Richard G. Hewlett and Francis Duncan, *A History of the United States Atomic Energy Commission: Volume II 1947/1952, Atomic Shield* (Washington, DC: US Atomic Energy Commission, 1972), pp. 7–8.

66. *Joint Committee on Atomic Energy, Confirmation of Atomic Energy Commission and General Manager, Jan. 27–31, Feb. 3–8, 10–12, 17–22, 24–26, Mar. 3–4, 1947* (Washington, DC: 1947), pp. 1–7; quoted in Hewlett and Duncan, *A History of the United States Atomic Energy Commission, Volume II*, p. 8.

67. Herbert F. York, *The Advisors: Oppenheimer, Teller, and the Superbomb* (San Francisco: W. H. Freeman and Co., 1976), p. 24.

68. The five included the current head of the theory division of the Los Alamos Laboratory, R. D. Richtmyer, his successor as head of theory, Carson Mark, Rolf Landshoff, the mathematician Stanislaw Ulam, who will play a conspicuous role in further developments, and a future Nobel laureate, Frederick J. Reines. The guests included Konopinski, George Placzek, Serber, Teller, von Neumann, and others; there were over thirty participants at the meeting.

69. Transcripts of a group interview with Edward Teller at Los Alamos National Laboratory, June 7, 1993, by Charles R. (Chuck) Hansen. I am grateful to Richard Garwin for a copy of the transcripts.

70. See, e.g., Boris S. Gorobets, *Krug Landau: Fizika voini i mira* (Landau's Circle: Physics of War and Peace) (Moscow: URSS, 2008), p. 103.

71. Edward Teller, "Atomic Physicists Have Two Responsibilities," *Bulletin of the Atomic Scientists* 3, no. 12 (1947): 355–56, 356.

72. Ibid.

73. York, *Advisors*, p. 24.

74. Ibid., p. 27.

75. Edward Teller, *Energy from Heaven and Earth* (San Francisco: W. H. Freeman and Co., 1979), p. 160.

76. Hewlett and Duncan, *Atomic Shield*, p. 196.

77. Teller, *Energy from Heaven and Earth*, p. 162.

78. Teller, *Memoirs*, p. 363, n. 8.

79. Hewlett and Duncan, *Atomic Shield*, p. 208.

80. Teller, *Energy from Heaven and Earth*, p. 161.

81. Harold Brown and Michael May, "Edward Teller in the Public Arena," *Physics Today* (August 2004): 51–53, 53.

82. E. A. Lobikov, *Sovremennaya fizika i atomnii proekt* (Modern Physics and

Atomic Project) (Moscow and Izhevsk: Institute of Computer Investigations, 2002), p. 122.

83. Lewis L. Strauss, *Men and Decisions* (New York: Doubleday & Co., 1962), p. 215.

84. Teller, *Memoirs*, p. 279, n. 5.

85. Strauss, *Men and Decisions*, p. 230.

86. Hewlett and Duncan, *Atomic Shield*, p. 369.

87. Teller, *Memoirs*, p. 279.

88. Hewlett and Duncan, *Atomic Shield*, p. 380.

89. Alvarez, *Alvarez*, p. 169.

90. Teller, *Memoirs*, pp. 280–82.

91. Hewlett and Duncan, *Atomic Shield,* pp. 375–76.

92. Interoffice Memorandum from Edward Teller to N. E. Bradbury, dated October 10, 1949, Los Alamos National Laboratory Records Center/Archives (unclassified).

93. Hewlett and Duncan, *Atomic Shield*, p. 380.

94. Letter in October 1949 from Edward Teller to Maria Goeppert Mayer, UCSD Library/Hoover Archives (No. 15 in 1949).

95. Hewlett and Duncan, *Atomic Shield*, p. 381.

96. Ibid., p. 382.

97. Seaborg was on a trip to Sweden that had been planned long before.

98. Glenn T. Seaborg, *A Chemist in the White House: From the Manhattan Project to the End of the Cold War* (Washington, DC: American Chemical Society, 1998), pp. 41–42. Seaborg could have raised his objections upon his return, during November and December, but he did not. He was a junior member of the GAC, at the beginning of a long and distinguished career in matters of nuclear policy during which he would deal with a total of ten American presidents. In time, he would develop into a seasoned diplomat in addition to being a world-renowned scientist; apparently he preferred to keep quiet for the duration of this debate.

99. For the three documents, see, for example, in York, *The Advisors,* appendix, pp. 150–59; the majority and minority opinions can also be found in Seaborg, *A Chemist in the White House*, pp. 42–43.

100. York, *The Advisors*, pp. 156–57; Seaborg, *A Chemist in the White House*, pp. 42–43.

101. Ibid.

102. Hewlett and Duncan, *Atomic Shield*, p. 385.

103. P. M. Stern with H. P. Green, *The Oppenheimer Case: Security on Trial* (New York: Harper & Row, 1969), p. 146, footnote.

104. York, *The Advisors*, p. 54.

105. Stern, *The Oppenheimer Case*, p. 138.

106. Strauss, *Men and Decisions*, pp. 211–12.

107. Ibid., p. 212.

108. Hargittai and Hargittai, *Candid Science VI,* "Vitaly L. Ginzburg," pp. 808–37.

109. From the transcript of the *Cold War* chat on CNN, March 21, 1999, moderated by *Cold War* associate editor Andy Walton.

110. Hargittai, *Candid Science*, "Kenneth S. Pitzer," pp. 438–47.

111. David Holloway, *Stalin and the Bomb: The Soviet Union and Atomic Energy 1939–1956* (New Haven and London: Yale University Press, 1994), pp. 317–18.

112. Teller, *Memoirs*, p. 286.

113. York, *The Advisors*, p. 34.

114. Alvarez, *Alvarez*.

115. Hewlett and Duncan, *Atomic Shield*, p. 391.

116. Ibid., p. 392.

117. Hans A. Bethe, "Robert Oppenheimer 1904–1967," *Biographical Memoirs of Fellows of the Royal Society* 14 (1968): 391–416, 404.

118. Hewlett and Duncan, *Atomic Shield*, p. 393.

119. Ibid.

120. Ibid., p. 394.

121. Edward Teller, "Back to the Laboratories," *Bulletin of the Atomic Scientists* 6, no. 2 (February 1950): 71.

122. John A. Wheeler with Kenneth Ford, *Geons, Black Holes & Quantum Foam: A Life in Physics* (New York: W. W. Norton & Co., 2000), p. 189.

123. York, *The Advisors*, p. 66.

124. Stern with Green, *The Oppenheimer Case*, p. 147.

125. D. Acheson, *Present at Creation: My Years in the State Department* (New York: W. W. Norton, 1969).

126. H. Urey, *Bulletin of the Atomic Scientists* 6, no. 3 (1950): 72–73, quoted in York, *The Advisors*, p. 71.

127. See, for example, the unsigned note in the *Bulletin of the Atomic Scientists* 6, no. 3 (March 1950): 66.

128. Hans Thirring, *Die Geschichte der Atombombe* (Vienna: Neues Österreich Zeitung–und–Verlagsgesellschaft, 1946).

129. Hans Thirring, "The Super Bomb," *Bulletin of the Atomic Scientists* 6, no. 3 (March 1950): 69–70.

## CHAPTER 6

1. W. F. Bynum and Roy Porter, eds., *Oxford Dictionary of Scientific Quotations* (Oxford: Oxford University Press, 2005), p. 125.

2. Ibid., p. 543.

3. Richard G. Hewlett and Francis Duncan, *A History of the United States Atomic Energy Commission, Volume II: Atomic Shield, 1947/1952* (University Park: Pennsylvania State University Press, 1969), p. 410.

4. Silvan S. Schweber, *In the Shadow of the Bomb: Bethe, Oppenheimer, and*

*the Moral Responsibility of the Scientist* (Princeton, NJ: Princeton University Press, 2000), p. 162.

5. Hans A. Bethe, *Scientific American* (April 1950).

6. Ibid.

7. Hans Bethe, "Edward Teller: A Long Look Back," review of Teller's *Memoirs*, *Physics Today* (November 2001): 55–56.

8. S. K. Allison, K. T. Bainbridge, H. S. [*sic*] Bethe, R. B. Brode, C. C. Lauritsen, F. W. Loomis, G. B. Pegram, B. Rossi, F. Seitz, M. A. Tuve, V. F. Weisskopf, and M. G. White, "Let Us Pledge Not to Use H-Bomb First," *Bulletin of the Atomic Scientists* 6, no. 3 (1950): 75.

9. See, in John A. Wheeler with Kenneth Ford, *Geons, Black Holes & Quantum Foam: A Life in Physics* (New York: W. W. Norton & Co., 2000), p. 272.

10. Allison et al., "Let Us Pledge Not to Use H-Bomb First," p. 75.

11. Ibid.

12. Albert Einstein's letter of January 30, 1950, to A. J. Muste, in the Albert Einstein Archives, Hebrew University, Jerusalem, AEA 60-636, quoted after Walter Isaacson, *Einstein: His Life and Universe* (New York: Simon & Schuster, 2007), p. 501.

13. Edward Teller with Judith Shoolery, *Memoirs: A Twentieth-Century Journey in Science and Politics* (Cambridge, MA: Perseus Publishing, 2001), p. 290.

14. Richard Rhodes, *Dark Sun: The Making of the Hydrogen Bomb* (New York: Touchstone, 1996), p. 465.

15. Letter of September 5, 1942, from Edward Teller to J. R. Oppenheimer, Hoover Archives.

16. István Hargittai and Magdolna Hargittai, *Candid Science VI: More Conversations with Famous Scientists*, "Vitaly L. Ginzburg," pp. 808–37.

17. Stanislaw M. Ulam, *Adventures of a Mathematician* (Berkeley: University of California Press, 1991), p. 210.

18. Hewlett and Duncan, *Atomic Shield*, p. 439.

19. George Gamow, *My World Line: An Informal Autobiography* (New York: Viking Press, 1970), p. 154.

20. Ulam, *Adventures*, p. 212.

21. Wheeler with Ford, *Geons*, p. 193.

22. Ulam, *Adventures*, pp. 209–24.

23. Magdolna Hargittai and István Hargittai, *Candid Science IV: Conversations with Famous Physicists*, "John A. Wheeler" (London: Imperial College Press, 2004), pp. 424–39.

24. Hargittai and Hargittai, *Candid Science VI*, "Wolfgang K. H. Panofsky," pp. 600–29.

25. Letter of March 21, 1950, from Edward Teller to Senator Millard Tydings, courtesy of Burtron Davis, Lexington, KY.

26. Leo Szilard, "The Sensitive Minority among Men of Science," dinner lecture at Brandeis University, Los Angeles, December 8, 1954, in George Marx, ed.,

*Leo Szilárd Centenary Volume* (Budapest: Eötvös Physical Society, 1998), pp. 176–84, 183.

27. Hargittai and Hargittai, *Candid Science VI*, "Matthew Meselson," pp. 40–61, 54–55.

28. Stephanie Young, "Something Resembling Justice: John Francis Neylan and the AEC Personnel Hearings at Berkeley, 1948–49," in *Reappraising Oppenheimer: Centennial Studies and Reflections*, edited by Cathryn Carson and David A. Hollinger, *Berkeley Papers in History of Science*, Volume 21 (Berkeley: Office for History of Science and Technology, University of California, Berkeley, 2005), pp. 217–51.

29. István Hargittai and Magdolna Hargittai, "'Doing Something Creative': Melvin Calvin," *Chemical Intelligencer* 6, no. 1 (2000): 52–55.

30. Carson Mark, interviewed by Kenneth W. Ford, February 24, 1995 (redacted transcripts with deletions made at the request of US Department of Energy in April 2009).

31. Ulam, *Adventures*, p. 214.

32. Hewlett and Duncan, *Atomic Shield*, p. 440.

33. Ibid.

34. Ibid., p. 441.

35. Luis W. Alvarez, *Alvarez: Adventures of a Physicist* (New York: Basic Books, 1987), p. 166.

36. Wheeler with Ford, *Geons*, p. 207.

37. Conversation with Richard L. Garwin, Scarsdale, NY, March 29, 2008.

38. Hewlett and Duncan, *Atomic Shield*, p. 527.

39. Teller knew the outcome before the official judgment was available, and he sent a telegram to Los Alamos, using no code, just saying, "It's a boy!" This is described in Hewlett and Duncan's official history of the Atomic Energy Commission (Hewlett and Duncan, *Atomic Shield*, p. 542). According to Teller's *Memoirs* (p. 352), he sent such a telegram after the Mike test on November 1, 1952, and he sent it from Berkeley, where he gathered the information about the explosion from a seismograph. Hewlett and Duncan based their account on documents, whereas Teller in this case had to rely on his memory about events well over forty years earlier. Unfortunately, no documents could be found in the archives of Los Alamos that would support or contravene either the Duncan–Hewlett or the Teller version (personal communication by e-mail from Los Alamos historian Alan Carr in September 2008). John Wheeler's description of the communications about the Mike test is consistent with Teller's version (Wheeler with Ford, *Geons*, p. 227). When Teller told his story to an audience in the 1990s, someone asked in the spirit of the time whether he would have sent a telegram saying "It's a girl!" had the test failed. (Teller, *Memoirs*, p. 352, n. 12.) In response to that question, Teller said in a 2002 interview: "I have to apologize for its sexist character." (Michael Lennick "Edward Teller," *American Heritage* [June/July 2005]: 54–63, 62.)

40. Edward Teller's memorandum in George A. (Jay) Keyworth's office on September 20, 1979.

41. Ulam, *Adventures*, p. 219.

42. Françoise Ulam, "Postscript to Adventures," in Ulam, *Adventures*, pp. 305–15, 311.

43. Ulam, *Adventures*, p. 220.

44. Teller's memorandum, September 20, 1979.

45. Wheeler with Ford, *Geons*, p. 210.

46. István Hargittai, "The Last Boat from Lisbon: Conversation with Peter Lax," *Mathematical Intelligencer* (2010), in press.

47. Ibid.

48. J. Rud Nielsen, "Memories of Niels Bohr," *Physics Today* (October, 22–30, 1963).

49. Hargittai and Hargittai, *Candid Science VI*, "Richard L. Garwin," pp. 480–517, 490.

50. Rhodes, *Dark Sun*, pp. 469–70.

51. Richard L. Garwin, private communication by e-mail, March 25, 2009.

52. Ibid.

53. Edward Teller, "A New Thermonuclear Device," LA-1230, Los Alamos Scientific Laboratory of the University of California, April 4, 1951.

54. Richard Polenberg, ed., *In the Matter of J. Robert Oppenheimer: The Security Hearing* (Ithaca, NY: Cornell University Press, 2002), pp. 110–11.

55. Polenberg, *In the Matter of J. Robert Oppenheimer*, p. 261.

56. Hargittai and Hargittai, *Candid Science VI*, "Richard Garwin," pp. 480–517.

57. Walter Sullivan, "Experts Doubt View That Atom Blast Could End All Life," *New York Times*, November 23, 1975.

58. Ulam, *Adventures*, pp. 310–11.

59. James D. Watson, *Girls, Genes, and Gamow: After the Double Helix* (New York: Alfred A. Knopf, 2002), p. 132.

60. Teller, *Memoirs*, pp. 99–100, slightly paraphrased.

61. James R. Shepley and Clay Blair Jr., *The Hydrogen Bomb: The Men, the Menace, the Mechanism* (New York: David McKay Co., 1954).

62. "The Atom: The Road Beyond," *Time*, April 12, 1954, pp. 21–24.

63. Teller, *Memoirs*, p. 368.

64. Robert C. Toth, "Teller Shows Consistency in Opposing Test Ban," *New York Times*, August 23, 1963.

65. Shepley and Blair, *Hydrogen Bomb*, p. 26.

66. Ibid., pp. 28–29.

67. Ibid., p. 40.

68. Ibid., p. 140.

69. Ibid., p. 146.

70. Ibid., p. 228.

71. In response to Kennedy's complimenting him by referring to the Shepley–Blair book, Teller recited a Gilbert and Sullivan ditty, which included the phrase "Scarce a word of it is true," according to a letter Teller published in *Science* 218 (1982): 1270.

72. Edward Teller, "The Work of Many People," *Science* 121 (February 25, 1955): 267–75.

73. Letter of November 16, 1954, from James Franck to Edward Teller, Special Collections Research Center, University of Chicago Library, James Franck Collection, Box 9, Folder 6.

74. Teller, *Memoirs*, p. 407, n. 6.

75. Ibid., p. 272.

76. Ibid., p. 273.

77. Ibid.

78. Ibid.

79. Letter of March 22, 1955, from Warren Weaver to Edward Teller, Hoover Archives.

80. Edward Teller with Allen Brown, *The Legacy of Hiroshima* (Garden City, NY: Doubleday & Co., 1962).

81. Ibid., p. 53.

82. Teller's memorandum, 1979.

83. Ibid.

84. Ibid.

85. Ibid.

86. Edward Teller, *Better a Shield Than a Sword: Perspectives on Defense and Technology* (New York: Free Press; London: Collier Macmillan Publishers, 1987).

87. Ibid., p. 77.

88. Ibid., p. 84.

89. Teller, *Memoirs*, p. 327.

90. Emilio Segrè, *A Mind Always in Motion: The Autobiography of Emilio Segrè* (Berkeley: University of California Press, 1993), p. 251.

91. Ibid.

92. Blumberg and Owens, *Energy and Conflict*, p. 374.

93. Teller, *Memoirs*, p. 405.

94. Conversation with Richard L. Garwin, Scarsdale, NY, March 29, 2008.

95. Ibid.

96. Ibid.

97. Hargittai and Hargittai, *Candid Science VI*, "Richard L. Garwin," p. 516.

98. Hans A. Bethe, "Comments on the History of the H-bomb," *Los Alamos Science* (Fall 1982): 43–53.

99. Peter Goodchild, *J. Robert Oppenheimer: Shatterer of Worlds* (London: BBC Books, 1980).

100. Bethe, "Comments on the History of the H-Bomb," p. 48.

101. Ibid., p. 51.

102. Ibid., p. 53.

103. Edward Teller, "Letters" section, *Science* 218 (1982): 1270. Teller's letter was in response to an article in *Science* based mostly on Bethe's article in *Los Alamos Science*: William J. Broad, "Rewriting the History of the H-bomb," *Science* 218 (November 19, 1982): 769–72.

104. Hans A. Bethe, "Letters" section, *Science* 218 (1982): 1270.

105. Hewlett and Duncan, *Atomic Shield*, pp. 542–45.

106. Bethe, "Comments on the History of the H-Bomb," p. 49.

107. Bethe, "Robert Oppenheimer 1904–1967," *Biographical Memoirs of Fellows of the Royal Society* 14 (1968): 391–416, 404.

108. István Hargittai, *Candid Science II: Conversations with Famous Biomedical Scientists*, "Marshall W. Nirenberg" (London: Imperial College Press, 2002), pp. 130–41.

109. Rudolf Peierls, *Bird of Passage: Recollections of a Physicist* (Princeton, NJ: Princeton University Press, 1985), pp. 199–200.

110. Conversation with John Nuckolls at Lawrence Livermore National Laboratory, April 23, 2009.

111. Transcripts of a group interview with Edward Teller at Los Alamos National Laboratory, June 7, 1993, by Charles R. (Chuck) Hansen. I am grateful to Richard Garwin for the transcripts of the interview.

112. Letter in midwinter 1949 from Edward Teller to Maria Goeppert Mayer, UCSD Library/Hoover Archives (No. 3 in 1949).

113. Conversation with Neal Snyderman at Stanford University, April 24, 2009.

114. E-mail message of April 4, 2009, from Gregory Canavan and subsequent e-mail messages.

115. Conversation with Siegfried Hecker at Stanford University on April 24, 2009.

116. Conversation with John Nuckolls.

117. Lowell Wood and John Nuckolls, "The Development of Nuclear Explosives," in *Energy in Physics, War and Peace: A Festschrift Celebrating Edward Teller's 80th Birthday*, edited by Hans Mark and Lowell Wood (Boston: Kluwer Academic Publishers, 1988), p. 317.

## CHAPTER 7

1. Magdolna Hargittai and István Hargittai, *Candid Science IV: Conversations with Famous Physicists*, "Edward Teller" (London: Imperial College Press, 2004), pp. 404–23, 412.

2. Richard G. Hewlett and Francis Duncan, *A History of the United States Atomic Energy Commission: Volume II 1947/1952, Atomic Shield* (Washington, DC: US Atomic Energy Commission, 1972), p. 541.

3. Ibid., p. 541.

4. Ibid., p. 569.

5. Edward Teller with Judith Shoolery, *Memoirs: A Twentieth-Century Journey in Science and Politics* (Cambridge, MA: Perseus Publishing, 2001), p. 334.

6. Hewlett and Duncan, *Atomic Shield*, pp. 569–70.

7. Ibid., p. 569.

8. Edward Teller with Allen Brown, *The Legacy of Hiroshima* (Garden City, NY: Doubleday & Co., 1962), pp. 54–55.

9. Edward Teller, "Back to the Laboratories," *Bulletin of the Atomic Scientists* 6, no. 2 (February 1950): 71.

10. Teller, *Memoirs*, p. 333.

11. Hewlett and Duncan, *Atomic Shield*, p. 570.

12. Theodore von Kármán with Lee Edson, *The Wind and Beyond: Theodore von Kármán, Pioneer in Aviation and Pathfinder in Space* (Boston: Little, Brown and Co., 1967), p. 270.

13. Richard Rhodes, "Running the Arms Race," *New York Times*, December 20, 1987.

14. Von Kármán, *The Wind and Beyond*, p. 301.

15. Ibid., p. 300.

16. Herbert F. York, *The Advisors: Oppenheimer, Teller, and the Superbomb* (San Francisco: W. H. Freeman and Co., 1976), p. 121.

17. Ibid., p. 122.

18. Christopher Anderson, "Are Two Labs Too Many?" *Science* 262 (October 8, 1993): 169.

19. István Hargittai and Magdolna Hargittai, *Candid Science VI: More Conversations with Famous Scientists*, "Wolfgang K. H. Panofsky" (London: Imperial College Press, 2006), pp. 600–29.

20. Hewlett and Duncan, *Atomic Shield*, p. 582.

21. Ibid., p. 581.

22. Ibid., p. 583.

23. Teller, *Memoirs*, p. 340.

24. Ibid.

25. Luis W. Alvarez, *Alvarez: Adventures of a Physicist* (New York: Basic Books, 1987), p. 178.

26. York, *The Advisors*, p. 127.

27. Ibid., p. 131.

28. Herbert F. York, *Making Weapons, Talking Peace: A Physicist's Odyssey from Hiroshima to Geneva* (New York: Basic Books, 1987).

29. Friedrich Schiller, *Fiesco* (*Fiesko*, in German, "Der Mohr hat seine Schuldigkeit getan, der Mohr kann gehen").

30. Ralph W. Emerson, *Essays: First Series. Self Reliance*, quoted in John Bartlett, *Familiar Quotations*, 14th ed. (Boston: Little, Brown and Co., 1968), p. 606.

31. York, *The Advisors*, p. 131.

32. Ibid., p. 132.

33. Teller, *Memoirs*, p. 343.

34. Ibid., p. 344.

35. York, *Making Weapons*, p. 10.

36. Herbert Childs, *An American Genius: The Life of Ernest Orlando Lawrence* (New York: E. P. Dutton & Co., 1968), p. 477.

37. Letter of September 2, 1954, from Ernest O. Lawrence to the regional office manager of the Atomic Energy Commission, Hoover Archives.

38. Herbert F. York, *Arms and the Physicist* (Woodbury, NY: American Institute of Physics, 1995), p. 10.

39. Thomas C. Reed, *At the Abyss: An Insider's History of the Cold War* (New York: Ballantine Books, 2004), p. 114.

40. York, *The Advisors*, p. 133

41. C. E. Leith, "The Computational Physics of the Global Atmosphere," in *Energy in Physics, War and Peace: A Festschrift Celebrating Edward Teller's 80th Birthday*, edited by Hans Mark and Lowell Wood (Dordrecht, Holland: Kluwer Academic, 1988), pp. 161–73.

42. Letter of February 5, 1953, from T. H. Johnson, director of the Division of Research, AEC, to E. O. Lawrence, Hoover Archives.

43. Letter of July 20, 2000, from Edward Teller to Energy Secretary Bill Richardson and NASA administrator Daniel S. Goldin, Hoover Archives.

44. Stanley A. Blumberg and Gwinn Owens, *Energy and Conflict: The Life and Times of Edward Teller* (New York: G. P. Putnam's Sons, 1976), p. 290.

45. Blumberg and Owens, *Energy and Conflict*, p. 287.

46. Letter in September 1950 from Edward Teller to Maria Goeppert Mayer, UCSD Library/Hoover Archives (No. 17 of 1950).

47. Letter of December 23, 1970, from Edward Teller to Herbert York, Hoover Archives.

48. York, *The Advisors*, p. 135.

49. Ibid.

50. Stanislaw M. Ulam, *Adventures of a Mathematician* (Berkeley: University of California Press, 1991), p. 222.

51. Teller, *Memoirs*, p. 359.

52. Lowell Wood and John Nuckolls, "The Development of Nuclear Explosives," in Mark and Wood, *Energy in Physics, War and Peace*, pp. 311–17, 316.

53. Duane C. Sewell, "The Branch Laboratory at Livermore during the 1950's," in Mark and Wood, *Energy in Physics, War and Peace*, pp. 319–26, 322.

54. James Sterngold, "Can This Guy Take the Heat? It Sure Looks That Way," *New York Times*, May 20, 2000.

55. Gregory Benford, "Reality Is Wilder: How a Terrifying Interview Led to Fun Doing Physics," *Nature* 432 (2004): 955.

56. Boyce Rensberger, "Troubled Laser Scientist Quitting Weapons Work: 'Star Wars' Innovator to Join Academia," *Washington Post*, September 11, 1986.

57. Reed, *At the Abyss*, p. 116.

58. Nicholas Metropolis, Arianna W. Rosenbluth, Marshall N. Rosenbluth, Augusta H. Teller, and Edward Teller, "Equation of State Calculations by Fast Computing Machines," *Journal of Chemical Physics* 21 (1953): 1087–92.

59. Carolin Middleton and Stephen B. Libby, "Exchanging Insights on Quantum Behavior: Teller's Contributions to Condensed-Matter Physics," *Science & Technology Review* (Lawrence Livermore National Laboratory) (May 2007): 26–27.

60. York, *The Advisors*, p. 82.

61. J. Robert Oppenheimer to James Conant, letter of October 21, 1949; quoted after Abraham Pais, *J. Robert Oppenheimer: A Life* (New York: Oxford University Press, 2006), p. 172.

62. Blumberg and Owens, *Energy and Conflict*, pp. 296–97.

63. Ibid., p. 296.

64. Stanley A. Blumberg and Louis G. Panos, *Edward Teller: Giant of the Golden Age of Physics* (New York: Charles Scribner's Sons, 1990), p. 142.

65. Teller, *Memoirs*, pp. 352–53.

66. Letter of May 25, 1943, from Robert Oppenheimer to Enrico Fermi, Hoover Archives.

67. J. Robert Oppenheimer, "Atomic Weapons and American Policy," *Foreign Affairs Quarterly* (July 1953).

68. Letter of July 1, 1977, from Edward Teller to Robert F. Bacher, Hoover Archives.

69. York, *Arms and the Physicist*, p. 123.

70. William J. Broad, *Teller's War: The Top-Secret Story Behind the Star Wars Deception* (New York: Simon & Schuster, 1992), p. 205.

71. Ibid., p. 124.

72. See, e.g., Josephine Anne Stein and Frank von Hippel, "Laboratories vs. A Nuclear Ban," *New York Times*, March 28, 1986; Fred Hiatt and Rick Atkinson, "Lab Creating New Generation of Nuclear Arms," *Washington Post*, June 9, 1986.

73. York, *The Advisors*, p. 84.

74. Richard Rhodes, *Dark Sun: The Making of the Hydrogen Bomb* (New York: Simon & Schuster, 1995), p. 541.

75. York, *Arms and the Physicist*, p. 12.

76. Sewell, "The Branch Laboratory at Livermore during the 1950's," p. 324.

77. Robert Lindsey, "Domestic Arms Race Pits Livermore vs. Los Alamos," *New York Times*, July 31, 1977.

78. Balazs Hargittai and István Hargittai, *Candid Science V: Conversations with Famous Scientists*, "Harold Agnew" (London: Imperial College Press, 2005), pp. 300–15.

79. Ibid., p. 303.

80. Private communication (by e-mail) from George (Jay) Keyworth, April 2008.

81. Conversation with Siegfried Hecker at Stanford University on April 24, 2009.

82. Robert B. Laughlin's autobiography in *Les Prix Nobel/The Nobel Prizes 1998: Nobel Prizes, Presentations, Biographies, and Lectures* (Stockholm: Nobel Foundation, 1999), pp. 52–71.

83. Richard More, "Conversations with Dr. Teller," printed notes as private communication in the framework of a conversation in Livermore, CA, April 22, 2009.

84. Conversations with Neal Snyderman at Stanford Guest House, Menlo Park, CA, April 24, 2009, and with Stephen Libby at Stanford Guest House, Menlo Park, CA, April 28, 2009.

85. Conversation with Ralph Moir at Lawrence Livermore National Laboratory, April 23, 2009.

86. Ralph W. Moir and Edward Teller, "Thorium-Fueled Underground Power Plant Based on Molten Salt Technology," *Nuclear Technology* 151 (2005): 334–40.

87. Teller, *Memoirs*, p. 483.

88. Alvin M. Weinberg, *Reflections on Big Science* (Cambridge, MA: MIT Press, 1967), pp. 89, 155.

89. Yuval Ne'eman, "Edward Teller, the Hungarian, American and Jew, Too," *Acta Physica Hungarica* 7 (1998): 155–65, 162.

90. Teller, *Memoirs*, p. 489.

## CHAPTER 8

1. Private communication (by e-mail) from George (Jay) Keyworth, April 2008.

2. Here is a non-exhaustive list in chronological order: Nuel Pharr Davis, *Lawrence & Oppenheimer* (New York: Simon & Schuster, 1968); Philip M. Stern with Harold P. Green, *The Oppenheimer Case: Security on Trial* (New York: Harper & Row, 1969); Herbert F. York, *The Advisors: Oppenheimer, Teller, and the Superbomb* (San Francisco: W. H. Freeman and Co., 1976); Peter Goodchild, *J. Robert Oppenheimer: Shatterer of Worlds* (London: BBC Books, 1980); Silvan S. Schweber, *In the Shadow of the Bomb: Bethe, Oppenheimer, and the Moral Responsibility of the Scientist* (Princeton, NJ: Princeton University Press, 2000); Gregg Herken, *Brotherhood of the Bomb: The Tangled Lives and Loyalties of Robert Oppenheimer, Ernest Lawrence, and Edward Teller* (New York: Henry Holt and Co., 2002); Jeremy Bernstein, *Oppenheimer: Portrait of an Enigma* (Chicago: Ivan R. Dee, 2004); David C. Cassidy, *J. Robert Oppenheimer and the American Century* (New York: Pi Press, 2005); Cathryn Carson and David A. Hollinger, eds., *Reappraising Oppenheimer: Centennial Studies and Reflections*, Berkeley Papers in History of Science, Vol. 21 (Berkeley: Office for History of Science and Technology, University of California, Berkeley, 2005); Priscilla J. McMillan, *The Ruin of J. Robert Oppenheimer and the Birth of the Modern Arms Race* (New York: Penguin Books, 2005); Abraham Pais with Robert P. Crease, *J. Robert Oppenheimer: A Life* (New York: Oxford University Press, 2006); Charles Thorpe, *Oppenheimer: The Tragic Intellect* (Chicago: University of Chicago Press, 2006); Silvan S. Schweber, *Einstein & Oppenheimer* (Cambridge, MA: Harvard University Press, 2008); Mark Wolverton, *A Life in Twilight: The Final Years of J. Robert Oppenheimer* (New York: St. Martin's Press, 2008). See also Barton J.

Bernstein, "In the Matter of J. Robert Oppenheimer," *Historical Studies in the Physical Sciences* 12 (1982): 195–252; Barton J. Bernstein, "The Oppenheimer Loyalty–Security Case Reconsidered," *Stanford Law Review* 42 (1990): 1383–1484.

3. The acclaimed movie *Gentleman's Agreement* exposed such discrimination as late as the years following World War II.

4. Among the Nobel laureates of the last decades of the twentieth century, there were quite a few examples of scientists who had had such experience of discrimination. See, e.g., István Hargittai, *Candid Science II: Conversations with Famous Biomedical Scientists*, "Arthur Kornberg" (London: Imperial College Press, 2002), pp. 50–71; István Hargittai, *Candid Science III: More Conversations with Famous Chemists*, "Herbert A. Hauptman" (London: Imperial College Press, 2003), pp. 292–317. These practices of discrimination are not quite widely known because the victims of such discrimination are often reluctant to speak about it. See, e.g., Jerome Karle's story in István Hargittai, *Our Lives: Encounters of a Scientist* (Budapest: Akadémiai Kiadó, 2004), pp. 73–74.

5. John S. Rigden, *Rabi: Scientist and Citizen* (New York: Basic Books, 1987), p. 33.

6. S. S. Schweber, *In the Shadow of the Bomb*, p. 54.

7. Cassidy, *J. Robert Oppenheimer*, p. 32.

8. Stern, *Oppenheimer Case*, p. 251.

9. Ibid.

10. Ibid., p. 23.

11. Victor Weisskopf, *The Joy of Insight: Passions of a Physicist* (New York: Basic Books, 1991), p. 115.

12. Victor Weisskopf, *The Privilege of Being a Physicist* (New York: W. H. Freeman and Co., 1989), p. 160.

13. Max Born, *My Life: Recollections of a Nobel Laureate* (New York: Charles Scribner's Sons, 1975), p. 229.

14. Ibid.

15. Abraham Pais, *Inward Bound* (New York: Oxford University Press, 1988), p. 367.

16. Hans Bethe, "J. Robert Oppenheimer, 1904–1967," *Biographical Memoirs of Fellows* (London: Royal Society, 1968), pp. 14, 391–416.

17. Quoted by Edward Teller in the transcripts of a group interview with Edward Teller at Los Alamos National Laboratory, June 7, 1993, by Charles R. (Chuck) Hansen. I am grateful to Richard Garwin for a copy of the transcripts.

18. Barton J. Bernstein, "The Puzzles of Interpreting J. Robert Oppenheimer, His Politics, and the Issues of His Possible Communist Party Membership," in Carson and Hollinger, *Reappraising Oppenheimer*, pp. 77–112, 111.

19. John Garrand and Carol Garrand, *Russian Orthodoxy Resurgent: Faith and Power in the New Russia* (Princeton, NJ: Princeton University Press, 2008), p. 49.

20. Ibid., p. 10.

21. See, e.g., Hans Krebs with Roswitha Schmid, *Otto Warburg: Cell Physiologist, Biochemist, and Eccentric* (Oxford: Clarendon Press, 1981), p. 91, after H. Fraenkel and R. Manvell, *Hermann Göring* (Hannover: Verlag für Literatur und Zeitgeschehn GmbH, 1964), p. 125.

22. Leslie R. Groves, *Now It Can Be Told: The Story of the Manhattan Project* (New York: Da Capo Press, 1983), p. 63.

23. Richard Polenberg, ed., *In the Matter of J. Robert Oppenheimer: The Security Clearance Hearing* (Ithaca, NY: Cornell University Press, 2002), p. 81.

24. Cassidy, *J. Robert Oppenheimer*, p. 225.

25. Edward Teller with Judith L. Shoolery, *Memoirs: A Twentieth-Century Journey in Science and Politics* (Cambridge, MA: Perseus Publishing, 2001), p. 163.

26. Max Born, *My Life & My Views* (New York: Charles Scribner's Sons, 1968), p. 40.

27. Polenberg, *In the Matter of J. Robert Oppenheimer*, p. 67.

28. Ibid., p. 360; this appears in the Personnel Security Board Reports.

29. Stern, *Oppenheimer Case*, p. 120.

30. Ibid., p. 74.

31. Ibid., p. 84.

32. K. C. Cole, *Something Incredibly Wonderful Happens: Frank Oppenheimer and the World He Made Up* (Boston and New York: Houghton Mifflin Harcourt, 2009), pp. 66–67.

33. Pais, *Oppenheimer: A Life*, p. 49.

34. Ibid., p. 53.

35. Ibid., p. 57.

36. Stanislaw M. Ulam, *Adventures of a Mathematician* (Berkeley: University of California Press, 1991), p. 224.

37. Stern, *Oppenheimer Case*, pp. 97–98.

38. Ibid., pp. 101, 103.

39. Ibid., pp. 127–28.

40. István Hargittai and Magdolna Hargittai, *Candid Science VI: More Conversations with Famous Scientists*, "Wolfgang K. H. Panofsky" (London: Imperial College Press, 2006), pp. 600–30, 623–24.

41. Stern, *Oppenheimer Case*, p. 179.

42. Ibid., p. 309.

43. Ulam, *Adventures*, p. 224.

44. Hargittai and Hargittai, *Candid Science VI*, "Donald Glaser," pp. 518–53, 549.

45. George Gamow, *Mr Tompkins in Paperback* (Cambridge: Cambridge University Press, 1965), p. 146.

46. Stern, *Oppenheimer Case*, p. 204.

47. Polenberg, *In the Matter of J. Robert Oppenheimer*, p. 277.

48. Stern, *Oppenheimer Case*, p. 205.

49. Polenberg, *In the Matter of J. Robert Oppenheimer*, pp. 304–307.

50. Steven H. Heimoff, "A Conversation with Edward Teller," *East Bay Express* 12, no. 47 (1990): 1, 12–14, 16, 18–19, 16.

51. Lewis Strauss, *Men and Decisions* (New York: Popular Library, 1963).

52. Stern, *Oppenheimer Case*, p. 305.

53. Polenberg, *In the Matter of J. Robert Oppenheimer*, p. 204.

54. Ibid., pp. 249–52.

55. Ibid., p. 252.

56. Ibid., pp. 253–54.

57. Teller, *Memoirs*, chapter 30 contains Teller's comments on his testimony in the Oppenheimer hearing.

58. FBI teletype dated June 12, 1952, decoded copy, Hoover Archives.

59. Letter of June 17, 1977, from Edward Teller to the Deputy Attorney General, Office of Privacy and Information Appeals, Washington, DC, Hoover Archives.

60. Office memorandum of May 3, 1954, from Charter Heslep to Lewis L. Strauss about Heslep's conversation with Edward Teller at Livermore on April 22, 1954, Hoover Archives.

61. Polenberg, *In the Matter of J. Robert Oppenheimer*, p. 256.

62. Ibid., p. 261.

63. Ibid., p. 262.

64. Ibid., p. 264.

65. Teller, *Memoirs*, chap. 30.

66. Polenberg, *In the Matter of J. Robert Oppenheimer*, p. 264.

67. Stern, *Oppenheimer Case*, p. 340.

68. See, e.g., in Teller, *Memoirs*, p. 601.

69. Stern, *Oppenheimer Case*, p. 381.

70. A. Pais, *Einstein Lived Here* (New York: Oxford University Press, 1994), p. 241. Yet it did not necessarily mean that Einstein much sympathized with Oppenheimer. When Einstein first heard about the proposed hearing, he burst into laughter and said, "The trouble with Oppenheimer is that he loves a woman who doesn't love him—the United States government." (Abraham Pais being quoted in R. Serber with R. P. Crease, *Peace & War: Reminiscences of a Life on the Frontiers of Science* [New York: Columbia University Press, 1998], pp. 183–84.)

71. Rigden, *Rabi*, p. 230, from an interview of Philip Morrison by John S. Rigden.

72. Polenberg, *In the Matter of J. Robert Oppenheimer*, p. 288.

73. Ibid., p. 232.

74. Ibid., p. 237.

75. Hans A. Bethe, "J. Robert Oppenheimer (1904–1967)," *Biographical Memoirs of the Royal Society* (London: Royal Society), pp. 391–416, 406.

76. Herbert Romerstein and Eric Breindel, *The Venona Secrets: Exposing Soviet Espionage and American Traitors* (Washington, DC: Regnery Publishing, 2000), p. 268.

77. Ibid., pp. 273–74.

78. Gertrude Weiss, according to Barton J. Bernstein in H. S. Hawkins, G. A. Greb, and G. Weiss Szilard, eds., *Toward a Livable World: Leo Szilard and the Crusade for Nuclear Arms Control* (Cambridge, MA: MIT Press, 1987), p. xliii.

79. Letter from John A. Wheeler to Edward Teller, July 2, 1954. I thank Judith Shoolery for a copy of this letter.

80. Robert F. Christy, interview by Sara Lippincott in Pasadena, CA, June 1994, http://resolver.caltech.edu/CaltechOH:OH_Christy_R.

81. Rigden, *Rabi*, p. 230.

82. George Johnson, *Strange Beauty: Murray Gell-Mann and the Revolution in Twentieth-Century Physics* (London: Jonathan Cape, 2000), p. 125.

83. Letter in July 1954 from Edward Teller to Maria Goeppert Mayer, UCSD Library/Hoover Archives (No. 3 in 1954).

84. Ibid.

85. Letter of October 18, 1954, from Edward Teller to James Franck; Special Collections Research Center, University of Chicago Library, James Franck Collection, Box 9, Folder 6. ("In den letzen Monaten ist es mir klar geworden, dass ich recht wenige Freunde habe.")

86. Letter of November 16, 1954, from James Franck to Edward Teller; Special Collections Research Center, University of Chicago Library, James Franck Collection, Box 9, Folder 6.

87. Ann Finkbeiner, *The Jasons: The Secret History of Science's Postwar Elite* (New York: Viking, 2006). The name "Jason" means nothing; it is not an acronym either. The initiator of the organization was Marvin Goldberger (we met him as Teller's critic in Chicago), and his wife, Mildred, suggested the name, which was a reference to Jason and the golden fleece. Jason is a hero in Greek mythology who performs a sort of mission impossible.

88. Private communication by e-mail from Alan B. Carr, laboratory historian, Los Alamos National Laboratory, September 2008.

89. Magdolna Hargittai and István Hargittai, *Candid Science IV: Conversations with Famous Physicists*, "John A. Wheeler" (London: Imperial College Press, 2004), pp. 424–39, 429.

90. Balazs Hargittai and István Hargittai, *Candid Science V: Conversations with Famous Scientists*, "Harold Agnew" (London: Imperial College Press, 2005), pp. 300–15, 304.

91. István Hargittai and Magdolna Hargittai, *Candid Science VI: More Conversations with Famous Scientists*, "Richard L. Garwin" (London: Imperial College Press, 2006), pp. 480–517.

92. Conversation with Richard Garwin, Scarsdale, NY, March 29, 2008.

93. Edward Teller, "Seven Hours of Reminiscences," *Los Alamos Science* (Winter/Spring 1983): 190–95, 190.

94. Ibid., p. 195.

95. Hargittai and Hargittai, *Candid Science VI*, "Freeman J. Dyson," pp. 440–77, 453.

96. Conversation with Freeman J. Dyson, Princeton, NJ, March 31, 2008.

97. Hargittai and Hargittai, *Candid Science VI*, "Freeman J. Dyson," p. 453

98. See, e.g., Gerald Horton's review of Stern, *The Oppenheimer Case* in the *New York Times*, January 4, 1970.

99. The award consisted of a gold medal and $50,000 in 1963 (as of 2010 the amount is $375,000).

100. John C. Devlin, "Teller Is Mute on Oppenheimer; Has TV Show Delete Question," *New York Times*, April 16, 1962.

101. Teller, *Memoirs*, p. 397.

102. Ibid., p. 392.

103. Ibid., p. 399.

104. Ibid., p. 396.

105. Ibid., p. 398.

106. Ibid., p. 399.

107. Ibid.

108. Letter in July 1954 from Edward Teller to Maria Goeppert Mayer, UCSD Library/Hoover Archives (No. 3 in 1954).

109. Clarence and Jane Larson's video interview with Edward Teller, 1984.

110. Hargittai and Hargittai, *Candid Science VI*, "Vitaly L. Ginzburg," pp. 808–37, 825.

111. Hargittai and Hargittai, *Candid Science V*, "Alexei A. Abrikosov," pp. 176–97, 190.

112. Heimoff, "A Conversation with Edward Teller," p. 19.

113. Michael Lennick, "Edward Teller," *American Heritage* (June/July 2002): 54–63, 63.

114. http://www.thememoryhole.org/fbi/teller_edward.htm. The file was obtained and posted by The Memory Hole.

115. Hargittai and Hargittai, *Candid Science IV*, "Laszlo Tisza," pp. 390–403.

116. See "Fusion of Identity Irks Non-Atomic Dr. Teller," *New York Times*, September 5, 1958.

117. Letter of September 23, 1958, from Edward Teller to Ludwig Teller, Hoover Archives.

118. Conversation with Wendy Teller, Naperville, IL, May 10, 2008.

119. E-mail message from Paul Teller, August 3, 2008.

120. Conversation with Janos Kirz, Budapest, January 16, 2008.

121. John H. Bunzel, "In Memoriam: Edward Teller: A Personal Remembrance," *Hoover Digest*, no. 1 (2004).


## CHAPTER 9

1. Lewis L. Strauss (then chairman of the US Atomic Energy Commission) was quoted in *Time* magazine's cover story, November 18, 1957.

2. István Hargittai and Magdolna Hargittai, *Candid Science VI: More Conversations with Famous Scientists*, "Donald A. Glaser" (London: Imperial College Press, 2006), pp. 518–53, 548.

3. István Hargittai, *Candid Science II: Conversations with Famous Biomedical Scientists*, "Edward B. Lewis" (London: Imperial College Press, 2002), pp. 350–63, 353–55.

4. Edward B. Lewis, "Leukemia and Ionizing Radiation," *Science* 125 (1957): 965–72.

5. Hermann J. Muller, "Artificial Transmutation of the Gene," *Science* 66 (1927): 84–87.

6. Another Nobel laureate, Bruce Merrifield (Chemistry 1984), told me about his related experience. As a teenager, he had an infection on his leg, and a dermatologist treated it with X-rays, but then the doctor used X-rays also on Merrifield's acne on his face, and with large doses. About fifteen years later tumors began showing up on Merrifield's face, and he suffered from them throughout his life. His warning was "Don't let your kids have X-ray treatment unless it's absolutely critical." Hargittai, *Candid Science II*, "Bruce Merrifield," pp. 206–19, 217.

7. Hargittai, *Candid Science II*, p. 353.

8. Herbert Childs, *An American Genius: The Life of Ernest Orlando Lawrence* (New York: E. P. Dutton & Co., 1968), p. 498.

9. John W. Finney, "Scientists May End Fall-Out in Bombs," *New York Times*, June 21, 1957.

10. William L. Laurence, "Construction of a 'Clean' H-Bomb Presents Formidable Problems for the Experts," *New York Times*, June 23, 1957.

11. Childs, *An American Genius*, p. 504.

12. John W. Finney, "U. S. Eliminates 95% of Fall-Out from the H-Bomb," *New York Times*, June 25, 1957.

13. Andrei Sakharov, *Memoirs*, translated from the Russian by Richard Lourie (New York: Alfred A. Knopf, 1990), p. 201.

14. Hargittai and Hargittai, *Candid Science VI*, "Wolfgang K. H. Panofsky," pp. 600–29, 611.

15. Ted Goertzel and Ben Goertzel with Mildred Goertzel and Victor Goertzel, *Linus Pauling: A Life in Science and Politics* (New York: Basic Books, 1995), p. 144.

16. Edward Teller, "The Compelling Need for Nuclear Tests," *Life*, 1958.

17. The text of their debate is available under the title "Fallout and Disarmament," http://osulibrary.oregonstate.edu/specialcollections/coll/pauling/catalogue/02/1958.

18. Thomas Hager, *Force of Nature: The Life of Linus Pauling* (New York: Simon & Schuster, 1995); Anthony Serafini, *Linus Pauling: A Man and His Science* (New York: Paragon House, 1989).

19. Hager, *Force of Nature*, p. 483.

20. "We Will Bury You!" *Time*, November 26, 1956. Khrushchev addressed this sentence toward Western diplomats at a reception in Moscow on November 18, 1956. The meaning of his words "Нравится вам или нет, но история на нашей стороне. Мы вас закопаем" was not something like "We will liquidate you and then bury you" but more like "Whether you like it or not, history is on our side, and we will survive you," or "We will be around to witness your funeral," or

530 NOTES TO PAGES 328–33

something similar. In any case, it was a harsh and undiplomatic statement, which could be easily interpreted to be even worse than it was.

21. Denis Brian, *Genius Talk: Conversations with Nobel Scientists and Other Luminaries* (New York: Plenum Press, 1995), p. 7.

22. Ibid., p. 28.

23. Hager, *Force of Nature*, p. 406.

24. Teller, *Memoirs*, p. 441.

25. Hager, *Force of Nature*, plate 16, between pp. 194 and 195.

26. Magdolna Hargittai and István Hargittai, *Candid Science IV: Conversations with Famous Physicists*, "Edward Teller" (London: Imperial College Press, 2004), pp. 404–23, 414.

27. Letter of May 29, 1958, from Edward Teller to John M. Richardson, Hoover Archives.

28. Letter of June 13, 1963, from Edward Teller to chairman-elect of the American Chemical Society, Arthur W. Adamson, Hoover Archives.

29. Allen Drury, "Teller Sees Peril If Atom Tests End," *New York Times*, April 17, 1958.

30. "New Issue Raised on Test Ban Pact," *New York Times*, March 18, 1960, referring to a statement by Teller on March 17 in San Francisco.

31. "Kennedy Confers with Experts on Nuclear Research," *New York Times*, December 1, 1961.

32. James Reston, "The Scientific Revolution and Democracy," *New York Times*, March 7, 1962.

33. Edward Teller in a letter to the *Times*, "Teller Advocates Testing: Physicist Says Soviet Explosions Make Our Effort Essential," *New York Times*, November 12, 1961.

34. Linus Pauling in a letter to the *Times*, "Pauling Opposes Tests: War Danger Held Increased by Further Weapon Development," *New York Times*, November 19, 1961.

35. Henry DeWolf Smyth, "This Science Knows—and This It Guesses," *New York Times*, May 13, 1962.

36. Edward Teller and Albert L. Latter, *Our Nuclear Future . . . Facts, Dangers and Opportunities* (New York: Criterion Books, 1958), p. 125.

37. Ibid., p. 124.

38. Valery N. Soyfer, *Lysenko and the Tragedy of Soviet Science* (New Brunswick, NJ: Rutgers University Press, 1994), p. 267.

39. Edward Teller with Allen Brown, *The Legacy of Hiroshima* (Garden City, NY: Doubleday & Co., 1962), pp. 180–81.

40. Sakharov, *Memoirs*, p. 198.

41. Ibid., pp. 200–203.

42. "Teller Test-Ban Warning," *New York Times*, February 1, 1963.

43. "Excerpts from Testimony by Teller," *New York Times*, August 15, 1963.

44. Robert C. Toth, "Teller Opposes Test Ban Treaty but 35 Nobel Laureates Ask Senate to Ratify Accord," *New York Times*, August 15, 1963.

45. Ibid.

46. E. W. Kenworthy, "President Vows U.S. Will Step Up Atomic Readiness: Disputes Teller on Danger to Security in Test Ban—Schriever Against Pact," *New York Times*, August 21, 1963.

47. Ibid.

48. E. W. Kenworthy, "Eisenhower for Test Ban, but Asks One Reservation: Says Treaty Must Make Clear Right of U.S. to Repel Aggression with Atom Arms—Truman Reaffirms Support," *New York Times*, August 27, 1963.

49. Teller, *Better a Shield*, p. 109.

50. D. S. Greenberg, "The Big Picture: House Committee Hears Views on Basic Problems of Science–Government Relations," *Science* 142 (November 8, 1963): 650–53.

51. Ibid., p. 652.

52. "The Nuclear Test Ban," *Nature* (September 28, 1963): 1219–21, 1220.

53. Robert C. Toth, "Teller Shows Consistency in Opposing Test Ban," *New York Times*, August 25, 1963.

54. Harold P. Furth, "Perils of Modern Living," *New Yorker*, November 10, 1956.

55. Letter of November 26, 1956, from Edward Teller to the editor of the *New Yorker*, Hoover Archives. Teller's letter appeared in the December 15, 1956, issue of the magazine.

56. Letter of January 31, 1963, from Edward Teller to Eugene P. Wigner, Hoover Archives.

57. Edward Teller and Eugene P. Wigner, "America Needs a Better Civil Defense Program," *New York Times*, July 22, 1979.

58. In 1969, I attended Wigner's lectures at the University of Texas at Austin, and he talked about the Budapest subway as designed for civil defense purposes.

59. Edward Teller, "Civil Defense Is Crucial," *New York Times*, January 3, 1984.

60. István Hargittai, "Otto Bastiansen (1918–1995)," *Chemical Intelligencer* 2, no. 2 (1996): 51.

61. "Pauling Welcomes Backing," *New York Times*, October 11, 1963.

62. Linus Pauling, "Science and Peace," Nobel lecture in Oslo, Norway, December 11, 1963.

63. István Hargittai and Magdolna Hargittai, *Candid Science VI: More Conversations with Famous Scientists*, "Seymour Benzer" (London: Imperial College Press, 2006), pp. 114–33, 116.

64. Lanouette, *Genius in the Shadows*, p. 411.

65. Helen S. Hawkins, G. Allen Greb, and Gertrud Weiss Szilard, eds., *Toward a Livable World: Leo Szilard and the Crusade for Nuclear Arms Control* (Cambridge, MA: MIT Press, 1987), pp. 238–50.

66. Jack Gould, "TV: Victims of Production Nonsense," *New York Times*, November 12, 1960.

67. Ibid.

68. Hawkins, Greb, and Weiss Szilard, *Toward a Livable World*, pp. 381–97.

69. Edward Teller, *The Reluctant Revolutionary* (Columbia: University of Missouri Press, 1964).

70. A. Cornelius Benjamin, foreword in Teller, *The Reluctant Revolutionary*, p. vi.

71. Teller, *The Reluctant Revolutionary*, p. 20.

72. Teller refers to *The Republic of Plato*, translated into English by John Llewelyn Davies and David James Vaugham (London: Macmillan & Co., 1921), pp. iii, 389.

73. Teller, *The Reluctant Revolutionary*, p. 16.

74. Ibid., p. 28.

75. A young member of the Senate Foreign Relations Committee, Democrat Joseph Biden met with Soviet Foreign Minister Gromyko and secured some concessions from him that the secretary of state and even the president had not been able to do.

76. Wolfgang K. H. Panofsky and Edward Teller, "Debate on SALT II," *Physics Today* (June 1979): 32–38.

77. *The Expanded Quotable Einstein*, collected and edited by Alice Calaprice (Princeton, NJ: Princeton University Press, 2000), p. 175; from "Atomic War or Peace," *Atlantic Monthly*, November 1945.

78. Edward Teller, "Deterrence? Defense? Disarmament? The Many Roads Toward Stability," in *Thinking about America: The United States in the 1990s*, edited by Annelise Anderson and Dennis L. Bark (Stanford, CA: Hoover Institution, 1988), pp. 21–32, 26.

79. Wolfgang K. H. Panofsky, "Against Deployment of 'Safeguard,'" *New York Times*, June 8, 1969.

80. Bryce Nelson, "ABM: Senators Request Outside Scientific Advice in Closed Session," *Science* 162 (1968): 1374–75.

81. See, e.g., Edward Teller, "Defense Is the Best Defense," *New York Times*, April 5, 1987.

82. Conversations with Judith Shoolery, Half Moon Bay, CA, 2004.

83. Edward Teller, *The Pursuit of Simplicity* (Malibu, CA: Pepperdine University Press, 1981), p. 136. Teller's interest in this story may have originated from the tragic event of the Mongolian invasion of Hungary in 1241–1242.

84. Herbert L. Abrams, "Surviving a Nuclear War Is Hardly Surviving," *New York Times*, February 27, 1983.

85. *Quotable Einstein*, p. 175; from "Atomic War or Peace," *Atlantic Monthly*, November 1945.

86. Ibid., p. 181, from an interview with Alfred Werner, *Liberal Judaism* (April–May 1949): 16, 12.

87. Ibid., p. 182, from a contribution to Eleanor Roosevelt's television program on the implications of the H-bomb, February 13, 1950, published in Albert Einstein, *Ideas and Opinions* (New York: Crown, 1954), pp. 159–61.

88. President Ronald Reagan's speech to the National Association of Evangelicals in Orlando, FL, March 8, 1983.

89. In view of considerable criticism toward ABC for having shown such a terrible scenario, I felt absurd because I did not find the story sufficiently terrible. See István Hargittai, *Our Lives: Encounters of a Scientist* (Budapest: Akadémiai Kiadó, 2004), p. 181.

90. R. P. Turco, O. B. Toon, T. P. Ackerman, J. B. Pollack, and Carl Sagan, "Nuclear Winter: Global Consequences of Multiple Nuclear Explosions," *Science* 222 (1983): 1283–92.

91. Carl Sagan, "Nuclear War and Climatic Catastrophe: Some Policy Implications," *Foreign Affairs* 62 (Winter 1983): 257–92.

92. Edward Teller, *Bulletin of the Atomic Scientists* (February 1947).

93. Edward Teller, "Widespread After-Effects of Nuclear War," *Nature* 310 (1984): 621–24.

94. Quoted after Stanley A. Blumberg and Louis G. Panos, *Edward Teller: Giant of the Golden Age of Physics* (New York: Charles Scribner's Sons, 1990), p. 181.

95. Russell Seitz, "The Melting of 'Nuclear Winter,'" *Wall Street Journal*, November 5, 1986.

96. Conversation with Freeman Dyson at the Institute for Advanced Study, Princeton, March 31, 2008.

97. Edward Teller, "Climatic Change with Nuclear War," *Nature* (1985): 318, 99.

98. R. P. Turco, O. B. Toon, T. P. Ackerman, J. B. Pollack, and Carl Sagan, "Climate and Smoke: An Appraisal of Nuclear Winter," *Science* 247 (1990): 166–76.

99. Hargittai and Hargittai, *Candid Science VI*, "Werner Arber," pp. 152–63, 163.

100. Letter of January 1949 from Edward Teller to Maria Goeppert Mayer, UCSD Library/Hoover Archives (No. 1 in 1949).

101. Letter of September 27, 1961, from Edward Teller to Senator Stuart Symington, Hoover Archives.

## CHAPTER 10

1. *Khochesh mira—bud' sil'nym* (in Russian, If you want peace—be strong) (Arzamas-16: Publisher of the Federal Nuclear Center of Russia, 1995).

2. Edward Teller with Judith L. Shoolery, *Memoirs: A Twentieth-Century Journey in Science and Politics* (Cambridge, MA: Perseus Publishing, 2001), p. 397.

3. Ibid., p. 400, in a letter to Maria Goeppert Mayer.

4. Ibid., p. 435, in a letter to Maria Goeppert Mayer around 1958.

5. James D. Watson and Francis H. C. Crick, "Molecular Structure of Nucleic Acids: A Structure for Deoxyribose Nucleic Acid," *Nature* 171 (1953): 737–38.

6. James D. Watson and Francis H. C. Crick, "Genetical Implications of the Structure of Deoxyribonucleic Acid," *Nature* 171 (1953): 964–67.

7. Letter of August 8, 2008, from James D. Watson to the author.

8. James D. Watson, *Genes, Girls, and Gamow: After the Double Helix* (New York: Alfred A. Knopf, 2002), p. 132.

9. Watson, *Genes, Girls, and Gamow*, plates 9 and 17 among the Gamow Memorabilia appended in unnumbered pages to the text of the book. In the British edition of the book (Oxford: Oxford University Press, 2001), these plates appear on pp. 142 and 155, respectively.

10. See, e.g., Lily E. Kay, *Who Wrote the Book of Life? A History of the Genetic Code* (Stanford, CA: Stanford University Press, 2000), pp. 144–46.

11. Horace Freeland Judson, *The Eighth Day of Creation: Makers of the Revolution in Biology*, expanded ed. (Cold Spring Harbor, NY: Cold Spring Harbor Laboratory Press, 1996), pp. 269–70.

12. Letter of September 25, 2000, from Kevin J. Fraser to Edward Teller and printout of an e-mail exchange between Dr. Fraser and Teller's assistant, Lois Nisbet, Hoover Archives.

13. V. Ya. Frenkel', "*Vstrechy*" (Meetings), in *Vospominaniya o I. E. Tamme* (Reminiscenses about I. E. Tamm), edited by E. L. Feinberg (Moscow: Nauka, 1986), pp. 269–95. Tamm's presentation was made in defiance of the then still dominant unscientific teachings of Lysenko.

14. Letter of June 27, 1949, from Edward Teller to Maria Goeppert Mayer, UCSD Library/Hoover Archives (No. 5 in 1949).

15. "Rickover Takes Cruise: Dr. Teller Accompanies Him on Nuclear Submarine," *New York Times*, August 15, 1960.

16. George C. Wilson, "Hyman Rickover Dies; Fostered Nuclear Navy: Maverick Admiral's Vision Forced Revolution in Strategy," *Washington Post*, July 9, 1986.

17. Tim Hilchey, "J. Carson Mark, 83, Physicist in Hydrogen Bomb Work, Dies," *New York Times*, March 9, 1997.

18. Transcripts of a group interview with Edward Teller at Los Alamos National Laboratory, June 7, 1993, by Charles R. (Chuck) Hansen. I am grateful to Richard Garwin for a copy of the transcripts.

19. Teller, *Memoirs*, p. 420.

20. Letter in February 1955 from Edward Teller to Maria Goeppert Mayer, UCSD Library/Hoover Archives.

21. Ann Finkbeiner, *The Jasons: The Secret History of Science's Postwar Elite* (New York: Viking, 2006), p. 34.

22. Richard G. Hewlett and Francis Duncan, *A History of the United States Atomic Energy Commission: Volume II 1947/1952, Atomic Shield* (Washington, DC: US Atomic Energy Commission, 1972), p. 541.

23. Freeman Dyson, *Disturbing the Universe* (New York: Harper & Row, 1979), p. 97.

24. Ibid., p. 98.

25. Teller, *Memoirs*, p. 457.

26. Ibid., p. 455.

27. Russell Porter, "Rockefeller Fund Prods U. S. on Aims: Report Asks Leadership to Seize Initiative in the 'Cold War' from Communists," *New York Times*, September 8, 1960.

28. Ibid., p. 8.

29. D. S. Greenberg, "The Big Picture: House Committee Hears Views on Basic Problems of Science–Government Relations," *Science* 142 (November 8, 1963): 650–53.

30. Letter of October 29, 1963, from Edward Teller to Henry Kissinger, Hoover Archives.

31. D. S. Greenberg, "Goldwater: An Effort to Evaluate the Effects That His Election Might Have on Scientific Activity," *Science* 145 (August 14, 1964): 685–87.

32. Stanley A. Blumberg, "Secrecy in Science: An Interview with Dr. Edward Teller," *Johns Hopkins Magazine* (Winter 1972): 8–16, 11.

33. Margalit Fox, "Tony Schwartz, Father of 'Daisy Ad' for the Johnson Campaign, Dies at 84," *New York Times*, June 17, 2008.

34. Teller was a member of the following panels: Panel I, Energy and Its Relationship to Ecology, Economics, and World Stability; Panel III, Raw Materials, Industrial Development, Capital Formation, Employment and World Trade; and Panel V, Change, National Security and Peace.

35. *Energy: A Plan for Action: A Report by Edward Teller to the Energy Panel of the Commission on Critical Choices for Americans* (New York: Commission on Critical Choices for Americans, 1975).

36. Edward Teller, *Power & Security: Critical Choices for Americans* (Lanham, MD: Lexington Books, 1976).

37. Colin Norman, "Washington See," *Nature* 251 (September 20, 1974): 179.

38. Edward Teller, "A New Look at War-Making," *New York Times*, July 7, 1957.

39. Letter of May 28, 1957, from Edward Teller to Henry A. Kissinger, Hoover Archives.

40. This was in the cover story of the November 18, 1957, issue of *Time* magazine. The title of the cover story was "Knowledge Is Power," taken from the sixteenth-century British philosopher Francis Bacon.

41. Letter of November 4, 1957, from Edward Teller to Senator Henry M. Jackson, Hoover Archives.

42. Letter of April 3, 1956, from Edward Teller to C. L. Brown, Hoover Archives.

43. "Dr. Teller Finds Block to Science Is 'Square,'" *New York Times*, February 16, 1957.

44. Grace Hechinger and Fred M. Hechinger, "X-Ray of the Scientific Mind," *New York Times*, October 18, 1959.

45. "Dr. Teller Cites Teaching 'Crisis': H-Bomb Physicist Proposes Better Science Courses—Pupil Talks Open Here," *New York Times*, October 12, 1960.

46. Private communication from Richard F. Weyand, Teller's son-in-law, July 31, 2008.

47. "Men of the Year," *Time*, January 2, 1961.

48. "Space Race Warning: Teller Says Russians Have Power to Shape World," *New York Times*, November 5, 1961, http://www.jfklibrary.org/Historical+Resources/Archives/Reference+Desk/Speeches/JFK/003POF03NationalNeeds 05251961.htm.

49. Ibid.

50. A CIA-operated U-2 spy plane was shot down over Soviet territory on May 1, 1960. The United States issued denials of spying and was humiliated when the Soviets revealed that the pilot, Francis Gary Powers, was alive and the wreckage of the plane with spying equipment was largely intact. Khrushchev used the incident to cancel a summit in Paris a couple of weeks later to which both he and President Eisenhower had come.

51. Letter of March 19, 1979, from Edward Teller to Dixy Lee Ray, Hoover Archives.

52. Letter of October 8, 1984, from Edward Teller to Tom Lantos, Hoover Archives.

53. Letter of August 18, 1980, from Edward Teller to Governor Ronald Reagan, Hoover Archives.

54. Letter of November 15, 1988, from Edward Teller to (President-Elect) George Bush, Hoover Archives.

55. Letter of July 23, 1999, from Edward Teller to Governor of Texas George W. Bush, Hoover Archives.

56. Edward Teller, "The Atom at Work," *New York Times*, December 9, 1956.

57. Edward Teller, "In the Eye of a Hurricane," *Progress in Scientific Culture* 7, no. 1 (1982): 1–2.

58. Isaiah 2:4.

59. Dan O'Neill, *The Firecracker Boys: H-bombs, Inupiat Eskimos and the Roots of the Environmental Movement* (New York: Basic Books, previously published by St. Martin's Press, 1994), p. 27.

60. Teller, *Memoirs*, p. 466.

61. Alvin Weinberg, "A magyar maffia Chicagóban" (The Hungarian mafia in Chicago), *Fizikai Szemle*: 93–96, 95.

62. "Dr. Teller Receives $50,000 Enrico Fermi Award," *New York Times*, December 4, 1962.

63. *New York Times*, December 11, 1961.

64. Edward Teller with Allen Brown, *The Legacy of Hiroshima* (Garden City, NY: Doubleday & Co., 1962), p. 84.

65. O'Neill, *Firecracker Boys*, p. 96.

66. Ibid., pp. 97–98.

67. Edward Teller, *Energy from Heaven and Earth* (San Francisco: W. H. Freeman and Co., 1979), p. 153.

68. Andrei Sakharov, *Memoirs*, translated from the Russian by Richard Lourie (New York: Alfred A. Knopf, 1990), p. 144.

69. O'Neill, *Firecracker Boys*, p. 84.

70. Ibid., p. 99.

71. "Underground Nuclear Blast Moved Mountain of Sand," *New York Times*, November 29, 1962.

72. Ibid., p. 14.

73. Peter Kihss, "A-Blasts Urged to Tap Gas Wells: Teller Says Plan Would Put New Sources Into Use," *New York Times*, January 11, 1967.

74. James P. Sterba, "H-Bombs Blast to Free Gas," *New York Times*, May 20, 1973.

75. James P. Sterba, "Oil Shale: Leasing to Begin on Vast Reserves in West," *New York Times*, January 6, 1974.

76. Hargittai Magdolna and István Hargittai, *Candid Science IV: Conversations with Famous Physicists*, "Yuval Ne'eman" (London: Imperial College Press, 2004), pp. 32–63.

77. Trevor Findlay, *Nuclear Dynamite: The Peaceful Nuclear Fiasco* (Brassey's Australia, 1990).

78. "Teller Consults Thais on Nuclear-Dug Canal," *New York Times*, January 14, 1973, p. 3.

79. Ernest Volkman, *Science Goes to War: The Search for the Ultimate Weapon, from Greek Fire to Star Wars* (New York: John Wiley & Sons, 2002), pp. 210–11.

80. Teller, *Memoirs*, p. 475.

81. Ibid., p. 474.

82. Stanley A. Blumberg and Gwinn Owens, *Energy and Conflict: The Life and Times of Edward Teller* (New York: G. P. Putnam's Sons, 1976), p. 417.

83. "'Father of H-Bomb' Agrees to Rally Scientific Talent," *New York Times*, December 31, 1965.

84. David Bird, "Teller Backs New Smog Curbs as a New Source of Broad Benefits," *New York Times*, September 14, 1967.

85. Robert C. Toth, "Teller Suggests Congress Guide Growth of Scientific Projects," *New York Times*, October 19, 1963.

86. R. L. Park, "Star Warrior on Sky Patrol: Edward Teller Wants to Nuke Asteroids," *New York Times*, March 25, 1992.

87. David Morrison and Edward Teller, "The Impact Hazard: Issues for the Future," in *Hazards Due to Comets and Asteroids*, edited by Tom Gehrels (Tucson and London: University of Arizona Press, 1994), pp. 1135–43.

88. Edward Teller, "Foreign Trip Report," October 31, 1994, Lawrence Livermore National Laboratory, Hoover Archives.

89. Hilary Rose, "Pangloss and Jeremiah in Science," *Nature* 229 (1971): 459–462, 460.

90. Letter of July 10, 1967, from Edward Teller to Leo Cherne, Hoover Archives. There is an attached memorandum, "Statement on Vietnamese Language."

91. Letter of February 28, 1968, from Edward Teller to Peter D. Lax, Hoover Archives.

538 NOTES TO PAGES 376–83

92. "Dissent Blooms at AAAS Circus," *Nature* 229 (1971): 81–82.

93. Philip M. Boffey, "AAAS Convention: Radicals Harass the Establishment," *Science* 171 (1971): 47–49.

94. "Physicists and Public Issues," *Nature* 217 (1968): 494–95.

95. Finkbeiner, *Jasons*, p. 106.

96. Following the Teller–Russell debate, one of Teller's fans wrote to him, saying among other things, "Dear Dr. Teller, you looked so wise, so sad, so calm, you spoke so firmly 'I know—I know'—from which source do you take your optimism that someone will survive on this globe even if 'a mad government' starts bilateral suicide? And if there are survivors, would it be worth it in a 'red' world under red world government?" Letter of February 28, 1960, from Margot J. Nicolaus to Edward Teller, Hoover Archives.

97. Eugene P. Wigner, *The Recollections of Eugene P. Wigner as Told to Andrew Szanton* (New York: Plenum Press, 1992), pp. 299–303.

98. Wigner with his outstretched banner is recorded in photograph.

99. Edward Teller, "How Many Secrecies?" *New York Times*, December 1, 1971.

100. Edward Teller, "Perilous Illusion: Secrecy Means Security," *New York Times*, November 13, 1960.

101. Ibid.

102. Letter of August 23, 1945, from Edward Teller to Maria Goeppert Mayer, Hoover Archives.

103. Arthur Schlesinger Jr., "The Secrecy Dilemma," *New York Times*, February 6, 1972.

104. Teller, "How Many Secrecies?"

105. Blumberg, "Secrecy in Science: An Interview with Dr. Edward Teller."

106. Ibid.

107. Teller liked to quote the following distinctions of four nations at the time of his youth: "In England, everything is permitted except a few things that are forbidden; in Prussia, everything is forbidden except a few things that are permitted; in Austro-Hungary, everything that is forbidden is permitted; and in Russia everything that is not forbidden is obligatory." See Edward Teller, *Better a Shield Than a Sword: Perspectives on Defense and Technology* (New York: Free Press/Macmillan, 1987), pp. 238–39.

108. "Classified Research," *Nature* 243 (1973): 6.

109. Ibid.

110. Teller, *Memoirs*, p. 431.

111. Edward Teller, "Kicking the Secrecy Habit," *New York Times*, May 27, 1973.

112. Ibid.

113. Wallace Turner, "Writer of H-Bomb Letter Expresses Surprise at Uproar," *New York Times*, September 19, 1979. The article makes reference to Charles R. (Chuck) Hansen.

114. Ibid.

115. Edward Teller, "Reopen the Skies," *New York Times*, April 25, 1971.

116. William F. Buckley, "How Much Is Secrecy Hurting the U.S.?" *Firing Line*, #511 (Columbia, SC: Southern Educational Communications Association).

117. Richard Lourie, *Sakharov: A Biography* (Hanover, NH: University Press of New England, 2002), p. 219.

118. Letter of January 14, 1982, from Edward Teller to Sidney Drell, Hoover Archives.

119. Letter of February 3, 1982, from Sidney Drell to Edward Teller, Hoover Archives.

120. Conversation with Sidney Drell at Stanford University, April 29, 2009.

121. Conversation with Stephen Libby at Stanford University, April 28, 2009.

122. Edward Teller, "Secret-Stealing, Then and Now," *New York Times*, May 14, 1999.

123. This silence prompted the dissident Soviet scientist Zhores Medvedev to investigate the accident and publish his findings: Zhores A. Medvedev, *Nuclear Disaster in the Urals* (New York: W. W. Norton and Co., 1979).

124. David Dickson, "End Nuclear Complacency—3 Mile Island Report," *Nature* 282 (November 8, 1979): 120–21.

125. D. S., "A Media Event for Science," *Science* 769 (August 30, 1974).

126. "Teller Tested in Hospital," *New York Times*, May 12, 1979.

127. Private communication (by e-mail) from George (Jay) Keyworth, April 2008.

128. Peter A. Bradford, "The 'Casualty' of Three Mile Island," *Washington Post*, October 25, 1979.

129. Matthew L. Wald, "Nuclear Agency Weighs Attack Threat at Plants," *New York Times*, July 2, 2008.

130. Letter of March 23, 1987, from Edward Teller to health official Robert Moffitt, Hoover Archives.

131. Letter of March 1, 2003, from Edward Teller to Senator Orrin G. Hatch, Hoover Archives.

132. Edward Teller, "Open Letter to the Citizens of the Year 2100," July 20, 1999, Hoover Archives.

133. John J. Miller, "Truth Teller," *National Review*, September 30, 2002, pp. 38–40, 38.


## CHAPTER 11

1. *Pravda*, according to a TASS report from London, February 11, 1967, p. 1, as quoted by Edward Teller, "Deterrence? Defense? Disarmament? The Many Roads Toward Stability," in *Thinking about America: The United States in the 1990s*, edited by Annelise Anderson and Dennis L. Bark (Stanford, CA: Hoover Institution, 1988), pp. 21–32, 26.

2. The full text of Ronald Reagan's speech on March 23, 1983, can be found in *Public Papers of the Presidents, Ronald Reagan, 1983, Book 1* (Washington, DC: US Government Printing Office, 1984), pp. 437–43.

3. Dwight D. Eisenhower's farewell address, January 1961. The full sentence: "Yet, in holding scientific research and discovery in respect, as we should, we must also be alert to the equal and opposite danger that public policy could itself become the captive of a scientific–technological elite." The full text of the speech can be found in *Public Papers of the Presidents, Dwight D. Eisenhower, 1960,* pp. 1035–40.

4. William F. Buckley, "How Much Is Secrecy Hurting the U.S.?" *Firing Line,* #511 (Columbia, SC: Southern Educational Communications Association, 1982), p. 9.

5. Ibid., p. 10.

6. Magdolna Hargittai and István Hargittai, *Candid Science IV: Conversations with Famous Physicists,* "Edward Teller" (London: Imperial College Press, 2004), pp. 404–23, 413–14.

7. Private communication (by e-mail) from George (Jay) Keyworth, April 2008.

8. Ibid.

9. Ibid.

10. Buckley, "How Much Is Secrecy Hurting the U.S.?" p. 9.

11. William J. Broad, *Teller's War: The Top-Secret Story Behind the Star Wars Deception* (New York: Simon & Schuster, 1992), p. 118.

12. Edward Teller with Judith Shoolery, *Memoirs: A Twentieth-Century Journey in Science and Politics* (Cambridge, MA: Perseus Publishing, 2001), p. 330.

13. Private communication (by e-mail) from George (Jay) Keyworth, April 2008.

14. See, e.g., Teller, "Deterrence? Defense? Disarmament? The Many Roads Toward Stability," p. 27.

15. Gregg Herken, *Cardinal Choices: Presidential Science Advising from the Atomic Bomb to SDI* (Stanford, CA: Stanford University Press, 2000), p. 211.

16. Jerome B. Wiesner and Kosta Tsipis, "Put 'Star Wars' before Panel," *New York Times,* November 11, 1986.

17. See William J. Broad, "'Star Wars' Gem Loses Its Luster," *New York Times,* February 13, 1990.

18. *Public Papers of the Presidents, Ronald Reagan, 1983, Book 1,* pp. 437–43.

19. Robert M. Gates, *From the Shadows: The Ultimate Insider's Story of Five Presidents and How They Won the Cold War* (New York: Simon & Schuster, 1996), p. 264.

20. Ibid., p. 265.

21. David Holloway, "Moscow's Reaction Should Be Familiar: It Used to Be Ours," *Washington Post,* April 3, 1983.

22. See, e.g., Philip M. Boffey, "Pressures Are Increasing for Arms Race in Space," *New York Times*, October 18, 1982.

23. A light pulse can be amplified by stimulating light emission from atoms that are at a higher energy state and emit light as its electrons are transferred to bring the atoms to a lower energy state, of which the lowest is the ground state. This amplification can only take place if the electron population is "inverted," that is, there are more electrons in the higher-energy state than in the lower.

24. Jeff Hecht, *Beam: The Race to Make the Laser* (New York: Oxford University Press, 2005).

25. Charles H. Townes, *How the Laser Happened: Adventures of a Scientist* (New York: Oxford University Press, 1999), p. 105, quoting the *New York Times*, July 8, 1960.

26. Ibid.

27. Ibid., pp. 166–67.

28. Ibid., p. 167.

29. Balazs Hargittai and István Hargittai, *Candid Science V: Conversations with Famous Scientists*, "Charles Townes" (London: Imperial College Press, 2005), pp. 94–137, 127.

30. Hans Mark, "The Airborne Laser from Theory to Reality: An Insider's Account," *Defense Horizons*, April 2002, pp. 1–6.

31. Edward Teller, "Lasers in Chemistry," in *Proceedings of the Robert A. Welch Foundation Conferences on Chemical Research. XVI. Theoretical Chemistry*, edited by W. O. Milligan (Houston: Robert A. Welch Foundation, 1973), pp. 205–28.

32. J. W. Boag, P. E. Rubinin, and D. Shoenberg, eds., *Kapitza in Cambridge and Moscow: Life and Letters of a Russian Physicist* (North-Holland, Amsterdam, 1990), p. 390.

33. Ibid., pp. 389–90.

34. Charles Mohr, "Scientists Dubious over Missile Plan: Some Consider It Technically Unworkable—Others View It as Strategic Danger," *New York Times*, March 25, 1983.

35. Ibid.

36. Stephen Budiansky, "President Reagan Opts for Anti-Missile Defence," *Nature* (1983): 302, 365.

37. Lillian Hoddeson and Vicki Daitch, *True Genius: The Life and Science of John Bardeen; The Only Winner of Two Nobel Prizes in Physics* (Washington, DC: Joseph Henry Press, 2002), p. 269.

38. Richard L. Garwin, "Reagan's Riskiness," *New York Times*, March 30, 1983.

39. Edward Teller, "Reagan's Courage," *New York Times*, March 30, 1983.

40. Ibid.

41. Edward Teller, *Better a Shield Than a Sword: Perspective on Defense and Technology* (New York: Free Press, 1987).

42. Letter of January 11, 1960, from Edward Teller to Henry Kissinger, Hoover Archives.

43. Letter of August 23, 1963, from Edward Teller to Donald Rumsfeld, Hoover Archives.

44. Philip M. Boffey, "Dark Side of 'Star Wars': System Could Also Attack," *New York Times*, March 7, 1985.

45. Ibid.

46. Ibid.

47. R. J. Smith, "Experts Cast Doubts on X-Ray Laser: The Jewel of the 'Star Wars' Missile Defense Program Fails to Glitter," *Science* 8 (November 1985): 646.

48. Art Buchwald, "Capitol Punishment: The Laser Brain Hunt," *Washington Post*, April 7, 1983.

49. Stephen Budiansky, "Star Wars: Strategic Weaknesses Made Plain," *Nature* 310 (1984): 530.

50. David B. Ottaway, "SDI Office Seeks Funds to Speed Atomic Tests: Program's 'Non-Nuclear' Nature Questioned," *Washington Post*, December 13, 1985.

51. Mary McGrory, "Senate Nonsense on Star Wars," *Washington Post*, June 9, 1985.

52. Ibid. John Kerry ran as the Democratic candidate for the presidency and lost against George W. Bush in 2004.

53. Wiesner and Tsipis, "Put 'Star Wars' before a Panel."

54. Hargittai and Hargittai, *Candid Science IV*, "Philip W. Anderson," pp. 586–601, 593–94.

55. Personal communication from Philip W. Anderson (by e-mail) on August 26, 2008.

56. Hargittai and Hargittai, *Candid Science IV*, p. 592.

57. Arno A. Penzias, "Sakharov and SDI," in *Andrei Sakharov: Facets of a Life*, edited by B. L. Altshuler (Gif-sur-Yvette, France: Edition Frontiers, 1991), pp. 507–16.

58. Ibid., p. 513.

59. Broad, *Teller's War*, pp. 193–94.

60. R. Jeffrey Smith, "Lab Officials Squabble Over X-ray Laser: Edward Teller Lobbies for $100 Million in New Research Funds, but Others Say It Isn't Warranted," *Science* 230 (1985): 923.

61. William J. Broad, "In from the Cold at a Top Nuclear Lab," *New York Times*, December 27, 1987. After the collapse of communism, the name of the city Gorky was reverted to its old name, Nizhny Novgorod.

62. See, e.g., Marcia Barinaga, "GAO Report Vindicates Teller but Critics Disagree," *Nature* 282 (1988): 334.

63. Flora Lewis, "A 'Star Wars' Cover-Up," *New York Times*, December 3, 1985.

64. Ibid.

65. Edward Teller, "Progress Made in Protective-Defense Research," *New York Times*, December 29, 1985.

66. Conversation with Freeman Dyson at the Institute for Advanced Study, Princeton, March 31, 2008.

67. Freeman J. Dyson, "Edward Teller 1908–2003," *Biographical Memoirs* (Washington, DC: National Academy of Sciences, 2007), pp. 1–20.

68. Private communication (by e-mail) from George (Jay) Keyworth, April 2008.

69. Michael R. Gordon, "'Star Wars' Final X-ray Laser Weapon Dies as Its Final Test Is Canceled," *New York Times*, July 21, 1992.

70. George C. Wilson, "Mirror-Reflected Laser Suggested to Shield Allies," *Washington Post*, March 3, 1986.

71. Memorandum of February 6, 1986, from Shirley Petty to Lowell Wood, Hoover Archives.

72. Mary McGrory, "A 'Star Wars' Gathering," *Washington Post*, June 24, 1986.

73. Rowland Evans and Robert Novak, "SDI: What Will Reagan Decide Now?" *Washington Post*, October 13, 1986.

74. Gates, *From the Shadows*, p. 263.

75. Frances FitzGerald, *Way Out There in the Blue: Reagan, Star Wars and the End of the Cold War* (New York: Simon & Schuster, 2000), p. 315.

76. Ibid., p. 327.

77. FitzGerald, *Way Out There in the Blue*; Jack F. Matlock Jr., *Reagan and Gorbachev: How the Cold War Ended* (New York: Random House, 2004); George P. Shultz and Sidney D. Drell, eds., *Implications of the Reykjavik Summit on Its Twentieth Anniversary: Conference Report* (Stanford, CA: Hoover Institution Press, 2007); George P. Shultz, Sidney D. Drell, and James E. Goodby, eds., *Reykjavik Revisited: Step Toward a World Free of Nuclear Weapons—A Summary Report of 2007 Hoover Institution Conference* (Stanford, CA: Hoover Institution Press, 2008); George P. Shultz, Steven P. Andreasen, Sidney D. Drell, and James E. Goodby, eds., *Reykjavik Revisited: Step Toward a World Free of Nuclear Weapons—Complete Report of 2007 Hoover Institution Conference* (Stanford, CA: Hoover Institution Press, 2008).

78. Quoted in FitzGerald, *Way Out There in the Blue*, p. 355 (James Schlesinger, "Reykjavik and Revelations: A Turn of the Tide?" *Foreign Affairs* 65 [1986]: 433–34).

79. Quoted in FitzGerald, *Way Out There in the Blue*, p. 337 (George Shultz, *Turmoil and Triumph* [New York: Charles Scribner's Sons, 1993], p. 716).

80. Teller, *Memoirs*, p. 510.

81. Richard Lourie, *Sakharov: A Biography* (Hanover, NH: University Press of New England, 2002), pp. 358–59. The French Maginot Line was a concrete fortification built in the 1930s on the border with Germany as a protection against a possible German attack. However, the Germans did not find it a real barrier, as they had first attacked Belgium and then invaded France at the early stage of World War II.

82. Elizabeth Kastor and Donnie Radcliffe, "The Night of the Peacemakers: After the Treaty, Toasts—and Gorbachev's Sentimental Sing-Along," *Washington Post*, December 9, 1987.

83. Barbara Gamarekian, "A State Dinner for the Gorbachevs: Front-Row Seat on World History: Traditions Fell—the Soviet Leader Wore a Blue Suit," *New York Times*, December 9, 1987.

84. Teller, *Memoirs*, p. 534.

85. Private communication from George Shultz (by e-mail), March 17, 2009.

86. Hugh Sidey, "Not Since Jefferson Dined Alone," *Time*, December 21, 1987, p. 22. The title is an obvious reference to President Kennedy's characterization of the White House dinner he gave for American Nobel laureates.

87. Letter of December 10, 1987, from Edward Teller to Ronald Reagan, Hoover Archives.

88. Teller wrote a note of congratulation to Oppenheimer when he heard about his Fermi Award. He singled out two pieces of Oppenheimer's activities; one, their joint work at Berkeley in 1942; and the other, Oppenheimer's contribution to the Baruch Plan for making nuclear matters international, right after the end of World War II. It was not obvious why Teller neglected to mention Oppenheimer's directorship at Los Alamos, while it was obvious why he omitted Oppenheimer's postwar activities. Undated letter from Edward Teller to Robert Oppenheimer, Hoover Archives. Oppenheimer graciously acknowledged the note of congratulation: Letter of April 23, 1963, from Robert Oppenheimer to Edward Teller, Hoover Archives.

89. Quoted in FitzGerald, *Way Out There in the Blue*, p. 143 (letter of November 24, 1982, from Caspar Weinberger to Daniel O. Graham).

90. Gregory Canavan and Edward Teller, "Strategic Defence for the 1990s," *Nature* 344 (1990): 699–704.

91. Letter of August 29, 1988, from Edward Teller to Ira Rosen, producer of *60 Minutes*, CBS News, Hoover Archives.

92. P.S. note in the letter of February 17, 1976, from Edward Teller to Rolland D. Headlee, Hoover Archives.

93. Sidney Blumenthal, "When Giants Meet: H-Bomb Fathers Sakharov & Teller's SDI Dialogue," *Washington Post*, November 17, 1988.

94. Hargittai and Hargittai, *Candid Science IV*, "Arno Penzias," pp. 272–85, 279.

95. Penzias, "Sakharov and SDI," p. 507.

96. Ibid., pp. 511–15.

97. Ibid., pp. 513–14.

98. Ibid., p. 515.

99. Edward Teller, "From Brilliant Pebbles to Brilliant Eyes," *New York Times*, June 23, 1991.

100. Robert L. Park, "Another 'Star Wars' Sequel," *New York Times*, February 15, 1999.

101. Alan Brinkley, "An Idea Whose Time Will Not Go: Ronald Reagan's Missile Shield Is Still Alive, Though It Shows No Sign of Consciousness," *New York Times*, April 16, 2000.

102. Ellen Pawlikowski, "The Airborne Laser (ABL) Program," *Defense Horizons*, April 2002, p. 4.

103. Peter J. Westwick, "The Strategic Offense Initiative? The Soviets and Star Wars," *Physics Today* (June 2008): 43–49.

104. Hargittai and Hargittai, *Candid Science IV*, p. 592.

105. Ibid., p. 279.

106. Ibid., p. 278.

107. Arthur H. Compton, *Atomic Quest: A Personal Narrative* (New York: Oxford University Press, 1956), p. 128.

108. Hargittai and Hargittai, *Candid Science IV*, p. 278.

109. Letter of August 6, 1985, from Edward Teller to President Reagan, Hoover Archives.

## CHAPTER 12

1. Leo Szilard, "Draft of a Statement about Edward Teller (August 23, 1963)," in *Toward a Livable World: Leo Szilard and the Crusade for Nuclear Arms Control*, edited by Helen S. Hawkins, G. Allen Greb, and Gertrud Weiss Szilard (Cambridge, MA: MIT Press, 1987), p. 406.

2. Oscar Wilde, *The Picture of Dorian Gray*, "The Preface," in *Complete Works of Oscar Wilde* (London and Glasgow: Collins, 1973), p. 17.

3. Bill Clinton's letter was printed in facsimile in Edward Teller, "Hiroshima—The Psychology of a Decision," *Fizikai Szemle* 49 (1995): 176–81, 181. This was in a special issue in English of the Hungarian magazine.

4. Fred Ausbach, "Singing Hungarian," in *The Manhattan Project: The Birth of the Atomic Bomb in the Words of Its Creators, Eyewitnesses, and Historians*, edited by Cynthia C. Kelly (New York: Black Dog & Leventhal, 2007), p. 165.

5. Herbert Childs, *An American Genius: The Life of Ernest Orlando Lawrence* (New York: E. P. Dutton & Co., 1968), p. 496.

6. Edward Teller with Judith Shoolery, *Memoirs: A Twentieth-Century Journey in Science and Politics* (Cambridge, MA: Perseus Publishing, 2001), p. 427.

7. Conversation with Balazs Rozsnyai at Lawrence Livermore National Laboratory, April 23, 2009.

8. Ede Teller, "I Came Home" (in Hungarian, "Hazajöttem"), *Fizikai Szemle* 41, no. 1 (1991): 1.

9. Ede Teller and László Zeley, *A biztonság bizonytalansága* (The Uncertainty of Security) (Budapest: Relaxa kft, 1991), pp. 14–17.

10. Teller's original message dated April 12, 2002, is documented in a memorandum of April 17 by Teller and is at the Hoover Archives.

11. "Teller Ede üzenete" (Edward Teller's Message), *Népszabadság* (September 15, 2003).

12. "Teller a barátjának diktált" (Teller Dictated to his Friend"), *Népszabadság* (September 16, 2003).

13. Laura Fermi, *Illustrious Immigrants: The Intellectual Migration from Europe 1930–41* (Chicago: University of Chicago Press, 1971), p. 53.

14. This was conveyed by one of the participants of the First Shelter Island Meeting, Robert Marshak; see James Gleick, *Genius: Richard Feynman and Modern Physics* (London: Abacus, 2003), p. 233.

15. Teller, *Memoirs*, p. 119.

16. Letter of December 3, 1957, from Rabbi Herschel Levin to Edward Teller, Hoover Archives.

17. Letter of December 13, 1957, from Edward Teller to Rabbi Herschel Levin, Hoover Archives.

18. Edward Teller's letter of January 17, 1995, to the Honorable Hazel O'Leary, secretary of energy, US Department of Energy, Washington, DC, 20545. I am grateful to Alexey Semenov (Yulii Khariton's grandson), Moscow, for a copy of this letter.

19. Yuli Khariton and Yuri Smirnov, "The Khariton Version," *Bulletin of the Atomic Scientists* (May 20–31, 1993).

20. Conversation with Siegfried Hecker at Stanford University, April 24, 2009.

21. Letter of February 9, 1999, from Edward Teller to Siegfried S. Hecker, Hoover Archives.

22. Edward Teller, "Foreign Trip Report," September 15, 1992, Lawrence Livermore National Laboratory, Hoover Archives.

23. Edward Teller, "Foreign Trip Report," October 31, 1994, Lawrence Livermore National Laboratory, Hoover Archives.

24. Letter of December 29, 1993, from Edward Teller to President Clinton, Hoover Archives.

25. Boris S. Gorobets, *Krug Landau: Fizika voini i mira* (Landau's Circle: Physics of War and Peace) (Moscow: URSS, 2008), p. 50.

26. Ibid., p. 20, referring to a publication by E. Andronikashvili, *Vospominaniya o zhidkom gelii* (Memoirs about Liquid Helium) (Tbilisi: Hanat Leba, 1980).

27. Hans A. Bethe, "Comments on the History of the H-Bomb," *Los Alamos Science* (Fall 1982): 43–53, 51. This was a slightly edited version of Bethe's article written in 1954, but was unclassified only in 1982.

28. Ibid.

29. Hans A. Bethe, "Edward Teller: A Long Look Back," *Physics Today* (November 2001): 55–56.

30. Steven Weinberg, private communication (by e-mail), February 7, 2009.

31. George Gamow, *Mr Tompkins in Paperback* (Cambridge: Cambridge University Press, 1965), p. 148.

32. Richard Garwin is quoting this from Teller's (!) open letter of January 19, 1987, to Hans Bethe; letter of March 23, 1987, from Richard L. Garwin to Edward Teller, Hoover Archives.

33. Steven Weinberg, private communication (by e-mail), February 7, 2009.

34. Jerome B. Wiesner and Kosta Tsipis, "Put 'Star Wars' before a Panel," *New York Times*, November 11, 1986.

35. Edward Teller, "The Atomic Scientists Have Two Responsibilities," *Bulletin of the Atomic Scientists* 3, no. 12 (1947): 355–56.

36. Eugene P. Wigner, *The Recollections of Eugene P. Wigner as Told to Andrew Szanton* (New York: Plenum Press, 1992), p. 219.

37. Stanley A. Blumberg and Louis G. Panos, *Edward Teller: Giant of the Golden Age of Physics* (New York: Charles Scribner's Sons, 1990), p. 2.

38. Gregg Herken, *Cardinal Choices: Presidential Science Advising from the Atomic Bomb to SDI* (Stanford, CA: Stanford University Press, 2000), p. 15.

39. Andrei Sakharov, *Memoirs*, translated from the Russian by Richard Lourie (New York: Alfred A. Knopf, 1990), p. 221.

40. Stanislaw M. Ulam, *Adventures of a Mathematician* (Berkeley: University of California Press, 1991), p. 224.

41. Lee Edson, "Scientific Man for All Seasons," *New York Times*, March 10, 1968.

42. George Gamow, *My World Line: An Informal Autobiography* (New York: Viking Press, 1970), p. 153.

43. David Javorsky, *The Lysenko Affair* (Chicago: University of Chicago Press, 1970), p. 149.

44. Peter D. Smith, *Doomsday Men: The Real Dr. Strangelove and the Dream of the Superweapon* (New York: St. Martin's Press, 2007); Peter Goodchild, *Edward Teller: The Real Dr. Strangelove* (London: Weidenfeld and Nicolson, 2004).

45. Ann Finkbeiner, *The Jasons: The Secret History of Science's Postwar Elite* (New York: Viking, 2006), p. 18.

46. Salman Rushdie, "Sakharov: The Courage of His Convictions," *Washington Post*, June 24, 1990.

47. Conversation with Freeman J. Dyson in Princeton, NJ, March 31, 2008.

48. Edward Teller, "Guest Comment: Military Applications of Technology—A New Turn," *American Journal of Physics* 59 (1991): 873.

49. Freeman J. Dyson, "Edward Teller 1908–2003," *Biographical Memoirs* (Washington, DC: National Academy of Sciences, 2007), pp. 2–20, 12.

50. Magdolna Hargittai and István Hargittai, *Candid Science IV: Conversations with Famous Physicists*, "Valentine Telegdi" (London: Imperial College Press, 2004), pp. 160–91, 185.

51. Freeman Dyson, *Disturbing the Universe* (London: Pan Books, 1979), pp. 84–93.

52. Ibid., p. 93.

53. Conversation with Freeman J. Dyson in Princeton, NJ, March 31, 2008.

54. H. S. Hawkins, G. A. Greb, and G. Weiss Szilard, eds., *Toward a Livable World: Leo Szilard and the Crusade for Nuclear Arms Control* (Cambridge, MA: MIT Press, 1987), p. 196.

55. Kameshwar C. Wali, *Chandra: A Biography of S. Chandrasekhar* (New Delhi, India: Penguin Books India, 1992), p. 93, note.

56. Edward Teller interview by an unidentified interviewer for the Academy of Achievement, Palo Alto, CA, September 30, 1990.

57. Ibid.

58. A. S. Cherniaev, *Shest' let s Gorbachevym* (Six Years with Gorbachev) (Moscow: Kul'tura, 1993), pp. 105–17, 112–13; quoted here after David Holloway, "The Soviet Preparation for Reykjavik: Four Documents," in Sidney D. Drell and George P. Shultz, eds., *Implications of the Reykjavik Summit on Its Twentieth Anniversary: Conference Report* (Stanford, CA: Hoover Institution, 2007), pp. 45–95, 47–48.

59. Duane C. Sewell, "The Branch Laboratory at Livermore during the 1950's," in *Energy in Physics, War and Peace: A Festschrift Celebrating Edward Teller's 80th Birthday*, edited by Hans Mark and Lowell Wood (Dordrecht, Holland: Kluwer Academic, 1988), p. 321.

60. Private communication from Ralph W. Moir at Lawrence Livermore National Laboratory, April 23, 2009.

61. We submitted our transcripts for his scrutiny—as was always our practice for the *Candid Science* book series—and he asked us to delete this statement. This was a departure from his usual approach to interviews, because he liked to insist on having the complete conversation appear or nothing at all. One other case known as an exception is described in Philip M. Stern with Harold P. Green, *The Oppenheimer Case: Security on Trial* (New York: Harper & Row, 1969), p. 454.

62. Edward Teller, *The Pursuit of Simplicity* (Malibu, CA: Pepperdine University Press, 1981), p. 62.

63. Bethe, "Edward Teller: A Long Look Back," p. 56.

64. István Hargittai, *The DNA Doctor: Candid Conversations with James D. Watson* (Singapore: World Scientific, 2007), pp. 38–40.

65. Edward Teller, "Open Letter to the Citizens of the Year 2100," http://wwwlhg.org/time/tellerletter.html.

66. Hargittai and Hargittai, *Candid Science IV*, "John A. Wheeler," pp. 424–39, 429.

# INDEX

Page numbers in **bold** indicate sidebars in the text
with detailed information.